T0390631

# NEW MATERIALS AND DEVICES ENABLING 5G APPLICATIONS AND BEYOND

# NEW MATERIALS AND DEVICES ENABLING 5G APPLICATIONS AND BEYOND

*Edited by*

Nadine Collaert
*IMEC, Heverlee, Belgium*

Elsevier
Radarweg 29, PO Box 211, 1000 AE Amsterdam, Netherlands
125 London Wall, London EC2Y 5AS, United Kingdom
50 Hampshire Street, 5th Floor, Cambridge, MA 02139, United States



ISBN: 978-0-12-822823-4

For Information on all Elsevier publications
visit our website at https://www.elsevier.com/books-and-journals

*Publisher:* Matthew Deans
*Acquisitions Editor:* Stephen Jones
*Editorial Project Manager:* Sara Greco
*Production Project Manager:* Manju Paramasivam
*Cover Designer:* Miles Hitchen

Typeset by MPS Limited, Chennai, India

# Contents

## 6. InP-based monolithic microwave integrated circuit technologies for 5G and beyond 179

Hiroshi Hamada



## 7. RF-MEMS for 5G: high performance switches and reconfigurable passive networks 205

Jacopo Iannacci



## 8. Antenna-in-package design considerations for millimeter-wave 5G 245

Bodhisatwa Sadhu, Duixian Liu and Xiaoxiong Gu



## 9. Circuits for 5G applications implemented in FD-SOI and RF/PD-SOI technologies 275

Lucas Nyssens, Martin Rack and Jean-Pierre Raskin



## 10. Power amplifiers monolithic microwave integrated circuit design for 5G applications 317

Xin Liu, Guansheng Lv, Dehan Wang and Wenhua Chen

# List of contributors


**Wenhua Chen** Department of Electronic Engineering, Tsinghua University, Beijing, P.R. China

**Nadine Collaert** IMEC, Heverlee, Belgium

**Xiaoxiong Gu** IBM Thomas J. Watson Research Center, Yorktown Heights, NY, United States

**Hiroshi Hamada** NTT Device Technology Labs, NTT Corporation, Atsugi-shi, Kanagawa-ken, Japan

**Jacopo Iannacci** Fondazione Bruno Kessler (FBK), Trento, Italy

**Duixian Liu** IBM Thomas J. Watson Research Center, Yorktown Heights, NY, United States

**Xin Liu** Department of Electronic Engineering, Tsinghua University, Beijing, P.R. China; School of Microelectronics, Xidian University, Xi'an, Shaanxi, P.R. China

**Guansheng Lv** Department of Electronic Engineering, Tsinghua University, Beijing, P.R. China

**Lucas Nyssens** Electronics and Applied Mathematics (ICTEAM), Université Catholique de Louvain, Institute of Information and Communication Technologies, Louvain-la-Neuve, Belgium

**Martin Rack** Electronics and Applied Mathematics (ICTEAM), Université Catholique de Louvain, Institute of Information and Communication Technologies, Louvain-la-Neuve, Belgium

**Jean-Pierre Raskin** Electronics and Applied Mathematics (ICTEAM), Université Catholique de Louvain, Institute of Information and Communication Technologies, Louvain-la-Neuve, Belgium

**Bodhisatwa Sadhu** IBM Thomas J. Watson Research Center, Yorktown Heights, NY, United States

**Jagar Singh** pSemi/ex-GlobalFoundries Analog and RF device Technologist, Technology Development Group, Malta, NY, United States

**Dehan Wang** ZTE Corporation, Shenzhen, P.R. China

**Josef Watts** IBM/GlobalFoundries, Modeling Group, Malta, NY, United States

CHAPTER

# 1

# Introduction to 5G applications and beyond

*Nadine Collaert*

**IMEC, Heverlee, Belgium**

## 1.1 Introduction

The wireless communication industry has been pushing the spectrum to ever higher frequencies, looking for the bandwidths needed to enable the data rates, latency, and energy efficiency required, and driven by the insatiable need for more data and more connectivity, especially now when society is facing tremendous challenges. Every 10 years there is a new generation that until now has been focused on delivering more connections and ever higher data rates, but with 5G the industry is also at an inflection point. For the first time, the vision encompasses the enablement of new use cases like for example, extended reality XR, vehicle-to-vehicle communication where an increasing amount of nonhumans will take place in the communication and interaction. Dreams of being able to map out an entire room and use an avatar, remotely controlled, to walk around and be able to perform tasks, including haptic feedback, might still sound like a thing of the future, but with the announcement like metaverse this might be become true even faster than most of us can imagine. These examples also show that the amount of autonomous nonhuman users connected to the networks using sustainable energy sources will start to overtake the human bandwidth consumption. These nonhumans have different bandwidth needs and the balance in energy cost might change quite significantly, for example, machines don't necessary need displays to take in the information needed, unlike humans who rely on input from all five senses.

In analogy to Moore's law for complementary metal-oxide semiconductor (CMOS) scaling, Edward A. Nielsen, a pioneering engineer in wireless communication, first proposed what is now known as "Nielsen's law of internet bandwidth" more than three decades ago. According to this law, data throughput (or speed) over broadband networks doubles every 21 months—a phenomenon that has held true since 1984 and shows no signs of changing anytime soon. The premise behind this rule is simple: with technological

*New Materials and Devices Enabling 5G Applications and Beyond*
**DOI: https://doi.org/10.1016/B978-0-12-822823-4.00001-7**

advancements come faster speeds at lower costs; as competition increases between service providers, they are incentivized to offer higher download/upload speeds for their customers in order to stay competitive or attract new users who demand high performance from technologies such as 5G and 6G communications systems. Thus far all evidence suggests that this trend will continue going forward which means increased connectivity across broader geographic locations along with improved signal strength throughout cities both large and small due to significant infrastructure upgrades being made today by telecommunication companies worldwide.

The maximum achievable data rate of a given system is proportional to its bandwidth B times the logarithm of the number of available frequency channels. In other words, doubling the bandwidth allows you to transmit twice as much information, but adding just one more channel only lets you add 10% more capacity. This relationship between data rate and bandwidth might seem counterintuitive at first, but it can be explained by Shannon's law, as seen below in (1.1), which defines an upper limit on how much information can be carried over a noisy communications channel. The larger the bandwidth (i.e., the greater the frequency range), the less noise there is within that range and therefore the higher the potential data rates become:

$$C = B\log_2\left(1 + \frac{S}{N}\right) \tag{1.1}$$

with $C$ being the highest error-free data speed in bps, $B$ the bandwidth of the channel, $S$ the average signal power, and $N$ the average interference power or noise over the bandwidth.

The availability of higher bandwidths has been the reason why the wireless communication industry has been pushing the operating frequencies higher. Table 1.1 shows an overview of the different frequency ranges currently defined and under consideration for wireless communication. The table shows the typical carrier frequency and bandwidth and the number of antennas that can be integrated on an example area of $1.5 \times 1.5\ \text{cm}^2$. Currently, 5G is being rolled out for the sub-7GHz frequency bands, with recent interest in the C-band (4–8 GHz). 3GPP, the group of standards organizations which develop protocols for mobile telecommunications, has defined these bands below 7.125 GHz as FR1, and they are also called the low frequency bands. It is the primary band for 5G. However,

**TABLE 1.1** Overview of the frequency spectrum defined and under consideration for wireless communication.

| Communication perspective | FR1 | "FR3" | FR2 | "FR4" | "FR5" |
|---|---|---|---|---|---|
| Typical range | 0.1–7 GHz | 7–24 GHz | 24–52 GHz | 57–72 GHz | 95–325 GHz |
| Example carrier frequency | 3.5 GHz | 10 GHz | 30 GHz | 60 GHz | 140 GHz |
| Typical bandwidth | 100 MHz | 200 MHz | 400 MHz | 2 GHz | 10 GHz |
| Wavelength | 8.6 cm | 3 cm | 1 cm | 5 mm | 2.1 mm |
| Number of antennas on 1.5 cm x 1.5 cm | 0 | 1 | 9 | 36 | 196 |

FR1 and FR2 are currently defined. The frequency ranges between quotation marks are not defined yet, but are being explored.

this frequency spectrum is quite congested with little amount of continuous bandwidth available. Therefore, higher frequencies are being considered. 3GPP has already defined frequency range FR2 encompassing the frequencies between 24.25 and 52.6 GHz. This is typically referred to as 5G-mmwave.

While there are already products addressing these low mm-wave bands (24–40 GHz), their adoption is far from ideal with challenges in energy efficiency and performance still to be addressed, first publications and white papers are already appearing on 6G (Chowdhury et al., 2020; Strinati et al., 2019; Yang et al., 2019). For 6G or 5.5G—as at this moment no standards exist for 6G—the ambition is to allow data rates up to 100 Gbps and higher, extreme low power and latency, instantaneous transfer of large amounts of data, Artificial Intelligence (AI) empowered networks, massive coverage, and new interfaces (e.g., biological).

For these applications even higher frequencies than FR2 are being considered, sub-THz frequencies above 100 GHz, with the D-band as a first target. Bandwidths higher than 10 GHz are available and they can enable potentially simpler modulation schemes, eventually leading to lower power consumption. Besides the higher bandwidths, the reduced wavelength $\lambda$, thereby allowing to integrate more antennas (pitch scaling with $\lambda/2$) on a given area, is another advantage. This allows to reduce the requirements in output power for the power amplifiers (PA) significantly as the antenna gain will increase.

Moreover, it allows to implement beamforming architectures. Beamforming is a technique used in wireless communication systems to control the directionality of the transmitted or received signal. It involves using multiple antennas at the transmitter or receiver to create a "beam" of radio frequency (RF) energy that is directed at a specific angle or location. This can be used to improve the signal-to-noise ratio, increase the range and capacity of the communication system, or reduce interference from other sources. In the antenna array, the phase of the signal at each antenna is shifted so that the beam can be directed into the desired direction, and away from unwanted directions. At the receive side, the signal of interest will hit each antenna of the array with a slight phase shift. Phase shifters are then used to align the phases of these signals and by combining these a strong signal is created. Waves that hit the array from another direction will not give rise to this constructive combination and they are rejected. As the free space path loss, atmospheric attenuation, scattering, and reflections increase as the frequencies go up, beamforming can be used to restore the loss.

Beamforming can be implemented using either analog or digital techniques. In analog beamforming, the phase and amplitude of the RF signals at each antenna are adjusted to create the desired beam pattern. In digital beamforming, the RF signals at each antenna are converted to a digital form and processed using digital signal processing (DSP) techniques to create the desired beam pattern. Digital beamforming allows for more flexibility and precision in controlling the beam pattern, but it requires more complex hardware and processing power. Beamforming can be used in a variety of wireless communication systems, including cellular networks, wireless local area networks, and satellite communication systems. It is particularly useful in systems that operate in crowded or cluttered environments, where interference is a major concern, or in systems that require high capacity or long range communication.

However, enabling a typical beamforming transceiver for mm-wave and sub-THz applications, as depicted in Fig. 1.1, shows many challenges, from the antenna all the way

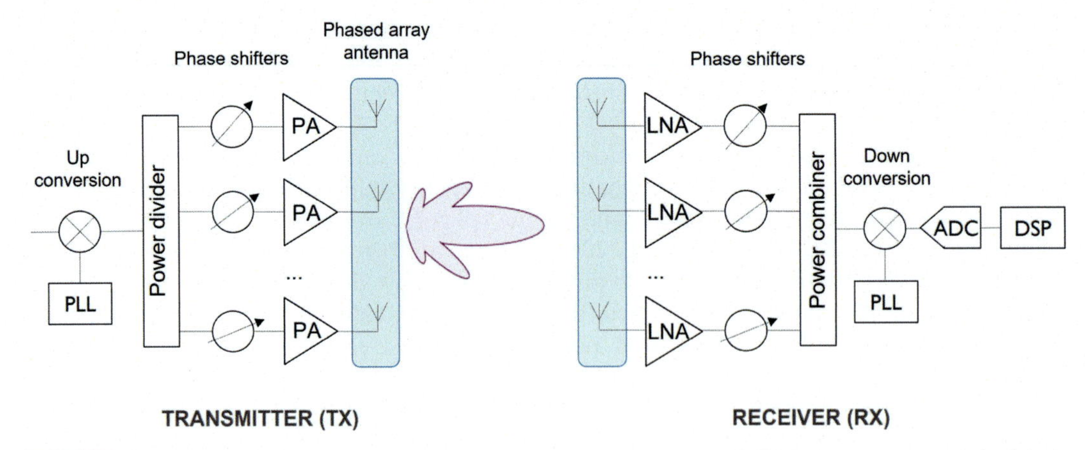


**FIGURE 1.1** Schematic view of a mm-wave beamforming transceiver; (left) the transmitter and (right) the receiver including the data convertors and digital baseband.

down to the baseband. The left-hand side of the figure shows a schematic view of the transmitter (TX) using a phased antenna array for the beamforming, and the right-hand side shows the receiver (RX). First of all, because of the higher bandwidth and data rates, faster and more complex DSP will be required. The analog-to-digital and digital-to-analog converters (ADC and DAC) are expected to have speeds up to 10GS/s and even higher. It is expected that the digital baseband and the ADCs and DACs will continue to be implemented in the most scaled CMOS.

The PAs, which are required to drive the antennas, are expected to be multi-stage and utilize new materials to deliver the output power and efficiency that is required. For generating the power, CMOS might not be the right technology choice. Compound semiconductor technologies like GaAs, GaN, and InP hold much greater promise than CMOS at these frequencies.

Different semiconductor materials can have different properties that can impact the performance of high-frequency power transistors made from them. For example, materials with high electron mobility and low resistivity can improve the speed and efficiency of transistors, which can in turn improve the power handling capacity and frequency response of the device. Similarly, materials with a high breakdown voltage and a low dielectric constant can improve the voltage and power handling capabilities of the device. The Johnson figure of merit (FOM) can be used to compare the suitability of different semiconductor materials for high-frequency power transistor applications by taking into account these and other performance factors. It can be a useful metric for evaluating the performance of RF devices and choosing the best material for a particular application. The Johnson FOM is the product of the charge carrier saturation velocity in the material and the electric breakdown field under the same conditions (Table 1.2).

Ultimately, it is not only about the intrinsic material properties but the possibility to use these materials in complex heterostructures for transistors like high electron mobility transistors (HEMT) or heterojunction bipolar transistors (HBT). These device architectures are quite popular in RF and microwave circuits because of their high electron mobility and

**TABLE 1.2** Comparison of material properties for different key semiconductor materials at 300K, amongst other the Johnson figure of merit, in this case normalized to the value for Si.

| Property | Si | GaAs | GaN | 4H-SiC | InGaAs (53% In) | InP |
|---|---|---|---|---|---|---|
| Bandgap energy (eV) | 1.12 | 1.4 | 3.39 | 3.26 | 0.75 | 1.3 |
| Breakdown field (MV/cm) | 0.3 | 0.4 | 3.3 | 3 | 0.2 | 0.5 |
| Electron mobility @300k ($cm^2/vs$) | 1400 | 8500 | 900 | 700 | 12,000 | 5400 |
| Saturated (peak) electron velocity ($10^7$ cm/s) | 1.0 (1.0) | 1.3 (2.1) | 1.3 (2.7) | 2.0 (2.0) | 0.77 (4.0) | 0.67 (2.5) |
| Relative dielectric constant | 11.8 | 12.8 | 9 | 10 | 13.1–14.1 | 12.5 |
| Thermal conductivity (W/mK) | 150 | 50 | 130 | 350 | 1.4 | 70 |
| Lattice constant (Å) | 5.43 | 5.65 | 3.19 | 3.07 | 5.87 | 5.87 |
| Johnson figure-of-merit (normalized to the Si value) | **1.0** | **2.7** | **27.5** | **20** | | **0.33** |

fast switching speeds. In a HEMT, the active region of the transistor is made from a layer of a compound semiconductor material, such as GaAs or InGaAs, which has a high electron mobility. This allows the HEMT to have fast switching speeds and high transconductance, making it suitable for high-frequency applications. HEMTs are also used in high-power and high-voltage applications because of their high breakdown voltage and low on-resistance.

HBTs are a type of bipolar transistors, which means that they use both electrons and holes as charge carriers. In a HBT, the base region is a thin layer of a different semiconductor material than the collector and emitter regions, and the base current controls the flow of current through the transistor. Just like HEMTs, HBTs are also used in high-frequency and high-power applications because they can operate at high frequencies and have a high gain, which makes them suitable for amplification and switching applications.

In general, while these compound semiconductors are extremely interesting to generate power at high frequencies, these circuits still need to be co-integrated with a RF-friendly CMOS technology which are typically older nodes of bulk CMOS or Fully Depleted Silicon-On-Insulator (FDSOI), or SiGe Bipolar CMOS (BiCMOS) for the calibration and the control needed to enable fast switching of the beam. SiGe BiCMOS technology involves the use of both SiGe bipolar transistors, which can deliver higher speed than Si bipolar transistors, and CMOS transistors on a single chip. Bipolar transistors are useful for high-speed and high-power applications, while CMOS transistors are low power and low noise.

Finally, the compactness of the mm-wave and sub-THz transceivers will require new packaging and new antenna designs. For mm-wave and sub-THz applications, this means a re-think of how the antenna is integrated in the package. For frequencies above 100 GHz, the interconnects between the chips or circuits as well as the passive components can be very lossy, and so the antennas should be put very close to the circuits. That's a challenge for the package. Advanced 3D IC techniques, where either dies or entire wafers are stacked on top of each other with short and well-defined interconnects and perhaps even new materials will need to be considered.

And in general, it requires careful co-design and co-optimization between the different ICs, the connections, the antenna, and the package from the beginning. This also calls for new electronic design automation tools that are able to do this co-design in an efficient way.

In this chapter, we will first give an overview of the different front-end technologies, with a specific focus on the PA. In a second section, we will zoom in on the compound semiconductors, in particular GaN and InP, and why these devices are so interesting. For the mm-wave applications we will focus on GaN, while for frequencies above 100 GHz, InP-based technologies will be chosen. And in a last technical section we will briefly highlight the need for heterogeneous integration, putting the compound semiconductor technologies and CMOS together to enable these highly performing and energy-efficient radios.

## 1.2 Overview front-end technologies

The most critical component of the RF front-end module (RF-FEM) that will require attention is the PA. Depending on the use case, system level studies show that 50% of the RF-FEM power consumption can be taken up by the PA. Moving to higher frequencies and thus higher data rates in general will also improve energy efficiency, but at the same an increase in power consumption as can be seen in Fig. 1.2. In general moving to higher frequencies and enabling higher data rates helps to improve the energy efficiency as can be seen in Fig. 1.2; however, we do see a significant increase in overall power consumption. For particular use cases like handsets there is in general a limit to the total (both TX as well as RX) power consumption, for example, 10 W.

But as the power consumption of the RF-FEM is dominated by the power consumption of the PA in the system, having the devices and technologies that come with high efficiency in a balanced and proper way is extremely important.

The typical FOM we will use in this document to compare different technologies are the unit current gain frequency $f_t$ and the maximum oscillation frequency $f_{max}$, and large signal metrics such as output power ($P_{out}$) and power added efficiency (*PAE*). They are defined as follows:

$$f_t = \frac{g_m}{2\pi\left(\left(C_{gs} + C_{gd}\right)\left(1 + g_{ds}R_{ds}\right) + g_m C_{gd} R_{ds}\right)} \tag{1.2}$$

$$f_{max} = \frac{f_t}{2\sqrt{g_{ds}\left(R_g + R_{ds}\right) + 2\pi f_t R_g C_{gd}}} \tag{1.3}$$

with $g_m$ the transconductance, $g_{ds}$ the output conductance, $R_g$ the gate resistance, $R_{ds}$ the drain-source resistance, $C_{gs}$ the gate-to-source capacitance, and $C_{gd}$ the gate-to-drain capacitance.

$$\mathrm{PAE} = \frac{P_{out} - P_{in}}{P_{DC}} \tag{1.4}$$

with $P_{out}$ the RF output power, $P_{in}$ the RF input power, and $P_{DC}$ the total DC power.

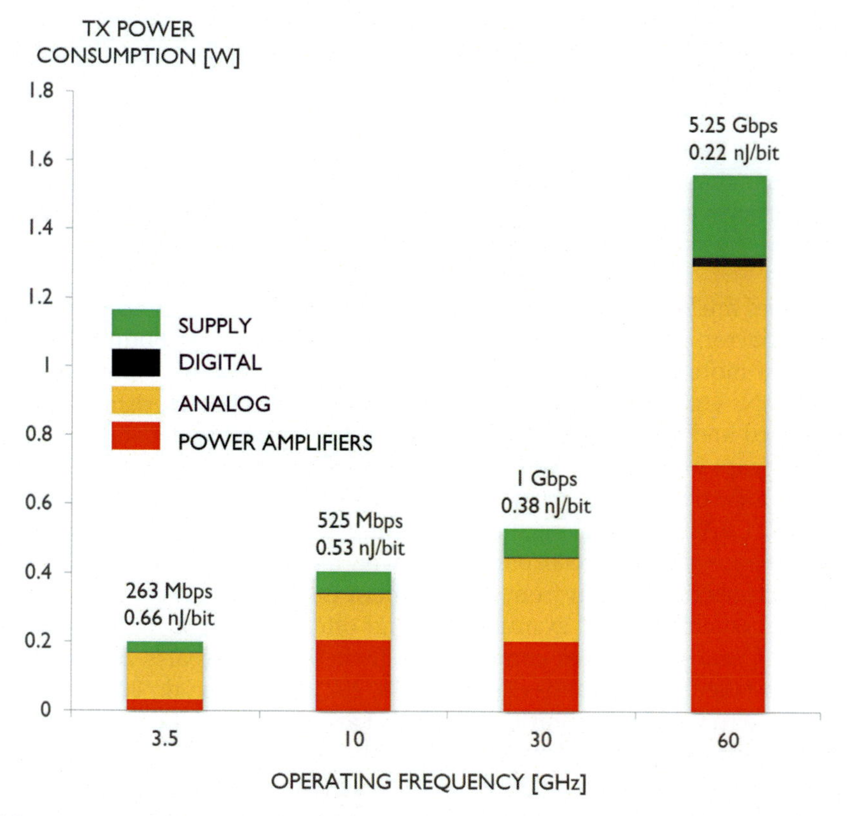


**FIGURE 1.2** Power consumption at the TX side as function of the operating frequency. A CMOS-only implementation has been considered and a 16-QAM modulation scheme. The figure shows how the contribution to the overall power consumption is mapped over the different parts of the transmitter: supply, digital, analog, and power amplifiers.

For receivers the noise figure NF, gain, linearity, and power consumption are important metrics. In the chapter, however we will focus mainly on the PA in the transmitter.

With generations of bulk CMOS downscaling we have seen consistent and gradual improvements in $f_t$ and $f_{max}$. However, this was no longer the case anymore when the Fin Field Effect Transistor (FinFET) was introduced in 2011 (Kavalieros et al., 2006; Wakayama, 2013). In order to keep the leakage under control when further scaling the gate length in CMOS, FinFET was introduced as a new device architecture to improve the short channel effects (SCE) in these scaled transistors. FinFETs are characterized by a gate that wraps around the channel region, formed in a fin shape. This allows to create fully depleted devices that have improved subthreshold swing and better short channel control. The Si fin is typically formed by recessing the Shallow Trench Isolation (STI) after definition of the active areas. Multiple fins are combined to define the total transistor width. As it is in use a 3D device, the additional benefit is that the fin height is an additional parameter that can be used to improve the current drivability and area efficiency as it is a parameter that does not appear in the layout of the devices. This device architecture was first introduced in 1998 (Hisamoto et al., 1998), with a predecessor dating back to even 1989 (Hisamoto et al., 1989).

The intrinsically higher parasitics resulted in a drop in $f_t$ for these early FinFET nodes. Parasitics such as gate-drain and gate-source capacitances and resistance are generally higher in FinFETs than in bulk CMOS transistors. Since then, many innovations have been introduced to improve the RF performance of these scaled CMOS devices, leading to particular RF-tailored versions of the core digital node, and this will be further detailed out in Chapter 3.

CMOS devices are made out of Si as starting material. Si is an example of a single element semiconductor, but other semiconductors exist like compound semiconductors which consist of more than one chemical element, like for example, III–V materials (InP, InAs, GaAs, GaN, etc.). They typically have a number of characteristics that favor their use in high-speed and high-power applications next to optical applications: wide bandgap or tunable bandgap, high carrier velocity, and often direct bandgaps. The additional benefit is that the properties of many can be adjusted by alloying them, for example, InGaAs, InAlAs, AlGaN, GaAsSb, and they allow to form complex heterostructures that can often be tuned to enhance the beneficial properties even more. HEMTs and HBTs are typically the device structures that benefit from the design of these heterostructures.

III–V devices combining, for example, InP and often used high indium content materials, show high $f_t$ and $f_{max}$ values compared to their Si counterparts as shown in Fig. 1.3 (data have been taken from Arabhavi et al., 2018, 2021; Carter et al., 2017; Chevalier et al., 2015; Europratice, n.d.; Heinemann et al., 2017; Lai et al., 2007; Lee et al., 2020; Manger et al., 2018; Mayeda et al., 2021; Micovic et al., 2017; Pekarik et al., 2021; Post et al., 2006; 2006; Razavieh et al., 2021; Singh et al., 2018; Teledyne website, n.d.; Then et al., 2021; Urteaga et al., 2011; Yadav et al., 2019; Yun et al., 2018). This is due to the III–V electron

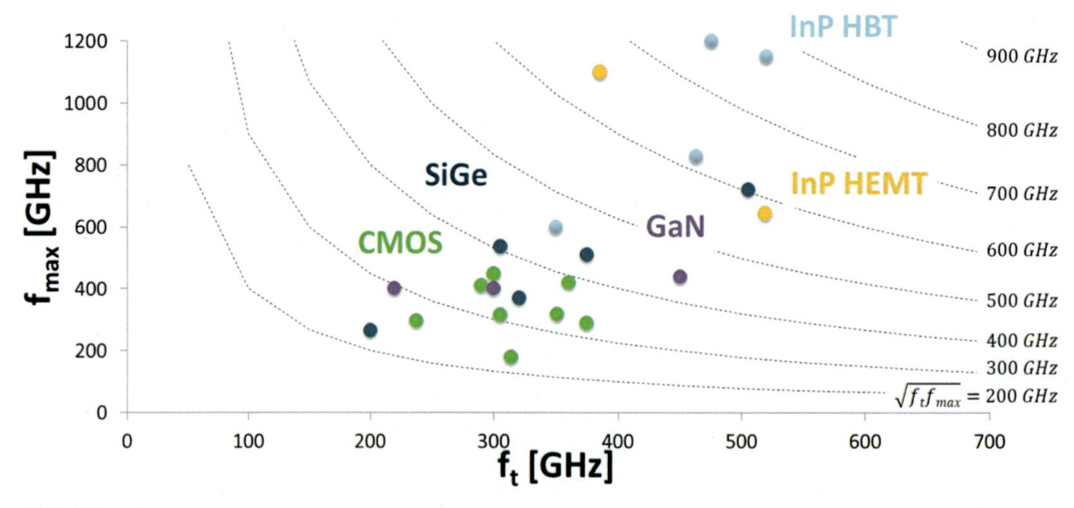


**FIGURE 1.3** $f_t$ as function of $f_{max}$ comparing different technologies: CMOS, GaN, SiGe HBT, InP HEMT, and InP HBT.

mobility being significantly higher than that of Si. What is even more interesting is that they typically combine this high $f_t$ and $f_{max}$ with the possibility to deliver high output power at much higher efficiencies than CMOS. This is shown in Fig. 1.4 where the output power of the PA is shown as function of the operating frequency, comparing different technologies (Collaert et al., 2020; Wang et al., n.d.).

For sub-6GHz and low mm-wave frequencies, medium to high-power applications, GaN is the better choice, while InP-based devices like HBTs and HEMTs work better for frequencies above 100 GHz.

Higher output power and efficiency means a lower number of elements are needed to drive the antennas and more energy-efficient systems can be enabled as shown in Fig. 1.5 where the required output power per PA is shown for two different cases: EIRP = 43 dBm (EIRP = equivalent isotropic radiated power) which corresponds to a user equipment (UE) case and EIRP = 65 dBm (base station case). Next to that, these compound semiconductors can be inherently solutions for use cases where form factor is a restriction, for example, mobile phones, glasses, or watches. As fewer PAs can be integrated in this case, higher performance is expected from these circuits.

While we have so far compared different semiconductor technologies using FOMs like $f_t$, $f_{max}$, $P_{out}$, and PAE, other things need to be considered like maturity of the technology, manufacturability, and cost-effectiveness.

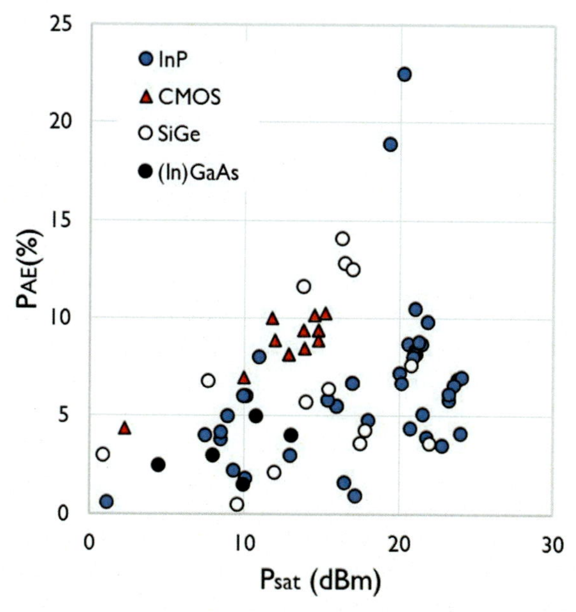


**FIGURE 1.4** Power added efficiency as function of saturated output power $P_{sat}$ for frequencies above 100 GHz comparing different technologies: InP, CMOS, SiGe, and (In)GaAs. *Source: From Collaert, N., Alian, A., Banerjee, A., Chauhan, V., Elkashlan, R. Y., Hsu, B., Ingels, M., Khaled, A., Vondkar Kodandarama, K., Kunert, B., Mols, Y., Peralagu, U., Putcha, V., Rodriguez, R., Sibaja-Hernandez, A., Simoen, E., Vais, A., Walke, A., Witters, L., … Waldron, N. (2020). From 5G to 6G: Will compound semiconductors make the difference? In:* 2020 IEEE 15th international conference on solid-state and integrated circuit technology, ICSICT 2020 – Proceedings. *Belgium: Institute of Electrical and Electronics Engineers Inc. http://ieeexplore.ieee.org/xpl/mostRecentIssue.jsp?punumber = 9278127. https://doi.org/10.1109/ICSICT49897.2020.9278253.*

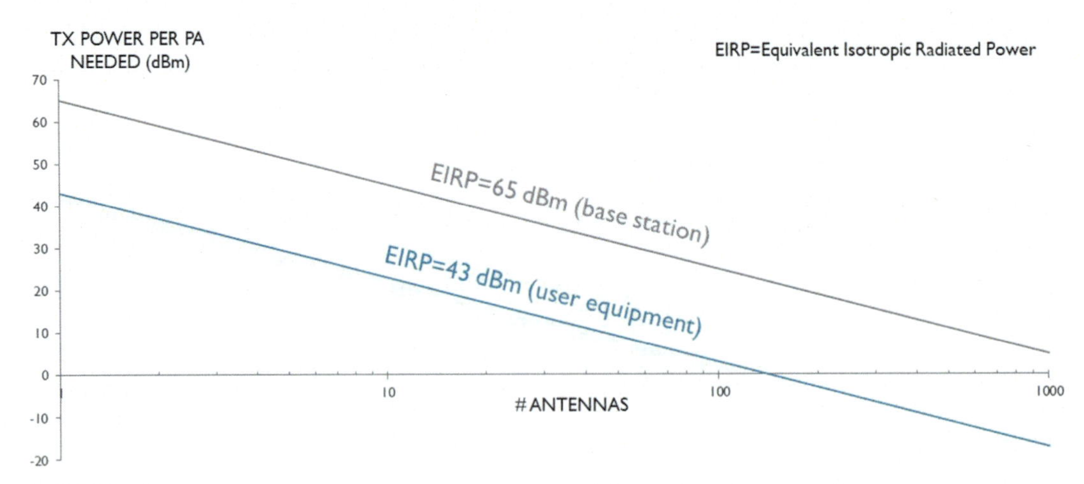


**FIGURE 1.5** Required output power per PA as function of the number of antennas for a user equipment case and a base station case.

In the next section, we will therefore give a short overview of the pros and cons of different semiconductor technologies for the RF-FEM.

1. Bulk CMOS and FinFET: The success of mobile devices can be largely attributed to the availability of low-cost integrated circuits and modules based on CMOS technology. The continued scaling of silicon-based technologies has allowed for economies of scale by upscaling to processing on 300 mm Si substrates. However, as mentioned before, scaled Si CMOS devices with short gate lengths, small fin width, and pitch in the case of FinFET are not well suited for RF operation due to the higher gate parasitics, decreased mobility, and in general degraded analog performance. Legacy nodes ($>$14 nm) on advanced engineered substrates, such as RF-SOI and FDSOI, are these days more widely used for high-end RF-FEM modules and RFIC System-on-Chip (SoC) applications up to the mm-wave frequency range, and they offer in general better trade-offs as compared to bulk CMOS.
2. FDSOI is an attractive solution for digital, analog, RF, and mixed signal circuits because of its excellent electrostatic behavior and efficient body biasing. SOI wafers consist of a thin crystalline Si (c-Si) layer (few nanometers up to tens of nanometers) bonded on a oxide layer that can also be scaled down to several tens of nanometers. Similar to FinFET, the channel thickness is therefore limited in thickness and allows the gate to have a much stronger control over the current in the channel, thereby limiting the leakage. The thin buried oxide also enables efficient back biasing using the carrier Si substrate as second gate. By etching away the Si layer and the thin buried oxide in the bulk contact areas, the underlying substrate can be contacted. A dedicated implant going through the channel layer and buried oxide into the Si substrate can be used to change the polarity of that gate and shift the workfunction of the back gate. This back gate can be used to increase the drive current, change the threshold voltage of the transistor, and reduce the power consumption of the transistor.

The ability of the FDSOI technology to operate at frequencies up to and beyond 100 GHz, with the same integration density as advanced bulk CMOS technologies, makes this an interesting candidate to be used across multiple 5G and even 6G applications (Sadlo et al., 2021). 28 nm FDSOI with $f_t$ and $f_{max}$ values above 300 GHz has been demonstrated and when combining this technology with high resistivity SOI (HR-SOI) enhanced RF switch performance can be shown while allowing at the same time the integration of high-quality passives, low-loss interconnects, and antenna-on-chip.

More details will be given in Chapters 2 and 9.

3. LDMOS (laterally diffused metal-oxide semiconductor) is a type of transistor technology that is widely used in high-power applications, such as in cellular base station amplifiers, broadcast transmitters, and microwave ovens. It is a type of RF PA technology that is designed to operate at high frequencies and deliver high levels of output power.

   One of the key advantages of LDMOS technology is its high-power density, which allows it to deliver high levels of output power in a compact package. LDMOS transistors can operate at frequencies ranging from a few megahertz to several gigahertz, making them suitable for a wide range of applications. Additionally, LDMOS technology is relatively low cost and can be manufactured using standard CMOS processes, which makes it attractive for high-volume production.

   However, there are also some disadvantages to LDMOS technology. For example, it can suffer from thermal runaway at high power levels, which can cause the device to overheat and fail. Additionally, LDMOS transistors have a relatively low breakdown voltage, which limits their ability to handle high voltage signals. Finally, LDMOS technology has a limited operating bandwidth, which can make it less suitable for some high-frequency applications. Therefore it's application is typically limited to frequencies below 6 GHz.

4. SiGe BiCMOS enables applications that require high RF/analog performance and high computational speed at low power by combining SiGe HBTs and CMOS into the same substrate. The SiGe HBTs allow to go to operating frequencies above 100 GHz and offer at the same time high linearity, low noise, and high output power. Both PA (Visweswaran et al., 2019) and low noise amplifier (LNA) have been demonstrated in this technology, showing excellent performance. The 130 nm node is appealing for a variety of applications mainly below 30 GHz, while the 55 nm node enables higher frequencies and larger digital integration. The latter node is especially well suited for the demanding circuits used in optical (fiber) communications, allowing an increase in the data rate up to 400 Gb/s and beyond. Wireless circuits up to 200 GHz can also be fabricated, covering both the highest 5G frequency bands and bands anticipated for 6G (Pallotta et al., 2021).

   In Chapter 5. more details will be provided regarding SiGe BiCMOS technologies.

5. GaAs is a common semiconductor used in RF applications, and can be found in many consumer, commercial, and military electronics to generate high-frequency signals. Besides these applications, there is high interest in GaAs for micro-LED. This material combines excellent electrical and thermal properties. Typical breakdown voltages are above 2.5 V (vs 1–1.8 V for Si) due to its high bandgap. Devices are made on native GaAs substrates with HR and dimensions even going up to 8 inch. Both HBT and

HEMT devices have been considered. GaAs HBTs typically target sub-6 GHz frequencies, while pseudo-morphic (pHEMT) and metamorphic HEMTs (mHEMT) achieved FOMs that are suitable for mm-wave frequencies as they use low In content InGaAs channels with higher mobility than GaAs. The difference in pHEMT and mHEMT lies in the way the heterostructure is grown.

Typically the In content of the channel material in GaAs HEMTs is lower than 53%. If In concentrations higher than 53% need to be used, the use of InP starting substrates is a better choice because of the increasing mismatch and therefore the thinner critical thickness. In pHEMT devices, the problem of lattice mismatch is solved by using a thin layer (typically AlGaAs), below the critical thickness, on top of the GaAs to make the heterostructure. In the case of mHEMT, a buffer layer is used between the GaAs substrate and the final channel layer. The buffer layer consists for example, of InAlAs where the In concentration is graded in such a way that it can match the lattice constant of both the GaAs substrate and the InGaAs channel.

In Chapter 10. MMIC design in GaAs technology will be discussed.

6. GaN has been deemed one of the most promising semiconductor materials for emerging RF applications because of its outstanding material properties, such as high carrier mobility, high breakdown voltage, and exceptional thermal stability. GaN is a semiconductor that can be easily grown on a wide variety of substrates. Applications for GaN include power semiconductors, lasers, optical amplifiers, and even high-efficiency solar cells. Especially GaN on silicon carbide (SiC) and GaN grown on Si are of high interest. The former one for infrastructure applications due to the excellent thermal conductivity of the SiC substrate. The latter option is currently under investigation for small cell infrastructure and applications that are more cost-sensitive like handsets. Challenges lie in upscaling the GaN growth to large areas, developing efficient manufacturing processes, reducing the impact of trapping, improving thermal dissipation, and reducing the losses coming from the substrate.

   Chapter 4 deals with aspects concerning technology optimization and challenges, while in Chapter 10, MMIC designs in GaN technology will be discussed.
7. InP HEMT (Mei et al., 2015) and HBT (Urteaga et al., 2017) have shown that $f_{max}$ numbers above 1 THz can be achieved, giving these technologies a clear edge over all the other technologies, especially for sub-THz and THz operating frequencies. Combining speed, high output power and efficiency, and low noise makes these transistors very interesting candidates for the frequencies targeted for 6G. However, from all the technologies, InP is the least mature, and suffers from a low level of integration, thereby still using small size substrates (up to 4 inch, although 6 inch wafers have been announced but are at this moment not widely used) and lab-like processes like lift-off, e-beam lithography, and Au-based metallization which make these devices hard to integrate in the overall RF-FEM module. Therefore, targeted R&D is needed to bring these technologies to the same level as Si-based technologies before they can be considered for next generation wireless applications. Chapters 6 and 5 deal with InP HEMT and InP HBT technologies, respectively.

Ultimately the choice of technology, in particular for the TX side, will largely depend on the frequency range and use case. For the digital part and parts of the beamforming chip, advanced CMOS will still be used. This could be different CMOS nodes where for the DSP and ADC/DAC the most scaled CMOS nodes will be used and where for the up/down conversion, phased locked loop, etc. more analog/RF friendly CMOS nodes will be used. In general, co-integration of the technology chosen for the PA and/or LNA, which can be different from CMOS, and the CMOS beamforming transceiver will be needed. Depending on the requirements, this might involve standard PCB technologies or more advanced approaches involving wafer-level heterogeneous integration as described in the last technical section of this chapter and in Chapter 8.

In the next section, we will particularly focus on the compound semiconductor technologies GaN and InP, and describe, next to the benefits and challenges, the progress towards upscaling and maturing these technologies.

## 1.3 Compound semiconductor devices

### 1.3.1 GaN devices

There are different flavors of GaN. GaN-on-GaN is the most expensive solution with substrates going up to 4 inch, but ultimately giving the best crystal quality. GaN-on-SiC is currently being considered for infrastructure applications due to the excellent thermal properties of the SiC substrates. The solution is more cost-friendly than GaN-on-GaN but still lacks the availability of large size SiC substrates to further reduce the cost. Therefore, GaN-on-Si is a very interesting option. The blanket growth of GaN on 200 mm (111) Si, so far mostly for high-power applications, has become a mature process. While achieving large-area GaN growth has been a major challenge for years, multiple techniques have been investigated to grow GaN, such as molecular beam epitaxy (MBE), metalorganic chemical vapor deposition (MOCVD), and plasma enhanced chemical vapor deposition (PECVD). From all these techniques, MOCVD and PECVD have proven to be the most scalable and manufacturable techniques for large-area GaN growth.

The starting material is typically a GaN seed crystal or a substrate, such as sapphire, silicon carbide, or silicon wafer, onto which the GaN layer is grown. The growth process involves the vaporization of a gallium-based compounds, such as trimethylgallium (TMG) or triethylgallium (TEG), and a nitrogen-based compound, such as ammonia or hydrazine, in a high-temperature, high-pressure reactor. The vaporized compounds react to form GaN, which deposits onto the seed crystal or substrate as an epitaxial layer. Another method being used is hydride vapor phase epitaxy (HVPE). HVPE has several advantages over other growth techniques like MOCVD and MBE. It allows for the growth of thick GaN layers at relatively low temperatures, making it suitable for use with temperature-sensitive substrates. HVPE also allows for the growth of high-quality GaN layers with low defect densities and good crystal quality. However, HVPE has some limitations, such as the need for high-pressure reactor equipment and the potential for contamination from the hydride compounds used in the growth process. Therefore, most commonly used methods to grow GaN layers and heterostructures still rely on MOCVD and MBE.

The lattice mismatch between GaN and SiC is 4%, but this can go up to 14% in the case of GaN and sapphire and even 17% in the case of GaN and Si. Typically, a buffer layer is grown to mitigate this difference in lattice constant. These buffer layers are typically several μms thick, especially when high breakdown voltages need to be ensured. Wafer bending, layer cracking, and wafer breakage are the main concerns, and careful optimization of the epi process is needed because of the thermal mismatch between the layers. The use of thick (1.15 mm thick) nonstandard starting Si substrates is required to solve these problems.

In power electronics Qromis Substrate Technology (QST) wafers are often used. These are engineered substrates with a CTE-matched (Coefficient of Thermal Expansion) poly-AlN core, featuring a thermal expansion that closely matches the thermal expansion of GaN materials. This allows for the fabrication of GaN devices with thicker buffer layers (needed for high voltage applications, e.g., 650 V) with higher yield since issues of wafer bending and breakage are omitted.

For RF applications, especially when looking to handheld devices where the requirements for breakdown voltage and bias conditions are much lower than in the case of high-power applications, the thickness of the buffer can be reduced significantly down to 2.5 μm and below, but requires the use of HR-Si substrates to reduce the substrate losses and improve the linearity.

When trying to grow GaN on a Si platform, it's important to consider how the substrate might affect the quality of not just passive components, but also the PA performance. This can be studied using coplanar transmission lines. Performance competing with best-in-class SOI technologies has been reported in Yadav et al. (2020). In the case of GaN-on-Si stacks, however, the direct relationship between RF losses and distortion that has been observed in SOI substrates does not seem to hold true. The hysteresis observed as described in Cardinael et al. (2021) is likely due to the long emission time constants of traps located inside the GaN buffer layers. This poses also many challenges for predicting substrate nonlinearity as a function of DC bias and it requires precise modeling of the buffer dynamic effects (Parvais et al., 2021; Rack & Raskin, 2017).

Current collapse or dynamic Ron is one of the challenges when designing RF GaN devices, especially for low bias conditions and high operating frequencies. This DC-RF dispersion can be attributed to defects in the barrier, surface traps, or bulk traps in the GaN structure. The surface traps can be controlled by the use of field plates. Bulk traps on the other hand come from the deep level dopants in the buffer to control the leakage and SCE. The role of C/O/H impurities and the acceptor/donor energy levels associated with other buffer defects has been discussed in Putcha et al. (2021). Next to that, charge carrier interaction between the 2D electron gas (2DEG) and defects in the barrier layer or the surface interface states also results in DC-RF dispersion. Getting a good insight into the mechanisms and having models that accurately describe the impact on device level and even circuit level is essential and a big part of today's modeling research (Parvais et al., 2021).

Because of the large power involved, the RF-FEM can heat up rapidly, and the performance of the technology will be limited by thermal effects. Most models do not account for the thin film and interfacial filtering effects. Using Monte Carlo Boltzmann Transport Equations (BTE) simulations with first-principles phonon dispersions and scattering rates (Carrete et al., 2017), the thickness dependent in-plane and cross-plane GaN, AlGaN, and AlN thermal conductivities can be determined, and used to assess the self-heating behavior of GaN devices, where not only the choice of material stack plays an important role, but also the device layout (e.g., number of gate fingers, finger width, S/D distance, etc.).

Often a trade-off between design for self-heating and electrical performance needs to be made, for example, scaling the thickness of the barrier and GaN channel to account for SCE often leads to an increase in self-heating.

Pushing the devices to higher frequencies and thus higher $f_t$ and $f_{\max}$ means scaling the device: gate length $L_g$ as well as the source-to-gate $L_{gs}$ and gate-to-drain distance $L_{gd}$. Lateral scaling always goes hand in hand with vertical scaling to control SCE. For RF devices special attention is needed to reduce parasitics as interconnect capacitances and inductances cannot be neglected at high frequencies. The cut-off frequency of the current gain $f_t$ and of the unilateral power gain $f_{\max}$ are the metrics used to benchmark the RF performance of transistors. However, in the context of the PA, $f_{\max}$ is more relevant and this parameter also depends on the particular layout of the transistor. The gate resistance $R_g$ is thereby an important parameter to take into account. Also, given the high level of output power involved in the RF-FEM, the nonlinear characteristics of the devices must be accurately taken into account.

Fig. 1.6 compares the efficiency and output power of GaN-on-Si and GaN-on-SiC devices at 28–40 GHz. Data have been taken from (Then et al., 2021; Lin et al., 2020; Mi et al., 2017; Moon et al., 2019; Ommic website, n.d.; Parvais et al., 2020; THAL archives, n. d.; Wang et al., 2018; Wui Then et al., 2020; Zhang et al., 2018). This figure shows that efficiencies and output power for GaN-on-Si are becoming quite comparable to GaN-on-SiC, even at reduced bias conditions ($V_{DD} < 15$ V). This makes them interesting candidates for battery-powered UE applications. Seeing the full potential of GaN-on-Si also relies on having good models that include all nonidealities to assess the performance at circuit level. Establishing a Design-Technology Co-Optimization loop, just like in the case of digital applications, will give insight if these nonidealities can be dealt with at circuit and system level by using techniques such as digital pre-distortion or if innovations at material and device level are needed for example, the use of N-polar GaN as a way to push the performance of the devices to even higher frequencies like 94 GHz (Liu et al., 2021). When it

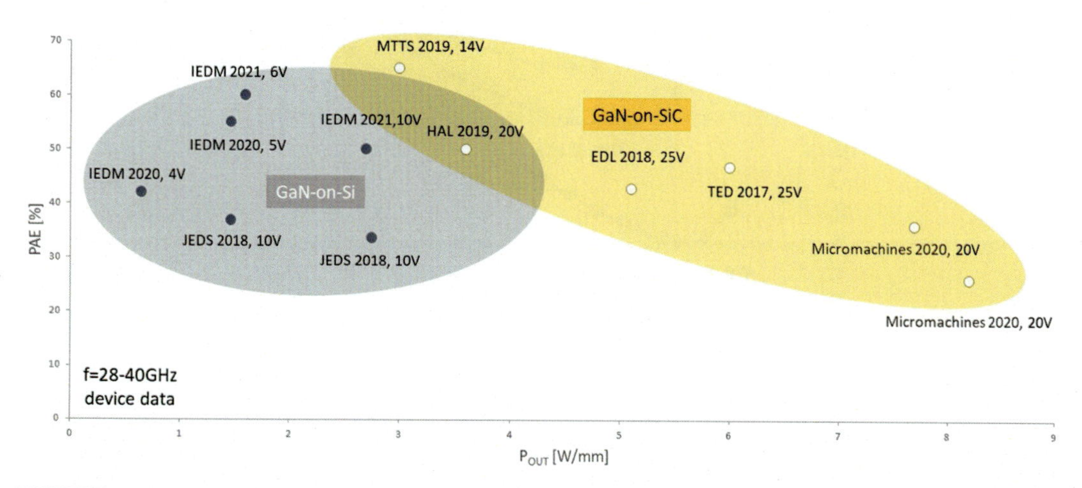


**FIGURE 1.6** **Benchmark** PAE versus $P_{out}$ comparing GaN-Si and GaN-SiC devices; the data is extracted from load-pull measurements at 28–40 GHz at different $V_{DD}$.

comes to GaN-on-Si circuit performance, recently Yu et al. (2022) reported PA and LNA results at 39 GHz showing $P_{sat}$, peak PAE, linear gain, and $OP_{1dB}$ of 25 dBm, 38.8%, 24.8 dB, and 19.8 dBm, respectively, and an area of only 0.079 $mm^2$, while the LNA results show 24.6 dB gain, 2.9 dB noise figure, and −11.4 dBm IIP3 at 38 GHz. This is the first demonstration of circuits enabled by a 300 mm GaN-on-Si platform.

## 1.3.2 InP-based devices

Using the model described in Desset et al. (2021), Fig. 1.7 illustrates how different PA technologies compare when sweeping the number of antennas for each of them, in order to optimize the total power. The comparison targets a UE device with an EIRP of 40 dBm and assumes 2 dBi antenna element gain. We compare three different cases: full implementation in CMOS, combining a CMOS beamforming transceiver, and a final amplification (PA) using either SiGe HBT or a third case where InP HBT is used for the PA.

For each technology, a trade-off is observed. Having a small number of antennas means the PA contributes more to the power consumption. But in a system with a lot of antennas, the analog chains dominate the power consumption. The differences seen in Fig. 1.7 between the three technologies directly relate to their difference in maximum efficiencies, that is, 6%, 9%, and 30% for CMOS, BiCMOS (SiGe), and InP, respectively. Fig. 1.7 illustrates that the optimum configuration for the CMOS-only implementation uses 48 antennas. In this case, 4.5 dBm per PA is assumed, while the InP technology uses only 20 antennas

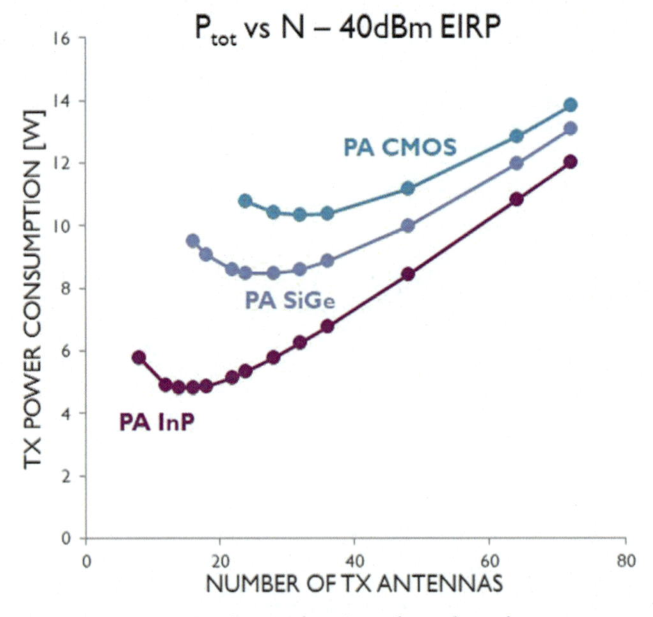


**FIGURE 1.7** Transmitter power consumption as function of number of antennas comparing three different cases at 140 GHz: a full CMOS implementation, a CMOS transceiver combined with SiGe HBT for the final power amplification stage and a InP HBT/CMOS implementation. EIRP = 40 dBm.

(with 12 dBm per PA), thanks to the higher efficiency and output power enabled by the InP PAs.

In general, the optimum power consumption $PW_{opt}$ and number of antennas $N_{TXopt}$ can be derived:

$$PW_{opt} = 2\sqrt{\frac{EIRP}{\eta} PW_{TX}} \tag{1.5}$$

$$N_{TXopt} = 2\sqrt{\frac{EIRP}{\eta} \frac{1}{PW_{TX}}}$$

where $\eta$ is the PA efficiency, EIRP is the equivalent isotropically radiated power, and $PW_{TX}$ is the power dissipated by each TX front-end element.

Three observations can be made when looking at Fig. 1.7. Firstly, by increasing the number $N_{TX}$ of antenna elements, the required PA output power per antenna is reduced, and so is the PA contribution to the total power budget, and all solutions converge to a similar power efficiency. For all the solutions, there is an optimum number of elements that allows to achieve the lowest and thus optimum power consumption. With respect to CMOS or BiCMOS, the InP PA implementation results in the lowest optimum number of antennas and power efficiency, thereby showing that this solution benefits from lower power consumption, reduced space, and reduced requirements on alignment of the TX antenna phases.

While this is an interesting solution, it requires a mature and cost-efficient InP technology that can also be co-integrated with CMOS.

The big challenge with using these semiconductor materials on a Si substrate is that the large lattice mismatch with Si leads to the origin of defects and suboptimal performance of the devices. But depending on the device we're trying to make, there are different ways to integrate these materials onto a Si substrate. In this section, we will discuss three different ways of upscaling.

Several approaches are being considered for the direct growth of III–V materials on Si, including both blanket and selective area growth (Kunert, Mols, et al., 2018). Blanket growth relies, just as in the case of GaN, on thick buffer layers to mitigate the difference in lattice constant. While threading dislocation densities (TDD) down to 1e6 $cm^{-2}$ have been shown for GaAs, for the high In-content layers numbers are still stuck in the 1e9 $cm^{-2}$ range (Huang et al., 2015). Nano-ridge engineering (NRE), which combines defect trapping with epitaxial lateral overgrowth, is the most effective way to reduce defectivity (Baryshnikova et al., 2020; Kunert, Langer, et al., 2018; Mols et al., 2021). The technique relies on creating high aspect ratio (AR) trenches in silicon. This can be done starting from a standard STI process commonly used in CMOS, after which the Si in the active layers is removed by either dry etch or wet etch. After that the Si surface is cleaned in-situ and the III–V layer is grown selectively by MOCVD. The part in the trench is highly defective, while the III–V area grown outside the trench is the lowly defective region which will be the core of the final device. As complex heterostructures need to be defined, a second template can be made on top of the STI before etching away the Si guiding the growth outside of the high AR trench and avoiding the sidewall growth of the different layers.

And example of a complex III–V heterostructure grown on a 300 mm Si wafer is shown in Fig. 1.8. In Baryshnikova et al. (2020), a more elaborate study is presented for GaAs nano-ridges (NRs). The study shows that TDD decreases as the width of the trenches in the Si decreases, as one would expect when AR trapping is used. In this case, for trench widths larger than 300 nm ($AR \leq 1$), the TDD is higher than 1e8 $cm^{-2}$. However, when the width is decreased to 100 nm, there is a significant drop in TDD, reaching numbers between 3e6 and 8e6 $cm^{-2}$. In cases where the trench width is only 80 nm wide ($AR = 3.75$), no signs of threading dislocations were found at all.

The other type of defects are planar defects (PD). The planar defect density (PDD) is defined as defects per NR trench length. The PDD increases with a reduction in trench width. For a large trench width, the PDD stays below 0.2 $\mu m^{-1}$ but rises sharply up to 0.5 $\mu m^{-1}$ for narrow trenches. Wafers that have been through different STI fabrication processes to fabricate the trenches can result in a spread of the PDD values, even when the MOCVD conditions are identical. If the seed nucleation is not optimized, it can result in a poorer quality deposition. However, it seems that the seed growth conditions improve for wider trenches with a larger pitch, as this results in a decrease of the PDD. However, impurities or $SiO_2$ residuals on the Si surface can also initiate PD formation. One possible

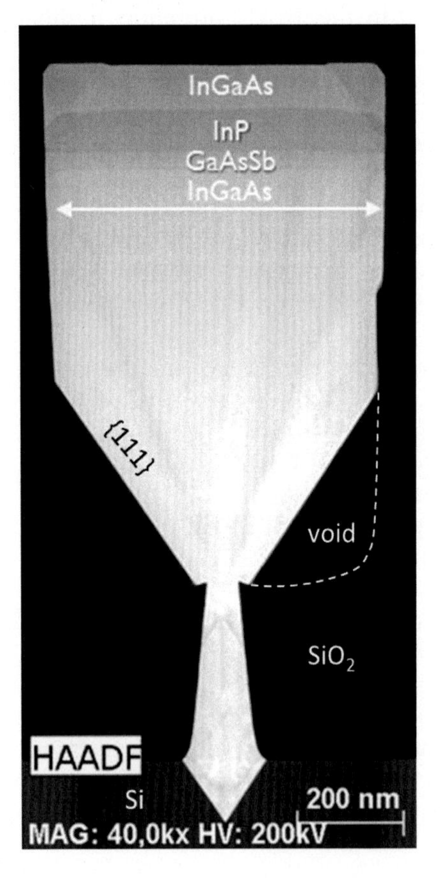


**FIGURE 1.8** TEM picture of a InGaAs/GaAsSb nano-ridge selectively grown on a 300 mm Si substrate.

reason for the observed PDD trend could be that wider trenches are cleaned more efficiently during the patterning process, as well a more efficient oxide removal process applied before growth. This is supported by the fact that a large spread in data for narrow trenches coming from different fabrication processes is seen. This suggests that the surface cleanliness of these wafers varies significantly. As many insights have been gained over the last years regarding the GaAs-on-Si growth, focus has moved to InP and InGaAs materials to enable the InP-based HBTs and HEMTs. The challenges are more significant, moving from 4% lattice mismatch to 8% lattice mismatch for InP and 53% InGaAs.

First demonstration that NRE can be a viable approach has been demonstrated in Vais et al. (2019) and Yadav et al. (2021) for GaAs HBTs. These publications not only demonstrate the defect reduction and first device data, but also show that the performance can be up to par with the blanket GaAs-on-Si approaches. However, different device optimization is needed. Using discrete NRs to build up the device is a departure from the standard HBT device layout which relies on big areas of III–V material. Several NRs need to be connected in parallel to achieve the same effective area. Even merged NRs are under study to be able to create locally larger areas of III–V as seed layers to start the growth of the complex heterostructures for the HBT devices (Vais et al., 2022).

The other approach relies on transferring InP tiles, with or without the active layers, onto a cost-effective and large size Si wafer using collective die-to-wafer bonding. This is the wafer reconstruction technology (Manda et al., 2019). The main advantage is that one can start from material with high crystal quality. The challenges with this process come from transferring the materials in a cost-efficient way and without introducing additional damage of the InP tiles. On top of that, in reality only the active layers are needed, and most of the InP tiles or seed can be removed or reused during transfer. Grinding, laser lift-off, microtransfer printing (Carter et al., 2019), and SmartCut (Ghyselen et al., 2022) are some of the techniques that are being considered, but controlling the warpage, caused by CTE mismatch between different materials in the reconstructed assembly, delamination, and planarization are just few of the other challenges that still need to be addressed.

Progress in maturing the small size substrates and processes should not be underestimated. Techniques as SmartCut used to enable 200 and 300 mm Si wafers with InP tiles can ultimately also be used to transfer a thin blanket layer of 2–4 inch InP onto an a same size Si wafer. This solves the issue of brittleness and cost as the donor InP substrate can be used several times, while the Si carrier wafer provides the mechanical strength in processing the wafers and improves yield.

Which approach finally will make it, not only depends on the performance of the devices that can be made using this technology but also the cost-effectiveness. Because of the upscaling, the direct growth and wafer reconstruction approaches are already more cost-effective than the options using small size InP substrates (up to 1000x more costly than 28 nm CMOS per $mm^2$). In the case of NRE, increasing the amount of useable area by NR pitch scaling and width increase are important parameters. For blanket approaches, bringing in reusability of the substrate leads to clear gains in cost, while for wafer reconstruction, both reusability as well as increasing the III–V density is important. Techniques like SmartCut and layer stack engineering which build in layers that can be selectively removed are quite interesting. All approaches will benefit from further cost reduction of

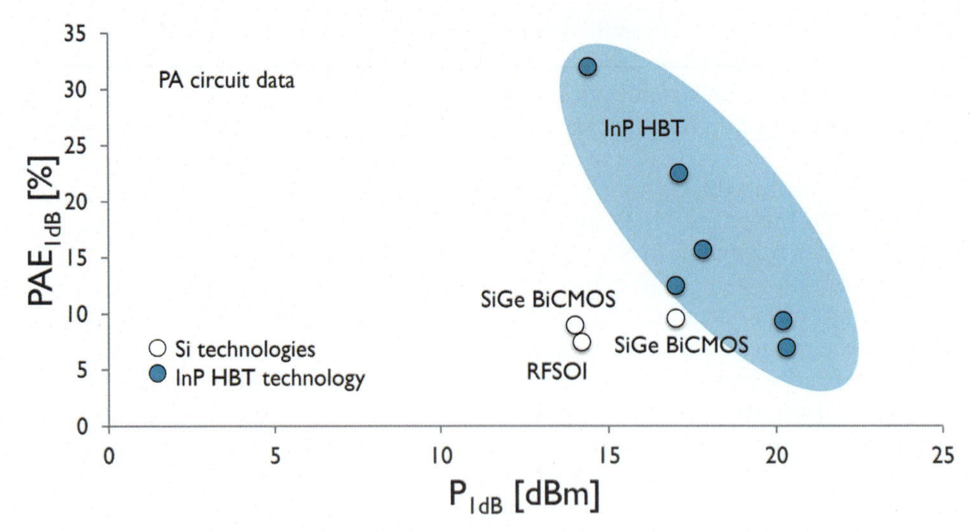


**FIGURE 1.9** **PA_InP** PA circuit data for different technologies; $PAE_{1db}$ versus $P_{1db}$ is compared for PAs designed in SiGe BiCMOS, RFSOI, and InP HBT at 140 GHz.

the III–V growth for the active layers and in the case of wafer reconstruction also cost reduction of the starting InP substrates.

What the technology can bring in terms of PA performance is clearly shown in Fig. 1.9 (data taken from: Ahmed et al., 2020; Banerjee & Wambacq, 2022; Griffith et al., 2019; Li & Rebeiz, 2021; Li et al., 2022; Ning et al., 2020; Visweswaran et al., 2019). In this figure the performance of InP PAs is compared to implementations in BiCMOS and RFSOI. The best trade-off in efficiency PAE and output power, here $P_{1dB}$, can be achieved for InP. The InP implementations are based on a 250 nm HBT process. Progress towards scaling the technology has been presented in Urteaga et al. (2011).

## 1.4 Heterogeneous integration

### 1.4.1 Introduction

In previous sections, we have seen that hybrid InP/CMOS technologies bring significant benefits in energy efficiency, performance, and footprint scaling. InP devices have been around for a while and have been used a lot in niche applications, mostly in the defense and military area. We have also shown in previous section that if InP devices are going to be used in the next generation of wireless communications, they need to be upscaled to improve their maturity, lower cost, and compatibility with heterogeneous integration options.

In this section, we will be looking at a few options for co-integrating these materials with CMOS. One way is monolithic integration where the compound semiconductor devices are placed next to the CMOS in the same substrate. Another way is 3D integration. The latter approach is becoming more popular because it allows to process and optimize different technologies independently and then put them together with wafer-to-wafer (W2W) or die-to-wafer (D2W) bonding and through silicon via (TSV) interconnections.

## 1.4.2 Different approaches

### *1.4.2.1 Chip-let and die-let*

While hybrid integration has been around for a while, mostly in the area of logic and memory, the co-integration with compound semiconductors for RF applications is still in its infancy. Chip-let and die-let approaches are rather being considered than wafer-level approaches.

In the chip-let approach, a small semiconductor chip that can be designed and manufactured separately from other chips is then connected together to form a larger SoC or a multi-chip module. Chip-let technology allows for the creation of more complex and powerful electronic devices by combining multiple smaller chips, each with its own specialized function, into a single integrated circuit. This can improve performance, reduce costs, and increase flexibility and scalability compared to traditional monolithic designs. Chip-let architectures can be used in a variety of applications, including smartphones, laptops, and high-performance computing systems.

While there are many advantages of blending different technologies together, there are also challenges related to interconnects, integration, performance optimization, and testing. Chip-lets need to be connected together in order to function as a single device, and often this requires the use of specific interconnects, such as wires or traces, which can add complexity and cost to the design. Chip-lets also need to be integrated into a larger system, which can be a complex process. It may be necessary to develop new tools (die pick-an-place) and processes to handle the integration of multiple chip-lets into a single system. Next, chip-lets may not always perform as well as a single, monolithic chip. There may be issues with signal integrity and power distribution that can impact overall performance.

Testing and verifying the performance of a chip-let-based system can be more complex than testing a single, monolithic chip. This can increase the time and cost of development. And finally, packaging chip-lets in a way that allows them to be easily integrated into a larger system can be a challenge. The packaging must protect the chip-lets and provide the necessary electrical and mechanical connections.

### *1.4.2.2 Wafer-level bonding*

One advantage of a W2W approach (3D integration) is that the integration can be done at the wafer level, so that the whole wafer is processed at once. This avoids the need to assemble a large number of individual dice, which could be a time-consuming and costly step, and avoid the challenges with chip-lets as mentioned above. There are many flavors of W2W bonding, depending on the material at the surface that needs to be bonded or the kind of bonding used: direct bonding, indirect bonding, anodic bonding, fusion bonding, and hybrid bonding.

Direct bonding involves bringing the two wafers into close contact and applying pressure to create a bond. This process can be used with wafers made of the same or different materials.

In indirect bonding an intermediate layer is used, such as a polymer or adhesive, to bond the two wafers together. This process can be used with wafers made of different materials and can provide improved adhesion compared to direct bonding.

Anodic bonding is a process that uses an electric current to create a bond between two wafers. This process is typically used to bond wafers made of different materials, such as silicon and glass.

Fusion bonding involves applying heat and pressure to the two wafers to create a bond. This process can provide strong, reliable bonds.

And finally hybrid bonding involves the creation of dielectric-to-dielectric and metal-to-metal bonds between the two surfaces or wafers.

Many innovations in the last decade have continuously pushed the performance of wafer-level bonding to smaller interconnect pitches and dimensions with increasing yield, for example, the conventional Cu/oxide based hybrid bonding scheme has faced significant technological challenges when scaling the Cu interconnects in size and pitch. Optimization of the Cu pad design using unequal sizes, the use of SiCN-to-SiCN dielectric bonding in combination with direct Cu-Cu bonding, and different surface topography for the top and bottom wafers can lead to further scaling of the interconnects as shown in Beyne et al. (2018).

Some of the challenges associated with this wafer-to-wafer bonding include the following:

1. Alignment: Precise alignment of the two wafers is critical for successful bonding. Any misalignment can result in defects or reduced performance due to loss of interconnects or to only partially connected Cu lines.
2. Cleanliness: The wafers must be extremely clean in order for the bonding process to work effectively. Any contaminants or particles can interfere with the bonding process and result in defects or failure in bonding.
3. Stress: The bonding process can create stress in the wafers, which can affect their performance. It may be necessary to use specialized techniques to mitigate stress in the bonded wafers like the use of backside layers that can compensate for the induced stress.
4. Thickness mismatch: If the two wafers have different thicknesses, it can be challenging to achieve a uniform bond. This can again result in defects or reduced performance.
5. Cost: Wafer-to-wafer bonding can be a complex and time-consuming process, which can increase the overall cost of manufacturing the device.

The other approach is D2W bonding. This process is similar to W2W bonding, but involves bonding a single die to a wafer rather than two wafers together. The process of D2W bonding typically involves aligning the die to the wafer using precision alignment equipment, applying a bonding material, such as epoxy or solder, and applying heat and pressure to complete the bonding process. D2W bonding can be used to create devices with a high degree of integration and functionality, but it can be a complex and time-consuming process, which can increase the overall cost of manufacturing the device. Cleanliness and alignment complexity are even more challenging than W2W.

In the case of mm-wave and sub-THz frequencies, having short and well-defined connections between different chips and circuits is important, especially between the antennas and the PAs. It would be ideal to have a 3D implementation that brings the antenna module closer to the PA driving the different antenna patches.

Another benefit is the reduced area, which is helpful for applications where the space is restricted. However, while 3D integration brings many benefits for RF applications at frequencies targeted for 6G, the main challenge will be the heat removal generated during operation, especially for the PA. The limited efficiency of the PAs, especially at higher frequencies, and the reduced area lead to a higher dissipated power density.

#### 1.4.2.3 *Interposer*

Another option is 2.5D or Si interposer/bridge, which connects the compound semiconductor die which contains the PA (and possibly low noise amplifier and switches) and the CMOS die using metal lines on a Si substrate that can be processed in a CMOS fab. Silicon interposers are typically made from silicon wafers and may include features such as micro-vias, which enable the connection of the different layers of the interposer, and passive components such as resistors and capacitors, which can be used to filter signals or provide voltage reference points. Silicon interposers can also be used to connect devices with different form factors or different materials.

Typically Si interposer technologies that have been developed are intended for use in high-density digital systems (Hellings et al., 2015). The essential modules are MIM decoupling capacitors (1–7 nF/mm$^2$), 0.5–2 μm thick Cu/oxide damascene, and $10 \times 100$ μm$^2$ or $5 \times 50$ μm$^2$ via-middle Cu TSVs.

In the next section, we will briefly review what is needed to make this approach suitable for RF applications.

### 1.4.3 Radio frequency interposer

The key to making this interposer technology work for RF applications is to reduce the metal pitch down to a few tens of micrometers and use low loss dielectrics.

In Sun et al. (2022) and Peeters et al. (2022), a first implementation of this RF-tailored interposer using wafer-level (300 mm) scale integration has been shown. This RF Si interposer, schematically shown in Fig. 1.10, uses a thick benzo-cyclobutene (BCB) layer and Cu-based redistribution layers (RDL).

A 1 μm thick ground plane, using a standard damascene Cu back end, is used to reduce the Si substrate loss and a 18 μm thick BCB layer to lower the parasitic capacitance to the ground shield. This also reduces the transmission line loss, while the 5 μm thick RDL enables low resistance. In Fig. 1.11, a schematic of the 300 mm integration flow is shown to fabricate the RF interposer. One also has to note that this is enabled using a standard Si substrate, not a HR-Si substrate. Avoiding the use of HR-Si is beneficial as, besides the higher cost of the HR-Si, additional surface passivation procedures are often needed since the RF characteristics of bulk HR-Si are severely undermined by unavoidable free charges at the Si-$SiO_2$ interface.

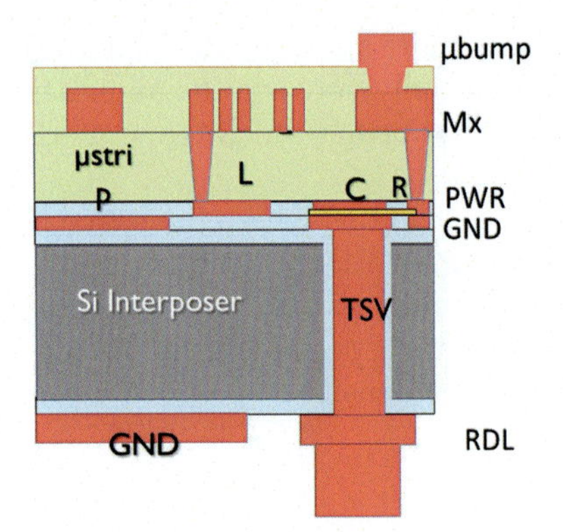


**FIGURE 1.10** Radio frequency (RF) interposer. A schematic view of a RF-tailored Si interposer, suitable for mm-wave and sub-THz frequencies. *Source: From Sun X., Slabbekoorn J., Sinha S., Bex P., Pinho N., Webers T., Velenis D., Miller A., Collaert, N., Van der Plas G., & Beyne E. (2022). Cost-effective RF interposer platform on low-resistivity Si enabling heterogeneous integration opportunities for beyond 5G. In:* Proceedings ECTC.

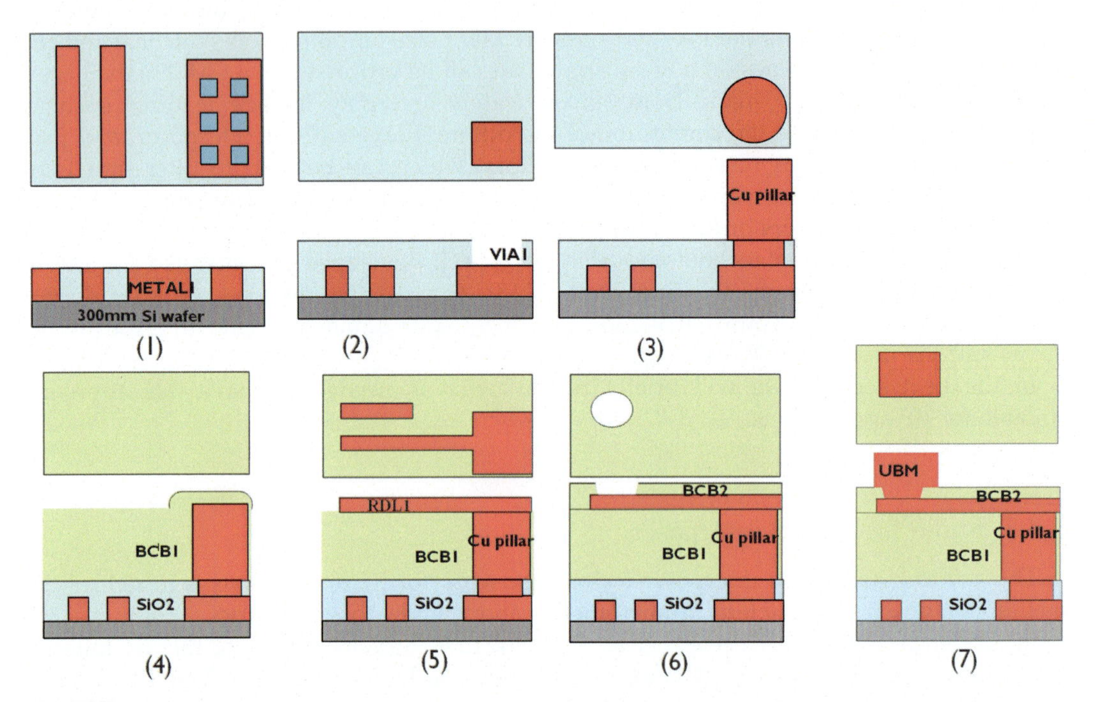


**FIGURE 1.11** Radio frequency (RF) interposer flow. Schematic presentation of the 300 mm RF interposer integration flow (top view and cross-section are shown): (1) Metal1 damascene (METDAM1), (2) via formation, (3) pillar definition, (4) BCB coating, (5) RDL layer formation, (6) passivation, (7) under bump metallization (UBM). *Source: From Sun X., Slabbekoorn J., Sinha S., Bex P., Pinho N., Webers T., Velenis D., Miller A., Collaert, N., Van der Plas G., & Beyne E. (2022). Cost-effective RF interposer platform on low-resistivity Si enabling heterogeneous integration opportunities for beyond 5G. In:* Proceedings ECTC.

Compared to widely used advanced printed circuit board (PCB) technologies like IC substrate with a modified semi-additive process (mSAP) and anylayer high density interconnect (HBI) PCB, the narrow μbump pitch (20–40 μm) on the RF interposer enables flip-chip performance up to 500 GHz as confirmed by simulations (Sun et al., 2022). A comparison between the different technologies is shown in Table 1.3.

TABLE 1.3 Comparison of the radio frequency Si interposer with two widely used printed circuit board technologies.

| | RF Si interposer | IC substrate + mSAP | Anylayer HDI PCB |
|---|---|---|---|
| Routing layers | 3–4 | 6–8 | 6–8 |
| Trace / space | 5 μm/5 μm | 15 μm/15 μm | 50 um/50 μm |
| Layer thickness | 5–20 μm | 10–30 μm | 50–250 μm |
| Bump pitch | 20–40 μm | 80–120 um | 200 um |
| Flip-chip Freq. | 500 GHz | 300 GHz | 170 GHz |
| Flip-chip area | 150 x 150 $\mu m^2$ | 300 x 300 $\mu m^2$ | 600 x 600 $\mu m^2$ |
| Flip-chip loss @150 GHz | 0.2 dB | 0.5 dB | 1 dB |
| Line loss @150 GHz | 0.3 dB/mm | 0.4 dB/mm | 0.2 dB/mm |

The technology is effective up to 110 GHz, validated by measurements, and as simulations have shown can be extended to the D-band and much higher frequencies. Moreover, the demonstration of high-Q inductors (up to 50 GHz) integrated on this RF interposer shows that passive components, normally integrated on the CMOS chip and thus taking up area, can successfully off-loaded to the Si interposer. The RF interposer also enables antennas above 100 GHz, which are challenging to implement on PCBs due to the small array pitch required (1 mm at 150 GHz) and the increased conductor losses caused by the large Cu roughness on PCB. These challenges are overcome by the RF interposer's fine-pitch RDL and negligible Cu roughness.

Overall, the superior RF performance and thermal management of RF Si interposers allows for them to be used as a heterogeneous platform for co-integrating mm-wave ICs in different technologies (de Kok et al., 2021; Ndip, et al., 2020).

Next steps include increasing the amount of RDL layers for routing of the RF and DC signals, demonstrating the TSVs and backside metallization. Ultimately, one could think of even integrating the antenna in the Si interposer as mentioned above, thereby having an antenna in package with the shortest possible connections to the circuits.

## 1.5 Summary

To achieve the increased throughput that future communication systems will require, wider bandwidths must be exploited. This can only come from using higher carrier frequencies, ranging from mm-wave to frequencies above 100 GHz. While the benefits are clear, many challenges need to be addressed. These challenges are related to identifying the right technologies for the different parts of the RF-FEM. While CMOS is in general the technology of choice, compound semiconductors like GaN and InP might be the better choice for especially the transmit side of the radio. System level studies have shown that co-integrating CMOS front ends with, for example, InP PAs promises savings of 2x or more in power consumption and array size in the frequency range where traditional CMOS technology becomes less efficient.

To make these compound semiconductors viable to be used in these future mass-volume applications, upscaling to a more cost-friendly and CMOS compatible platform is needed. Driven by the success of GaN-on-Si for power electronics lots of attention is focused on optimizing these devices for high-frequency applications. This GaN-on-Si technology must thereby be seen as a more cost-friendly option as compared to GaN-on-SiC, now widely considered mostly for infrastructure.

InP, envisioned as a must-have for frequencies above 100 GHz, is the least mature of all non-Si RF technologies, and the road to upscaling that technology is still in an exploratory phase with many options still on the table.

Finally, as the ultimate 6G architecture is envisioned to be hybrid, heterogeneous integration is becoming an important topic. Techniques going from monolithic integration to 2.5D interposer to 3D integration are being considered. Each approach has its advantages and challenges. In this work, we have lifted out the RF Si interposer technology as it shows the promise to enable the co-integration of CMOS and compound semiconductors for frequencies even up to 500 GHz while offering a more generic solution that can be used across different technologies.

## References

Ahmed, A. S. H., Seo, M., Farid, A. A., Urteaga, M., Buckwalter, J. F., & Rodwell, M. J. W. (2020). A 140GHz power amplifier with 20.5dBm output power and 20.8% PAE in 250-nm InP HBT technology. In: *IEEE MTT-S international microwave symposium digest* (pp. 492–495). United States: Institute of Electrical and Electronics Engineers Inc. Available from https://doi.org/10.1109/IMS30576.2020.9224012.

Arabhavi, A. M., Ciabattini, F., Hamzeloui, S., Fluckiger, R., Popovic, T., Han, D., Marti, D., Bonomo, G., Chaudhary, R., Ostinelli, O., & Bolognesi, C. R. (2021). InP/GaAsSb double heterojunction bipolar transistor technology with fMAX = 1.2 THz. In: *Technical digest – International electron devices meeting*, IEDM (pp. 11.4.1–11.4.4). Switzerland: Institute of Electrical and Electronics Engineers Inc. Available from https://doi.org/10.1109/IEDM19574.2021.9720644.

Arabhavi, A. M., Quan, W., Ostinelli, O., & Bolognesi, C. R. (2018). Scaling of InP/GaAsSb DHBTs: A simultaneous f-T /f-MAX = 463/829 GHz in a 10 mu m long emitter. In: *2018 IEEE BiCMOS and compound semiconductor integrated circuits and technology symposium*, BCICTS 2018. Switzerland: Institute of Electrical and Electronics Engineers Inc. http://ieeexplore.ieee.org/xpl/mostRecentIssue.jsp?punumber = 8536738.

Baryshnikova, M., Mols, Y., Ishii, Y., Alcotte, R., Han, H., Hantschel, T., Richard, O., Pantouvaki, M., Van Campenhout, J., Van Thourhout, D., Langer, R., & Kunert, B. (2020). Nano-ridge engineering of GaSb for the integration of InAs/GaSb heterostructures on 300 mm (001) Si. *Crystals, 10*(4), 330. Available from https://doi.org/10.3390/cryst10040330.

Banerjee, A., Wambacq, P. (2022). A 130GHz Two-Stage Common-Base Power Amplifier in 250nm InP, 2022 47th International Conference on Infrared, Millimeter and Terahertz Waves (IRMMW-THz), Delft, Netherlands, 1-2. https://doi.org/10.1109/IRMMW-THz50927.2022.9895854.

Beyne, E., Kim, S. W., Peng, L., Heylen, N., De Messemaeker, J., Okudur, O. O., Phommahaxay, A., Kim, T. G., Stucchi, M., Velenis, D., Miller, A., & Beyer, G. (2018). Scalable, sub 2μm pitch, Cu/SiCN to Cu/SiCN hybrid wafer-to-wafer bonding technology. In: *Technical digest – International electron devices meeting*, IEDM (pp. 32.4.1–32.4.4). Belgium: Institute of Electrical and Electronics Engineers Inc. Available from https://doi.org/10.1109/IEDM.2017.8268486.

Cardinael, P., Yadav, S., Zhao, M., Rack, M., Lederer, D., Collaert, N., Parvais, B., & Raskin, J. P. (2021). Impact of III-N buffer layers on RF losses and harmonic distortion of GaN-on-Si Substrates. In: *European solid-state device research conference* (pp. 303–306). Editions Frontieres Belgium. http://ieeexplore.ieee.org/xpl/conferences.jsp. Available from https://doi.org/10.1109/ESSDERC53440.2021.9631822.

Carrete, J., Vermeersch, B., Katre, A., van Roekeghem, A., Wang, T., Madsen, G. K. H., & Mingo, N. (2017). almaBTE: A solver of the space–time dependent Boltzmann transport equation for phonons in structured materials. *Computer Physics Communications, 220*, 351–362. Available from https://doi.org/10.1016/j.cpc.2017.06.023, http://www.elsevier.com/wps/find/journaldescription.cws_home/706710/description#description.

Carter, A. D., Urteaga, M. E., Griffith, Z. M., Lee, K. J., Roderick, J., Rowell, P., Bergman, J., Hong, S., Patti, R., Petteway, C., Fountain, G., Ghosel, K., & Bower, C.A. (2019). Si/InP heterogeneous integration techniques from the wafer-scale (hybrid wafer bonding) to the discrete transistor (micro-transfer printing). In: *IEEE SOI-3D-subthreshold microelectronics technology unified conference*, S3S 2018. United States: Institute of Electrical and Electronics Engineers Inc. Available from https://doi.org/10.1109/S3S.2018.8640196, http://ieeexplore.ieee.org/xpl/mostRecentIssue.jsp?punumber = 8636845.

Carter, R., Mazurier, J., Pirro, L., Sachse, J. U., Baars, P., Faul, J., Grass, C., Grasshoff, G., Javorka, P., Kammler, T., Preusse, A., Nielsen, S., Heller, T., Schmidt, J., Niebojewski, H., Chou, P. Y., Smith, E., Erben, E., Metze, C., ... Rice, B. (2017). 22nm FDSOI technology for emerging mobile, Internet-of-Things, and RF applications. In: *Technical digest – International electron devices meeting*, IEDM (pp. 2.2.1–2.2.4). Germany: Institute of Electrical and Electronics Engineers Inc. Available from https://doi.org/10.1109/IEDM.2016.7838029.

Chevalier, P., Avenier, G., Ribes, G., Montagné, A., Canderle, E., Céli, D., Derrier, N., Deglise, C., Durand, C., Quémerais, T., Buczko, M., Gloria, D., Robin, O., Petitdidier, S., Campidelli, Y., Abbate, F., Gros-Jean, M., Berthier, L., Chapon J. D., ... Borot, B. (2015). A 55 nm triple gate oxide 9 metal layers SiGe BiCMOS technology featuring 320 GHz fT / 370 GHz fMAX HBT and high-Q millimeter-wave passives. In: Technical digest – International electron devices meeting, IEDM 2015- no. February. (pp. 3.9.1–3.9.3). France: Institute of Electrical and Electronics Engineers Inc. Available from https://doi.org/10.1109/IEDM.2014.7046978.

Chowdhury, M. Z., Shahjalal, M., Ahmed, S., & Jang, Y. M. (2020). 6G Wireless communication systems: Applications, requirements, technologies, challenges, and research directions. *IEEE Open Journal of the Communications Society, 1*, 957–975. Available from https://doi.org/10.1109/ojcoms.2020.3010270.

Collaert, N., Alian, A., Banerjee, A., Chauhan, V., Elkashlan, R. Y., Hsu, B., Ingels, M., Khaled, A., Vondkar Kodandarama, K., Kunert, B., Mols, Y., Peralagu, U., Putcha, V., Rodriguez, R., Sibaja-Hernandez, A., Simoen, E., Vais, A., Walke, A., Witters, L., ... Waldron, N. (2020). From 5G to 6G: Will compound semiconductors make the difference? In: *2020 IEEE 15th international conference on solid-state and integrated circuit technology, ICSICT 2020 – Proceedings*. Belgium: Institute of Electrical and Electronics Engineers Inc. Available from https://doi.org/10.1109/ICSICT49897.2020.9278253, http://ieeexplore.ieee.org/xpl/mostRecentIssue.jsp?punumber = 9278127.

Desset, C., Collaert, N., Sinha, S., & Gramegna, G. (2021). InP/CMOS co-integration for energy efficient sub-THz communication systems. In: *2021 IEEE globecom workshops, GC Wkshps 2021 – Proceedings*. Belgium: Institute of Electrical and Electronics Engineers Inc. Available from https://doi.org/10.1109/GCWkshps52748.2021.9682092, http://ieeexplore.ieee.org/xpl/mostRecentIssue.jsp?punumber = 9681915.

de Kok, M., Smolders, A. B., & Johannsen, U. (2021). A review of design and integration technologies for D-band antennas. *IEEE Open Journal of Antennas and Propagation, 2*, 746–758. Available from https://doi.org/10.1109/OJAP.2021.3089052.

Europratice. (n.d). <https://europractice-ic.com/wp-content/uploads/2019/12/GLOBALFOUNDRIES-EUROPRACTICE-v1-2020-01-01.pdf>.

Ghyselen, B., Navone, C., Martinez, M., Sanchez, L., Lecouvey, C., Montmayeul, B., Servant, F., Maitrejean, S., & Radu, I. (2022). Large-diameter III–V on Si substrates by the smart cut process: The 200 mm InP film on Si substrate example. *physica status solidi (a), 219*(4)2270010. Available from https://doi.org/10.1002/pssa.202270010.

Griffith, Z., Urteaga, M., & Rowell, P. (2019). A 140-GHz 0.25-W PA and a 55-135 GHz 115-135 mW PA, high-gain, broadband power amplifier MMICs in 250-nm InP HBT. In: *IEEE MTT-S international microwave symposium digest* (pp. 1245–1248). United States: Institute of Electrical and Electronics Engineers Inc. Available from https://doi.org/10.1109/mwsym.2019.8701019.

HAL archives. (n.d). <https://hal.archives-ouvertes.fr/hal-0235688>.

Heinemann, B., Rucker, H., Barth, R., Barwolf, F., Drews, J., Fischer, G. G., Fox, A., Fursenko, O., Grabolla, T., Herzel, F., Katzer, J., Korn, J., Kruger, A., Kulse, P., Lenke, T., Lisker, M., Marschmeyer, S., Scheit, A., Schmidt, D., ... Wolansky, D. (2017). SiGe HBT with fx/fmax of 505 GHz/720 GHz. In: *Technical digest – International electron devices meeting*, IEDM (pp. 3.1.1–3.1.4). Germany: Institute of Electrical and Electronics Engineers Inc. Available from https://doi.org/10.1109/IEDM.2016.7838335.

Hellings, G., Scholz, M., Detalle, M., Velenis, D., De Ten Broeck, M. D. P., Roda Neve, C., Li, Y., Van Huylenbroek, S., Chen, S. H., Marinissen, E. J., La Manna, A., Van Der Plas, G., Linten, D., Beyne, E., & Thean, A. (2015). Active-lite interposer for 2.5 & 3D integration. In: *Digest of technical papers – Symposium on VLSI technology* (pp. T222–T223). Belgium: Institute of Electrical and Electronics Engineers Inc. Available from https://doi.org/10.1109/VLSIT.2015.7223647.

Hisamoto, D., Kaga, T., Kawamoto, Y., & Takeda, E. (1989). A fully depleted lean-channel transistor (DELTA)–A novel vertical ultra thin SOI MOSFET. In: *Technical digest – International electron devices meeting* (pp. 833–836). Publ by IEEE undefined.

Hisamoto, D., Lee, W.-C., Kedzierski, J., Anderson, E., Takeuchi, H., Asano, K., King, T.-J., Bokor, J., & Hu, C. (1998). A folded-channel MOSFET for deep–sub-tenth micron era. *IEDM*. Available from https://doi.org/10.1109/IEDM.1998.746531.

Huang, M. L., Chang, S. W., Chen, M. K., Fan, C. H., Lin, H. T., Lin, C. H., Chu, R. L., Lee, K. Y., Khaderbad, M. A., Chen, Z. C., Lin, C. H., Chen, C. H., Lin, L. T., Lin, H. J., Chang, H. C., Yang, C. L., Leung, Y. K., Yeo, Y. C., Jang ,S. M., ... Diaz, C. H. (2015). In0.53Ga0.47As MOSFETs with high channel mobility and gate stack quality fabricated on 300 mm Si substrate. In: *Digest of technical papers – Symposium on VLSI technology* (pp. T204–T205). Taiwan: Institute of Electrical and Electronics Engineers Inc. Available from https://doi.org/10.1109/VLSIT.2015.7223675.

Kavalieros, J., Doyle, B., Datta, S., Dewey, G., Doczy, M., Jin, B., Lionberger, D., Metz, M., Rachmady, W., Radosavljevic, M., Shah, U., Zelick, N., & Chau, R. (2006). Tri-gate transistor architecture with high-k gate dielectrics, metal gates and strain engineering. In: *Digest of Technical Papers – Symposium on VLSI Technology*. United States 50-51 07431562.

Kunert, B., Langer, R., Pantouvaki, M., Campenhout, J. V., & Thourhout, D. V. (2018). Gaining an edge with nano-ridges. *Compound Semiconductor, 24*(5), 36–41.

Kunert, B., Mols, Y., Baryshniskova, M., Waldron, N., Schulze, A., & Langer, R. (2018). How to control defect formation in monolithic III/V hetero-epitaxy on (100) Si? A critical review on current approaches. *Semiconductor Science and Technology, 33*(9)093002. Available from https://doi.org/10.1088/1361-6641/aad655.

Lai, R., Mei, X. B., Deal, W. R., Yoshida, W., Kim, Y. M., Liu, P. H., Lee, J., Uyeda, J., Radisic, V., Lange, M., Gaier, T., Samoska, L., & Fung, A. (2007). Sub 50 nm InP HEMT device with Fmax greater than 1 THz. In: *Technical digest – International electron devices meeting*, IEDM (pp. 609–611). United States. Available from https://doi.org/10.1109/IEDM.2007.4419013.

Lee, H. J., Callender, S., Rami, S., Shin, W., Yu, Q., & Marulanda, J. M. (2020). Intel 22nm low-power FinFET (22FFL) process technology for 5G and beyond. In: *Proceedings of the custom integrated circuits conference*. United States: Institute of Electrical and Electronics Engineers Inc. https://doi.org/10.1109/CICC48029.2020.9075914.

Li, S., & Rebeiz, G.M. (2021). A 130-151 GHz 8-way power amplifier with 16.8-17.5 dBm Psat and 11.7-13.4% PAE using CMOS 45 nm RFSOI. In: *Digest of papers – IEEE radio frequency integrated circuits symposium* (pp. 115–118). United States: Institute of Electrical and Electronics Engineers Inc. Available from https://doi.org/10.1109/RFIC51843.2021.9490507.

Li, X., Chen, W., Li, S., Wu, H., Yi, X., Han, R., & Feng, Z. (2022). A 110-to-130GHz SiGe BiCMOS doherty power amplifier with slotline-based power-combining technique achieving >22 dBm saturated output power and >10% power back-off efficiency. In: *Digest of technical papers – IEEE international solid-state circuits conference* (pp. 316–318). China: Institute of Electrical and Electronics Engineers Inc. Available from https://doi.org/10.1109/ISSCC42614.2022.9731552.

Lin, Y. C., Chen, S. H., Lee, P. H., Lai, K. H., Huang, T. J., Chang, E. Y., & Hsum, H.-T. (2020). Gallium nitride (GaN) high-electron-mobility transistors with thick copper metallization featuring a power density of 8.2 W/mm for Ka-band applications. *Micromachines, 11*. Available from https://doi.org/10.3390/mi11020222.

Liu, W., Romanczyk, B., Guidry, M., Hatui, N., Wurm, C., Li, W., Shrestha, P., Zheng, X., Keller, S., & Mishra, U. K. (2021). 6.2 W/Mm and record 33.8% PAE at 94 GHz from N-polar GaN deep recess MIS-HEMTs with ALD Ru gates. *IEEE Microwave and Wireless Components Letters, 31*(6), 748–751. Available from https://doi.org/10.1109/LMWC.2021.3067228, https://ieeexplore.ieee.org/servlet/opac?punumber=7260.

Manda, S., Zaizen, Y., Hirano, T., Iwamoto, H., Matsumoto, R., Saito, S., Maruyama, S., Minari, H., Hirano, T., Takachi, T., Fujii, N., & Yamamoto, Y. (2019). High-definition visible-SWIR InGaAs image sensor using Cu-Cu bonding of III-V to silicon wafer. In: *Technical digest – International electron devices meeting*, IEDM. Japan: Institute of Electrical and Electronics Engineers Inc. Available from https://doi.org/10.1109/IEDM19573.2019.8993432.

Manger, D., Liebl, W., Boguth, S., Binder, B., Aufinger, K., Dahl, C., Hengst, C., Pribil, A., Oestreich, J., Rohmfeld, S., Rothenhaeusser, S., Tschumakow, D., & Boeck, J. (2018). Integration of SiGe HBT with f -T = 305 GHz, f -max = 537 GHz in 130nm and 90nm CMOS. In: *IEEE BiCMOS and compound semiconductor integrated circuits and technology symposium*, BCICTS 2018 (pp. 76–79). Germany: Institute of Electrical and Electronics Engineers Inc. Available from https://doi.org/10.1109/BCICTS.2018.8550922, http://ieeexplore.ieee.org/xpl/mostRecentIssue.jsp?punumber = 8536738.

Mayeda, J. C., Lie, D. Y. C., & Lopez, J. (2021). A highly efficient 18-40 GHz linear power amplifier in 40-nm GaN for mm-Wave 5G. *IEEE Microwave and Wireless Components Letters*, *31*(8), 1008–1011. Available from https://doi.org/10.1109/LMWC.2021.3085241, https://ieeexplore.ieee.org/servlet/opac?punumber = 7260.

Mei, X., Yoshida, W., Lange, M., Lee, J., Zhou, J., Liu, P. H., Leong, K., Zamora, A., Padilla, J., Sarkozy, S., Lai, R., & Deal, W. R. (2015). First demonstration of amplification at 1 THz Using 25-nm InP high electron mobility transistor process. *IEEE Electron Device Letters*, *36*(4), 327–329. Available from https://doi.org/10.1109/LED.2015.2407193.

Mi, M., Ma, X.-H., Yang, L., Lu, Y., Hou, B., Zhu, J., Zhang, M., Zhang, H.-S., Zhu, Q., Yang, L.-A., & Hao, Y. (2017). Millimeter-wave power AlGaN/GaN HEMT using surface plasma treatment of access region. *IEEE Transactions on Electron Devices*, *64*(12), 4875–4881. Available from https://doi.org/10.1109/ted.2017.2761766.

Micovic, M., Brown, D. F., Regan, D., Wong, J., Tang, Y., Herrault, F., Santos, D., Burnham, S. D., Tai, J., Prophet, E., Khalaf, I., McGuire, C., Bracamontes, H., Fung, H., Kurdoghlian, A.K., & Schmitz, A. (2017). High frequency GaN HEMTs for RF MMIC applications. In: *Technical digest – International electron devices meeting*, IEDM (pp. 3.3.1–3.3.4). United States: Institute of Electrical and Electronics Engineers Inc. Available from https://doi.org/10.1109/IEDM.2016.7838337.

Mols, Y., Vais, A., Yadav, S., Witters, L., Vondkar, K., Alcotte, R., Baryshnikova, M., Boccardi, G., Waldron, N., Parvais, B., Collaert, N., Langer, R., & Kunert, B. (2021). Monolithic integration of nano-ridge engineered InGaP/GaAs HBTs on 300 mm Si substrate. *Materials*, *14*(19). Available from https://doi.org/10.3390/ma14195682, https://www.mdpi.com/1996-1944/14/19/5682/pdf.

Moon, J. S., Wong, J., Grabar, B., Antcliffe, M., Chen, P., Arkun, E., Khalaf, I., Corrion, A., & Post, T. (2019). Novel high-speed linear GaN technology with high efficiency. In: *IEEE MTT-S international microwave symposium digest* (pp. 1130–1132). United States: Institute of Electrical and Electronics Engineers Inc. Available from https://doi.org/10.1109/mwsym.2019.8700832.

Ndip, I., Andersson, K., Kosmider, S., Le, T. H., Kanitkar, A., Van Dijk, M., Senthil Murugesan, K., Maas, U., Loher, T., Rossi, M., Jaeschke, J., Ostmann, A., Aschenbrenner, R., Schneider-Ramelow, M., & Lang, K.D. (2020). A novel packaging and system-integration platform with integrated antennas for scalable, low-cost and high-performance 5G mm wave systems. In: *Proceedings – Electronic components and technology conference* (pp. 101–107). Germany: Institute of Electrical and Electronics Engineers Inc. Available from https://doi.org/10.1109/ECTC32862.2020.00029.

Ning, K., Fang, Y., Rodwell, M., & Buckwalter, J. (2020). A 130-GHz power amplifier in a 250-nm InP process with 32% PAE. In: *Digest of papers – IEEE radio frequency integrated circuits symposium* (pp. 195–198). United States: Institute of Electrical and Electronics Engineers Inc. Available from https://doi.org/10.1109/RFIC49505.2020.9218351.

Ommic website. (n.d). <https://www.ommic.com/>.

Pallotta, A., Roux, P., Del Rio, D., Sevillano, J. F., Pirbazari, M. M., Mazzanti, A., Ermolov, V., Lamminen, A., Saily, J., Frecassetti, M., Moretto, M., & De Cos, J. (2021). SiGe:BiCMOS technology is enabling D-band link with active phased antenna array. In: *2021 Joint European Conference on Networks and Communications and 6G Summit*, EuCNC/6 G Summit 2021 (pp. 496–501). Italy: Institute of Electrical and Electronics Engineers Inc. Available from https://doi.org/10.1109/EuCNC/6GSummit51104.2021.9482432, http://ieeexplore.ieee.org/xpl/mostRecentIssue.jsp?punumber = 9482408.

Parvais B., Elkashlan R., Yu H., Sibaja-Hernandez A., Vermeersch B., Putcha V., Cardinael P., Rodriguez R., Khaled A., Alian A., Peralagu U., Zhao M., Yadav S., Gramegna G., Driessche J. V., & Collaert N. (2021). Transistor modelling for mm-Wave technology pathfinding. In: *International conference on simulation of semiconductor processes and devices, SISPAD* (pp. 247–250). Belgium: Institute of Electrical and Electronics Engineers Inc. Available from https://doi.org/10.1109/SISPAD54002.2021.9592530.

Parvais, B., Alian, A., Peralagu, U., Rodriguez, R., Yadav, S., Khaled, A., Elkashlan, R. Y., Putcha, V., Sibaja-Hernandez, A., Zhao, M., Wambacq, P., Collaert, N., & Waldron, N. (2020). In *GaN-on-Si mm-wave RF devices integrated in a 200mm CMOS Compatible 3-Level Cu BEOL, IEDM*. Institute of Electrical and Electronics Engineers Inc.

Peeters, M., Sinha, S., Sun, X., Desset, C., Gramegna, G., Slabbekoorn, J., Bex, P., Pinho, N., Webers, T., Velenis, D., Miller, A., Collaert, N., Van Der Plas, G., Beyne, E., Huynen M., & Broucke, R. (2022). (Why do we need) Wireless heterogeneous integration (anyway?). In: *Digest of technical papers – Symposium on VLSI technology* (pp. 256–257). Belgium: Institute of Electrical and Electronics Engineers Inc. Available from https://doi.org/10.1109/VLSITechnologyandCir46769.2022.9830480.

Pekarik, J., Jain, V., Kenney, C., Holt, J., Khokale, S., Saroop, S., Johnson, J. B., Stein, K., Ontalus, V., Durcan, C., Nafari, M., Nesheiwat, T., Saudari, S., Yarmoghaddam, E., Chaurasia, S., & Joseph, A. (2021). SiGe HBTs with fT/fMAX ~375/510GHz integrated in 45nm PDSOI CMOS. In: *2021 IEEE BiCMOS and compound semiconductor integrated circuits and technology symposium*, BCICTS 2021. United States: Institute of Electrical and Electronics Engineers Inc. Available from https://doi.org/10.1109/BCICTS50416.2021.9682454, http://ieeexplore.ieee.org/xpl/mostRecentIssue.jsp?punumber = 9682201.

Post, I., Akbar, M., Curello, G., Gannavaram, S., Hafez, W., Jalan, U., Komeyll, K., Lin, J., Lindert, N., Park, J., Rizk, J., Sacks, G., Tsai, C., Yeh, D., Bai, P., & Jan, C. H. (2006). A 65nm CMOS SOC technology featuring strained silicon transistors for RF applications. In: *Technical digest – International electron devices meeting*, IEDM. United States. Available from https://doi.org/10.1109/IEDM.2006.346816.

Putcha, V., Cheng, L., Alian, A., Zhao, M., Lu, H., Parvais, B., Waldron, N., Linten, D., & Collaert, N. (2021). On the impact of buffer and GaN-channel thickness on current dispersion for GaN-on-Si RF/mmWave devices. In: *IEEE international reliability physics symposium proceedings*. Belgium: Institute of Electrical and Electronics Engineers Inc. Available from https://doi.org/10.1109/IRPS46558.2021.9405139, http://ieeexplore.ieee.org/xpl/conhome.jsp?punumber = 1000627.

Rack, M., & Raskin, J. P. (2017). RF harmonic distortion modeling in silicon-based substrates including non-equilibrium carrier dynamics. In: *IEEE MTT-S international microwave symposium digest* (pp. 91–94). Belgium: Institute of Electrical and Electronics Engineers Inc. Available from https://doi.org/10.1109/MWSYM.2017.8058737.

Razavieh, A., Chen, Y., Ethirajan, T., Gu, M., Cimino, S., Shimizu, T., Hassan, M. K., Morshed, T., Singh, J., Zheng, W., Mahajan, V., Wang, H. T., & Lee, T. H. (2021). Extremely-low threshold voltage FinFET for 5G mmWave applications. *IEEE Journal of the Electron Devices Society*, *9*, 165–169. Available from https://doi.org/10.1109/JEDS.2020.3046953, http://ieeexplore.ieee.org/servlet/opac?punumber = 6245494.

Sadlo, S., De Matos, M., Cathelin, A., & Deltimple, N. (2021). One stage gain boosted power driver at 184 GHz in 28 nm FD-SOI CMOS. In: *Digest of papers – IEEE radio frequency integrated circuits symposium* (pp. 119–122). France: Institute of Electrical and Electronics Engineers Inc. Available from https://doi.org/10.1109/RFIC51843.2021.9490441.

Singh, J., Ciavatti, J., Sundaram, K., Wong, J. S., Bandyopadhyay, A., Zhang, X., Li, S., Bellaouar, A., Watts, J., Lee, J. G., & Samavedam, S. B. (2018). 14-nm FinFET technology for analog and RF applications. *IEEE Transactions on Electron Devices*, *65*(1), 31–37. Available from https://doi.org/10.1109/TED.2017.2776838.

Strinati, E. C., Barbarossa, S., Gonzalez-Jimenez, J. L., Ktenas, D., Cassiau, N., & Maret, L. (2019). Cedric dehos: 6G: The next frontier: From holographic messaging to artificial intelligence using subterahertz and visible light communication. *IEEE Vehicular Technology Magazine*, *14*, 1556–6080.

Sun, X., Slabbekoorn, J., Sinha, S., Bex, P., Pinho, N., Webers, T., Velenis, D., Miller, A., Collaert, N., Van der Plas, G., & Beyne, E. (2022). Cost-effective RF interposer platform on low-resistivity Si enabling heterogeneous integration opportunities for beyond 5G. In: *Proceedings ECTC*.

Teledyne website. (n.d). http://www.teledyne-si.com/products-and-services/scientific-company/mm-wave-and-thz-pa-chips.

Then, H. W., Radosavljevic, M., Koirala, P., Thomas, N., Nair, N., Ban, I., Talukdar, T., Nordeen, P., Ghosh, S., Bader, S., Hoff, T., Michaelos, T., Nahm, R., Beumer, M., Desai, N., Wallace, P., Hadagali, V., Vora, H., Oni, A., … Fischer, P. (2021). Advanced scaling of enhancement mode high-K gallium nitride-on-300mm-Si(111) transistor and 3D layer transfer GaN-silicon finfet CMOS integration. In: *Technical digest – International electron devices meeting*, IEDM (pp. 11.1.1–11.1.4). United States: Institute of Electrical and Electronics Engineers Inc. Available from https://doi.org/10.1109/IEDM19574.2021.9720710.

Urteaga, M., Hacker, J., Griffith, Z., Young, A., Pierson, R., Rowell, P., Seo, M., & Rodwell, M. J. W. (2017). A 130 nm InP HBT integrated circuit technology for THz electronics. In: *Technical digest – International electron devices meeting*, IEDM (pp. 29.2.1–29.2.4). United States: Institute of Electrical and Electronics Engineers Inc. Available from https://doi.org/10.1109/IEDM.2016.7838503.

Urteaga, M., Pierson, R., Rowell, P., Jain, V., Lobisser, E., & Rodwell, M. J. W. (2011). 130 nm InP DHBTs with ft >0.52THz and fmax >1.1THz. In: *Device research conference – Conference digest*, DRC (pp. 281–282). United States. Available from https://doi.org/10.1109/DRC.2011.5994532.

Vais, A., Alcotte, R., Ingels, M., Wambacq, P., Parvais, B., Langer, R., Kunert, B., Waldron, N., Collaert, N., Witters, L., Mols, Y., Hernandez, A. S., Walke, A., Yu, H., Baryshnikova, M., Mannaert, G., & Deshpande, V. (2019). First demonstration of III-V HBTs on 300 mm Si substrates using nano-ridge engineering. In: *Technical digest – International electron devices meeting*, IEDM. Belgium: Institute of Electrical and Electronics Engineers Inc. Available from https://doi.org/10.1109/IEDM19573.2019.8993539.

Vais, A., Yadav, S., Mols, Y., Vermeersch, B., Kodandarama, K. V., Baryshnikova, M., Mannaert, G., Alcotte, R., Boccardi, G., Wambacq, P., Parvais, B., Langer, R., Kunert, B., & Collaert, N. (2022). III-V HBTs on 300 mm Si substrates using merged nano-ridges and its application in the study of impact of defects on DC and RF performance. In: *European solid-state device research conference* (pp. 261–264). Belgium: Editions Frontieres. Available from https://doi.org/10.1109/ESSDERC55479.2022.9947124, http://ieeexplore.ieee.org/xpl/conferences.jsp.

Visweswaran, A., Vignon, B., Tang, X., Brebels, S., Debaillie, B., & Wambacq, P. (2019). A 112-142 GHz power amplifier with regenerative reactive feedback achieving 17 dBm peak Psat at 13% PAE. In: *ESSCIRC 2019 – IEEE 45th European solid state circuits conference* (pp. 337–340). Belgium: Institute of Electrical and Electronics Engineers Inc. Available from https://doi.org/10.1109/ESSCIRC.2019.8902764, http://ieeexplore.ieee.org/xpl/mostRecentIssue.jsp?punumber = 8895615.

Wakayama, M. H. (2013). Nanometer CMOS from a mixed-signal/RF perspective. In: *Technical digest – International electron devices meeting*, IEDM (pp. 17.4.4). United States. Available from https://doi.org/10.1109/IEDM.2013.6724648.

Wang, H., Choi, K., Abdelaziz, B., Eleraky, M., Lin, B., Liu, E., Liu, Y., Jalili, H., Ghorbanpoor, M., Chu, C., Huang, T.-Y., Sasikanth Mannem, N., Park, J., Lee, J., Munzer, D., Li, S., Wang, F., Ahmed, A.S., Snyder, C., Nguyen, H.T., Duffy Smith, M.E. Power Amplifiers Performance Survey 2000-Present [Online]. Available: https://ideas.ethz.ch/Surveys/pa-survey.html.

Wang, W., Yu, X., Zhou, J., Chen, D., Zhang, K., Kong, C., Kong, Y., Li, Z., & Chen, T. (2018). Improvement of power performance of GaN HEMT by using quaternary InAlGaN barrier. *IEEE Journal of the Electron Devices Society*, *6*(1), 360–364. Available from https://doi.org/10.1109/JEDS.2018.2807185, http://ieeexplore.ieee.org/servlet/opac?punumber = 6245494.

Wui Then, H., Radosavljevic, M., Desai, N., Ehlert, R., Hadagali, V., Jun, K., Koirala, P., Minutillo, N., Kotlyar, R., Oni, A., Qayyum, M., Rode, J., Sandford, J., Talukdar, T., Thomas, N., Vora, H., Wallace, P., Weiss, M., Weng, X., & Fischer, P. (2020). Advances in research on 300mm gallium nitride-on-Si(111) NMOS transistor and silicon CMOS integration. In: Technical digest – International electron devices meeting, IEDM (pp. 27.3.1–27.3.4). United States: Institute of Electrical and Electronics Engineers Inc. Available from https://doi.org/10.1109/IEDM13553.2020.9371977.

Yadav S., Cardinael P., Zhao M., Vondkar K., Khaled A., Rodriguez R., Vermeersch B., Makovejev, S., Ekoga, E., Pottrain, A., Waldron, N., Raskin, J.P., Parvais, B., & Collaert, N. (2020). Substrate RF losses and non-linearities in GaN-on-Si HEMT technology. In: *Technical digest – International electron devices meeting*, IEDM (pp. 8.2.1–8.2.4). Belgium: Institute of Electrical and Electronics Engineers Inc. Available from https://doi.org/10.1109/IEDM13553.2020.9371893.

Yadav S., Chow W. H., Bellaouar A., Wong J. S., Chen T., Sekine S., Schwan C., Chin M. S., Workman G., & Chew K. W. J. (2019). Demonstration and modelling of excellent RF switch performance of 22nm FD-SOI technology for millimeter-wave applications. In: *European solid-state device research conference* (pp. 170–173). Editions Frontieres India. Available from https://doi.org/10.1109/ESSDERC.2019.8901823, http://ieeexplore.ieee.org/xpl/conferences.jsp.

Yadav, S., Vais, A., Elkashlan, R. Y., Witters, L., Vondkar, K., Mols, Y., Walke, A., Yu, H., Alcotte, R., Ingels, M., Wambacq, P., Langer, R., Kunert, B., Waldron, N., Parvais, B., & Collaert, N. (2021). DC and RF characterization of nano-ridge HBT technology integrated on 300 mm Si substrates. In: *EuMIC 2020 – 2020 15th European microwave integrated circuits conference* (pp. 89–92). Belgium: Institute of Electrical and Electronics Engineers Inc. http://ieeexplore.ieee.org/xpl/mostRecentIssue.jsp?punumber = 9337289.

Yang, P., Xiao, Y., Xiao, M., & Li, S. (2019). 6G Wireless communications: Vision and potential techniques. *IEEE Network*, *33*(4), 70–75. Available from https://doi.org/10.1109/MNET.2019.1800418, https://ieeexplore.ieee.org/xpl/mostRecentIssue.jsp?punumber = 65.

Yu, Q., Then, H. W., Thomson, D., Chou, J., Garrett, J., Huang, I., Momson, I., Ravikumar, S., Hwangbo, S., Latorre-Rey, A., Roy, A., Radosavljevic, M., Beumer, M., Koirala, P., Thomas, N., Nair, N., Vora, H., Bader, S., Rode, J., ... Rami, S. (2022). 5G mmWave power amplifier and low-noise amplifier in 300mm GaN-on-Si technology. In: *Digest of technical papers – Symposium on VLSI technology* (pp. 126–127). United States: Institute of Electrical and Electronics Engineers Inc. Available from https://doi.org/10.1109/VLSITechnologyandCir46769.2022.9830383.

Yun, D. Y., Jo, H. B., Son, S. W., Baek, J. M., Lee, J. H., Kim, T. W., Kim, D. H., Tsutsumi, T., Sugiyama, H., & Matsuzaki, H. (2018). Impact of the source-to-drain spacing on the DC and RF characteristics of InGaAs/InAlAs high-electron mobility transistors. *IEEE Electron Device Letters*, *39*(12), 1844–1847. Available from https://doi.org/10.1109/LED.2018.2876709.

Zhang, Y., Wei, K., Huang, S., Wang, X., Zheng, Y., Liu, G., Chen, X., Li, Y., & Liu, X. (2018). High-temperature-recessed millimeter-wave AlGaN/GaN HEMTs with 42.8% power-added-efficiency at 35 GHz. *IEEE Electron Device Letters*, *39*(5), 727–730. Available from https://doi.org/10.1109/led.2018.2822259.

CHAPTER

# 2

# FD-SOI and RF-SOI technologies for 5G

*Lucas Nyssens, Martin Rack and Jean-Pierre Raskin*

**Electronics and Applied Mathematics (ICTEAM), Université Catholique de Louvain, Institute of Information and Communication Technologies, Louvain-la-Neuve, Belgium**

## 2.1 Introduction

Driven by the ever-increasing demand of more bandwidth for faster telecommunication with increased amount of data transfer, the radio-frequency (RF) electronic world has known a tremendous growth in the last two decades. The succession of various telecommunication standards (3G, 4G, and now 4G LTE) has required mobile phones to be compatible with multiple frequency bands. To support a whole range of network configurations, the front-end module is becoming increasingly more complex, with RF switch banks becoming an essential block with stringent requirements (Didier & Desbonnets, 2017; Skyworks, 2011). RF-SOI (silicon-on-insulator) technology offering high performance at low cost has become the mainstream technology for RF switch banks in mobile applications, by steadily replacing gallium-arsenide and silicon-on-sapphire technologies to be virtually present in all modern-day smartphones (Rack & Raskin, 2019). The success of partially depleted (PD) SOI technology in today's RF market has been enabled thanks to significant technological improvements at each level of the integrated circuit: at transistor level, and also at back end of line (BEOL) and, in particular, substrate levels.

At transistor level, the continuous downscaling of CMOS technology for increased integration density when combined with several technical boosters such as k-high and metal gate stack, channel thinning, strain engineering, and so on leads to improved DC, analog, and RF performances, widening its range of applications. The aggressive downscaling has enabled operation at millimeter-wave (mm-wave) frequencies, which, coupled to low-cost integration of digital and analog circuitry, has made Si-based technology a serious contender of III–V technologies for RF and mm-wave applications. With the gate length ($L_g$) reduction below the 100-nm node, new device architectures were necessary to keep short channel effects (SCE) under control. Two new types of devices have been introduced, PD

*New Materials and Devices Enabling 5G Applications and Beyond*
DOI: https://doi.org/10.1016/B978-0-12-822823-4.00002-9

SOI and fully depleted (FD) SOI (or FD-SOI). While only those two devices are going to be discussed throughout this chapter, other device architectures exist or are under development to further scale down the gate length beyond the 20-nm node keeping controlled SCE, such as FinFET (Jan et al., 2012; Raskin, 2016a), gate-all-around nanowire (Bangsaruntip et al., 2013; Barraud et al., 2012; Hur et al., 2013; Veloso et al., 2016), nanosheet MOSFET (Bu, 2017; Veloso et al., 2019), etc.

At BEOL and substrate levels, strong technological efforts have also been made. Indeed, the electrical properties of the physical substrate have a strong impact on device and circuit performance, since electric fields from active and passive devices penetrate inside the substrate, potentially inducing significant losses, distortion, and unwanted crosstalk. The development of trap-rich substrates featuring low-loss, high-linearity, low-crosstalk, is one of the main reasons for the commercial success of the RF-SOI technology in the RF industry. Furthermore, the presence of one or several thick metal layers in the BEOL has enabled the design of high-quality (low-loss) on-chip passives (particularly when coupled with the trap-rich substrate) required in order to achieve high-performance RF and mm-wave circuits.

The emergence of 5G and Internet-of-Things will lead to another strong growth in the RF industry that must be supported by a technology with low-cost and high-volume manufacturability. In addition, new transceiver architectures are considered to support new 5G functionalities (beamforming, multiple input multiple output), hence further increasing the front-end module complexity, which means a higher level of flexibility and configurability too. The SOI platform, fulfilling the demands of high-volume, low-cost manufacturability, and integrating high-performance RF functions along with digital, analog, and mixed-signal circuitry is an outstanding contender to meet the technological challenges presented by 5G.

This chapter is organized as follows. First, a brief introduction to SOI technology is presented in Section 2.1 (Introduction to SOI technology). Then, Section 2.2 (PD-SOI and FD-SOI devices) focuses on the transistor level, describing the behavior and characteristics of PD-SOI and FD-SOI devices and showing their RF performance. Section 2.3 (Passives in SOI technology) discusses about substrate and BEOL aspects. The parasitic effects associated to the substrate are briefly introduced and the trap-rich substrate solution is described. Finally, still in Section 2.3 (Passives in SOI technology), we discuss about the impact of substrate and BEOL on passives at mm-wave frequencies, with examples of passives in commercial SOI technologies. The present chapter focuses on SOI technology at a more fundamental level, that is, on transistors and material aspects. Instead, a subsequent chapter is devoted to SOI-based circuits, and provides a description and review of circuits in SOI technology targeting 5G mm-wave applications.

## 2.2 Introduction to silicon-on-insulator technology

The SOI technology consists of a relatively thin layer of silicon—where the active devices are implemented—on top of a dielectric (buried oxide, or BOX, in silicon dioxide) laying above a handle silicon substrate. The main advantages (summarized in Fig. 2.1)

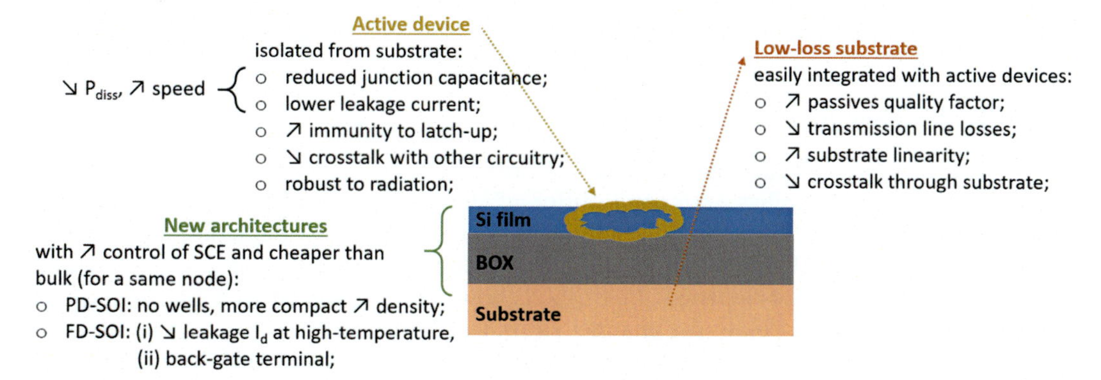


**FIGURE 2.1** Advantages of silicon-on-insulator (SOI) technologies. Summary of SOI technologies' advantages over bulk CMOS.

gained by SOI technology with respect to bulk CMOS are well known (Flandre et al., 2001; Colinge, 2004; Raskin, 2016b):

- The active devices are isolated from the substrate implying (1) reduced parasitic junction capacitances, (2) lower leakage current, (3) better immunity to latch-up phenomena, (4) decreased crosstalk between different circuits integrated on the same chip. The first two reductions lead to increased speed and lower power consumption ($P_{\text{diss}}$). Drain leakage current reduction in FD-SOI devices is particularly consequential at high temperature, because of (1) a reduction in drain junction area and (2) a change in the physical mechanism of leakage current. In bulk CMOS, the leakage current is dominated by carrier diffusion in the quasi-neutral region surrounding the drain junction, which is suppressed in FD-SOI devices and for which generation–recombination mechanisms in the depleted region dominate the leakage current. Therefore, the leakage current increases with temperature as the square of/ linearly with the intrinsic carrier concentration ($n_i$) in bulk/FD-SOI devices, respectively (Flandre, 1995).
- SOI circuits are particularly robust against radiation, as electron–hole pairs are generated in the thick handle substrate instead of inside the thin active film.
- PD-SOI CMOS does not need wells, or deep isolation trenches, unlike bulk CMOS to isolate devices from one another. The devices layout can then be made more compact, therefore increasing density for the same technology node.
- Low-loss high-resistivity (HR) Si substrate can be easily integrated as the substrate underlying the dielectric layer. Having an HR substrate has many advantages in terms of increased passives quality factor, reduced propagation losses along transmission lines, better linearity, lower crosstalk, etc. which will be detailed in a subsequent section.
- The ability to support new device architectures: the PD-SOI and FD-SOI MOSFET, which are both more efficient and cheaper (fewer process steps) than bulk CMOS compared with the same node.

## 2.3 Partially depleted-silicon-on-insulator and fully depleted-silicon-on-insulator devices

### 2.3.1 Technology device description

In planar SOI technologies, there are two distinct families of devices: PD-SOI and FD-SOI MOSFETs. These two families are mainly differentiated by the top silicon layer thickness as shown in Fig. 2.2. In FD-SOI, the silicon film is made so thin that the depletion region below the inversion channel extends down to the buried oxide. The entire Si film is depleted from free carriers, whence fully depleted. In PD-SOI, however, the silicon film is thicker, such that the depletion region below the channel stops within the silicon film, leaving an undepleted (neutral) region above the BOX called body. The existence of this body in PD-SOI MOSFET is at the origin of floating body effects affecting the device electrical behavior. Floating body effects (FBE) lead to undesired device behavior such as kink-effect, unusual subthreshold slope, etc. (Kazemi Esfeh et al., 2016; Makovejev et al., 2014) and can therefore be harmful in circuits, leading to circuit instabilities, frequency-dependent delay time, or

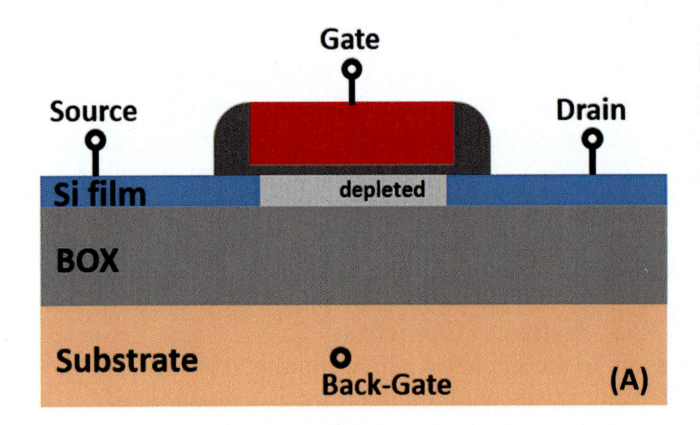


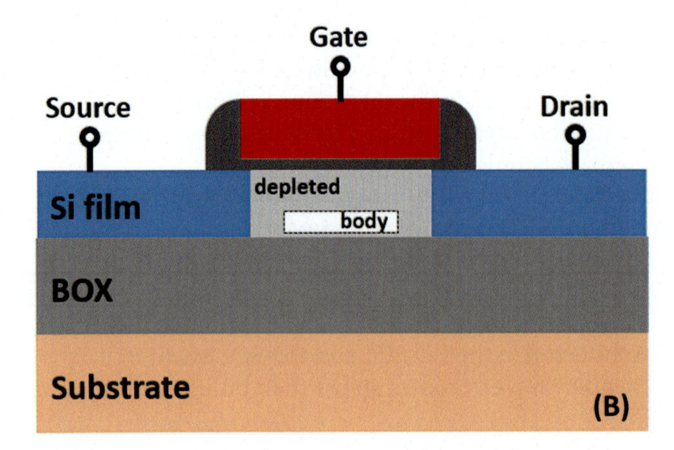


FIGURE 2.2 Sketch of FD-SOI and PD-SOI MOSFETs to scale. Schematic view of the FD-SOI (A) and PD-SOI (B) MOSFETs. The figures are not to scale. In case of ultra-thin BOX, the FD-SOI usually has an additional back-gate terminal.

pulse stretching (Vandana, 2013). FBE can be completely suppressed under DC conditions by fixing the body voltage by either connecting it to a terminal (source, gate) or controlling it as an additional one. Nevertheless, the dynamic control of the body still cannot be easily guaranteed in the GHz range because of the parasitic body resistance (Kilchytska et al., 2005; Lederer et al., 2004, 2005).

However, SCE become hard to control even with PD-SOI below the 40 nm technology. In order to minimize the detrimental impacts of the SCE, one solution is thinning down the channel Si of the transistor, and thus move to FD-SOI devices. In order to limit SCE, the channel thickness must be approximately 1/4 of the channel length (Raskin, 2016a). No channel doping is required since the Si film is entirely depleted thanks to its thin dimension. This architecture naturally enhances the electrical performances of the MOSFET with almost ideal subthreshold slope, reduced drain induced barrier lowering, and lower output conductance. Not necessitating halo implant and channel doping induces a great improvement in variability compared with bulk CMOS (Sugii et al., 2010). Furthermore, in ultra-thin body and BOX (UTBB) FD-SOI, the front-side transistor's electrostatics can be controlled from the doped region underneath the thin BOX (called back gate or body region). The front-gate threshold voltage ($V_{\mathrm{th}}$) can be tuned by applying a voltage on the back gate, which opens many perspectives in circuit as will be discussed further in this chapter. This device can be seen as an asymmetrical double gate transistor, with one gate at the front side (the regular one) and the other being the back gate with the BOX as the back-side gate oxide.

### 2.3.2 RF figures of merit

Furthermore, beside the great DC figures of merit (FoMs), both PD-SOI and FD-SOI technologies feature excellent analog and RF properties. The two main RF FoMs are the cutoff frequency ($f_t$) and the maximum oscillation frequency ($f_{\mathrm{max}}$). $f_t$ and $f_{\mathrm{max}}$ are the frequencies at which the short-circuited output current gain and maximum available gain, respectively, are unity. They are general indicators of how much gain a transistor can amplify at a given frequency, or until which frequency a transistor can be used to amplify a signal ($\sim 1/5$ to $1/3$ of $f_t$ & $f_{\mathrm{max}}$ as a rule of a thumb). They are also found in other more specific FoMs, such as the noise figure (see the next section). Benefitting from gate length reduction and controlled parasitics scaling, $f_t$ and $f_{\mathrm{max}}$ above 300 GHz have been demonstrated. SOI technologies currently exhibit the largest $f_t$ & $f_{\mathrm{max}}$ pair among all CMOS technologies, comparable with BiCMOS and some III–V technologies as reported in Table 2.1.

Commonly used (and rather simplified) expressions of $f_t$ and $f_{\mathrm{max}}$ are (Nyssens et al., 2020)

$$f_t \approx \frac{g_m}{2\pi\left(C_{gs}+C_{gd}\right)} \approx \frac{g_{m,i}}{2\pi\left(1+g_{m,i}R_s\right)\left(C_{gs,i}+C_{gd,i}+C_{gs,e}+C_{gd,e}+C_{gs,par}+C_{gd,par}\right)}, \tag{2.1}$$

$$f_{\mathrm{max}} \approx \frac{f_t}{2\sqrt{2\pi f_t R_g C_{gd}+\left(R_s+R_g\right)g_d}}. \tag{2.2}$$

Gate length scaling is the main force leading to increased values of $f_t$ & $f_{\mathrm{max}}$ as clearly shown in Fig. 2.3. Ideally, the transconductance increases linearly with $1/L_g$, whereas the gate capacitance decreases linearly with $L_g$; thus, $f_t$ should increase quadratically with $1/L_g$. Due to

**TABLE 2.1** Largest $f_t$ & $f_{max}$ reported values for various technologies.

| Technology | $f_t$ (GHz) | $f_{max}$ (GHz) | Technology | $f_t$ (GHz) | $f_{max}$ (GHz) |
|---|---|---|---|---|---|
| 32 nm bulk CMOS (VanDerVoorn et al., 2010) | 420 | 260 | SiGe HBT hein(Heinemann et al., 2016)<br>55 nm BiCMOS (Chevalier et al., 2014) | 505[a]<br>326[b] | 720[a]<br>376[b] |
| 22FDX (Zhao et al., 2021) | 365 | 413 | GaAs (Nguyen et al., 1988) InAs/InGaAs (Chang et al., 2008) | 152 360 | 230 380 |
| 45RFSOI (Ong et al., 2018) | 296 | 342 | 20 nm GaN HEMT (Shinohara et al., 2013) | 454 or 310 | 444 or 582 |
| 22 nm FinFET (Lee et al., 2018)<br>12 nm FiNFET (Razavieh et al., 2021) | 280<br>332 | 450<br>346 | 25 nm InP pHEMT (Mei et al., 2015)<br>130 nm InP HBT (Urteaga et al., 2011)<br>25 nm InP HEMT (Jo et al., 2019) | 610<br>520<br>703 | 1500<br>1100<br>820 |

[a]*Achieved from a purely (not industrial) HBT process.*
[b]*Achieved from an industrial BiCMOS process.*

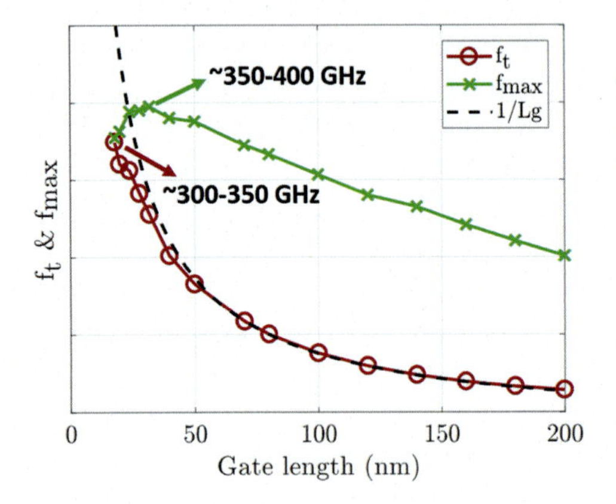


**FIGURE 2.3** $f_t$ and $f_{max}$ trends versus gate length. $1/L_g$ $f_t$ (red, circles) and $f_{max}$ (green, crosses) versus gate length. $1/L_g$ function in dashed line. Typical peak values for advanced CMOS technologies are in the 300–350 GHz in $f_t$ and 350–400 GHz range in $f_{max}$.

extrinsic elements (series resistances and extrinsic capacitances) the increase is not so fast and approximately a $1/L_g$ scaling of $f_t$ is observed. $f_{max}$ improvement is slower for deeply scaled transistors, because of a larger gate resistance (being proportional to $1/L_g$).

### 2.3.3 Extrinsic parasitics minimization

Despite the very large $f_t$ & $f_{max}$ advertised, care should be taken when implementing a real circuit, because such values correspond to the behavior of the transistor at the first metal layer, whereas, in an actual circuit, the transistor will be integrated with additional interconnects linking the passives (usually in the top metal layers) to the transistor's first metal layer. Such interconnects should comply to the electromigration rules and add as less parasitics as possible (mainly capacitive). In any case, these interconnects decrease the $f_t$ & $f_{max}$ of a usable transistor, which are representative of an actual circuit design. With proper care, $f_t$ & $f_{max}$, including the interconnects, above 200 GHz (up to 250–300 GHz) can be achieved (Gao et al., 2020; Inac et al., 2014; Nyssens, Rack, Wane, et al., 2022; Torres et al., 2020), enabling operation up to 60 GHz without any problem ($\sim f_t$ & $f_{max}$ / (3–5)) and up to 100 GHz with lower gain per stage.

To understand the limitations and trade-offs in $f_t$ & $f_{max}$, it is important to know the origin of those parasitics and their relative impact on the high-frequency behavior of the transistor. Fig. 2.4A shows the small-signal equivalent circuit, and Fig. 2.4B and C illustrate

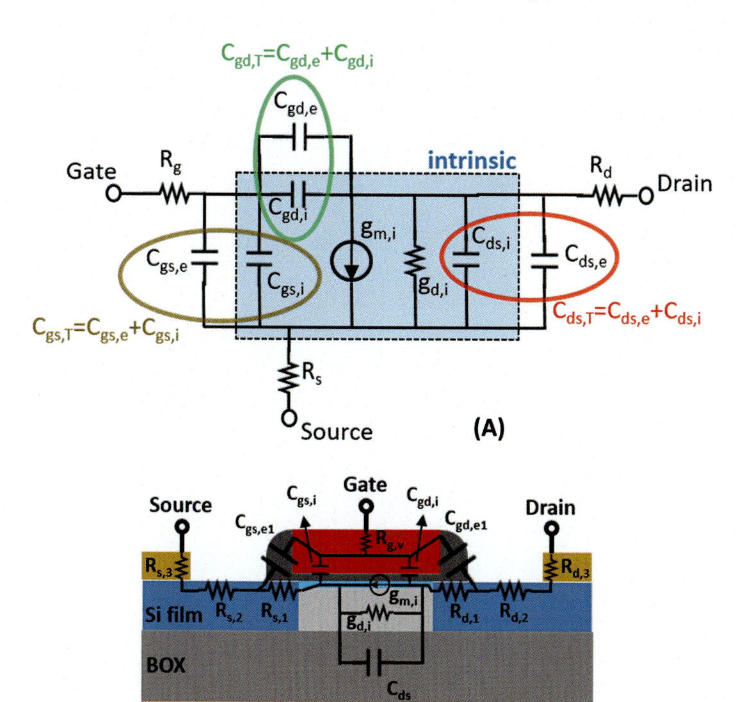


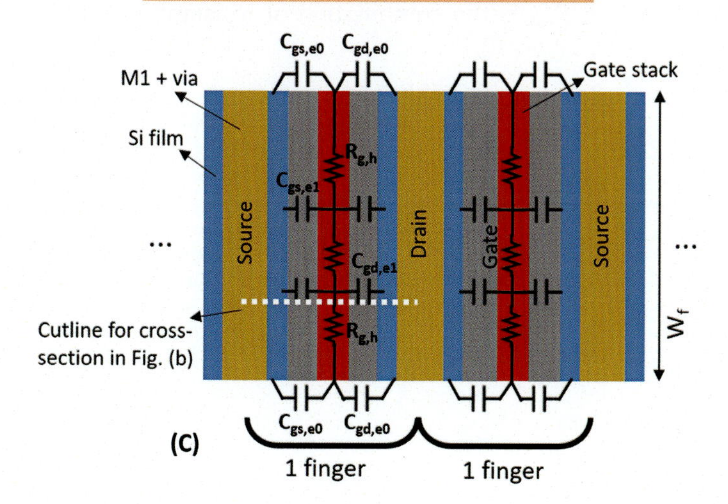


**FIGURE 2.4** MOSFET equivalent circuit and vertical and horizontal sketches with extrinsic parasitics. (A) Small-signal equivalent circuit of a MOSFET. (B) Schematic of a MOSFET's cross-section with equivalent circuit parameters. (C) Schematic of a MOSFET's top view along with equivalent circuit parameters. The elements related to the substrate are secondary parameters and have not been considered for simplicity.

the origin of each component on the MOSFET structure. The extrinsic components are usually the sum of several contributions (as for the extrinsic resistances, parasitic capacitances) and have been represented in a more detailed way in Fig. 2.4B and C. The intrinsic transconductance ($g_{m,i}$) and gate capacitances ($C_{gd,i}$ and $C_{gs,i}$) are clearly related to the technology, and there is not much a designer can do about it. $R_s$ increases slightly when the interconnects are added, but it is mainly dominated by the region underneath M1: silicidation + raised source (for UTBB FD-SOI MOSFET only) + doped region (Nyssens et al., 2020) and therefore related to technology as well. The principal design parameter that affects the $f_t$ value is the finger width ($W_f$). The intrinsic capacitances are purely proportional to $W_f$, while the extrinsic ones contain a linear term and an offset value due to edge effect ($C_{gd/s,e1}$ and $C_{gd/s,e0}$ terms in Fig. 2.4C): $C_{gd/s,e} = C_{gd/s,e1}\ W_f + C_{gd/s,e0}$. Therefore, $f_t$ is reduced for low $W_f$ values due to the increasing importance of edge effect ($C_{gs/d,e0}$) compared with the linear terms in $W_f$: $C_{gd/s,e1}$ and $C_{gd/s,i}$ as shown in Fig. 2.5. Further adding interconnects increases mainly the parasitic capacitances $C_{gd/s,par}$ and results in degraded $f_t$ & $f_{max}$.

$f_{max}$ is much more affected by layout design. The same considerations in terms of $g_{m,i}$, $C_{gd/s,i}$, $R_s$ (similarly for $R_d$), $C_{gd/s,e}$ hold. The major influence on $f_{max}$ that was not present in $f_t$ is the gate resistance ($R_g$). An analytical expression of $R_g$ is given by Lederer et al. (2018) and is applied to $R_g$ extraction as shown in Fig. 2.6:

$$R_g = R_{wire} + \frac{1}{N_f}\left( R_{ext} + \underbrace{\underbrace{\frac{R_{sil} W_f}{3 L_g N_c^2}}_{\equiv R_{g,h}} + \underbrace{\frac{R_v}{L_g W_f}}_{\equiv R_{g,v}}}_{\equiv R_{g,finger}} \right). \tag{2.3}$$

$N_f$ is the number of fingers. $R_{wire}$ is the effective resistance of a metal wire or polysilicon wire that connects the $N_f$ fingers together. $R_{ext}$ is the combination of resistances external

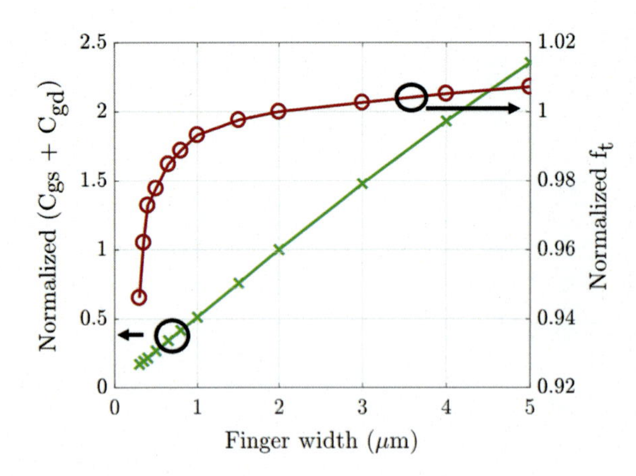


**FIGURE 2.5** $C_{gg}$ ($=C_{gs} + C_{gd}$) and $f_t$ trends versus finger width. Normalized total $C_{gg}$ ($=C_{gs} + C_{gd}$) and $f_t$ for different finger widths. A decrease in $f_t$ is observed at low $W_f$ due to the increased importance of the $W_f$ -independent extrinsic gate capacitance to the total gate capacitance. $C_{gg}$ and $f_t$ are normalized in this figure with respect to their values for $W_f = 2\ \mu m$. $f_t$ is typically in the 300–350 GHz range in advanced CMOS technologies.

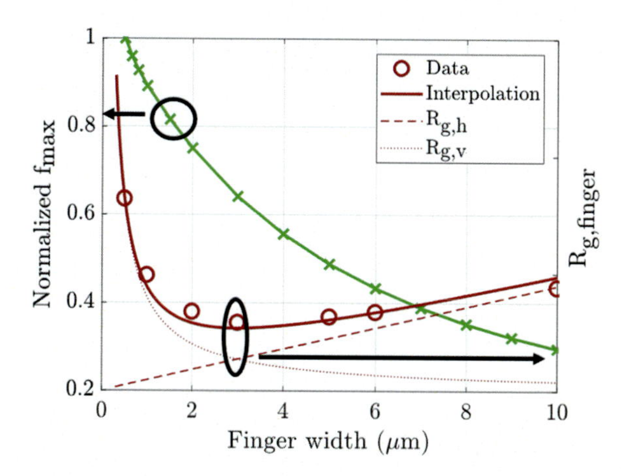


**FIGURE 2.6** $f_{max}$ and $R_{g,finger}$ trends versus finger width. Normalized $f_{max}$ and typical $R_{g,finger}$ versus finger width curve. In solid red line, interpolation of data points (circles) with Eq. (2.3). Resulting $R_{g,h}$ and $R_{g,v}$ components in dashed and dotted lines, respectively. $f_{max}$ is normalized in this figure with respect to its peak value, achieved for $W_f = 0.5$ μm. Peak $f_{max}$ is typically in the 350–400 GHz range in advanced CMOS technologies.

from the FET region to the wire connecting the different fingers. It includes (1) the horizontal expansion of the gate stack beyond the finger towards the vias (not shown in Fig. 2.4C) and (2) the M1 to gate stack via resistance (in case of metal wire in M1). $R_{ext}$ does not depend on finger width. $R_{sil}$ and $R_v$ are the sheet silicide and the vertical resistivity, respectively. $N_c$ is the number of gate contacts per finger ($N_c = 1$ or 2).

The finger gate resistance ($R_{g,finger}$) is a complex function of $W_f$, with two contributions, a vertical ($R_{g,v}$) and a horizontal ($R_{g,h}$) gate resistance. For low $W_f$, the vertical resistance dominates, such that $R_{g,finger}$ increases. Similarly, for large $W_f$, the horizontal resistance dominates and $R_{g,finger}$ linearly increases. So, there is an optimum in $W_f$ that depends on the sheet and vertical resisivities, which can be reduced by fabrication process optimization (Lederer et al., 2018). Nevertheless, the total gate resistance $R_g$ also depends on the total number of fingers. For a constant effective width ($W_{tot} = N_f W_f$), the contribution of $R_{g,v}$ is constant, while the contributions from $R_{g,h}$ and $R_{ext}$ are decreasing for low $W_f$. $R_{wire}$ slowly (if no polysilicon wire) increases for increased $N_f$, thus lower $W_f$, but it remains small with respect to the gain in $R_{g,finger}$, such that the total $R_g$ shows a decreasing behavior with lower values of $W_f$, and thereby an increasing trend in $f_{max}$ as shown in Fig. 2.6.

However, in practice, to achieve a desired total transistor width ($W_{tot}$) in an RF circuit, it is necessary to duplicate several times a FET instance (with a certain finger width and a number of fingers), because the equivalent number of fingers could be (1) out of process design kit (PDK) range, (2) so large that the wire gate resistance term or (3) distributed effects could become important. For very large transistor width (required in power amplifiers (PAs) to drive a large output current), parasitics due to interconnects degrade even more the $f_t$ & $f_{max}$, thus yielding a trade-off between total transistor width and maximum achievable gain by the transistor. Such trade-off usually results in an optimal choice of a $W_f$ larger than its minimum value.

Another important aspect about these RF FoMs is their dependence on bias. Indeed, the reported values of $f_t$ & $f_{max}$ above are all given at a certain bias yielding peak values. However, in an RF circuit, the FETs are not necessarily going to be biased at peak $f_t$ & $f_{max}$ values. In a low-noise amplifier, it is important to keep power consumption low, hence working with a limited bias current, trading-off with gain and noise. In a PA, the bias will determine the working class. The peak $f_t$ & $f_{max}$ bias in CMOS technology approximately corresponds to class A operation, with a low efficiency, which could be traded-off for some gain. The PD-SOI and FD-SOI (particularly UTBB) still have high $f_t$ and $f_{max}$ close to threshold voltage as shown in Fig. 2.7, with low drain current bias, thereby easing the trade-offs or achieving performant circuits in extreme conditions (El-Aassar & Rebeiz, 2020).

### 2.3.4 Back gate as RF tuning knob in UTBB fully depleted-silicon-on-insulator technology

With the additional back-gate terminal, the UTBB FD-SOI technology offers a unique feature with several implications. As mentioned above, the front-gate electrostatics can be controlled by the back-gate voltage. This is not a new feature per se, as body biasing can also be used in planar bulk technology. The differentiator is the tuning range offered by UTBB: bulk CMOS has a lower body factor (25 mV/V) and the effective safe body-bias voltage range spans from $\sim -0.3$ to $\sim 0.3$ V, yielding a $V_{th}$ modulation of $\pm 75$ mV (Cathelin, 2017). Fig. 2.8 shows a schematic of two flavors of UTBB nMOSFET: regular $V_{th}$ (RVT) and low $V_{th}$ (LVT). Thanks to the electrical isolation by the BOX of the source and drain regions from the back gate, the only limiting parasitic diodes are those between the N-back gate and P-substrate (for LVT, flipped well) and between the P-back gate and deep N-well (for RVT, conventional well). These are Zener-type diodes with an opening voltage of 3 V, which, together with the breakdown voltage of the thin BOX, limits the maximum magnitude of the back-gate voltage to 3 V (for 28 nm FD-SOI) or 2 V (for 22 nm FD-SOI). Overall, with a much larger body factor of $\sim 85$ mV/V, the $V_{th}$ tuning range spans from $-0.3$ to 3 V (2 V for 22 nm FD-SOI) for LVT nMOSFETs (forward body bias, FBB) and from $-3$ V ($-2$ V for 22 nm FD-SOI) to 0.3 V for RVT nMOSFETs (reverse body bias, RBB), yielding a total $V_{th}$ variation range of $\sim 280$ mV ($\sim 200$ mV for 22 nm FD-SOI) as shown in Fig. 2.9 (Cathelin, 2017).

This wide $V_{th}$ tuning range does not come with performance degradation penalty, as the RF FoMs are essentially shifted due to different bias condition (see Fig. 2.10). Improvement in nonlinearity can also be obtained by applying an appropriate back-gate

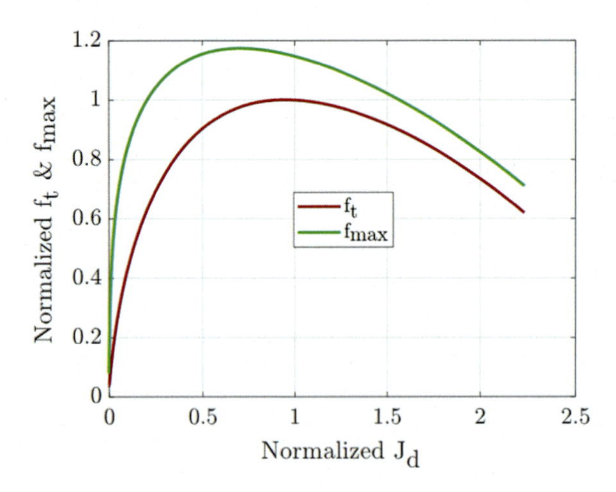


**FIGURE 2.7** $f_t$ and $f_{max}$ versus $J_d$. Normalized $f_t$ and $f_{max}$ versus normalized drain current density. $f_t$ and $f_{max}$ have a steep increase close to the threshold voltage, at low drain current, such that large values of $f_t$ and $f_{max}$ can be achieved with a low power consumption. $f_{max}$ and $f_t$ are both normalized in this figure with respect to the peak $f_t$ value. $J_d$ is normalized with respect to its value yielding the peak $f_t$. Peak values $f_t$ and $f_{max}$ are typically in the 300–350 GHz and 350–400 GHz range, respectively, in advanced CMOS technologies.

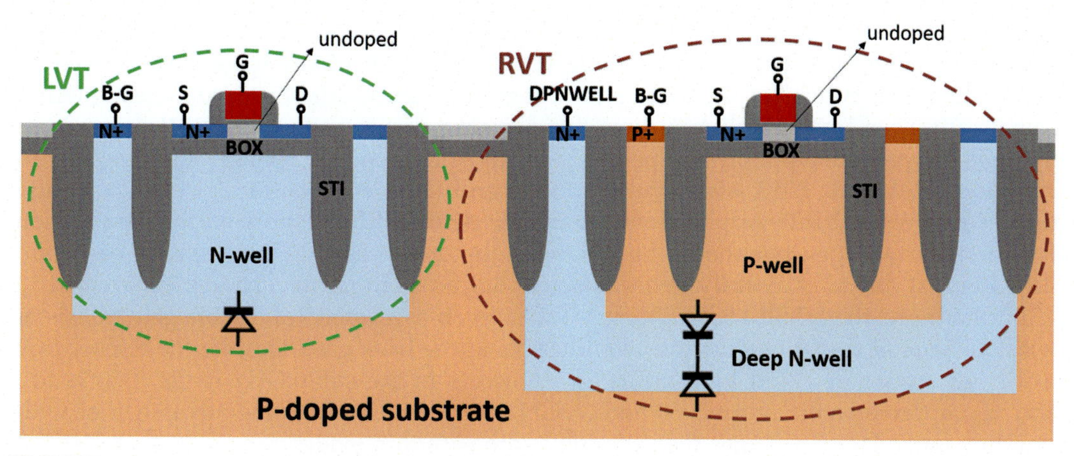


**FIGURE 2.8** LVT and RVT MOSFETs cross-section. Cross-section view of LVT (left) and RVT (right) nMOSFETs with back-gate terminals (B–G) and wells, deep N-well, and its terminal (DNWELL), p-doped substrate and diodes.

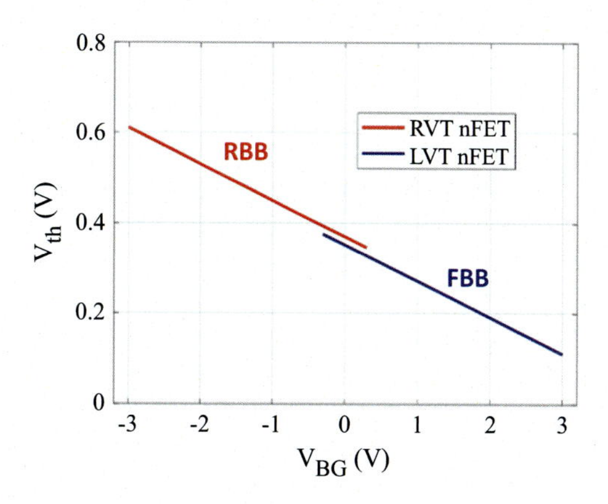


**FIGURE 2.9** $V_{th}$ tuning range with back-gate bias in fully depleted-silicon-on-insulator (FD-SOI) $V_{th}$ tuning with back-gate bias variation with reverse body bias (RBB) in RVT nFETs or forward body bias in LVT nFETs, in the 28 nm FD-SOI technology. Values for the 22 nm FD-SOI technology vary a bit, with a lower tuning range due to lower allowed $V_{BG}$ value (2 V instead of 3 V due to a thinner BOX thickness) and a different $V_{th}$ value at $V_{BG} = 0$ V.

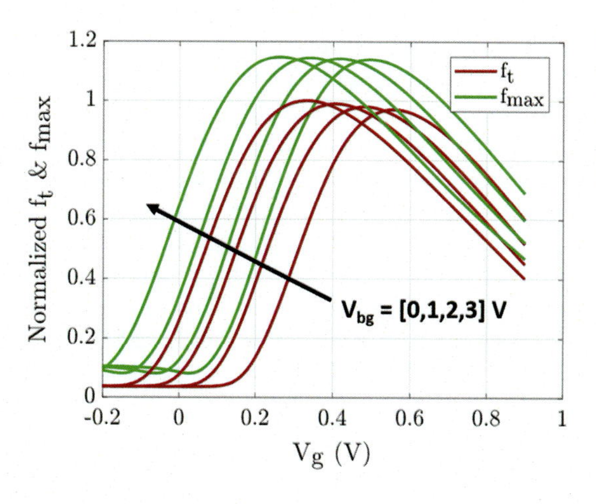


**FIGURE 2.10** $f_t$ and $f_{max}$ versus gate voltage at different back-gate voltages. Normalized front-gate $f_t$ and $f_{max}$ versus gate voltage at different back-gate voltages. $f_{max}$ and $f_t$ are both normalized in this figure with respect to the peak $f_t$ value among all $V_{bg}$ biases.

bias condition as demonstrated in Kazemi Esfeh, Kilchytska et al. (2017). In general, the back-gate terminal can be seen as a tuning knob, with a potentially huge impact on circuit design. Dynamic control of the back-gate voltage, also called adaptive body-biasing (ABB), has already been implemented in digital and analog circuits, allowing to (1) dynamically trade-off power consumption for speed, (2) to adjust for process, voltage, and temperature variations (Bol et al., 2019; Moursy et al., 2021) and (3) to compensate for aging degradation (Huard et al., 2022). Application of dynamic ABB will be detailed in the next section. It is in its early stages, but should become soon mature, as several studies have been published in that direction recently (more information can be found in Chapter 9).

Furthermore, the UTBB MOSFET can be seen as an asymmetrical double gate transistor, with one gate at the front side and the other being the back gate. Despite the thicker BOX oxide, the back gate could be used as an RF access in the sub-6 GHz range, thanks to $f_t$ and $f_{max}$ approximately around 80 and 30 GHz, respectively, with the advantage of withstanding much larger voltages than the front gate with thin oxide.

## 2.4 Passives in silicon-on-insulator technology

### 2.4.1 Substrate impact

The physical properties of the underlying substrate have a strong impact on device and circuit performance, particularly at RF and mm-wave frequencies. At high frequencies, the thin dielectric layers isolating the metal interconnects and transistors from the substrate have a low impedance by their capacitive nature, resulting in a significant portion of the electric field propagating into the substrate, inducing losses, nonlinearity, and increasing the complexity of the parasitic coupling modeling. Standard resistivity silicon substrates ($\sim 10\ \Omega$cm, cfr Fig. 2.11A) have strong RF losses, nonlinearity, and interdevice coupling and degrade the quality factor of passives [inductors, coplanar waveguide (CPW) lines] (Rack & Raskin, 2019). Overall, they are of poor performance for RF and mm-wave circuits. Moving to HR substrates ($>3$ k$\Omega$cm, cfr Fig. 2.11B) is of little benefit, as the parasitic surface conduction (PSC) phenomenon counteracts the large substrate resistivity and still exhibits large (although lower than a standard Si substrate) losses and nonlinearity. This PSC effect originates in the presence of fixed charges (generally positive) trapped in the BOX and generated during manufacturing that induces a thin sheet of highly conductive free carriers (generally electrons) at the silicon–BOX interface. Several solutions have been proposed to counter this PSC effect. The most famous and spread technique is the trap-rich (TR) substrate (Lederer & Raskin, 2005) illustrated in Fig. 2.11C. In such substrates, a layer rich of traps is created with polysilicon deposited on top of the HR Si substrate, at the interface with the BOX. The amount of traps is so high that the Fermi level is pinned in this layer close to the mid gap, resulting in an extremely low amount of free carriers and therefore a high local resistivity. The TR substrate exhibits very low RF losses, enhanced quality factor of passives (Lederer & Raskin, 2008; Shim et al., 2013), high linearity (Kazemi Esfeh, Makovejev, et al., 2017; Roda Neve & Raskin, 2012), interdevice capacitive-like coupling at RF frequencies (Ben Ali et al., 2011).

A major advantage of the SOI topology is to enable the separation of the intrinsic device from the bulk substrate. This flexibility allows to engineer the intrinsic device and the

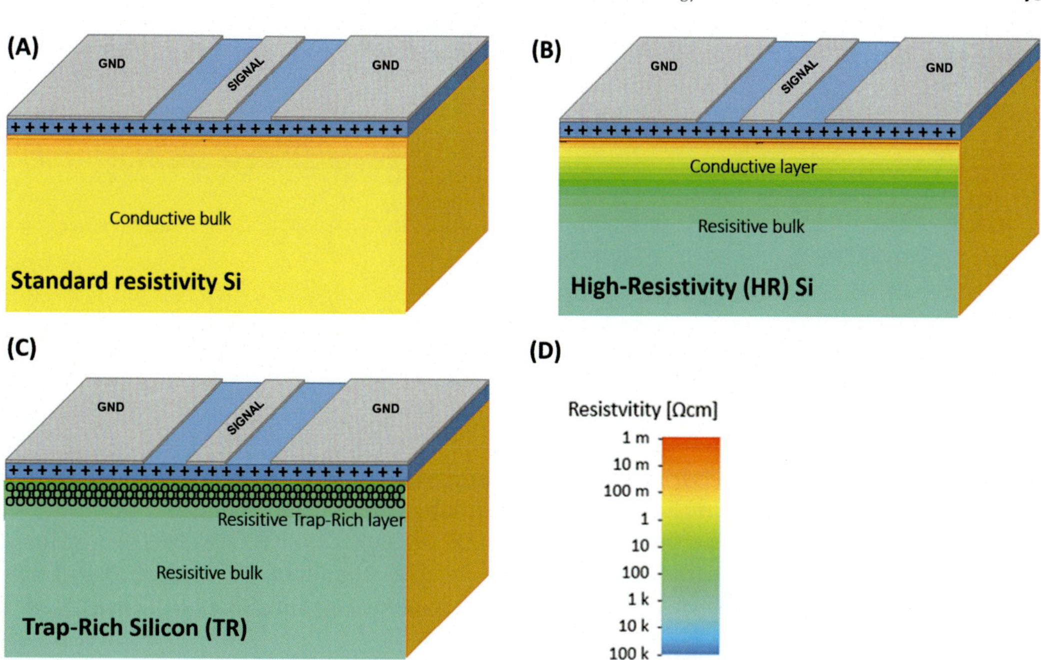

FIGURE 2.11 Schematic representation of a standard Si substrate, high-resistivity Si substrate, trap-rich Si substrate with a CPW line on top. Schematic representation of a standard Si substrate (A), high-resistivity Si substrate (B), trap-rich Si substrate (C), with a CPW line on top. A representative profile of the local resistivity is given inside the substrates with the scale present in (D). Fixed charges (in " + " symbols inside the BOX) attract free carriers at the Si-oxide interface inducing a thin conductive layer that mitigates the benefits of HR substrates. This conductive sheet is suppressed thanks to the introduction of a trap-rich layer in the TR Si substrate, yielding a high-quality RF substrate.

substrate separately (to first order) and achieve a combination of the best of two worlds: benefitting from the advantages of SOI at the device level as already discussed above and from a high RF quality substrate with low losses and high linearity. TR is the benchmark Si substrate for SOI technologies with best high-frequency performance. It is however not compatible with UTBB FD-SOI technology, as the defect-rich poly-silicon layer degrades the isolation of the back-gate contact of FD-SOI devices with the substrate's junctions, therefore increasing the reverse diode current. Nevertheless, other solutions compatible with the UTBB FD-SOI process have been successfully demonstrated to counter the PSC effect. The first solution is based on a smart array of shallow P- and N-implants in the substrate (Rack et al., 2019a, 2019b). The second one consists of a local post-CMOS process porosification step of the silicon substrate that yields ultra-high RF performance, with great high linearity and low-loss performance (Scheen et al., 2020).

### 2.4.2 Millimeter-wave back end of line

Any RF circuit includes passives as matching network. Although at low RF frequencies, part (or the entirety) of the matching network can be implemented off-chip, at mm-wave

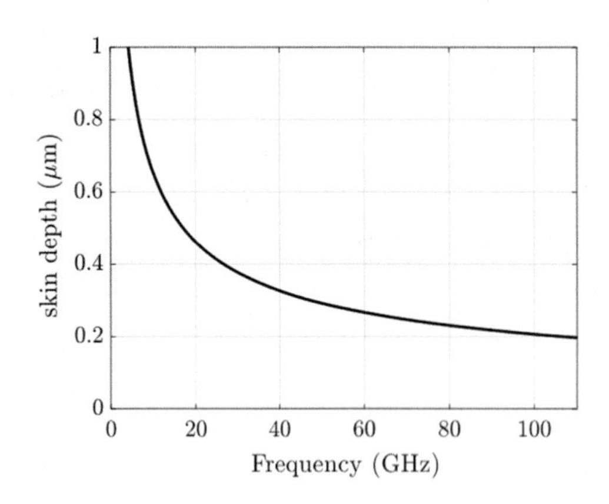


**FIGURE 2.12** Skin depth in a metal as a function of frequency. Skin depth of Cu as a function of frequency.

frequencies, the matching network is usually completely on-chip to reduce the losses and circuit variability arising from large packaging interconnects (wire bonding, bumps, etc.) compared to wavelength. The passives need to be of high quality (low loss) in order to achieve high-performance mm-wave circuits. Passive elements are either distributed (based on transmission lines) or lumped (mainly capacitors and inductors). Losses in passive devices have two main origins: (1) resistive, metallic losses and (2) losses in a poorly resistive substrate[1] (Nieuwoudt & Massoud, 2005). By using high RF quality substrates, compatible with SOI technologies, such as the TR substrate described above, substrate losses are either reduced to such an extent that they become negligible with respect to metallic losses or become insignificant in any case. Metallic losses can be reduced by using highly conductive metal lines, for instance by using copper (Cu) instead of aluminum (Al), and increasing their thickness. Metallic loss is anyway intrinsically limited at mm-wave frequencies due to skin effect. Fig. 2.12 shows the skin depth in a Cu line as a function of frequency. Due to the skin depth decreasing with frequency, metallic losses increase and are usually dominant at high frequency.

In technologies targeting mm-wave applications such as the 22FDX and 45RFSOI technologies from GlobalFoundries, specific BEOL, rich of thick Cu metal layers, are offered (cfr Fig. 2.13, Gao & Rebeiz, 2019; Li et al., 2018). Having several thick metal layers of high conductivity has two advantages. The first and most obvious outcome is to reduce metallic losses and enable more flexibility in passive designs thanks to multiple thick metal lines. Secondly, the upper thick metal layers are farther away from the substrate, thus reducing the portion of electric field penetrating into the substrate and thereby reducing the losses dissipated in poor RF quality substrates. So, even a rich BEOL with a poor RF performance substrate, such as the one offered by 22FDX, enables pretty high-quality RF and millimeter-wave passives.

Capacitors can be either implemented with MOSFETs by shorting the drain and source terminals together or with metal-oxide-metal (MOM) capacitors. The first type of

[1] There might also be losses due to Foucault currents in inductors, but these can only be induced in RF if the inductor is implemented on top of a highly doped Silicon substrate, with resistivity below 1 Ωcm [ref below], which is not standard/found in CMOS technology nowadays to the best of the authors' knowledge.

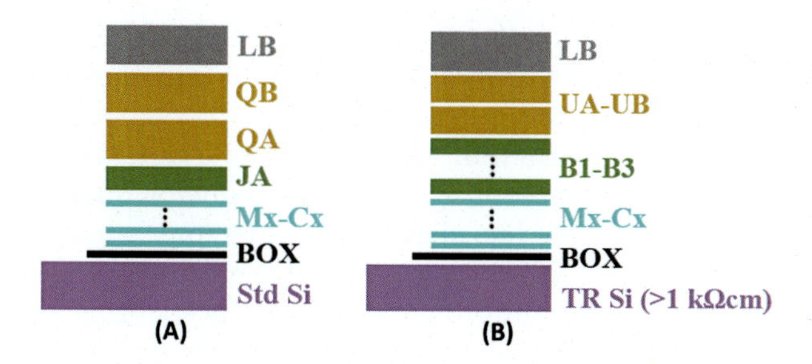


**FIGURE 2.13** Mm-wave back end of line (BEOL) from 2 silicon-on-insulator technologies. Mm-wave BEOL from the 22FDX technology (A) and from the 45RFSOI technology (B). The figures are indicative and not to scale.

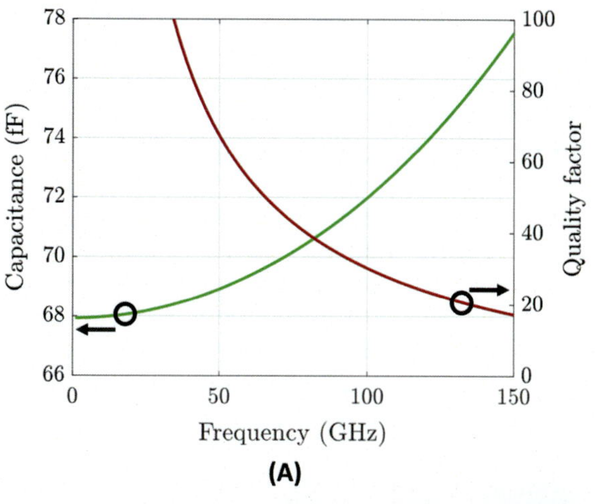


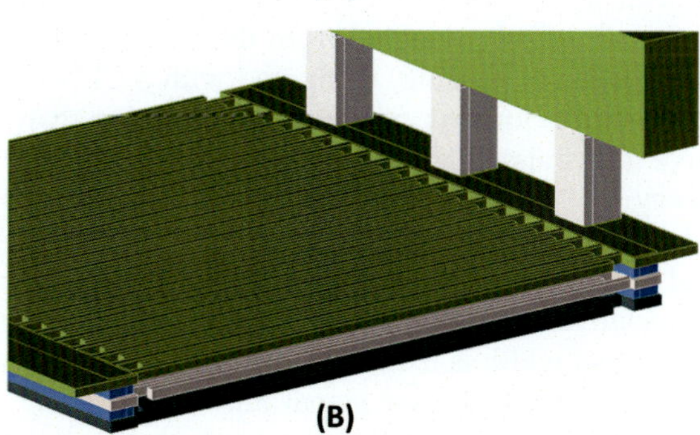


**FIGURE 2.14** Example of integrated metal-on-metal capacitor. Example from of a 3.3 V, 5.5 × 6 μm, 68 fF MOM capacitor in 22FDX, with 1 fF of shunt capacitance a quality factor above 20 up to 135 GHz. (A) Capacitance and quality factor versus frequency, (B) capacitor layout.

capacitors (shorted source-drain FET) results in highly dense capacitances, but with limited applied voltage (due to oxide breakdown), high nonlinearity (due to active device) and poor quality factor at mm-wave frequencies (due to large series resistances). MOM capacitors with interdigitated fingers enable sufficiently large values of capacitance per unit area (from 2.5 to 7.6 fF/$\mu m^2$ according to operating voltage), with good linearity and high-quality factor. For instance, 22FDX technology offers a 3.3 V series MOM capacitor of 68 fF with an area of 5.5 × 6 μm, a quality factor above 20 up to 135 GHz, and shunt capacitances of 1 fF (see Fig. 2.14, Gao et al., 2020).

Spiral inductors can still be found in mm-wave circuits up to ~40–50 GHz (Li et al., 2021; Nyssens, Rack, Wane, et al., 2022; Parlak & Buckwalter, 2011) and spiral transformers above 100 GHz (Tang et al., 2020). These elements take a great advantage of rich BEOL thanks to (1) less lossy metallic lines and (2) fewer portion of electric field in a potentially lossy substrate, as mentioned above. A large improvement of the spiral inductor quality factor moving from a poor to high RF quality substrate has already been demonstrated at sub-6 GHz (Gianesello et al., 2007; Lederer & Raskin, 2008; Liu et al., 2017; Nyssens, Rack, Schwan, et al., 2022). At mm-wave frequencies, the improvement is reduced. This is explained by the fact that as frequency increases, the size of lumped elements decreases and, for a fixed distance between the spiral inductor and substrate, the portion of electric field flowing inside the lossy substrate is reduced, thus yielding the lumped element less sensitive to the substrate nature. An example of the improvement from Nyssens, Rack, Nabet, et al., (2022) and Nyssens, Rack, Schwan, et al. (2022) is shown in two simulated inductors (for operation at ~6 GHz—called RF—and 39 GHz—called mm-wave) implemented in the mm-wave BEOL of 22FDX in Fig. 2.15. Simulations have been carried out with ADS momentum on different substrate stacks for which sketches are given in Fig. 2.15D–F. Although the improvement in mm-wave spiral inductor's quality factor is not as large as in RF (up to ~30% increase for the RF inductor), it is still improved by ~13% at best. Nevertheless, the mm-wave inductor design could be optimized for low-loss substrate. Indeed, stacking all thick metal layers available in the BEOL reduces metallic losses and bring the inductor closer to the substrate, hence increasing the electric field concentration inside the semiconductor. In this way, the quality factor of the modified mm-wave spiral inductor benefits from an up to 52% improvement as long as it is integrated on a low-loss substrate.

A mm-wave BEOL with multiple thick metal layers also enables implementing thin-film microstrip (TFMS) lines with reasonable losses and entirely shielded from the substrate. Such an element benefits from high linearity (Kazemi Esfeh et al., 2018) compared with a CPW line and does not suffer from a poor RF quality substrate. However, it offers less flexibility than a CPW line, mainly when a highly inductive transmission line is needed (to implement inductance-like elements at very high frequency >60 GHz). Indeed, the characteristic impedance ($Z_c$) of the line is mainly a function of the ratio of signal line width and signal to ground lines distance. The distance between the signal and ground lines is fixed in a TFMS line (thickness determined by number and thickness of metal/dielectric layers), while it can be increased in a CPW line. Therefore, for a constant $Z_c$, the signal width is fixed in a TFMS line and can be increased in a CPW line, thus resulting in a wider and less lossy signal line. Losses along a CPW line (as well as nonlinearity) can be substantially improved by using a high RF quality substrate, such as the one used in the 45RFSOI technology (TR substrate). Fig. 2.16 shows the total propagation losses of different transmission lines (TFMS and CPW) implemented on several substrates (Nyssens, Rack, Nabet et al., 2022). For a fair comparison, all lines have been designed to achieve a $Z_c = 50\ \Omega$. Dimensions are reported on the transmission lines sketches. When only a lossy substrate is available (such as the standard or HR with PSC), the TFMS line is clearly less lossy and should be favored due to the strong substrate losses along the CPW line. Nevertheless, when a low-loss substrate is available, CPW lines with a reasonable cross-section dimension can outperform TFMS lines yielding reduced total losses thanks to negligible substrate losses and lower metallic losses from the wider signal line.

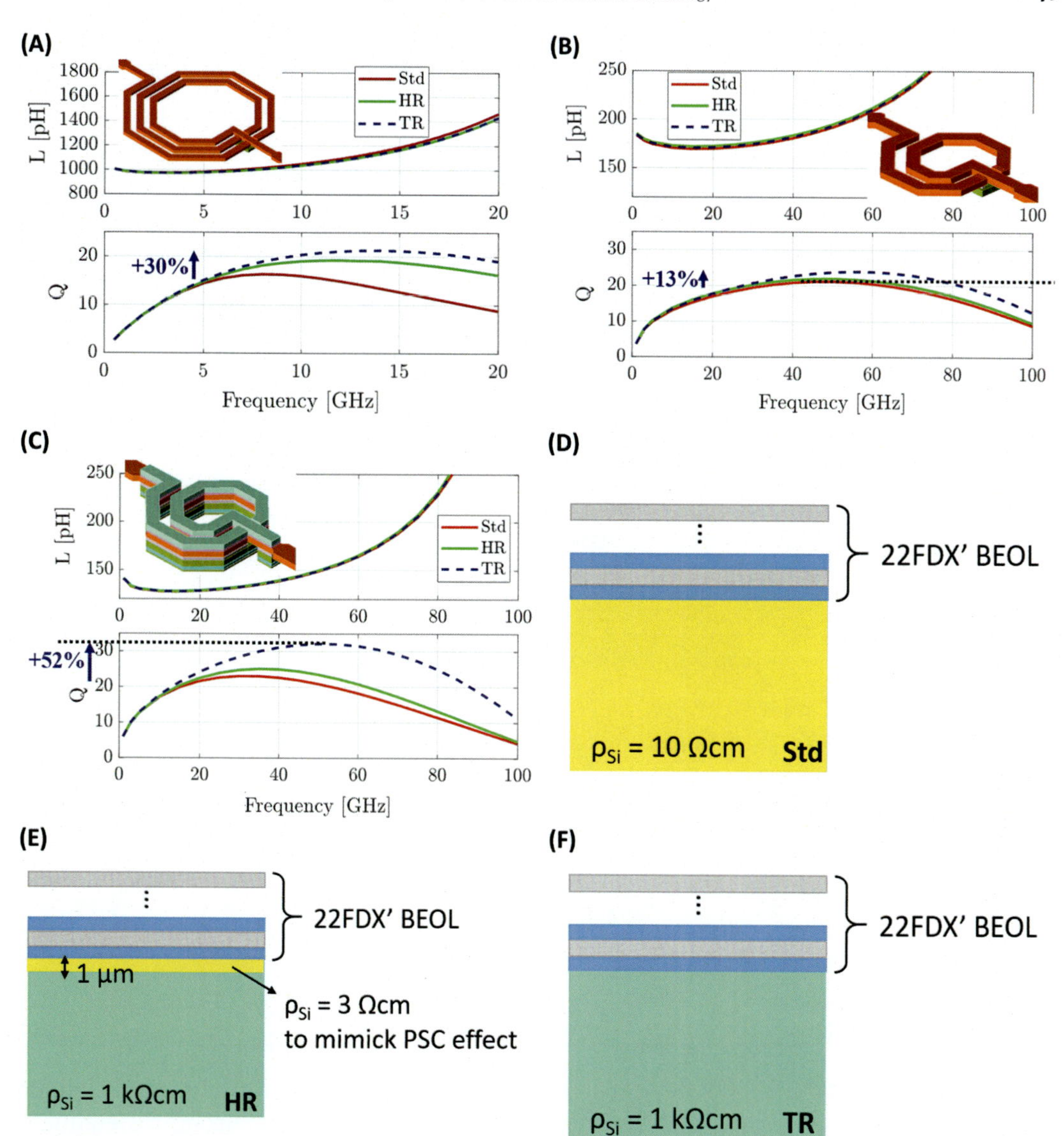


**FIGURE 2.15** Simulations of inductors on several substrates. Inductance and quality factor of a typical spiral inductor used for ~6 GHz [(A), called RF] and 39 GHz operation [(B), called mm-wave and (C)], implemented in the 22FDX' back end of line presented in Figure 5.13(A). The third (C) inductor has the dimensions as the mm-wave inductor, but is implemented by stacking several metal layers. Layout of the inductors is presented in the inset of each figure. In green and dashed blue lines different simulations performed by replacing the standard substrate to a representative stack mimicking a high-resistivity substrate and TR substrate, respectively. Schematic views of the standard, HR with PSC, and TR substrate stacks used by the simulator are displayed in (D), (E), and (F), respectively.

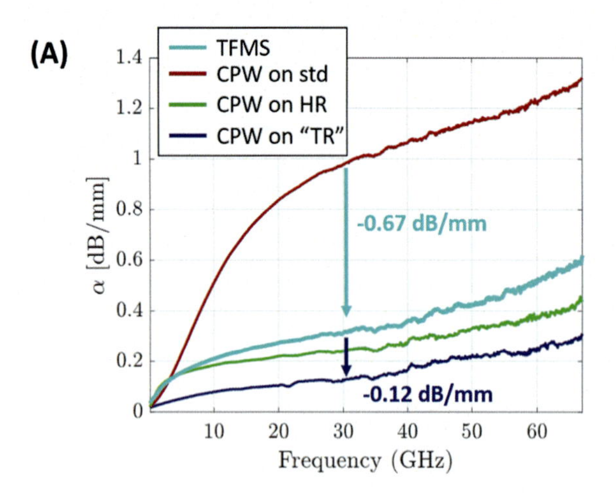


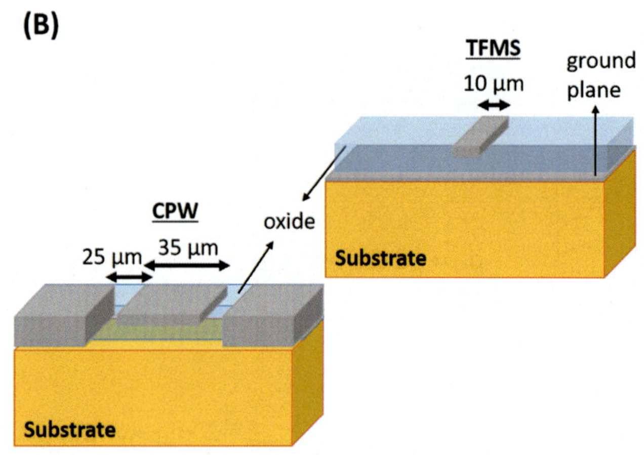


**FIGURE 2.16** Comparison of propagation losses for different types of transmission lines. (A) Measured propagation losses of TFMS and CPW lines on different substrates: (i) std, (ii) HR, and (iii) a low-loss HR substrate with a special passivation technique yielding results similar to a TR substrate. (B) Sketch of the CPW and TFMS lines along with their cross-section dimensions. For a fair comparison, the CPW and TFMS dimensions were designed to achieve a 50 Ω characteristic impedance, while minimizing the TFMS propagation losses.

## 2.5 Conclusion

While FD-SOI and PD-SOI device architectures are major innovations in the digital world compared with bulk CMOS, their RF performance is really a strong differentiator with best pair of $f_t$ & $f_{max}$ reported among all CMOS technologies. Moreover, their intrinsic ability to perform well even under harsh environment, such as high temperature, radiation, etc. makes them particularly suited to the growing demand of automotive or spatial applications. Although reported $f_t$ & $f_{max}$ values above 300 and 400 GHz for $f_t$ and $f_{max}$, respectively, are valid for the FET at the first metal layer (M1), adding interconnects for a realistic circuit implementation can still yield $f_t$ & $f_{max}$ in the 250 to 300 GHz range with a careful geometry-aware design. Indeed, finger width has been shown to have a strong effect on $f_t$ and $f_{max}$ through the gate capacitance and gate resistance dependences on finger width. The UTBB FD-SOI technology in particular, beside its outstanding digital, analog, and RF performances, has a powerful advantage with its additional back-gate access.

It acts as a fine-tuning knob, modifying the bias point and therefore RF characteristics without degradation. This additional feature opens new perspectives to RF circuits, some of which will be discussed in a subsequent chapter (Chapter 9).

Another great advantage of SOI technologies is the possibility to engineer the device and substrate separately thanks to the presence of the buried oxide that isolates under static condition at least the two components. Substrate engineering solutions, such as the TR Si substrate, showed great improvements in lumped and distributed passive losses, linearity, coupling, etc. Furthermore, BEOLs rich in thick Cu metal layers have been developed, enabling high-quality factor capacitors, inductors, and low-loss transmission lines, crucial to implement competitive circuits at mm-wave frequencies.

This chapter highlights some of the benefits and performances of SOI technologies at device level. The performances shown so far enable SOI technologies as an attractive choice for 5G applications. A subsequent chapter (Chapter 9) focuses on circuits and demonstrates that SOI technologies already meet the requirements to implement 5G transceivers.

## References

Bangsaruntip S., Balakrishnan K., Cheng S.-L., Chang J., Brink M., Lauer I., Bruce R. L., Engelmann S. U., Pyzyna A., Cohen G. M., Gignac L. M., Breslin C. M., Newbury J. S., Klaus D. P., Majumdar A., Sleight J. W., & Guillorn M. A. (2013). Density scaling with gate-all-around silicon nanowire MOSFETs for the 10 nm node and beyond. In *2013 IEEE international electron devices meeting* (pp. 20.2.1–20.2.4). https://doi.org/10.1109/IEDM.2013.6724667.

Barraud, S., Coquand, R., Casse, M., Koyama, M., Hartmann, J.-M., Maffini-Alvaro, V., Comboroure, C., Vizioz, C., Aussenac, F., Faynot, O., & Poiroux, T. (2012). Performance of omega-shaped-gate silicon nanowire MOSFET with diameter down to 8 nm. *IEEE Electron Device Letters, 33*(11), 1526–1528. Available from https://doi.org/10.1109/LED.2012.2212691.

Ben Ali, K., Roda Neve, C., Gharsallah, A., & Raskin, J.-P. (2011). Ultrawide frequency range crosstalk into standard and trap-rich high resistivity silicon substrates. *IEEE Transactions on Electron Devices, 58*(12), 4258–4264. Available from https://doi.org/10.1109/TED.2011.2170074.

Bol D., Schramme M., Moreau L., Haine T., Xu P., Frenkel C., Dekimpe R., Stas F., & Flandre D. (2019). 19.6 A 40-to-80MHz Sub-4μW/MHz ULV Cortex-M0 MCU SoC in 28nm FDSOI with dual-loop adaptive back-bias generator for 20μs wake-up from deep fully retentive sleep mode. In *2019 IEEE international solid- state circuits conference - (ISSCC)*. (pp. 322–324). https://doi.org/10.1109/ISSCC.2019.8662293.

Bu H. (2017). 2022 8 20 5 Nanometer transistors inching their way into chips. https://www.ibm.com/blogs/think/2017/06/5-nanometer-transistors/.

Cathelin, A. (2017). Fully depleted silicon on insulator devices CMOS: The 28-nm node is the perfect technology for analog, RF, mmW, and mixed-signal system-on-chip integration. *IEEE Solid-State Circuits Magazine, 9*(4), 18–26. Available from https://doi.org/10.1109/MSSC.2017.2745738.

Chang E. Y., Kuo C.-I., Hsu H.-T., & Chang C.-Y. (2008). InAs/In1-xGaxAs composite channel high electron mobility transistors for high speed applications. In *2008 European microwave integrated circuit conference* (pp. 198–201). https://doi.org/10.1109/EMICC.2008.4772263.

Chevalier P., Avenier G., Ribes G., Montagné A., Canderle E., Céli D., Derrier N., Deglise C., Durand C., Quémerais T., Buczko M., Gloria D., Robin O., Petitdidier S., Campidelli Y., Abbate F., Gros-Jean M., Berthier L., Chapon J. D., … Borot B. (2014). A 55 nm triple gate oxide 9 metal layers SiGe BiCMOS technology featuring 320 GHz fT / 370 GHz fMAX HBT and high-Q millimeter-wave passives. In *2014 IEEE international electron devices meeting* (pp. 3.9.1–3.9.3). https://doi.org/10.1109/IEDM.2014.7046978.

Colinge, J.-P. (2004). *Silicon-on-insulator technology: Material to VLSI. 3*. New York, NY, New York: Springer. Available from https://doi.org/10.1007/978-1-4419-9106-5.

Didier C., & Desbonnets E. (2017). 2022 9 20 RF-SOI wafer characterization. https://www.soitec.com/media/files/soitec_whitepaper_rfesi_product_characterization_300mm_v2.pdf.

El-Aassar, O., & Rebeiz, G. M. (2020). Design of low-power sub-2.4 dB mean NF 5G LNAs using forward body bias in 22 nm FDSOI. *IEEE Transactions on Microwave Theory and Techniques*, *68*(10), 4445–4454. Available from https://doi.org/10.1109/TMTT.2020.3012538.

Flandre, D. (1995). Silicon-on-insulator technology for high temperature metal oxide semiconductor devices and circuits. *Materials Science and Engineering: B.*, *29*(1-3), 7–12. Available from https://doi.org/10.1016/0921-5107(94)04018-Y.

Flandre, D., Raskin, J.-P., & Vanhoenacker, D. (2001). SOI CMOS transistors for RF and microwave applications. *International Journal of High Speed Electronics and Systems*, *11*(4), 1159–1248. Available from https://doi.org/10.1142/S0129156401001076.

Gao L., & Rebeiz G. M. (2019). A 24-43 GHz LNA with 3.1-3.7 dB noise figure and embedded 3-pole elliptic high-pass response for 5G applications in 22 nm FDSOI. In *2019 IEEE radio frequency integrated circuits symposium (RFIC)* (pp. 239–242). https://doi.org/10.1109/RFIC.2019.8701782.

Gao, L., Wagner, E., & Rebeiz, G. M. (2020). Design of E- and W-band low-noise amplifiers in 22-nm CMOS FD-SOI. *IEEE Transactions on Microwave Theory and Techniques*, *68*(1), 132–143. Available from https://doi.org/10.1109/TMTT.2019.2944820.

Gianesello F., Gloria D., Raynaud C., Montusclat S., Boret S., & Touret P. (2007). Integrated inductors in HR SOI CMOS technologies: On the economic advantage of SOI technologies for the integration of RF applications. In *2007 IEEE international SOI conference* (pp. 119–120) https://doi.org/10.1109/SOI.2007.4357881.

Heinemann B., Rücker H., Barth R., Bärwolf F., Drews J., Fischer G. G., Fox A., Fursenko O., Grabolla T., Herzel F., Katzer J., Korn J., Krüger A., Kulse P., Lenke T., Lisker M., Marschmeyer S., Scheit A., Schmidt D., … Wolansky D. (2016). SiGe HBT with fx/fmax of 505 GHz/720 GHz. In *2016 IEEE international electron devices meeting (IEDM)* (pp. 3.1.1–3.1.4). https://doi.org/10.1109/IEDM.2016.7838335.

Huard V., Jacquet F., Mhira S., Jure L., Montfort O., Louvat M., Zaia L., Bertrand F., Acacia E., Caffin O., Belhadj H., Durand O., Exibard N., Bonnet V., Charvier A., Bernardi P., & Cantoro R. (2022). Runtime test solution for adaptive aging compensation and fail operational safety mode. In *2022 IEEE international reliability physics symposium (IRPS)* (pp 8 A.2-1–8 A.2-4). https://doi.org/10.1109/IRPS48227.2022.9764590.

Hur S.-G., Yang J.-G., Kim S.-S., Lee D.-K., An T., Nam K.-J., Kim S.-J., Wu Z., Lee W., Kwon U., Lee K.-H., Park Y., Yang W., Choi J., Kang H.-K., & Jung E. (2013). A practical Si nanowire technology with nanowire-on-insulator structure for beyond 10nm logic technologies. In *2013 IEEE international electron devices meeting* (pp 26.5.1–26.5.4). https://doi.org/10.1109/IEDM.2013.6724698.

Inac, O., Uzunkol, M., & Rebeiz, G. M. (2014). 45-nm CMOS SOI technology characterization for millimeter-wave applications. *IEEE Transactions on Microwave Theory and Techniques*, *62*(6), 1301–1311. Available from https://doi.org/10.1109/TMTT.2014.2317551.

Jan C.-H., Bhattacharya U., Brain R., Choi S.-J., Curello G., Gupta G., Hafez W., Jang M., Kang M., Komeyli K., Leo T., Nidhi N., Pan L., Park J., Phoa K., Rahman A., Staus C., Tashiro H., Tsai C., … Bai P. (2012). A 22nm SoC platform technology featuring 3-D tri-gate and high-k/metal gate, optimized for ultra low power, high performance and high density SoC applications. In *2012 International electron devices meeting* (pp. 3.1.1–3.1.4). https://doi.org/10.1109/IEDM.2012.6478969.

Jo, H.-B., Yun, D.-Y., Baek, J.-M., Lee, J.-H., Kim, T.-W., Kim, D.-H., Tsutsumi, T., Sugiyama, H., & Matsuzaki, H. (2019). Lg = 25 nm InGaAs/InAlAs high-electron mobility transistors with both fT and fmax in excess of 700 GHz. *Applied Physics Express*, *12*(5)054006. Available from https://iopscience.iop.org/article/10.7567/1882-0786/ab1943/pdf.

Kazemi Esfeh, B., Kilchytska, V., Barral, V., Planes, N., Haond, M., Flandre, D., & Raskin, J.-P. (2016). Assessment of 28 nm UTBB FDSOI technology platform for RF applications: Figures of merit and effect of parasitic elements. *Solid-State Electronics*, *117*, 130–137. Available from https://doi.org/10.1016/j.sse.2015.11.020.

Kazemi Esfeh B., Kilchytska V., Parvais B., Planes N., Haond M., Flandre D., & Raskin J.-P. (2017). Back-gate bias effect on UTBB-FDSOI non-linearity performance. In *2017 47th European solid-state device research conference (ESSDERC)* (pp. 148–151). https://doi.org/10.1109/ESSDERC.2017.8066613.

Kazemi Esfeh B., Makovejev S., Allibert F., & Raskin J.-P. (2017). A SPDT RF switch small- and large-signal characteristics on TR-HR SOI substrates. In *2017 IEEE SOI-3D-subthreshold microelectronics technology unified conference (S3S)* (pp. 1–3). https://doi.org/10.1109/S3S.2017.8308737.

Kazemi Esfeh, B., Rack, M., Ben Ali, K., Allibert, F., & Raskin, J.-P. (2018). RF small- and large-signal characteristics of CPW and TFMS lines on trap-rich HR-SOI substrates. *IEEE Transactions on Electron Devices*, *65*(8), 3120–3126. Available from https://doi.org/10.1109/TED.2018.2845679.

Kilchytska, V., Lederer, D., Collaert, N., Raskin, J.-P., & Flandre, D. (2005). Accurate effective mobility extraction by split C-V technique in SOI MOSFETs: Suppression of the influence of floating-body effects. *IEEE Electron Device Letters, 26*(10), 749–751. Available from https://doi.org/10.1109/LED.2005.855408.

Lederer, D., Flandre, D., & Raskin, J.-P. (2004). AC behavior of gate-induced floating body effects in ultrathin oxide PD SOI MOSFETs. *IEEE Electron Device Letters, 25*(2), 104–106. Available from https://doi.org/10.1109/LED.2003.822658.

Lederer, D., Flandre, D., & Raskin, J.-P. (2005). High frequency degradation of body-contacted PD SOI MOSFET output conductance. *Semiconductor Science and Technology, 20*, 469–472. Available from https://doi.org/10.1088/0268-1242/20/5/025.

Lederer D., Jain S., Saroop S., Kumar A., & Freeman G. (2018). 45nm PD SOI FET gate resistance optimization for mmw applications. In *2018 IEEE SOI-3D-subthreshold microelectronics technology unified conference* (S3S) (pp. 1–3). https://doi.org/10.1109/S3S.2018.8640156.

Lederer, D., & Raskin, J.-P. (2005). New substrate passivation method dedicated to HR SOI wafer fabrication with increased substrate resistivity. *IEEE Electron Device Letters, 26*(11), 805–807. Available from https://doi.org/10.1109/LED.2005.857730.

Lederer, D., & Raskin, J.-P. (2008). RF performance of a commercial SOI technology transferred onto a passivated HR silicon substrate. *IEEE Transactions on Electron Devices, 55*(7), 1664–1671. Available from https://doi.org/10.1109/TED.2008.923564.

Lee H.-J., Rami S., Ravikumar S., Neeli V., Phoa K., Sell B., & Zhang Y. (2018). Intel 22nm FinFET (22FFL) process technology for RF and mm wave applications and circuit design optimization for FinFET technology. In *2018 IEEE international electron devices meeting (IEDM)* (pp. 14.1.1–14.1.4). https://doi.org/10.1109/IEDM.2018.8614490.

Li C., El-Aassar O., Kumar A., Boenke M., & Rebeiz G.M. (2018). LNA Design with CMOS SOI process-l.4dB NF K/Ka band LNA. In *2018 IEEE/MTT-S international microwave symposium - IMS* (pp. 1484–1486). https://doi.org/10.1109/MWSYM.2018.8439132.

Li S., Huang T.-Y., Liu Y., Yoo H., Na Y., Hur Y., & Wang H. (2021). A millimeter-wave LNA in 45nm CMOS SOI with over 23dB peak gain and sub-3dB NF for different 5G operating bands and improved dynamic range. In *2021 IEEE radio frequency integrated circuits symposium (RFIC)* (pp. 31–34). https://doi.org/10.1109/RFIC51843.2021.9490455.

Liu, S., Zhu, L., Allibert, F., Rad, I., Zhu, X., & Lu, Y. (2017). Physical models of planar spiral inductor integrated on the high-resistivity and trap-rich silicon-on-insulator substrates. *IEEE Transactions on Electron Devices, 64*(7), 2775–2781. Available from https://doi.org/10.1109/TED.2017.2700022.

Makovejev, S., Kazemi Esfeh, B., Andrieu, F., Raskin, J.-P., Flandre, D., & Kilchytska, V. (2014). Assessment of global variability in UTBB MOSFETs in subthreshold regime. *Journal of Low Power Electronics and Applications, 4*(3), 201–213. Available from https://doi.org/10.3390/jlpea4030201.

Mei, X., Yoshida, W., Lange, M., Lee, J., Zhou, J., Liu, P.-H., Leong, K., Zamora, A., Padilla, J., Sarkozy, S., Lai, R., & Deal, W. R. (2015). First demonstration of amplification at 1 THz Using 25-nm InP high electron mobility transistor process. *IEEE Electron Device Letters, 36*(4), 327–329. Available from https://doi.org/10.1109/LED.2015.2407193.

Moursy Y., Raupp Da Rosa T., Jure L., Quelen A., Genevey S., Pierrefeu L., Grand E., Winkler J., Park J., Pillonnet G., Huard V., Bonzo A., & Flatresse P. (2021). 35.2 A 0.021mm2 PVT-aware digital-flow-compatible adaptive back-biasing regulator with scalable drivers achieving 450% frequency boosting and 30% power reduction in 22nm FDSOI technology. In *2021 IEEE international solid- state circuits conference (ISSCC)* (pp. 492–494). https://doi.org/10.1109/ISSCC42613.2021.9365782.

Nguyen L. D., Tasker P. J., Radulescu D. C., & Eastman L. F. (1988). Design, fabrication, and characterization of ultra high speed AlGaAs/InGaAs MODFETs. In *Technical digest., international electron devices meeting* (pp 176–179). https://doi.org/10.1109/IEDM.1988.32783.

Nieuwoudt A., & Massoud Y. (2005). Efficient modeling of substrate eddy currents for integrated spiral inductor design automation. In *2005 48th Midwest symposium on circuits and systems* (vol. 2, pp. 1823–1826). https://doi.org/10.1109/MWSCAS.2005.1594477.

Nyssens, L., Halder, A., Kazemi Esfeh, B., Plane, N., Flandre, D., Kilchytska, V., & Raskin, J.-P. (2020). 28-nm FD-SOI CMOS RF figures of merit down to 4.2 K. *IEEE Journal of the Electron Devices Society, 8*, 646–654. Available from https://doi.org/10.1109/JEDS.2020.3002201.

Nyssens L., Rack M., Nabet M., Schwan C., Zhao Z., Lehmann S., Herrmann T., Henke D., Kondrat A., Soonekindt C., Koch F., Kache T., Kini D.P., Zimmerhackl O., Allibert F., Aulnette C., Lederer D., & Raskin J.-

P. (2022). PN junctions interface passivation in 22 nm FD-SOI for low-loss passives. In *2022 24th International microwave and radar conference (MIKON)* (pp. 1–4). Available from https://ieeexplore.ieee.org/document/9924803.

Nyssens, L., Rack, M., Schwan, C., Zhao, Z., Lehmann, S., Hermann, T., Allibert, F., Aulnette, C., Lederer, D., & Raskin, J.-P. (2022). Impact of substrate resistivity on spiral inductors at mm-wave frequencies. *Solid-State Electronics, 194*108377. Available from https://doi.org/10.1016/j.sse.2022.108377.

Nyssens L., Rack M., Wane S., Schwan C., Lehmann S., Zhao Z., Lucci L., Lugo-Alvarez J., Gaillard F., Raskin J.-P., & Lederer D. (2022). A 2.5-2.6 dB noise figure LNA for 39 GHz band in 22 nm FD-SOI with back-gate bias tunability. In *17th European microwave integrated circuits conference (EuMIC)* (pp. 60–63). https://doi.org/10.23919/EuMIC54520.2022.9923552.

Ong S. N., Lehmann S., Chow W. H., Zhang C., Schippel C., Chan L. H. K., Andee Y., Hauschildt M., Tan K. K. S., Watts J., Lim C. K., Divay A., Wong J. S., Zhao Z., Govindarajan M., Schwan C., Huschka A., Bellaouar A., Loo W., ... Harame D. (2018). A 22nm FDSOI technology optimized for RF/mmWave applications. In *2018 IEEE radio frequency integrated circuits symposium (RFIC)* (pp. 72–75). https://doi.org/10.1109/RFIC.2018.8429035.

Parlak M., & Buckwalter J. F. (2011). A 2.9-dB noise figure, Q-band millimeter-wave CMOS SOI LNA. In *2011 IEEE custom integrated circuits conference (CICC)* (pp. 1–4). https://doi.org/10.1109/CICC.2011.6055321.

Rack, M., Nyssens, L., & Raskin, J.-P. (2019a). Low-loss si-substrates enhanced using buried PN junctions for RF applications. *IEEE Electron Device Letters, 40*(5), 690–693. Available from https://doi.org/10.1109/LED.2019.2908259.

Rack M., Nyssens L., & Raskin J.-P. (2019b). Silicon-substrate enhancement technique enabling high quality integrated RF passives. In *2019 IEEE MTT-S international microwave symposium (IMS)* (pp. 1295–1298). https://doi.org/10.1109/MWSYM.2019.8701095.

Rack, M., & Raskin, J.-P. (2019). (Invited) SOI technologies for RF and millimeter wave applications. *ECS Transactions, 92*(4), 79–94. Available from https://doi.org/10.1149/09204.0079ecst.

Raskin, J.-P. (2016a). FinFET and UTBB for RF SOI communication systems. *Solid-State Electronics, 125*, 73–81. Available from https://doi.org/10.1016/j.sse.2016.07.004.

Raskin J.-P. (2016b). SOI technologies from microelectronics to microsystems - Meeting the more than moore roadmap requirements. International Journal of High Speed Electronics and Systems, 25(1–2), 1–26. https://doi.org/10.1142/S0129156416400048.

Razavieh, A., Chen, Y., Ethirajan, T., Gu, M., Cimino, S., Shimizu, T., Hassan, M. K., Morshed, T., Singh, J., Zheng, W., Mahajan, V., Wang, H. T., & Lee, T. H. (2021). Extremely-low threshold voltage FinFET for 5G mmWave applications. *International Journal of High Speed Electronics and Systems, 9*, 165–169. Available from https://doi.org/10.1109/JEDS.2020.3046953.

Roda Neve, C., & Raskin, J.-P. (2012). RF harmonic distortion of CPW lines on HR-Si and trap-rich HR-Si substrates. *IEEE Transactions on Electron Devices, 59*(4), 924–932. Available from https://doi.org/10.1109/TED.2012.2183598.

Scheen, G., Tuyaerts, R., Rack, M., Nyssens, L., Rasson, J., Nabet, M., & Raskin, J.-P. (2020). Post-process porous silicon for 5G applications. *Solid-State Electronics, 168*107719. Available from https://doi.org/10.1016/j.sse.2019.107719.

Shim, Y., Raskin, J.-P., Roda Neve, C., Rais-Zadeh, M., & MEMS, R. F. (2013). Passives on high-resistivity silicon substrates. *IEEE Microwave and Wireless Components Letters, 23*(12), 632–634. Available from https://doi.org/10.1109/LMWC.2013.2283857.

Shinohara, K., Regan, D. C., Tang, Y., Corrion, A. L., Brown, D. F., Wong, J. C., Robinson, J. F., Fung, H. H., Schmitz, A., Oh, T. C., Kim, S. J., Chen, P. S., Nagele, R. G., Margomenos, A. D., & Micovic, M. (2013). Scaling of GaN HEMTs and schottky diodes for submillimeter-wave MMIC applications. *IEEE Transactions on Electron Devices, 60*(10), 2982–2996. Available from https://doi.org/10.1109/TED.2013.2268160.

Skyworks. (2011). 2022 10 17 White paper. Choosing the Right RF switches for smart mobile device applications. https://www.microwavejournal.com/ext/resources/BGDownload/3/2/RF_SwitchesForSmartMobileDeviceApplications.pdf?1612490380.

Sugii, N., Tsuchiya, R., Ishigaki, T., Morita, Y., Yoshimoto, H., & Kimura, S. (2010). Local Vth variability and scalability in silicon-on-thin-BOX (SOTB) CMOS with small random-dopant fluctuation. *IEEE Transactions on Electron Devices, 57*(4), 835–845. Available from https://doi.org/10.1109/TED.2010.2040664.

Tang, X., Nguyen, J., Medra, A., Khalaf, K., Visweswaran, A., Debaillie, B., & Wambacq, P. (2020). Design of D-band transformer-based gain-boosting class-AB power amplifiers in silicon technologies. *IEEE Transactions on Circuits and Systems I: Regular Papers, 67*(5), 1447–1458. Available from https://doi.org/10.1109/TCSI.2020.2974197.

Torres, F., Cathelin, A., & Kerhervé, E. (2020). *The fourth terminal. Benefits of body-biasing techniques for FDSOI circuits and systems millimeter-wave power amplifiers for 5G applications in 28 nm FD-SOI technology. 1st* (pp. 169–222). Cham: Springer. Available from https://doi.org/10.1007/978-3-030-39496-7.

Urteaga M., Pierson R., Rowell P., Jain V., Lobisser E., & Rodwell M. J. W. (2011). 130nm InP DHBTs with ft >0.52THz and fmax >1.1THz. In *69th Device research conference* (pp. 281–282). https://doi.org/10.1109/DRC.2011.5994532.

Vandana, B. (2013). Study of floating body effect in SOI technology. *International Journal of Modern Engineering Research, 3*(3), 1817–1824.

VanDerVoorn P., Agostinelli M., Choi S.-J., Curello G., Deshpande H., El-Tanani M. A., Hafez W., Jalan U., Janbay L., Kang M., Koh K.-J., Komeyli K., Lakdawala H., Lin J., Lindert N., Mudanai S., Park J., Phoa K., Rahman A., ... Jan C.-H. (2010). A 32nm low power RF CMOS SOC technology featuring high-k/metal gate. In *2010 Symposium on VLSI technology* (pp. 137–138). https://doi.org/10.1109/VLSIT.2010.5556201.

Veloso, A., Cho, M. J., Simoen, E., Hellings, G., Matagne, P., Collaert, N., & Thean, A. (2016). (Invited) Gate-all-around nanowire FETs vs. triple-gate FinFETs: On gate integrity and device characteristics. *ECS Transactions, 72*(2), 85–95. Available from https://doi.org/10.1149/07202.0085ecst.

Veloso A., Huynh-Bao T., Matagne P., Jang D., Horiguchi N., Ryckaert J., & Mocuta D. (2019). Nanowire & nanosheet FETs for ultra-scaled, high-density logic and memory applications. In *2019 Joint international EUROSOI workshop and international conference on ultimate integration on silicon (EUROSOI-ULIS)* (pp. 1–4). https://doi.org/10.1109/EUROSOI-ULIS45800.2019.9041857.

Zhao Z., Lehmann S., Oo W. L., Sahoo A.K., Syed S., Le Q. H., Huynh D. K., Chohan T., Utess D., Kleimaier D., Wiatr M., Kolodinski S., Mazurier J., Hoentschel J., Knorr A., Cahoon N., & Kneitz S. (2021). 22FDSOI device towards RF and mmWave applications. In *2021 IEEE BiCMOS and compound semiconductor integrated circuits and technology symposium (BCICTS)* (pp. 1–6). https://doi.org/10.1109/BCICTS50416.2021.9682480.

CHAPTER

# 3

# Radio frequency FinFET bulk silicon technology

*Jagar Singh*[1] *and Josef Watts*[2]

**[1]pSemi/ex-GlobalFoundries Analog and RF device Technologist, Technology Development Group, Malta, NY, United States [2]IBM/GlobalFoundries, Modeling Group, Malta, NY, United States**

## 3.1 Introduction

For operating frequencies up to 6 GHz, the radio frequency (RF) complementary metal-oxide semiconductor (CMOS) technology has become the process of choice to fabricate wireless transceiver chips with integrated radios in high-volume production (Baldwin et al., 2003; Lakdawala et al., 2013; Li et al., 2017; Wu et al., 2018). Owing to superior scalability, and the rich infrastructure of design tools and intellectual property available, these RF integrated circuits (ICs) enable a variety of applications such as WiFi, ZigBee, Security Cameras, and Bluetooth as shown in Fig. 3.1 in conjunction with the low-power (LP) and high-performance (HP) computational circuits (Lee et al., 2019, 2021; Sell et al., 2018; Singh et al., 2017; Yoon & Baek, 2020). However, millimeter-wave (mmWave; 30–300 GHz) regime circuit manufacturing has been the realm of technologies such as heterojunction silicon germanium (SiGe) vertical NPN bipolar, UTBSOI (ultra-thin body silicon on insulator) CMOS, and bipolar-CMOS (Bi-CMOS), as operation in the mmWave regime requires faster transistors and high-quality passives (Alvin et al., 2018; Carter et al., 2017).

On the other hand, recent advancements in the FinFET technology beyond the conventional scaling below 22 nm, the new low gate resistance device architecture, such as the contact over an active gate, offers cost-effective and competitive mmWave device performance solutions (Razavieh et al., 2020; Yang et al., 2017). Thus, it provides a unique opportunity to bring LP and high-density logic devices and circuits closer to mmWave devices for SoC (system on chip) solutions (Yeap, 2013). This chapter compares in comprehensive detail RF FinFET and RF planar CMOS technology, in addition to a brief description of the alternative RF technologies, including the SiGe vertical heterojunction bipolar transistor (HBT) NPN and the UTBSOI CMOS.

*New Materials and Devices Enabling 5G Applications and Beyond*
**DOI: https://doi.org/10.1016/B978-0-12-822823-4.00003-0**

**FIGURE 3.1** The 5G circuit's application examples include WiFi, ZigBee, Security Cameras, BlueTooth, etc.

## 3.2 SiGe NPN heterojunction bipolar transistor

The SiGe bipolar technology has been an early choice for a wide range of wireless applications, especially power amplifier circuits. The smaller device size and lower cost compared to GaAs technology have led to widespread market adoption. The bipolar transistor is based on a different physical principle than the field effects transistor (FET). These SiGe bipolar junction transistors (BJT) are like Si BJTs, with the difference in germanium presence in the base region as shown in Fig. 3.2. An optimum Ge presence reduces the bandgap of the SiGe base region material. The bandgap offset at the emitter-base junction leads to improved performance, especially the beta, by reducing the hole current. Further bandgap engineering with a graded Ge profile in the base region enhances device performance. These mechanisms provide very low on-resistance, high current capability, and excellent linearity, which is particularly well suited for power amplifiers (Alvin et al., 2018; Kissinger et al., 2021).

## 3.3 UTBSOI MOSFET

The silicon on insulator (SOI) structure consists of a thin silicon and buried oxide layer on a bulk silicon substrate. The thin silicon ($t_{Si}$) layer is the base of the most of FET fabrication steps and device performance attributes. It is isolated by the buried oxide ($B_{OX}$) insulating layer from a silicon substrate as shown in Fig. 3.3. The thin silicon body ($t_{Si}$) of the device is fully depleted during the on-state operation and limits the drain field penetration toward the source side. As a result, it provides better electrostatic control against the short channel effects (SCEs) like drain-induced barrier lowering (DIBL) compared to the conventional bulk Si planar FETs. The latest generation of FDSOI (fully depleted silicon on insulator) technology, namely 12 nm UTBSOI with the 6 to 8 nm ultra-thin Si body, enables superior electrostatic scaling and lower

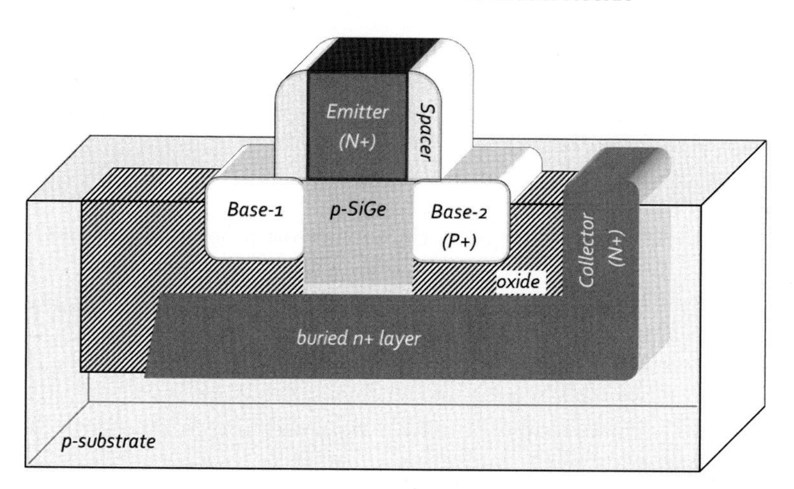


**FIGURE 3.2** A basic three-dimensional (3D) schematic of the NPN SiGe heterojunction bipolar transistor.

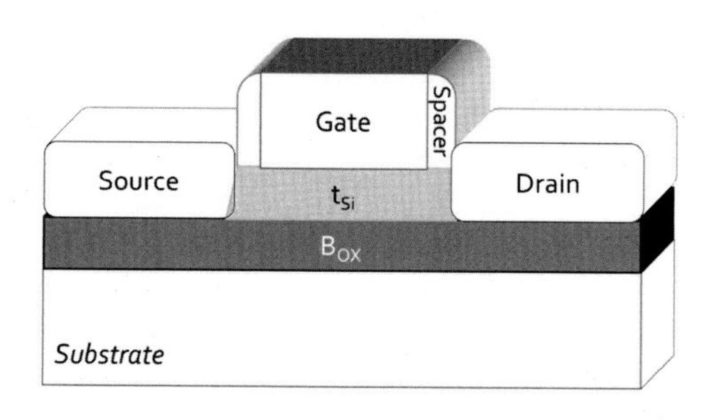


**FIGURE 3.3** A schematic view of the ultra-thin body silicon on insulator (UTBSOI) MOSFET.

source and drain (S/D) junction parasitic capacitance compared to the predecessor 22 nm FDSOI technology node (Carter et al., 2017; Choi et al., 2000; Muralidhar et al., 2018).

The reduced junction parasitic capacitance of UTBSOI MOSFET structure compared to bulk Si gives higher unity current gain transit frequency ($F_t$) and unity maximum power gain operating frequency ($F_{max}$). However, the presence of buried oxide degrades heat dissipation from the device channel region to the silicon substrate because of increased thermal resistance. This leads to more intrinsic SOI MOSFET self-heating and performance and reliability degradation. It precipitates an impact on the circuit efficiency and limits use in high-power applications such as power amplifiers (PAs).

In contrast, the buried oxide provides electrical isolation from the substrate which reduces cross-talk, nonlinearities, and substrate losses at RF frequencies which are key advantages of the SOI structures compared to bulk-Si (Ali et al., 2014; Raskin et al., 1997). The electrical insulating characteristics of the buried oxide layer enable transistor stacking to meet the requirement of the higher voltage handling capability without the body effect seen in the bulk CMOS ( Joseph et al., 2013). Stacking of FETs is an important technology feature for the RF switch and power amplifier designs on the leading-edge technologies because the RF peak voltage is typically higher than the FET breakdown voltage. Also, the backend passive devices, for example, MIM and MOM capacitor and inductor built over the $B_{OX}$ layer show improved device matching behavior, which is of particular importance for the ease of design and higher performance mmWave RF ICs (Cathelin, 2017). These factors make this an attractive process choice for mmWave circuit designs (Ong et al., 2018). However, compared to advanced FinFET technology nodes, in particular 12 nm and below, UTBSOI lags in terms of logic density and cost per die due to an expensive SOI wafer.

## 3.4 Radio frequency complementary metal-oxide semiconductor technology

In the traditional planar MOSFET, the conduction channel forms in the silicon region under the active gate area to allow carrier transport from source to drain contact. It spreads on the silicon surface, which is covered by the thin gate oxide first and then gate material as shown in Fig. 3.4A. On the other hand, in the three-dimensional (3D) FinFET case, the active channel forms on the three sides of the fully depleted Si region because the gate wraps around three sides as illustrated in Fig. 3.4B.

The planar CMOS has been the dominant manufacturing technology to build ICs in the semiconductor industry. It has been continuously scaled for more than five decades to serve the industry's growing needs. However, Moore's law scaling of planar CMOS ended with a 22 nm node (Jan et al., 2012). This is because the electrostatic effect of the drain on the channel becomes uncontrollable at shorter channel lengths ($L_g$). The bulk of the planar CMOS production has been in the 28 nm node and above (Wu et al., 2015). A very large design ecosystem exists for 28 nm planar technology, enabling shorter design cycles and faster time to market. Large-scale production and relative technological simplicity make for very competitive pricing. The main motivation of scaling is to have faster devices and lower cost (Hz-bit-function) for logic circuit applications as more circuit dies can be built in the same area or more functions packed into each die. As scaling continued to progress, the operating frequency of the device also improved as shown in Fig. 3.5A (Jan et al., 2010; Morifuji et al., 1999; Woerlee et al., 2001; Yang et al., 2017; Yeap, 2013). This is mainly attributed to the enhanced transconductance ($G_m$) with gate length reduction as shown in Fig. 3.5B. The parameter $F_t$ in these figures represents the unity current gain transit frequency.

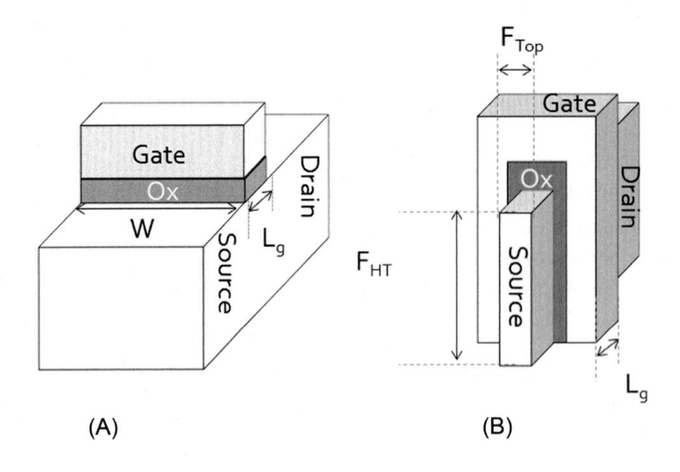


**FIGURE 3.4** A schematic showing of bulk silicon (A) the planar MOSFET and (B) single fin FinFET device.

Only in the last two decades, interest in building RF applications is CMOS increased, as the $F_t$ has become competitive with SiGe and SOI RF technology (see Fig. 3.6). The RF CMOS has been adapted extensively for RF circuit applications primarily below the 6 GHz range. However, at 28 or even 22 nm, the $F_t$ start to support some of the mmWave circuit applications (Ong et al., 2018), and as Fig. 3.6 shows, further reduction in channel length ($L_g$) does not increase $F_t$ because of increasing parasitics. However, migration to the 3D FinFET architecture led to a reduction in the peak $F_t$ performance primarily due to the rise in the external parasitic capacitance such as $C_{gs}$, $C_{gd}$, etc., despite an improved transconductance ($G_m$) as shown in (b). The planar CMOS scaling didn't improve the self-gain ($G_m \times R_{out}$) as can be seen in Fig. 3.7. The analog and RF designers are frequently forced to choose nonminimum $L_g$. The move to FinFET technology greatly increased self-gain, mainly due to enhanced electrostatic control over the channel region leading to a lowering of the DIBL and higher $R_{out}$.

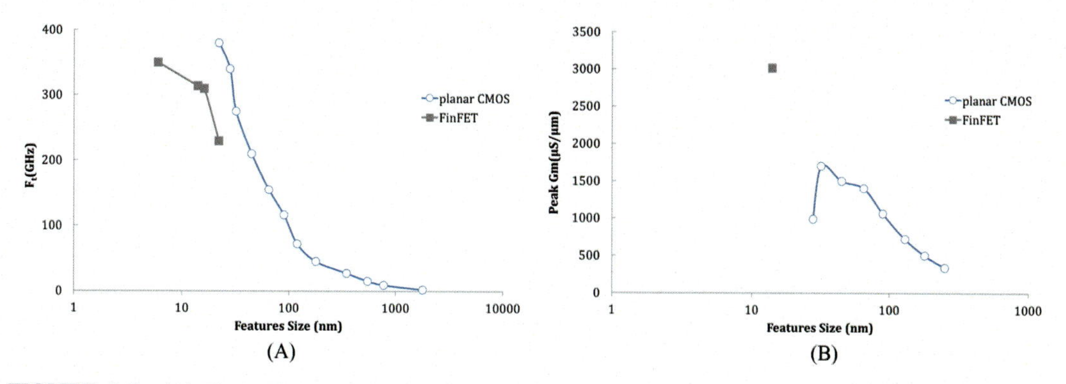


FIGURE 3.5 (A) The peak $F_t$ versus technology node feature size (data available down to 14 nm technology only), (B) the peak Gm versus technology node feature size (data available down to 14 nm technology only).

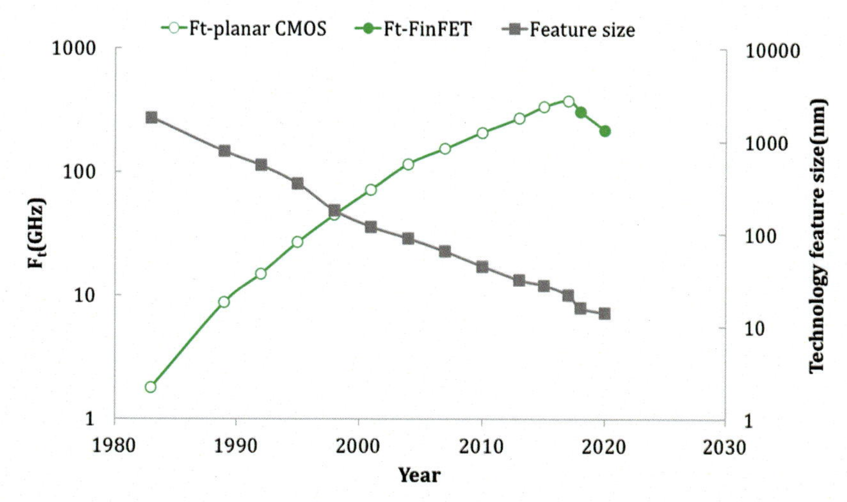


FIGURE 3.6 The $F_t$ history of radio frequency complementary metal-oxide semiconductor and FinFET devices over the past 30 years of period.

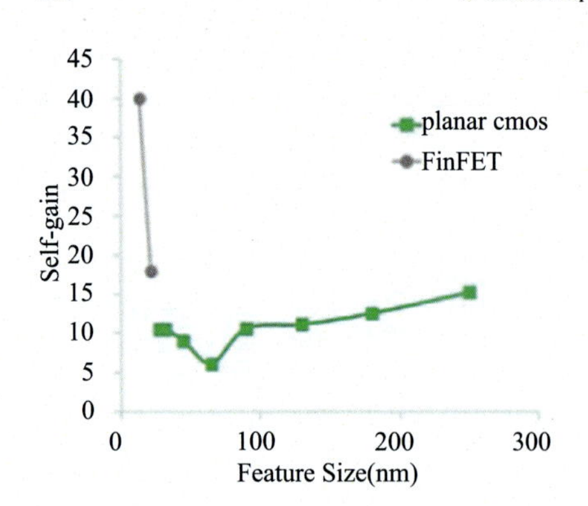


**FIGURE 3.7** Shows the self-gain of the complementary metal-oxide semiconductor (planar and FinFET) devices versus the technology feature size.

Actually, an improvement from the 3D FinFET architecture is quite noteworthy three to four times even as the device scaling enters the FinFET era (<22 nm).

For the 6 GHz Wi-Fi SoC applications, the FinFET RF technology next node migration from 14 nm to 8 or 6 nm offers the benefits of 35% area scaling and 30% power reduction while maintaining a decent ~30% $F_t$ improvement as shown in Fig. 3.8. On the other hand, scaling gives rise to parasitics including the increased gate resistance from narrow line widths and fringing capacitance from the shrinking spacing of gate to source and drain, metal to metal, etc., which are detrimental to the device RF performance. With just scaling, communications ICs tend to see degraded RF characteristics, such as lower gain and output power versus frequency with increased power consumption. To overcome the performance bottlenecks, additional process elements such as new material or stress layers are introduced to enhance the channel mobility. Similarly wider space or liner width of backend layers are introduced to the process platforms to minimize the parasitic capacitance (Lee et al., 2020). The criteria of RF technology selection now include not only the superior performance but also the area and power scaling (Samsung, 2021; TSMC, 2021).

## 3.5 Radio frequency FinFET

The FinFET geometry provides the best electrostatic device scaling because of the two parallel gates and results in an enhanced logic performance. This has led to wide adoption in the mainstream of logic CMOS technology manufacturing down to feature size 3 nm and beyond (Mocuta et al., 2018). On the mobile 5G application side, the increasing mobile data traffic demands a larger bandwidth and higher operating frequencies. To support such needs, the RF SoC circuits like transceivers and modems are seeing ever-increasing logic circuit complexity and performance requirements. Also, these battery-driven mobile devices put power as a critical parameter above any other. These three variables power, performance, and area (PPA) form an important logic metric at which FinFET CMOS excels (Jeong et al., 2017). The RF FinFET-based optimized RF SoC designs are an attractive choice for designers in the <6 GHz RF application space (Lee et al., 2020, 2021; Singh et al., 2017).

The critical $F_t$, $F_{max}$ parameters across the RF technologies are listed in Table 3.1. The FinFET technology offers competitive RF performance while holding the advantage of

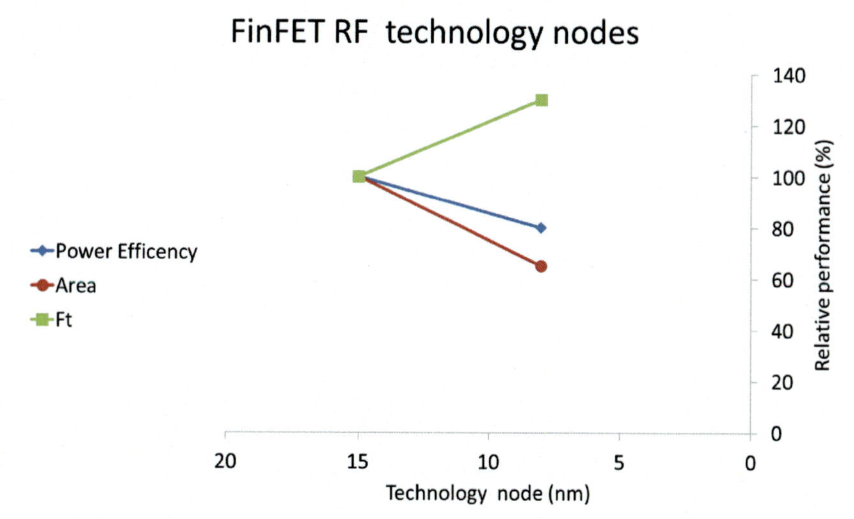


**FIGURE 3.8** Shows FinFET's next node progression from 14 to 8 or 6 nm.

**TABLE 3.1** A comparison of the radio frequency device parameters of in-production FinFET, UTBSOI, and SiGe heterojunction bipolar transistor technology.

| Parameters | Units | 12 nm RF FinFET (Singh et al., 2018) | 22 nm UTB SOI FET (Carter et al., 2017) | 28 nm HiKMG planar CMOS (Chew et al., 2015) | 90 nm SiGe HBT (Alvin et al., 2018) |
|---|---|---|---|---|---|
| Supply voltages ($V_{dd}$) | V | 0.8 | 0.8 | 1.05 | 1.8 |
| Peak $F_t$ (N/P) | GHz | 340/285 | 350/242 | 310/185 | 310 (npn) |
| Peak $F_{max}$ (N/P) | | 180/140 | 371/288 | 161/104 | 370 (npn) |

scaling. The digital contents portion of RF SoC is also complemented by the continuous node scaling of the FinFET era to fit in a smaller footprint. Progression to the next FinFET technology node also brings the power saving of nearly ~30% as shown in Fig. 3.8. The FinFET RF technology continues to lead the adoption of next-generation WiFi RF SoC designs in high-volume production. A very short channel of ultra-deep submicron FinFETs provides a high enough $F_t$ and $F_{max}$ for mmWave applications. Coupled with innovations at the circuit and architecture level, the advanced node FinFET is now well positioned to be the key technology that enables other RF and mmWave circuit applications beyond the 5G SoCs (Callender et al., 2018; Lee et al., 2019).

The FinFET benefits of superior scalability and low power continue to improve down to 3 nm nodes and below (Chiang et al., 2021; Mocuta et al., 2018). Extending these advantages to circuits like analog and RF requires an additional back end of line (BEOL) process and layout optimization to achieve low-wiring parasitic at a lower cost with more and better BEOL passive devices. Integrating these requires a deeper understanding of new materials, processes, process integration, layouts, architecture, reliability aspects, etc. As a matter of fact, the reliability of the devices emerged as a key challenge to qualifying for high-volume production. In this chapter, we will have a brief look at some general RF FinFET material and device challenges and solutions for 5G RF SoC circuits (Callender et al., 2018). Recent microwave band circuit application for multiband/multimode transmitters and receivers has been rapidly evolving. The increasing digital design contents and digitally assisted RF circuit developments have favored the introduction of both CMOS and FinFET technology to the mmWave applications.

## 3.6 Radio frequency planar MOSFET versus FinFET

The width of the planar device is the width of the active region in the plane of the wafer, which is a surface channel region. Conversely, the FinFET width ($W_{Fin}$) consists of the three side-channel regions, that is, two heights of active vertical fin sidewalls and one fin top surface width. This can be written as

$$W_{Fin} = 2 \times F_{HT} + F_{Top} \tag{3.1}$$

where $W_{Fin}$ is the device width of FinFET, $F_{HT}$ is the fin height, and $F_{Top}$ is the width of the fin top channel surface. Table 3.2 shows an example of 14 nm FinFET width parameters in comparison with the equivalent planar MOSFET. The width ($W_{Fin}$) of the FinFET device is >1.5 times larger than the planar one, therefore delivering a current of a similar higher ratio from the same footprint device. As can be seen in Eq. (3.1), the active fin height ($F_{HT}$) makes a significant contribution to the $W_{Fin}$. It is one of the popular knobs to increase the total $W_{Fin}$, while keeping the device footprint the same on a given FinFET technology node. Actually, this has the potential to produce a next-generation device without migrating to the next lithography node (Natarajan et al., 2015).

As mentioned earlier, in the FinFET architecture, the gate wraps around the three sides of the fin provides better electrostatic control of the gate over the channel due to the thin and fully depleted fin Si body. Fig. 3.9 shows the linear $I_dV_g$ behavior of the planar MOSFET (device 2) and two FinFET devices one with higher threshold voltage ($V_t$)

**TABLE 3.2** Comparing the 14 nm FinFET versus planar MOSFET device width parameters for the same planar area on the chip.

| Parameter | 14 nm FinFET value (Singh et al., 2018) | Equivalent planar MOSFET value |
|---|---|---|
| $F_{HT}$ (nm) | 35 | NA |
| $F_{Top}$ (nm) | 5 | |
| $W_{Fin}$ (nm) | 75 | 48[a] |
| Ratio ($W_{Fin}$/W) | 1.5625 | |

[a]*Equal to fin pitch.*

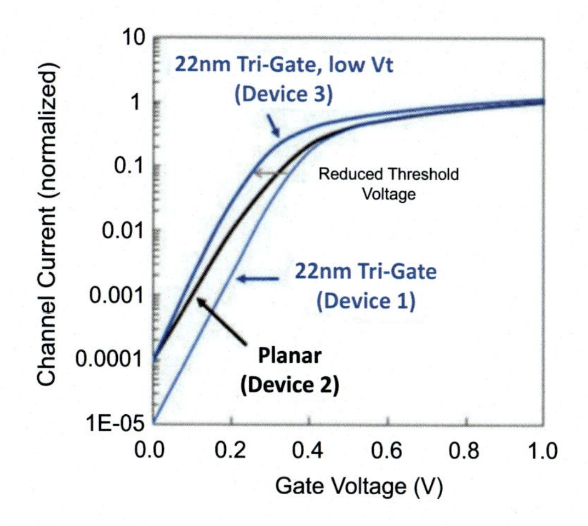


**FIGURE 3.9** A linear DC drain current versus gate voltage characteristics ($I_d V_g$) comparison from the 3D FinFET (Tri-Gate) and planar MOSFET. *Source: From Lee et al. (2019).*

(device 1) and the other with lower $V_t$ (device-3). The leakage current $I_{OFF}$, of the FinFET (device 1) is reduced by an order of magnitude as compared to planar (device 2), while still maintaining the same drive current strength ($I_{ON}$) (Lee et al., 2019). The improved subthreshold behavior effectively allows FinFET architecture to offer a lower $V_t$ device (device 3) without increasing the $I_{OFF}$ and thus achieving extreme $L_g$ scaling to a few nanometers. It shows the true capability of FinFET to produce higher current at the same $V_g$ and scaling opportunities can boost it further. This implies an improved transconductance $\left(G_m = \frac{\Delta I_d}{\Delta V_g}\right)$ for low voltage designs by comparing devices 2 and 3 of Fig. 3.9.

The FinFET geometry also improves the output impedance. In a short channel planar bulk-Si FET, there is a strong electrostatic coupling between the drain and the channel through the body of the FET. This makes the threshold voltage ($V_t$) drop as the drain voltage ($V_d$) increases and results in a lower output impedance in the saturation region, thereby creating nonlinearity in amplifiers (Razavi, 1999). In the FinFET, this is greatly reduced by the thinness of the fin. This is seen in the flattened slope of the $I_d V_d$ curve in the saturation region of operation in Fig. 3.10 despite being at an extremely short channel device. The thin fin body almost eliminates the effect of back body bias on the FinFET threshold voltage because the geometry prevents significant capacitive coupling between the channel and substrate. This is shown in Fig. 3.11 with an example of P-type devices (Zhang et al., 2021). In some analog and RF applications, FinFET technology provides better isolation to substrate noise from neighboring circuits compared to planar and may not require the use of triple well isolation.

The output conductance $\left(G_{ds} = \frac{\Delta I_d}{\Delta V_d}\right)$ is now lower, providing better intrinsic self-gain $(G_m/G_{ds})$, and less dependence on $V_d$. As a result, FinFET technology provides better linearity, a higher gain with a smaller device footprint, and improved substrate isolation offering an unique opportunity to reinvent some of the basic analog and HP RF circuit designs like amplifiers, current mirrors, etc.

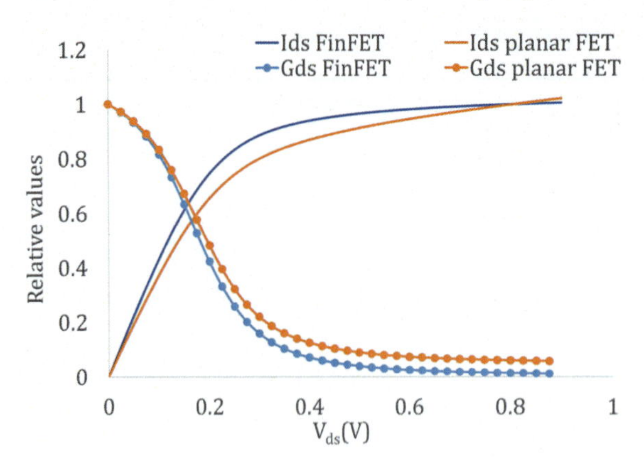


**FIGURE 3.10** The $I_d V_d$ and $G_{ds}$ characteristics comparison of planar and FinFET devices.

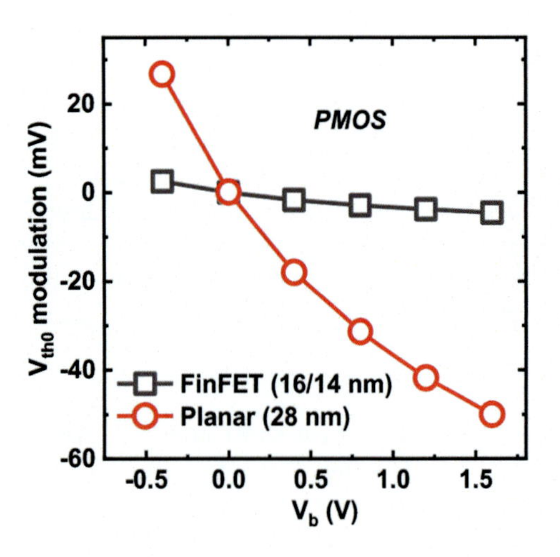


**FIGURE 3.11** The comparison of body effect between FinFET and planar devices. *Source: From Zhang et al. (2021).*

Note that Fig. 3.9 plots data based on first-generation FinFETs and the discussed performance gains are further improved in second- and third-generation FinFETs. This is advantageous for digital and low-frequency analog design, while these metrics are also important for RF and mmWave designs, at high frequencies parasitic effects become more important. Unfortunately, RF FinFET devices suffer from higher parasitics $R_g$, $C_{gs}$, $C_{gd}$, and fringing capacitance compared to planar MOSFET. For that, we need to look at both the small-signal figures-of-merit (FoMs); $G_m$, $F_t$, $F_{max}$, and the influence of BEOL, which will be discussed in this chapter. However, in the following section, a brief introduction to FinFET process flow is presented emphasizing the critical steps, which are of importance to the RF FinFET behavior in comparison with the planar CMOS technology.

## 3.7 The radio frequency FinFET fabrication process flow

As mentioned earlier, the FinFET is a 3D structure evolution from earlier planar MOSFET to meet the next-generation device performance specifications. The fin is an active part of the device to transport the carriers from source to drain. To extract the maximum FinFET architecture benefits the $W_{Fin}/W$ ratio must be >1; hence, a taller fin height is preferred. However, it brings additional process complexity and adverse impact on parasitics, particularly the fringing field capacitance and gate resistance. The fin formation, work function (WF) gate material, wrapping gate, S/D cavity, contact formation, etc. are innovated to meet the rising 3D structure, process, and performance challenges. In this section, the key processing steps of the 12 nm equivalent FinFETs technology are described below in comparison to the previous generation 28 nm planar CMOS technology.

**Fin formations:** The FinFET processing starts with a bulk-Si wafer similar to planar CMOS technology; however, it begins to diverge in the first few steps as shown in Fig. 3.12A. After depositing the thin pad oxide and silicon nitride (hard mask) layer, amorphous silicon (mandrel) layer, oxide, and silicon nitride (second hard mark) layer, these layers are patterned, creating mandrels @ *2X* fin pitch. Subsequently, an oxide layer is deposited for sidewall spacer formation over the patterned mandrel amorphous Si lines. It is followed with an oxide dry etch to form a narrow-width sidewall spacer, which marks the location of the fins. After removing the mandrel, the spacers are left as a hard mask to transfer these patterns down to the Si. The vertical fins are formed by etching away the Si between the fins using a dry etch process, as shown in the single fin schematic Fig. 3.12B. Typically, it is followed by the n-well and p-well implantation and shallow trench isolation processing.

**Polysilicon dummy gate processing**: In the gate last fabrication approach, dummy gates are formed by polysilicon (poly-Si) film deposition, patterning, and etching before the source and drain (S/D) regions processing (Kesapragada et al., 2010). These dummy poly-Si line patterns are created at the eventual location of the gates to hold the spot and allow spacer and S/D processing on either side as shown in Fig. 3.12B.

**Low-k dielectric material spacer formation:** In the following step, a thin low-k dielectric film is deposited using a process like CVD (chemical vapor deposition) covering the dummy poly-Si patterns. It is dry etched to form a low-k oxide spacer along the sidewall of these poly-Si lines. The low-k material sidewalls reduce the gate-to-source ($C_{gs}$) and gate-to-drain ($C_{gd}$) parasitic capacitance, which improves device RF performance.

**Source and drain (S/D) module**: As stated earlier that FinFET is a 3D structure, the active channels form along the top and sidewalls of fins as depicted in Fig. 3.4. To produce a drive current effectively from the bottom portion of the fin, highly doped N+ or P+ S/D regions reaching the bottom of the active vertical fin region are essential. To achieve this, deep cavities are etched onto S/D Si fins regions and filled with epitaxially grown, highly in-situ doped N+ SiP (phosphorus-doped Si) material for N-type and P+ SiGe for P-type FinFET. The stoichiometry of SiGe is capable of exerting compressive stress on the channel of p-type FinFET as desired to improve the hole carrier mobility. The channel stress or strain techniques boost the $G_m$ of the device and are a popular process improvement knob even in next-generation FinFET technologies (Samsung, 2021). The epi-Si is also grown larger than the initial fin Si volume, creating the diamond shape overhead, and a

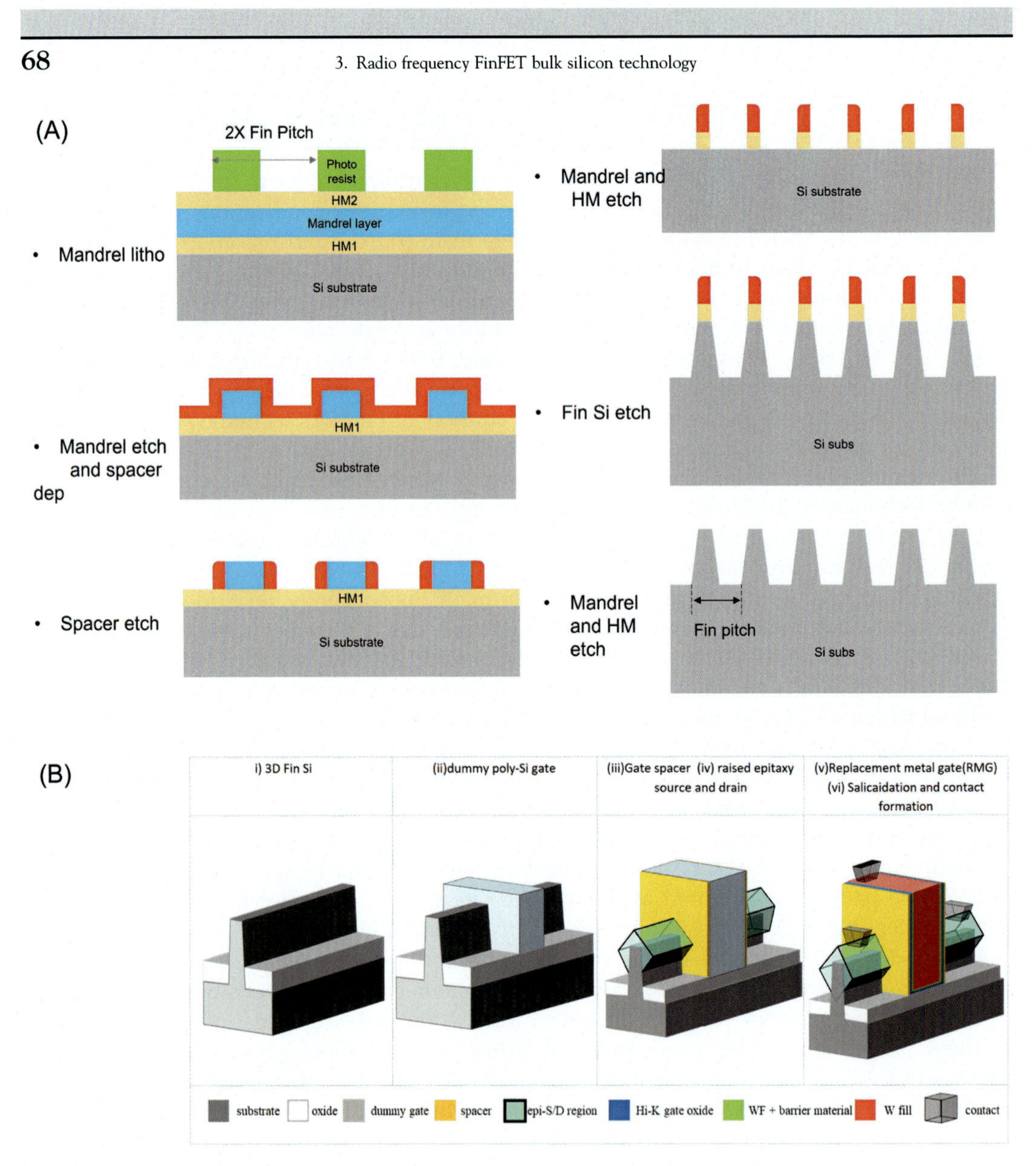


**FIGURE 3.12** Illustrating the (A) Si fins and (B) FinFET device fabrication steps using the single fin schematic.

resulting cross-section of the S/D regions as shown in Fig. 3.12B. This is beneficial because it reduces the contact resistance due to the larger surface area. However, it causes an extra $C_{gs}$ and $C_{gd}$ parasitic capacitance, which is unfavorable to RF device performance.

**Replacement metal gate (RMG) processing**: Post raised S/D formation, the dummy poly-Si gate material is etched away exposing the fin and leaving behind the spacer side-walls. A thin interfacial (native) oxide is naturally grown over the exposed Si fin regions. Afterward, a thin high-κ $HfO_2$ gate dielectric is deposited covering all three sides of the fin.

In the next step, a thin layer of TiN (refractory transition film) is deposited as a barrier to prevent the subsequent gate material penetration through the thin high-κ gate oxide layer. To define and modulate the threshold voltage ($V_t$) of the device, a thin layer material, whose WF close to the conduction band of Si (~4.1 eV) is needed. For the case of the N-type FinFET devices, a thin TiAlC layer is deposited. The presence of Al (aluminum) in the stack influence $V_t$ even though it is not in direct contact with the $HfO_2$. A barrier metal (TiN) is deposited again as a capping layer before the tungsten (W) gate fill. The final gate stack consists of interfacial oxide (IL)—high-κ dielectric ($HfO_2$)—TiN—TaN—TiAl—final TiN and W as shown in column 2 of Fig. 3.13 (Jones, 2017). Similarly, for P-type FinFET, a metal gate with a WF close to the Si valence band (~5 eV) is desired. The WF is tuned by adjusting the parameters of the TiN capping layer, including the Ti to N ratio, the thickness, and deposition techniques (Erben et al., 2018). A detailed 2D x-sectional view of the RMG processing sequential steps of both N and P-type FinFET is shown in Fig. 3.13.

**Salicaidation and contact formations**: After the RMG gate formation, a thick passivation low-temperature TEOS (tetraethyl orthosilicate) oxide film is deposited. It is planarized using the CMP (chemical mechanical polishing) process to allow a higher resolution litho patterning and etch to take place as needed for the narrowly spaced metal pitch and contact formation. Post planarization, small contact regions are opened over the gate, source, and drain regions. Subsequently, a silicide material such as tungsten silicide (WSi), CoSi, TiSi, or nickel silicide (NiSi) is deposited and annealed to form low-resistivity contacts.

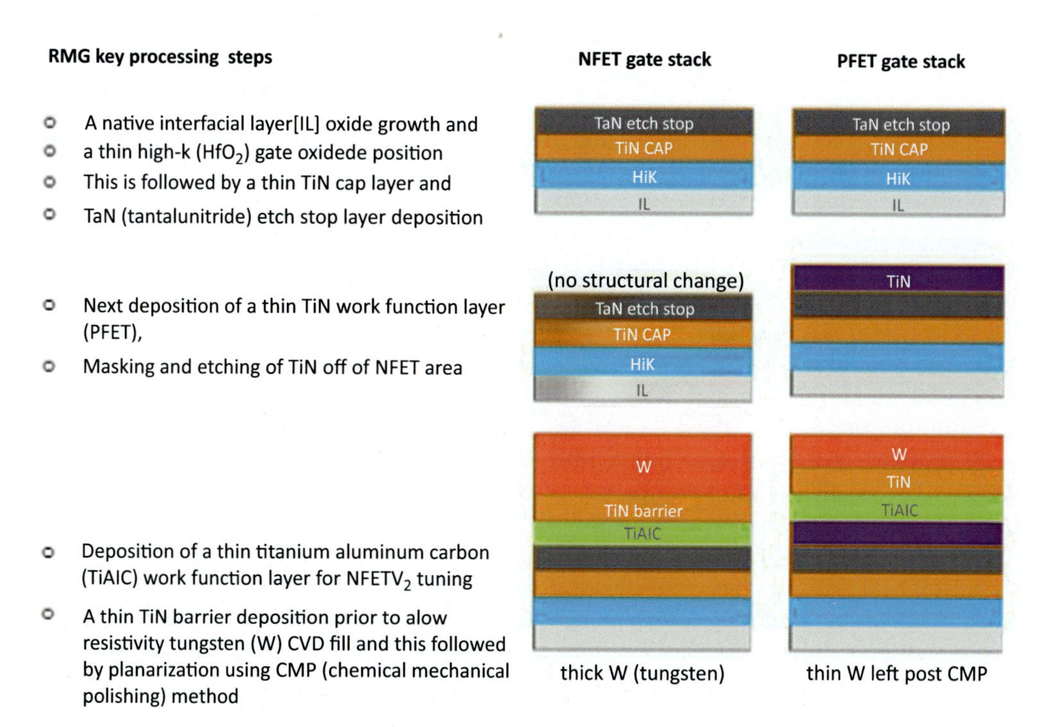


**FIGURE 3.13** Illustrating key RMG processing steps for both NFET and PFET gate stack.

**TABLE 3.3** Selected process design rules for Intel's 22 nm SoC, 14 nm SoC, and 22FFL process technologies (Sell et al., 2018).

| Parameter | Unit | 12 nm RF FinFET | 14 nm SoC | 22 nm SoC | 22FFL |
|---|---|---|---|---|---|
| $F_{HT}$ | nm | 38 | 42 | 34 | 42 |
| $F_{Top}$ (or width) | | 6 | 8[a] | 8 | 8[a] |
| $F_P$ | | 48 | 42 | 60 | 45 |
| Gate poly pitch (CPP) | | 84 | 70/84 | 90/108 | 108/144 |
| Min metal pitch | | 64 | 52 | 80 | 90 |
| Cell height | | – | 399 | 840 | 540 |

[a]*Indicate estimated values.*

**BEOL processing:** Following the processing of contacts, wiring layers are added. For logic circuit applications, a high density is paramount and in fine-metal line pitch processes either metal first or second layer is used. On the contrary, for RF and mmWave designs, density is often less important, and a relaxed metal pitch may be used to reduce cost while improving resistance and electromigration reliability. In addition to these, in most cases, a reduced-layer backend metal stack is favored. In one such case, Intel had presented a new "22FFL" RF FinFET process-centric technology, with relatively a loose gate and metal pitch, while adopting advanced 14 nm fin processing features as shown in Table 3.3 (Sell et al., 2018). Subsequently, a conventional backend layers processing is followed to maintain compatibility with the logic platform and to achieve lower processing costs by module, tool sharing, etc. until the last metal processing. A ~3.0 μm ultra-thick metal (UTM) is deposited and patterned to meet the demand of the low sheet resistance metal layer for high current handling and also for high-performance inductor and transformer designs.

## 3.8 FinFET device structures

Fig. 3.14A and B shows a single-side gate contact (SGC) layout and 2D profile of RF FinFET. The final x and y-direction cross-sectional TEM (transmission electron microscopy) images of 14 nm FinFET are shown in Fig. 3.14C and D. The FinFET device active area is patterned as multi-fins with a fixed 48 nm fin pitch in the case of 14 nm FinFET technology. After Si processing, the effective unit width ($W_{eff}$) corresponds to twice the fin height plus fin top, which equals ~75 nm compared to 48 nm equivalent planar FET footprint as indicated in Table 3.2 and also in Fig. 3.14C. In this micrograph, it is clear how the FinFET achieves a wider (75 nm) active device width than the planar FET width. It can also be observed that the RMG layers wrap around fins while filling a narrow gap left in between fins (Fig. 3.14C). The wrapping layers include high-κ gate ($HfO_2$) oxide, WF, barrier material, and low-resistance W fills as discussed in the previous RMG section. The corresponding RMG layer's resistivity is listed in Table 3.4, to understand basically which contributes more to gate resistance ($R_g$) and hence to RF parameters like $F_{max}$. This geometry gives rise to the lateral gate resistance ($R_g$). Also, a shorter $L_g$ at 14 nm technology node leaves a confined

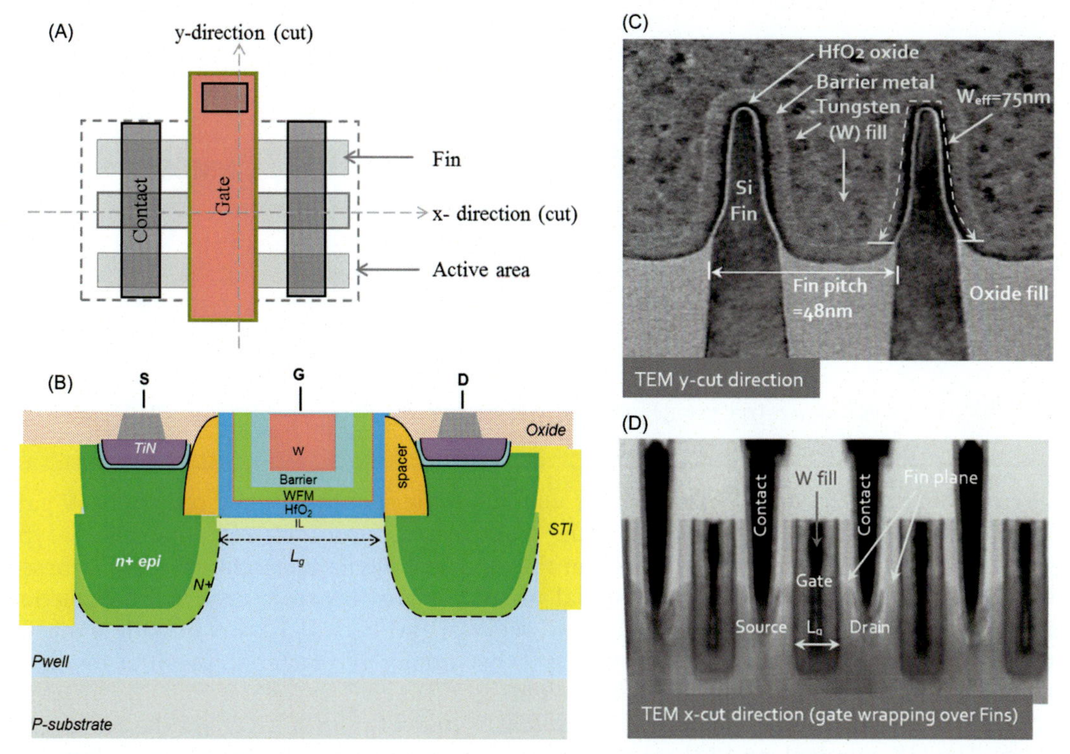


**FIGURE 3.14** A 14 nm FinFET device (A) layout schematic with three fins, (B) the drawing of x-cut along the fin top, (C) the TEM image along the y-cut direction with RMG material wrapping fin, and (D) the x-direction cut TEM image showing the small volume of W material fill inside a 14 nm long gate length ($L_g$) gate.

**TABLE 3.4** The typical resistivity numbers of the typical gate stack material used in the 12 or 14 nm FinFET technology.

| Stack layers | Material | Resistivity ($\mu\Omega \times$ cm) |
|---|---|---|
| Barrier | TiN | 200 |
| Etch stop layer | TaN | 120 |
| Work function (WF) | TiAlC | 2000 |
| Metal fills | W | 20–80 |

space to fill with low resistivity W material post the high-κ gate oxide and WF metal deposition (Fig. 3.14D), hence causing a higher vertical $R_g$ component. Thus, an overall gate resistance sees a dramatic rise at a scaled $L_g$ and narrow fin pitch.

In the P-type FinFET RMG gate stack (Fig. 3.13), an additional film of TiN appears due to the masking cap layer used over the P-type region for the first N-type layer processing. Now this film stays as a part of the final p-type gate stack and regulates WF tuning as

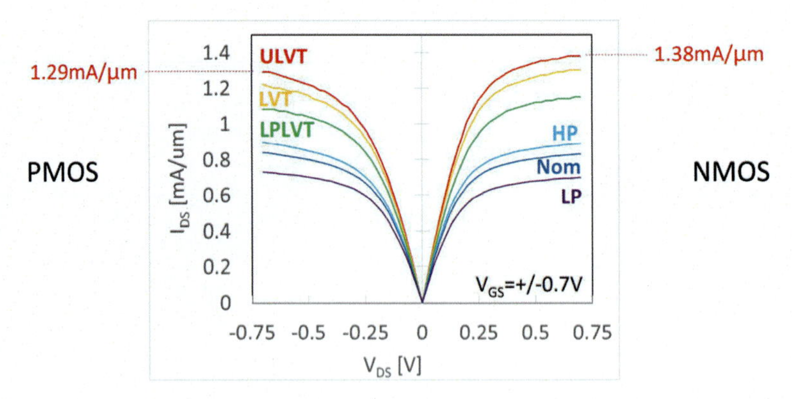


**FIGURE 3.15** Logic devices type example on 22FFL Intel technologies. *Source: From Sell et al. (2018).*

discussed earlier. However, it reduces the volume of the subsequent low resistivity W gate fill material in the gate canyons and is the main reason for a higher $R_g$ than the N-type gate stack. The detailed influence of $R_g$ on the electrical device characteristics is discussed in the following sections. The WF material properties are chosen to ensure full depletion of fins during the device operation. The low, regular, and high $V_t$ flavors of both N and P-type FinFET devices are further achieved by either tuning WF material, pwell or nwell, or halo implants, etc. as shown in Fig. 3.15 (Sell et al., 2018).

Three different $L_g$: 32 nm (ULVT), 36 nm (LVT), and 44 nm (LPLVT) are the HP device types. The HP, nominal, and LP are the low-leakage transistors with the same $L_g$, but having a different WF metal to increase $V_t$. All six of these logic transistors share are footprint compatible, with a 108 nm CPP (contacted poly pitch). The leakage current of the RF device in many circuits applications like local oscillators is less critical than logic and other circuits favored the use of ultra-low $V_t$ devices (Razavieh et al., 2021; Sell et al., 2018). So far, logic technologies are steered for superior NFET device performance because of higher $e^-$ mobility. As a consequence, the PFET has historically taken a role of a secondary device. At 14 nm and advanced FinFET technology nodes, the hole mobility has shown remarkable improvements due to <110> Si fin surface properties and SiGe channel stressors used in S/D regions (Samsung, 2021; TSMC, 2021). Now it has become comparable and exceeds the $e^-$ mobility in some cases; therefore, a fresh look is required to appropriately use these devices for RF and mmWave circuit applications (Ding et al., 2021).

## 3.9 Radio frequency device parasitics

One of the key differences between the logic device and the RF device is the rigorous consideration of parasitic resistance, capacitance, and inductance in the RF and mmWave device and circuit designs. As planar CMOS migrated to 3D FinFET several changes impacted parasitics, (1) the active channel region is patterned with multiple narrow vertical fins, (2) the gate processing is modified to cover three sides of fins, and (3) S/D processing added innovative cavity and epitaxy processing techniques to minimize the S/D parasitics. The

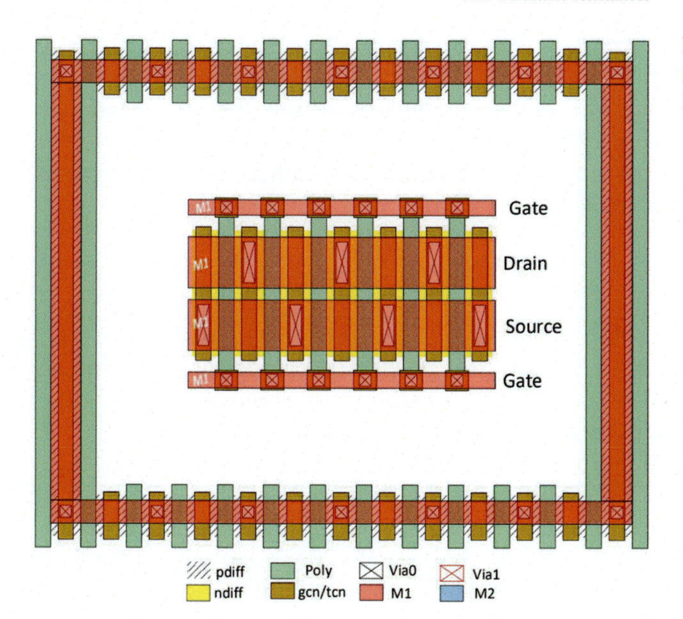


**FIGURE 3.16** High-performance-radio frequency transistor footprint (22FFL). *Source: From Lee et al. (2020).*

parasitic components overall see a noteworthy shift both in terms of severity and complexity (see Fig. 3.13). In one example, an optimized RF device layout with gaurdring is shown Fig. 3.16 (Lee et al., 2020). The guardring of the HP-RF transistor is moved away from the active region of a transistor, and therefore its parasitic contribution to the RF performance degradation is negligible. In the following sections, device process, structure, and layout-related parasitics components are discussed.

## 3.10 Parasitics resistances

### 3.10.1 Gate resistance ($R_g$)

The gate structure is a central device element of a MOSFET, which plays an important role in determining the actual RF device performance >1 GHz operating frequency. In the last two decades, the gate processing of CMOS logic devices has seen major changes in terms of material and structures. For the conventional planar CMOS, the gate material was doped poly-Si (non-silicided) with a sheet resistance ranging from 0.5–4 kΩ/□. It was adequate then to model the gate resistance as a simple transmission line as in Eq. (3.2) (Chen et al., 2014; Jin et al., 1998; Wachnik et al., 2013).

$$R_g = R_{sh} \times \left( \frac{W}{3L \times NF} \right) \tag{3.2}$$

where $R_{sh}$ is the sheet resistance, $W$ is the gate width, $L$ is the gate length, $NF$ is the number of gate fingers in RF device structure, and the factor 1/3 (or 1/12 for double-sided gate contacts) accounts for the distributed load of the gate capacitance (Jin et al., 1998).

Theoretically, there should also be a term for contact resistance, but compared to the poly-Si sheet resistance, this term was small enough to neglect. With the salicidation process, the $R_{sh}$ is reduced by a factor of 10 or more. But the gate current now travels horizontally mainly in the silicide and then travels vertically from the silicide to the gate dielectric. Accurate modeling in this case requires modeling of both the horizontal and vertical resistance plus the contact resistance as described in Eq. (3.3).

$$R_g = \frac{R_c}{NF} + R_{sh}\left(\frac{W}{3L \times NF}\right) + \frac{R_v}{L \times W \times NF} \tag{3.3}$$

where $L$, $W$, and $NF$ are the same as in Eq. (3.2) and $R_c$ is the gate contact resistance, $R_{sh}$ is the horizontal sheet resistance, and $R_v$ is the vertical resistance coefficient. For a planar gate, the vertical resistance coefficient is simple bulk resistivity times the vertical height of the gate. Except for very narrow (small $W$) devices, the horizontal term dominates.

As the technology scaled below 65 nm, the polydepletion effect became increasingly severe, and quantum mechanical tunneling through the gate oxide set a lower limit on the gate dielectric thickness. This motivated the evolution of high-κ dielectrics and metal stacks with materials of low resistivity (Wachnik et al., 2013). However, the $R_g$ of high-κ metal gate stack is typically higher than the PolySiON gate material (Chew et al., 2015).

## 3.10.2 FinFET gate resistance ($R_g$)

FinFET gate stack consists of several layers to set the WF which are much more resistive than the bulk tungsten of the gate (Jeong et al., 2017) (see Fig. 3.13). In RF operation, gate metal layers with different resistivity act as transmission lines of different characteristics. For simplicity of derivation, gate materials are considered into two groups high resistivity (including barrier metal) and low resistivity (W fill, etc.). The signal current passing through the FinFET gate (Fig. 3.17) sees first $R_c$ the resistance of via contact and $R_{int}$ the interface contact resistance. $R_h$ is the horizontal resistance component to the current flow plane across the fins direction and $R_v$ is the vertical resistance down between the fins. The $R_v$ term now includes the resistance of the interface materials that are required to set the WF of the gate to a value that gives the desired $V_t$. The terms in Eq. (3.3) for $R_g$ are still applicable in this case although the numerical coefficients change and an additional term is needed for the resistance of the WF setting layers.

$$R_g = \frac{R_c}{NF} + R_{sh}\left(\frac{F_p \times N_{\mathrm{Fin}}}{3L \times NF}\right) + \frac{R_v^{'}}{L \times N_{\mathrm{Fin}} \times NF} + \frac{R_{\mathrm{int}}}{L \times W_{\mathrm{Fin}} \times N_{\mathrm{Fin}} \times NF} \tag{3.4}$$

where $W_{\mathrm{Fin}}$ is the electrical width of one fin as calculated in Table 3.2. $F_p$ is the pitch of the fins (distance from the centerline of one fin to the centerline of the next) and $N_{\mathrm{Fin}}$ is the number of fins in a device. Compared to Eq. (3.3) $R_v$ is replaced by $R_v^{'}$ reflecting the fact that it no longer includes the interface resistance and a change in units. $R_v$ has units ohm-m$^2$, while $R_v^{'}$ has units of ohm-m. The value of $R_v^{'}$ depends on the width of the space between fins but since this is a constant for a given process offering It is convenient to incorporate it into the constant and leave it out of the equation. $R_{\mathrm{int}}$ has been introduced

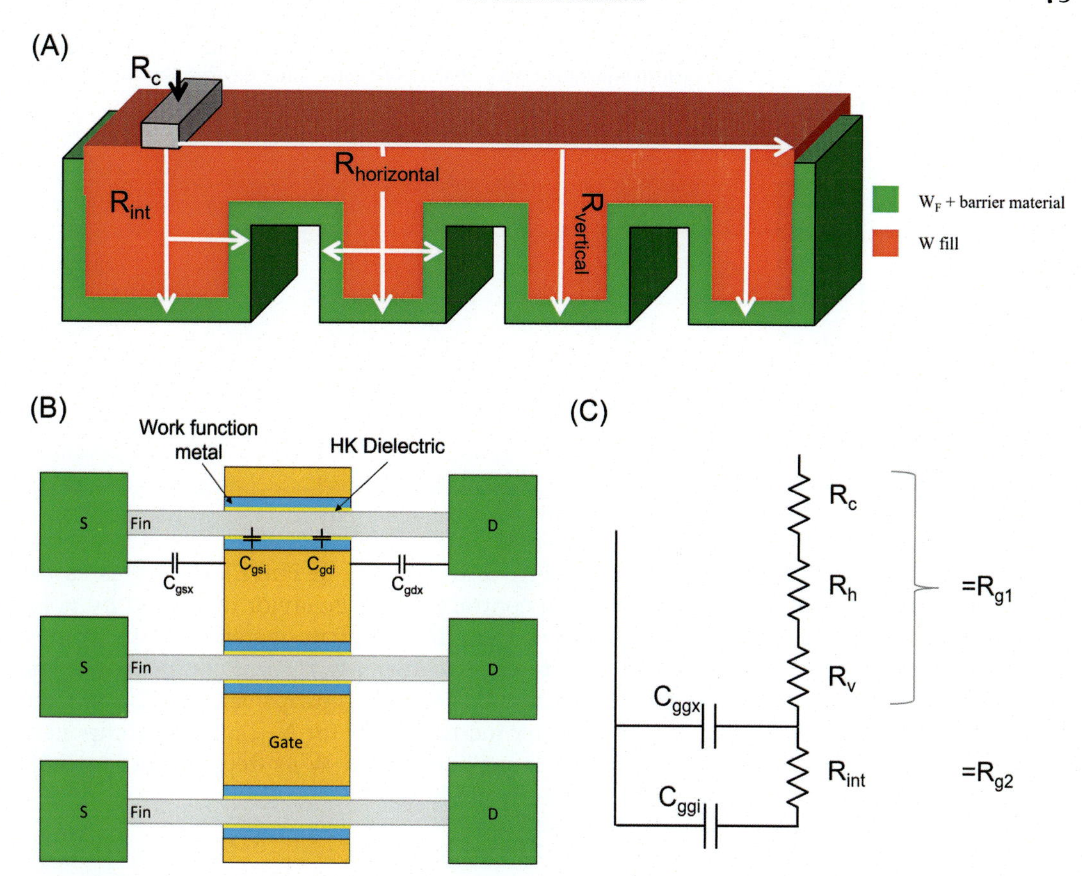


**FIGURE 3.17** (A) Gate horizontal and vertical resistance components in the FinFET RMG wrap-around gate structure along the width or y-cut corresponding to Fig. 3.14A, (B) top view cross-section through the fin. (C) The schematic shows the impedance components looking into the gate.

for the interface resistance since for some purposes it may need to be consider as a separate resistance as discussed below. The other symbols have the same meaning as in Eq. (3.3). One effect of $R_g$ is to insert an RC delay into signals applied to the gate of the FET. RF is more demanding of low $R_g$ because the slew rates are higher. In a FinFET, the gate capacitance scales linearly with $W_{Fin} \times N_{Fin} \times NF$ (the total device width). Multiplying Eq. (3.4) by $W_{Fin} \times N_{Fin} \times NF$ gives us the geometric scaling of this delay constant.

$$C_g R_g \sim R_c \times W_{\text{Fin}} \times N_{\text{Fin}} + R_{sh}\left(\frac{F_p \times N_{\text{Fin}}^2 \times W_{\text{Fin}}}{3L}\right) + \frac{R_v' \times W_{\text{Fin}}}{L} + \frac{R_{\text{int}}}{L} \tag{3.5}$$

The delay from contact resistance ($R_c$), $N_{Fin}$ and $R_h$ scale with $N_{Fin}^2$, while $R_v$ and $R_{int}$ don't scale with $N_{Fin}$. None of the components scale with NF. As technology scales down to a smaller node, the space to fill tungsten material continues to shrink (Jeong et al., 2017; Singh et al., 2018), and this results in the overall $R_g$ increase and overshadowing the benefit of the higher $G_m$. Unlike devices with unsilicided poly gates, $R_g$ in FinFET shows a

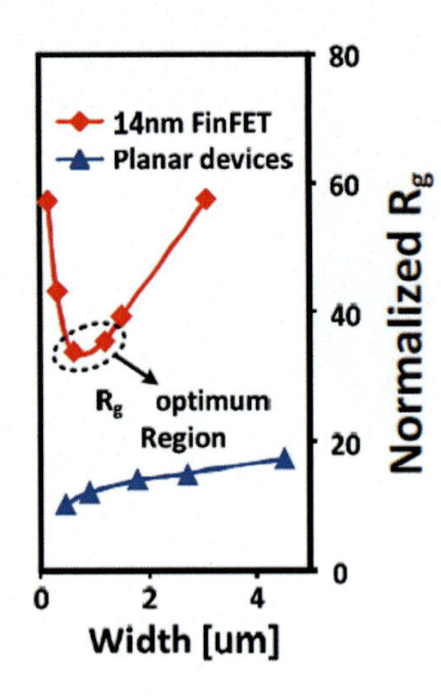


FIGURE 3.18 The gate resistance as a function of fins for 14 nm FinFETs and width equivalent h for 28 nm poly gate planar devices. *Source: From Jeong et al. (2017).*

nonlinear relation with channel width ($W$) (Fig. 3.18). The $R_g$ of FinFET decreases as the fin number is increased from 2 to 8, which is contrary to the behavior displayed by poly-gate devices. It is because of the 3D nature of the "fin," the gate material wraps around and goes to the bottom portion, which generates an additional vertical component of the $R_g$. As channel width increases from the narrowest, the total $R_g$ drops because $R_v$ contribution to total $R_g$ starts dropping. With the further increase of $W$ (or $N_{Fin}$), the $R_h$ component becomes stronger, and the $R_g$ relation turns more linear with $W$ as depicted in Fig. 3.18. The higher $R_g$ increases the input power for a given output.

## 3.10.3 The raised source and drain parasitic resistance ($R_s$, $R_d$)

In the planar CMOS technologies, the junction depth ($x_j$) of S/D regions is being used as one of the key parameters to reduce the SCEs, especially the DIBL. On the other hand, in the FinFET technology, narrowing the fin thickness is another parameter available to restrict the SCEs. A lower limit of the fin thickness is set by device operation. As we know, during the ON-state, a current enters via the source contact passes through the thin fins, and exits at the drain contact. The contact area and Si volume on the fins are reduced compared to planar FETs (Fig. 3.19), therefore leading to a higher $R_s(/R_d)$ (Lu, 2011). It is aggravated further as the fin is thinned to control the SCEs (Kedzierski et al., 2003). This gives rise more to $R_{s,\ d}$, which limits the device's performance. It has been captured in the following simple drain current ($I_{dsat}$) expression.

$$I_{dsat} = \frac{I_{dsat0}}{1 + \frac{I_{dsat0} \times (R_s + R_{ext} + R_c)}{(V_{gs} - V_t)}} \tag{3.6}$$

where $I_{dsat0}$ is the drain current in the absence of parasitic source resistance ($R_s$). Fig. 3.19 is a 2D schematic of the raised S/D structure showing the $R_c$, $R_s$, $R_d$, $R_{ext}$, and $R_{ch}$ regions of the FinFET devices along the current path (Lu, 2011). For the symmetric devices, $R_c$ and $R_{ext}$ resistance are similar on both source and drain ends. $R_c$ is the combined resistance due to the raised

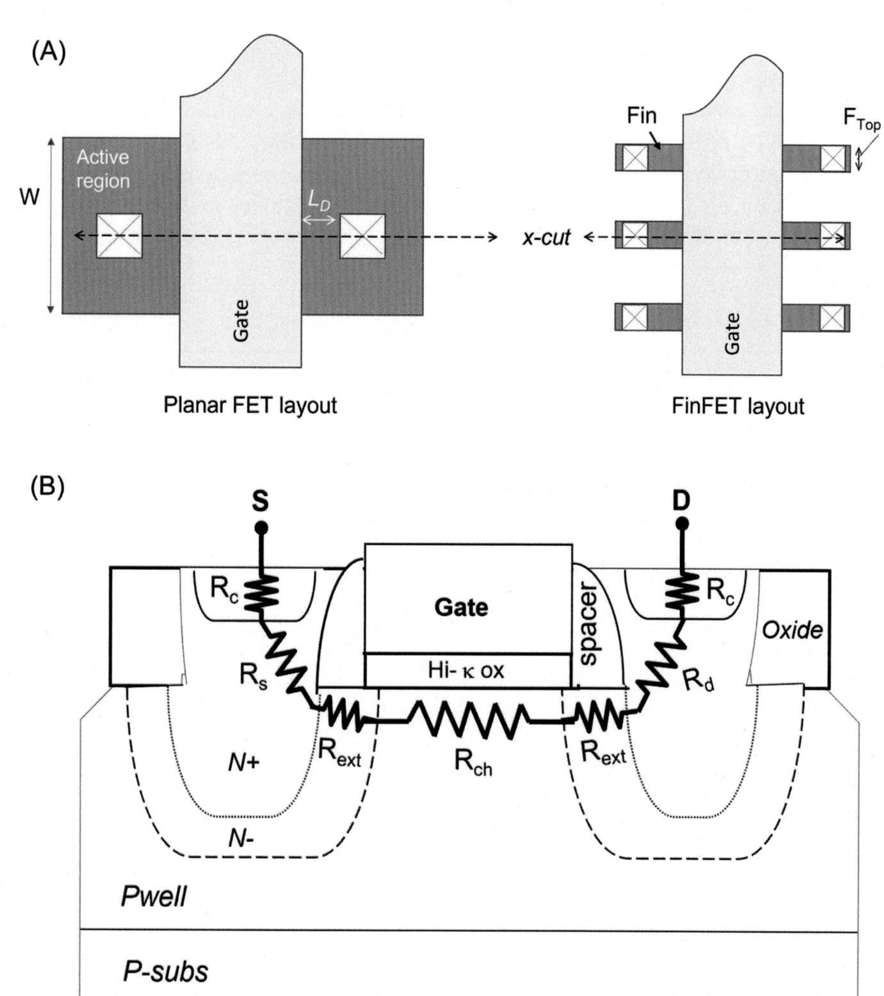


**FIGURE 3.19** Comparison of (A) a top view layout of planar FET versus FinFET and (B) shows the parasitic Rc, Rs, Rd, Rext, and Rch resistance component in 2D FinFET schematic along the channel.

S/D region bulk resistivity and the silicon/silicide interface resistance. The $R_s$ or $R_d$ is the resistance due to current spreading from the raised S/D region to the entrance to the fin, and the $R_{ext}$ is the bias-dependent resistance in the thin S/D extension region under the spacer.

The total resistance is $R_{(total)} = R_{ch} + R_{parasitic}$, where $R_{parasitic} = R_{ext} + R_{extrinsic}$. The extrinsic parasitics resistances include both source and drain side and contact resistance; $R_{extrinsic} = R_s + R_d + 2 \times R_c$. The contact resistance $R_c = \frac{\rho_c}{A}$, which scales inversely with the area, and $\rho_c$ is a specific contact resistivity. Compared to planar FET, the S/D contact area of FinFET is reduced as contact lands on the fin top's (see top view layout in Fig. 3.19A). A common practice to reduce parasitic resistance $R_s$, $R_d$, and $R_c$ is to increase the contact surface area and/or Si volume at S/D regions by epitaxially growing highly doped Si over these narrow fin regions.

Now, these regions can provide a larger surface area for landing the S/D contacts, hence a lower parasitic resistance, and also create a low resistance path for current to reach an even lower portion of the vertical fin channel as shown in Fig. 3.19B. Despite this reduction, the effect of parasitic resistance on device performance remains substantial. Also, there is a tradeoff with the $R_s$ ($/R_d$) reduction because the raised epi region causes a consequential increase in the fringing parasitic capacitances ($C_{gs}$ and $C_{gd}$), which is one of the primary factors limiting the device's RF performance on the FinFET platform despite having superior $G_m$ (Singh et al., 2018).

## 3.11 FinFET parasitic capacitance

Similar to parasitic resistances, the parasitic capacitances substantially trend worsening with planar FET migration to 3D FinFET. These parasitic capacitances can be subdivided mainly into two categories: (1) intrinsic and (2) extrinsic capacitances. The intrinsic capacitances are associated with the inversion layer of the channel and are intrinsic to the device's operation. On the other hand, extrinsic capacitance creates AC signal paths independent of the desired FET operation and degrades the device performance.

Physically the FinFET is symmetric to the interchange of source and drain, so $C_{gs}$ and $C_{gd}$ arise from the same physical structures as shown in Fig. 3.20. Electrically some of the capacitances are bias dependent, so $C_{gs}$ and $C_{gd}$ are not always of equal value. The circuit impacts are also different in detail, but both RF and logic performance degrade. The parasitic capacitances related to FinFET structure are discussed in the following sections often with reference to the planar counterpart.

**Overlap capacitance:** The overlap capacitances ($C_{gs,ov}$, $C_{gd,ov}$) arise from the overlap of the gate via gate dielectric onto the source and drain extension regions along the device width direction as illustrated in Fig. 3.20B. If the overlap area increases due to the S/D epitaxy, implant, and or the gate dimension processing changes, then the overlap capacitance increases proportionally. The overlap capacitance mainly depends upon the device's geometry, and largely it remains bias independent due to heavy doping ($>1e19/cm^3$) of the S/D regions.

**Inner fringe capacitance:** The inner fringe capacitance ($C_{gs,if}$, $C_{gd,if}$) results when the inversion layer is absent or incomplete. It is the result of e-field lines from the gate, through the depleted channel region to the S/D extensions. They are strongly bias-dependent bias. In a FinFET, these capacitances are small because the thin fin body allows little space for fringing fields.

**Through spacer capacitance:** A significant portion of $C_{gs}$ and $C_{gd}$ parasitic capacitances are through the spacer as depicted in Fig. 3.20A with notation $C_{gs,of}$ (often called outer fringe capacitance). They occur from the vertical sides of the gate to the fin outside the channel region and to the epi-S/D regions. The raised epi-S/D regions increase the through spacer capacitance compared to a planar device (without raised S/D). With device scaling, the contact pitch is also reduced, which leads to a smaller gap between the epitaxy Si and the gate edge and increased $C_{gs,of}/C_{gd,of}$. Increasing the spacer thickness or CPP decreases the parasitic capacitance, but this typically causes an increase S/D parasitic resistance via $R_{ext}$ component. One of the favored spacer formation techniques is the use of a low-k dielectric material with or without air gaps to reduce the parasitic capacitance while having minimal or no impact on the parasitic resistance.

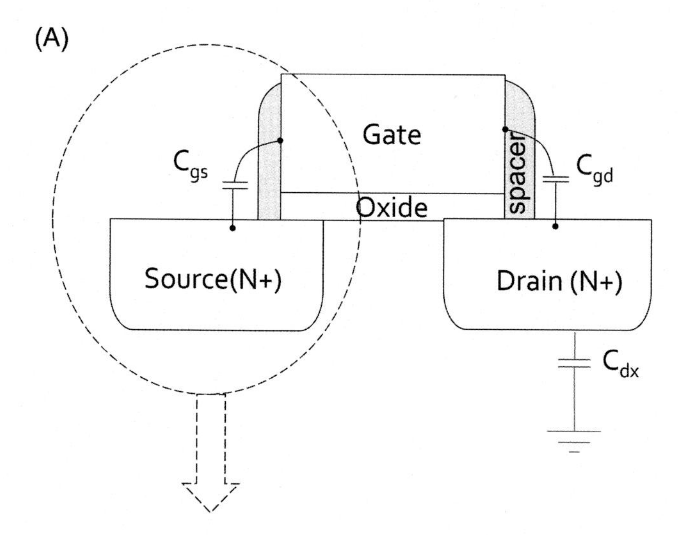


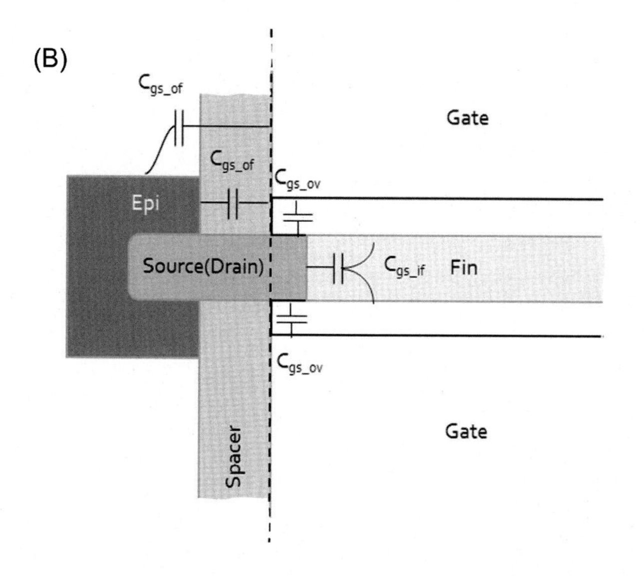


**FIGURE 3.20** The FinFET parasitic capacitance 2D illustration along the x-cut direction, (A) the extrinsic $C_{gs}$ and $C_{gd}$ components, (B) a cross-section through a plane parallel to the wafer on the source side representation.

**Capacitance to contacts:** There is also capacitance from the source and drain contacts to the gate, and between the contacts themselves ($C_{gs,cc}/C_{gd,cc}$). The FinFET architectures are mainly optimized for density and CMOS logic performance by packing gates as close together as possible. This is not optimum for RF performance but these logic platforms are well suited for circuits that have both logic and RF contents. In such a scenario, the main RF performance gain options lie with the layout optimizations. To achieve a better RF performance, the ground rules need to allow freedom to optimize RF layouts differently than logic layouts. Using simple approaches like 2*X* CPP and optimized contact density, high frequency and mmWave devices with UTM layers can be designed with minimum change to logic platforms processings.

### 3.11.1 Source to drain parasitic capacitance ($C_{sd}$)

The source to drain capacitance consist of two parts: (1) from the doped source to drain body region ($C_{sd,if}$), (2) part above silicon from the contact to contact ($C_{sd,cc}$), via1 to via1 ($C_{sd,vv}$), and metal1 to metal1 ($C_{sd,mm}$) levels. The parasitic $C_{sd}$ capacitance gets worse with channel length ($L_g$) scaling as the source-to-drain gap reduces as shown in Fig. 3.21. In the case of a planar FET, the $C_{sd}$ is proportional to junction depth, while in 3D FinFET it depends upon the fin width ($F_w$). For simplicity in the above silicon parasitic capacitance, the second-order capacitance from contact to via or metal to via is being ignored. The use of 2X CPP space can reduce the above silicon parasitic capacitances by optimizing contact pitch as shown in Fig. 3.22.

### 3.11.2 Gate-to-substrate overlap capacitance ($C_{gx}$)

The gate-to-bulk Si overlap capacitances is due to the gate regions extending over the substrate region. The value of this capacitance is small compared to device intrinsic capacitance ($C_{gg,i}$) because the field oxide is very thick compared to the effective thickness of the gate dielectric. In the case of the FinFET, it is associated with the region between the fins, as the number of fins increases, this capacitance value increases as shown in Fig. 3.20B. It also depends upon the fin pitch value, which determines the area of the capacitance.

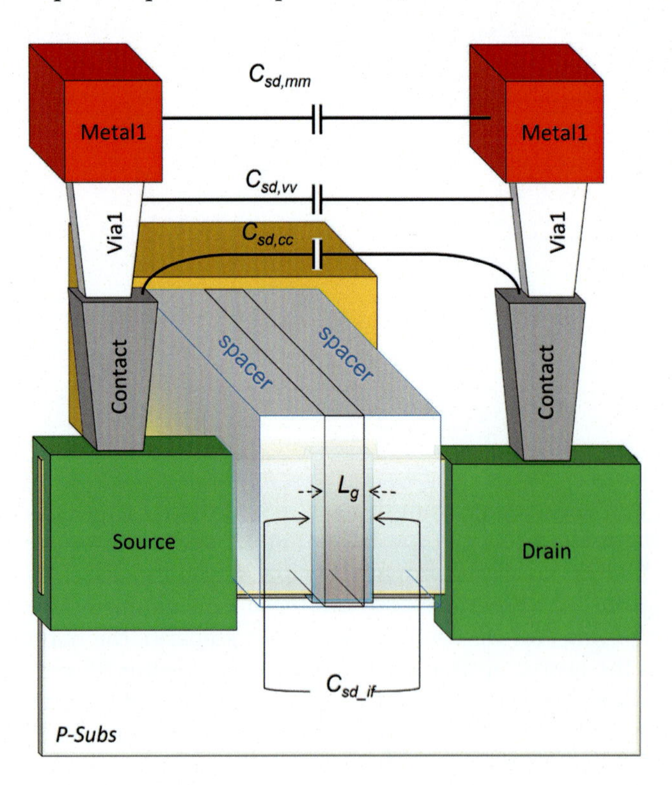


**FIGURE 3.21** The illustration of the source to drain parasitic capacitances from both (i) the doped source to drain region ($C_{sd,if}$), (ii) part above silicon from the contact to contact levels.

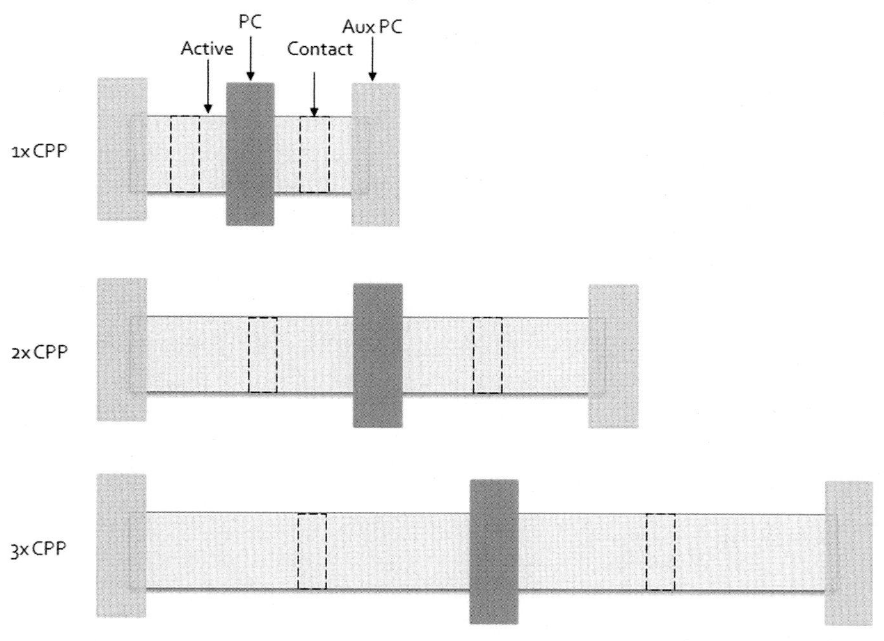


FIGURE 3.22 Showing the illustration example of 14 nm FinFET technology *1X, 2X, 3X* contacted poly pitch device cases on an active silicon area with contact area, gate poly (PC), and dummy auxiliary (Aux PC) on sides.

### 3.11.3 Source and drain to substrate capacitance ($C_{sx}$, $C_{dx}$)

In the case of the planar FET, this capacitance is across the N-P junction of the source and drain diffusions into the substrate as shown in Fig. 3.20A. This parasitic capacitance has both sidewall and bottom portions. In the case of FinFET, only the bottom of the fin contributes to the junction capacitance, so the area is much smaller and the sidewall is greatly reduced by the oxide surrounding the fins. The epi-Si raised S/D regions are isolated on sides with a thick oxide, so it contributes very little capacitance. The junction capacitance depends on the depletion width which is bias-dependent.

These parasitics capacitance has been one of the main reasons for lower FinFET RF performance as can be seen from earlier $F_t$ versus feature (scaling). On a logic technology platform, one of the common ways to minimize the parasitic capacitances ($C_{gs}$ and $C_{ds}$) is a device layout approach of increasing the CPP to two or three times the minimum value (Fig. 3.22). For the mmWave device applications, these spaces can be increased further. The increased space moves the contacts further from the source and drain greatly reduces capacitance. This increases the source and drain series resistance ($R_s$, $R_d$) but low resistivity salicidation over the S/D region minimizes this effect. Also, the space between the active gates allows wider wiring, which improves the wiring resistance and heat dissipation. This benefits the device and circuit reliability, particularly the electromigration safety margin. A qualitative comparative trend with planar MOSFET is summarized in Table 3.5.

## 3.12 FinFET radio frequency device figures-of-merit

Despite higher parasitics, the CMOS technology continues to advance below the 5 nm node, to meet the demand of the logic circuits mainly in terms of PPA. The single-chip integration of RF and communication designs with the microprocessor cores on a common CMOS SoC platform becomes increasingly attractive due to the PPA scaling. Overall circuits perform better compared to previous technology nodes (Fig. 3.8). However, parasitics, particularly capacitances associated with 3D FinFET architecture, tend to limit the higher frequency response of the circuit. To accurately predict the frequency response, these capacitances need to be estimated in detail (Figs. 3.20, 3.23) using simple representation and FoM. The unity current gain frequency $F_t$ can be simply represented by

$$F_t = \frac{G_m}{2\pi \times C_{gg}} \tag{3.7}$$

where $G_m$ ($= \Delta Id/\Delta Vg$) is the transconductance and $C_{gg} = C_{gs} + C_{gx} + C_{gd}$ is the total gate capacitance.

Here $F_t$ includes the effect of $R_s$ and $R_d$

$$F_t = \frac{G_m}{C_{gg} + C_{gd}G_m(R_s + R_d)}$$

**TABLE 3.5** A qualitative trend of parasitic capacitances between the planar versus FinFET devices.

| Parameter | FinFET (<22 nm node) compared to planar (>22 nm node) MOSFET |
|---|---|
| $C_{gg,i}$ | ↑(increased because of greater effective width) |
| $C_{gs}$, $C_{gd}$ | ↑(increased mainly due to through space capacitance to epi) |
| $C_{sd}$ | ↓(inner fringing reduced due to small fin cross-sectional area), = above silicon at the same pitch |
| $C_{sx}$, $C_{dx}$ | ↓(Fin N+ or P+ area to subs, thick field oxide under epitaxy) |

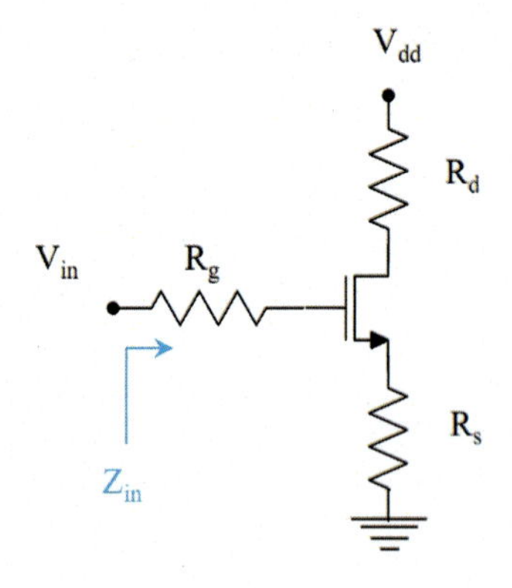


**FIGURE 3.23** The small-signal MOSFET schematic showing the source and drain resistance components.

These capacitance parameters are explained in the previous section. At high frequencies the impedance of $C_{gs}$ is small; therefore, the input impedance is simply $Z_{in} = R_g + R_s$. In this case, the $F_{max}$ can be represented by

$$F_{max} = \frac{1}{2}\frac{F_t}{\sqrt{2\pi \times F_t \times C_{gd}(R_g + R_s) + \frac{R_g + R_s}{r_o}}} \tag{3.8}$$

The numerator $F_t$ component consists of $(R_s + R_d)$ term, while $F_t$ denominator is without $(R_s + R_d)$ term.

## 3.13 The 3D FinFET small signal model

The 3D nature of the FinFET complicates the small signal model. Fig. 3.24 shows the standard small signal model of a planar FET. In this model, there is a single path for the signal from the gate to the source (the source is assumed to be tied to the substrate, so this path includes capacitance from the gate directly to the substrate). In a FinFET, a significant fraction of the gate-to-source capacitance is distributed along the gate, effectively bypassing part of the $R_g$. The gate path is more like a lossy transmission line than the single lumped R and C of the standard small-signal model. The 3D gate makes measured $R_g$ a function of the distribution of capacitances between extrinsic and intrinsic regions. To better model this path, we represent it by a π-network plus an additional resistor as shown in Fig. 3.25. Adding the extra node gives a physically correct model for $R_g$ looking into both

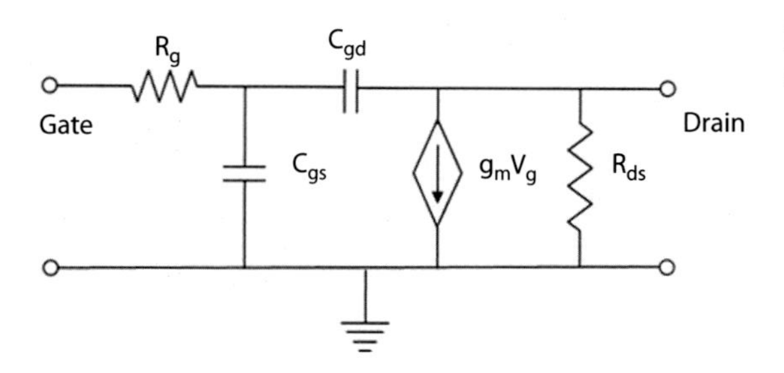


FIGURE 3.24 A small signal model of the equivalent $R_g$ for a standard topology.

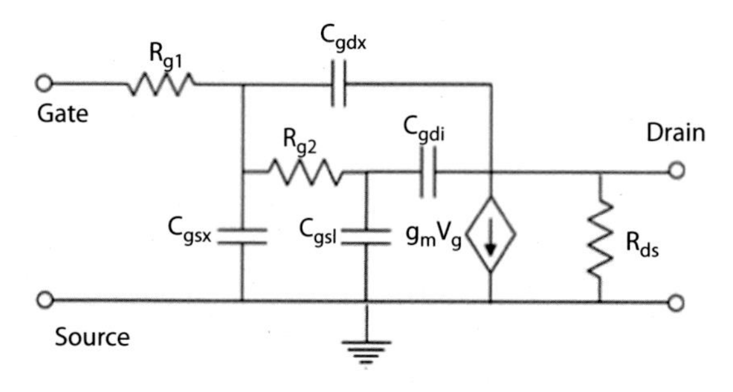


FIGURE 3.25 Showing the small signal model of 3D FinFET using approximated the RC distribution of the π−network.

gate and the drain. As a matter of fact, the $R_g$ is most important for its impact on input power. The resistor $R_{g1}$ represents the resistance that gate current sees, primarily the gate contact, but also part of the horizontal resistance. $C_{gsx}$, $C_{gd\ x}$, $R_{g2}$, and $C_{gsi}$, $C_{gd\ i}$ form the π-network which represents the remainder of the resistance and all of the parasitic capacitance of the gate.

Physically the effect of the distributed nature of resistance and capacitance is that much of the signal goes through the external capacitance and sees only a part of the $R_g$. Since $R_g \ll \frac{1}{\omega C_{gg}}$ the circuit sees the expected total gate capacitance but not the expected gate resistance which is important for $F_{max}$ and power gain. For a planar FET, the $R_g$ can be measured using Y-parameters looking into the gate Eq. (3.9) or looking into the drain Eq. (3.10). For the model in Fig. 3.24, the two measurements give the value of $R_g$. But for the model in Fig. 3.25, Eq. (3.9) gives a value lower than the total physical resistance, and Eq. (3.10) gives an even lower value.

$$R_g = R_e[1/Y11] \tag{3.9}$$

$$R_g{}' = \frac{\text{Re}[Y12]}{Im[Y11] \times \text{Re}[Y12]} \tag{3.10}$$

$$Y11 = \frac{I_1}{V_1, \text{with } V_2 = 0V} \tag{3.11}$$

$$Y12 = \frac{I_1}{V_2, \text{with } V_1 = 0V} \tag{3.12}$$

Using the topology of the figure, we obtain Y-parameters as

$$Y11 = \omega^2 C_{ggi}\left(C_{ggx}R_{g1} + C_{ggi}R_{g2}\right) + j\omega C_{gg} \tag{3.13}$$

$$Y12 = \omega^2\left(C_{gd}C_{gg}R_{g1} + C_{gd2}C_{gg2}R_{g2}\right) + j\omega C_{gd} \tag{3.14}$$

Using this expression for Y11 in Eq. (3.9) we get a value for $R_g$ equal to $R_{g1}$ plus a fraction of $R_{g2}$.

$$R_g = R_{g1} + R_{g2}\left(\frac{C_{ggi}}{C_{ggx} + C_{ggi}}\right)^2 \tag{3.15}$$

The value of $R_g$ looking into the drain ($R_g^{'}$) can be expressed using Eq. (3.10),

$$R_g^{'} = R_{g1} + R_{g2}\frac{C_{gdi}C_{ggi}}{\left(C_{gdi} + C_{gdx}\right)\left(C_{ggi} + C_{ggx}\right)} \tag{3.16}$$

Since the $\frac{\text{Re}[Y12]}{Im[Y12]} > \frac{\text{Re}[Y11]}{Im[Y11]}$, $R_g^{'}$ gives a more reliable measurement of $R_g$ for process monitoring purposes. For simulation, the model of Fig. 3.25 is more accurate but has an extra node which may be too large a computational burden for simulating large circuits. Since we are generally interested in simulating the input power consumed by $R_g$ if the model in Fig. 3.24 is used to reduce simulation runtime, $R_g$ is the appropriate number to put in the model. This number will be smaller than $R_g^{'}$. Output power is determined by $R_{ds}$ and output circuit impedance.

## 3.14 FinFET radio frequency silicon results

After establishing a leading logic CMOS technology below $\leq$22 nm node, the FinFET devices show excellent RF behavior. It is mainly attributed to the superior transconductance ($G_m$), despite having higher fringing field parasitic capacitance (Jeong et al., 2017; Sell et al., 2018; Singh et al., 2018). Also, the current flows through the vertical fins, leaving a potential space for carrier interactions at RF frequencies. The RF data presented in this section is the measurement of device-under-test units using a two-port S-Parameter technique and Vector Network Analyzer 40 GHz (Singh et al., 2017). The pad and the interconnect parasitic were de-embedded by using the conventional open-short de-embedding technique at the ground metal level (or first metal) to accurately estimate the intrinsic RF device performance (Cho & Burk, 1991; Koolen et al., 1992). For the current ($I_d$) normalization and comparison with the planar device, a width calculation method listed in Table 3.2 was employed. The FinFET uses the width quantization of multiple fins of the same height due to process design.

The unity current gain cutoff frequency ($F_t$) and unity power gain maximum oscillation frequency ($F_{max}$) of both N and P-type FinFETs channel length ($L_g = 14$ nm), number of fins ($N_{Fin}$) = 12, and foot width = $N_{Fin} \times F_p = 12 \times 48$ nm have been extracted from S-parameters measurements and are shown in Fig. 3.26. It can be clearly seen that FinFET achieves higher peak $F_t$ due to superior transconductance than 28 nm planar high-κ metal gate N and PFETs as listed in Table 3.5.

A significant increase in P-type FinFET peak $F_t$ is also observed and now for the first time at the 14 nm tech node, it is comparable to N-type FinFET devices. In this 14 nm FinFET technology, the P-type FinFET device architecture uses SiGe boron-doped S/D epitaxy stressor to enhance the silicon channel hole mobility in the RMG gate-wrapped silicon fins. The channel stress components in the 14 nm FinFET are significantly improved compared to previous node planar FETs, which can also be translated to high $G_m$ and $F_t$. Although, in the case of both N and P-type FinFET, the $F_t$ and $F_{max}$ roll-off at a much faster rate than the planar FETs, it is mainly higher parasitic capacitances.

The $F_{max}$ of both N and P-type FinFETs is also higher than that of the 28 nm counterparts as can be seen in Fig. 3.26. However, an adverse influence of the $R_g$ can be seen on the $F_{max}$ performance. Fig. 3.27A shows a $F_{max}$ trend with device $L_g$ scaling. In a typical device scaling, as the $L_g$ of the FET is reduced $g_m$, $F_t$, and $F_{max}$ all increase. On the contrary, for FinFETs the $R_g$ increases, especially at $L_g < 20$ nm. In addition, the parasitics ($C_{gs}$ and $C_{gd}$) capacitance increases due to poly pitch which is scaled with $L_g$. These effects combine to reduce $F_{max}$ as $L_g$ decreases.

In Fig. 3.27B, a self-gain ($G_m/G_{ds}$) of 14 nm FinFETs and 28 nm FETs are plotted against the normalized current density ($J_d$) with the device width. Due to a thin fin channel body, as mentioned earlier a FinFET device experienced complete depletion of the channel region during the operation, thus providing better electrostatic control over the channel region against the SECs. When scaling to a shorter $L_g$ higher $G_m$ and lower $G_{ds}$ compared to planar devices, which can be observed from 14 nm FinFET self-gain compared to 28 nm planar FETs. Despite a higher gain performance on the scaled device, the $R_g$ remains a challenge for FinFET technology to optimize further in the journey of higher $F_{max}$.

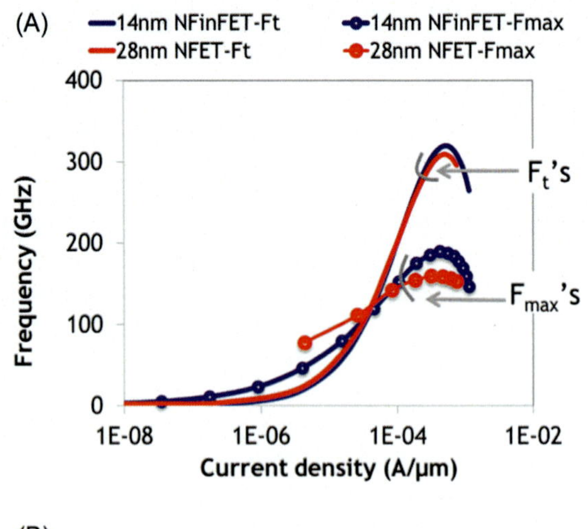


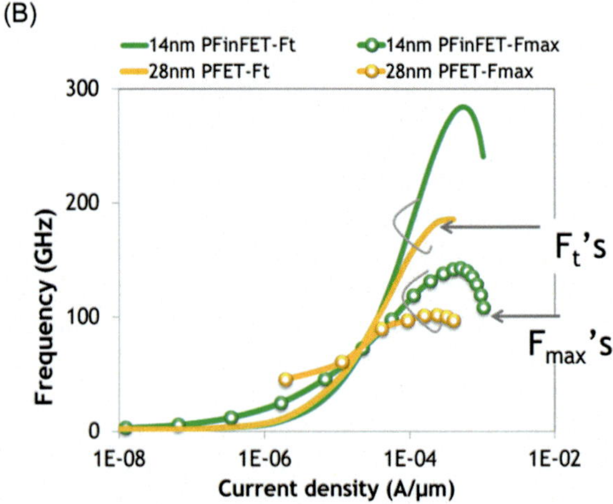


**FIGURE 3.26** A metal-1(ground metal) level cut-off frequency ($F_t$) and maximum oscillation frequency ($F_{max}$) versus current density (Jd-normalized with device width) of a single side gate contacted (A) N-type FinFET (B) and P-type FinFET channel length $L_g = 14$ nm, (NFin = 12) device in comparison with the 28 nm high-κ metal gate counterpart.

Fig. 3.28A shows the $F_{max}$ improvement achieved by using double side gate contact (DGC) layout. To make this comparison, the de-embedding and RF structure connections were modified compared to the design used for Fig. 3.26, mainly to fit the DGC strategies. From Fig. 3.28B, significant $F_{max}$ improvements of 1.26 and 1.40 times that of SGC devices can be achieved for N and P-type FinFETs DGC structure, respectively. A new FoM parameter ($F_t \times G_m/I_d$) is introduced to standardize the intrinsic RF performance across technologies as shown in Table 3.6. It can be observed that the 14 nm FinFET technology intrinsic performance outperforms 28 nm planar FETs while maintaining a smaller device footprint. Key DC and AC parameters for RF FinFET technology and planar MOSFET technology are compared in Table 3.6.

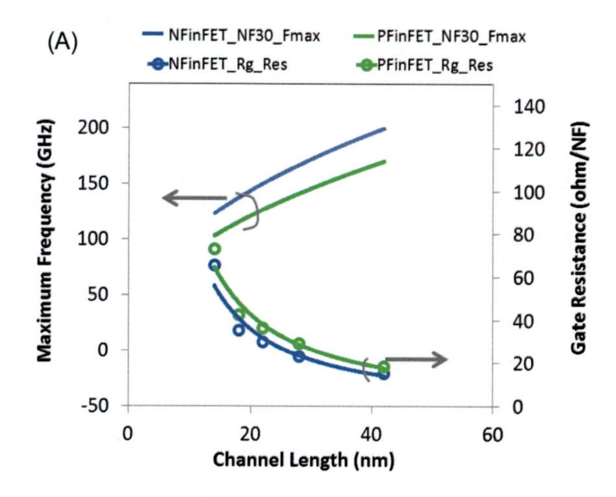


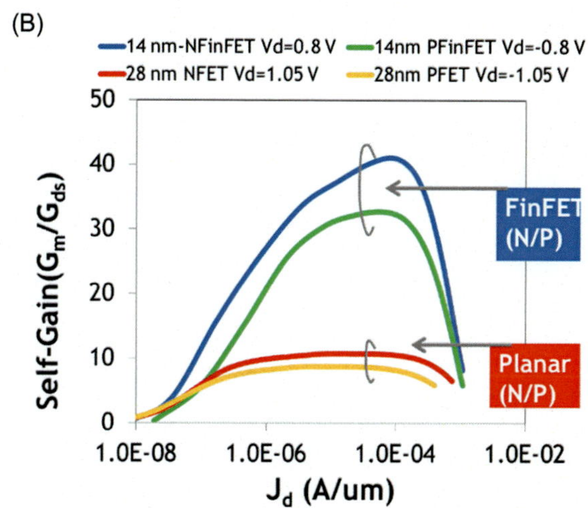


**FIGURE 3.27** (A) A maximum oscillation frequency ($F_{max}$) and gate resistance ($R_g$) relationship with channel length ($L_g$), (B) self-gain (Gm/Gds) of the N and P-type devices from 14 nm FinFET and 28 nm planar technologies, respectively.

## 3.15 Analog transistors

Similar to logic FETs, the analog devices also see the benefits of 3D FinFET architecture and deliver an improved $G_m \times R_{out}$ as shown in Fig. 3.29 (Sell et al., 2018). In the conventional planar CMOS scaling, the logic device gets faster, while its intrinsic analog performance ($G_m/G_{ds}$) degrades from one technology node to the next node as shown in Fig. 3.29A. With 3D migration, for the first time, the analog transistor sees an improved self-gain with $L_g$ scaling. The analog transistor at $L_g = 160$ nm delivers an outstanding $R_{out} = 0.68\text{M}\Omega$ from a 6 fins device (Fig. 3.29B). As a result, an enhanced analog circuit performance is expected.

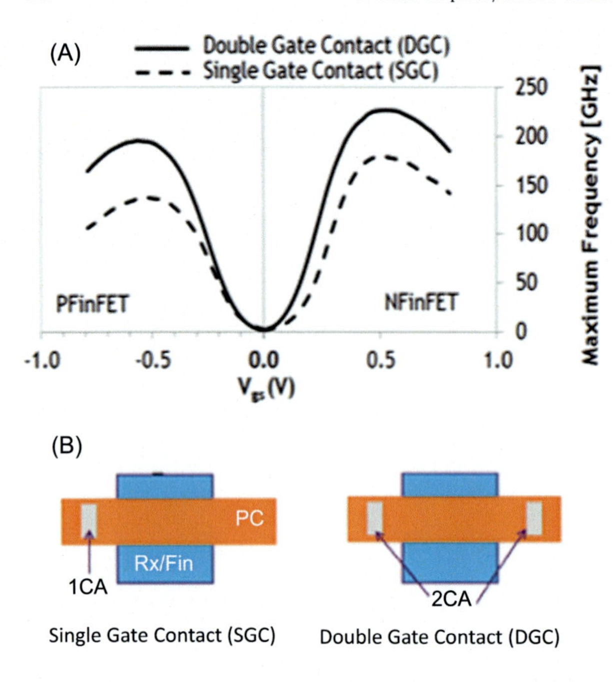


**FIGURE 3.28** (A) The layout of the single-side gate contact (SGC) and double-side gate contact (DGC) device structure. (B) A comparison of the SGC and DGC $F_{max}$ performance versus gate voltage ($V_{gs}$) of the N and PFinFET devices.

## 3.16 Radio frequency high-voltage (I/O) FETs

The high-voltage (HV) devices are also called I/O (input/output) interface FETs. These devices use thick gate oxide (35 Å) to support higher gate voltage ($V_g = 1.8$ V) operation. To support the same voltage on the drain an lightly doped drain (LDD) implant, separate from the S/D implant is typically used to increase junction breakdown, and reduce HCI. The RF characteristics of both N and P-type IO FETs from 14 nm tech platform are shown in Fig. 3.30. The IO N/PFinFETs exhibit an excellent peak $F_t$ (50.1/53.5 GHz) and $F_{max}$ (200/160 GHz). It can be seen that $F_{max}$ is higher than $F_t$ due to the lower $R_g$ benefit of a longer channel length ($L_g = 150$ nm). In Fig. 3.14C the gate current flows into the page through the tungsten in the gate (narrow dark region); although the thicker gate oxide decreases this region the greater gate length more than compensates resulting in lower $R_g$. In the orthogonal direction (y-cross-section, Fig. 3.14B), the gate pitch set the space available for vertical gate current flow, so a tighter gate pitch means more gate resistance. Owing to the higher voltage operation of these I/O FETs, a high current can be drawn despite having longer $L_g$, thereby providing an opportunity to support the cellular and RF power amplifier designs. A detail of such devices across the FinFET technology platform is summarized in Table 3.7.

**TABLE 3.6** Comparison of key core device design and DC/AC parameters.

| Technology | 14 nm FinFET[1] (Singh et al., 2018) | | 22FFL (Sell et al., 2018) | | 28 nm high-κ metal gate (Chew et al., 2015) | |
|---|---|---|---|---|---|---|
| Device flavor | Super Low $V_t$ | | High Perf | | Low Power | |
| $L_g{}^a$[nm] | 14 (Si = 24 nm) | | 32 | | 30 | |
| CPP[b][nm] | 78 | | 108 | | 126 | |
| $N_{fin} \times NF$ (or W) | 12 × 16 | | 6 × 16 | | 3 × 16 | |
| $V_{dd}$(V) | 0.8 | | 1.0 | | 1.0 | |
| FET type | N | P | N | P | N | P |
| $I_{dsat}$[μA/μm] | 1523[c] | 1433[c] | 1380[d] | 1290[d] | 686 | 406 |
| $G_{msat}$[μS/μm] | 3017[c] | 2748[c] | – | – | 985 | 395 |
| *peak* $F_t$ (GHz) | 314 | 285 | 309 | 282 | 310 | 185 |
| *peak* $F_{max}$ (GHz) | 227 | 195 | 452 | 282 | 161 | 104 |
| $F_t.G_m/I_d$[GHz/V] | 2650 | 2053 | 2840 | 2300 | 2000 | 1150 |
| $S_{vg}$ ($V^2\mu m^2$/Hz)@1KHz | 17e-15 | 35e-15 | 20e-15[e] | 36e-15[e] | 171e-15 | 106e-15 |
| $NF_{min}$ (dB) | <0.4@6 GHz[f] | – | <1.7[g]@50 GHz | <1.5[g]@50 GHz | 0.4[h]@6 GHz | 0.53[h]@6 GHz |

[a]*$L_g$-design channel length, final Si $L_g$ is longer for 14 nm case (~24 nm).*
[b]*CPP: contacted poly (gate) pitch.*
[c]*Normalized to footprint $W_{design}$.*
[d]*Measured@$V_{gs}$ = ±0.7 V.*
[e]*Extrapolated values.*
[f]*nfin = 40 (Razavieh et al., 2020)*
[g]*Values @50 Hz.*
[h]*Measured @$V_g = V_d = V_{dd}$.*

## 3.17 Gain-power efficiency

One of the additional benefits of the FinFET technology is the improved device gain per power dissipation efficiency, as FinFET devices can drive stronger drain current with reduced SCEs. Fig. 3.31 illustrates the gain-power efficiency FoM (GPFoM) of FinFET and planar from 32 nm process technology nodes for comparison purposes (Lee et al., 2019).

GPFoM is defined as

$$\mathrm{GPFoM} = \frac{U.G_m}{I_d}\,[dB/V] \tag{3.17}$$

FinFET devices offer about 600 dB/V improvement over planar technologies in the gain-power efficiency at 30 GHz. Note that Mason's gain (U or Unilateral gain) is used for the FoM to accommodate the performance metrics in the mmWave frequency range. The current density reaching the peak FoM is the optimum bias condition for the maximum gain-power efficiency.

(A)

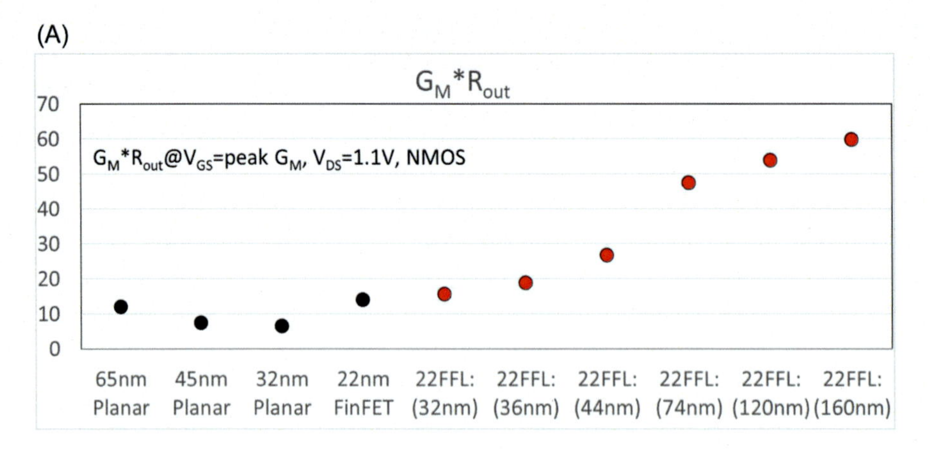


(B)

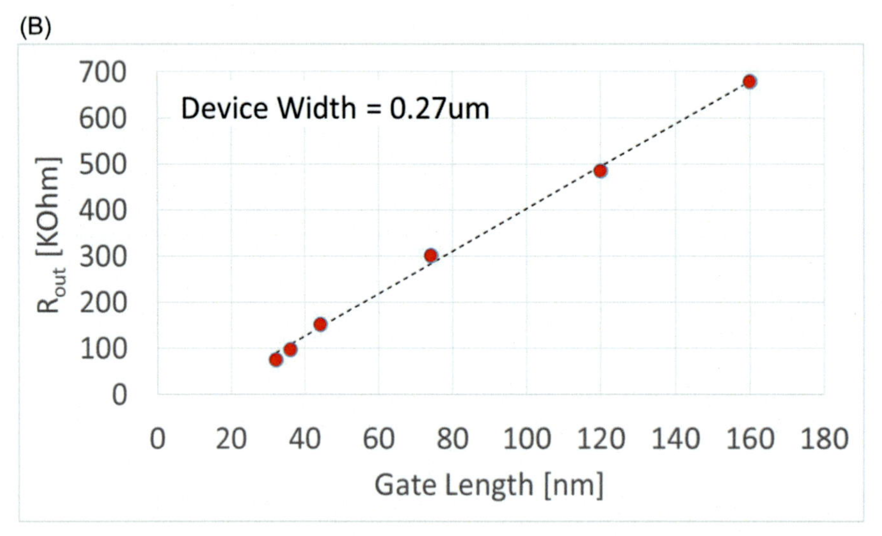


**FIGURE 3.29** The figure shows (A) self-gain (Gmx$R_{out}$) versus technologies features (B) the $R_{out}$ versus gate length of the 22FFL FinFET @ Vds = $V_d/2 = 0.425$ V and $V_{gs} = V_t + 0.2$ V.

## 3.18 Substrate network

At higher frequencies, the substrate region acts as an RC network with the source, drain, and gate all capacitively coupled to the substrate. AC current injected into the substrate is conducted to substrate contacts. In an ideal situation, the substrate potential equals the potential at the substrate contact (usually ground) because the substrate resistance is zero. However, in the real case, the substrate resistance is finite, and there is an AC signal on the substrate under the device (Kushwaha et al., 2018). This signal increases with frequency. It couples to the channel region of FET reducing the gain and introducing distortion. It can further couple to the source and drain terminals and neighboring

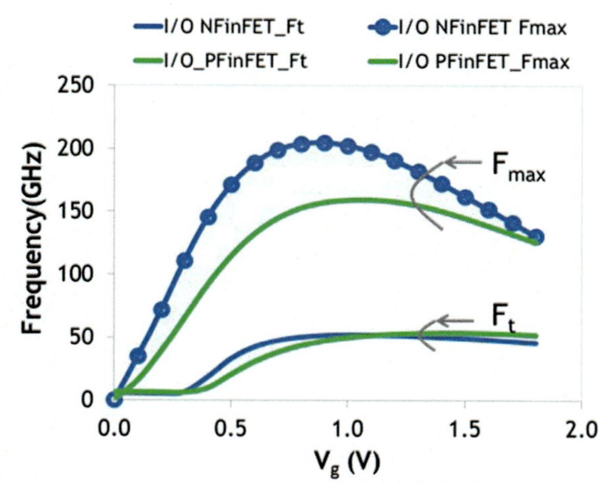


**FIGURE 3.30** The $F_t$ and $F_{max}$ of 1.8 V I/O of $L_g = 150$ nm N, P-type FinFET versus $V_g$ (gate voltage) characteristics. *Source: From Singh et al. (2018).*

**TABLE 3.7** The thick oxide N and P-type FinFET DC/AC parameters across the technology platform.

| Technology | 14 nm FinFET | | 22FFL | |
|---|---|---|---|---|
| Device | NFinFET | PFinFET | NFinFET | |
| $V_{dd}$(V) | 1.8 | | | |
| $L_g$[nm] | 150 | | 120 | 160 |
| CPP†[nm] | 260 | | 216 | 270 |
| $F_t$[GHz] | 50.1 | 53.5 | 63 | 45 |
| $F_{max}$[GHz] | 200 | 160 | 226 | 167 |

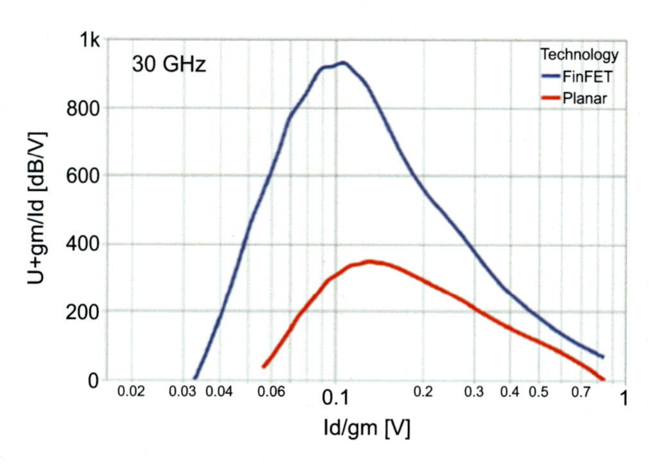


**FIGURE 3.31** Gain-power efficiency FoM of 22FFL FinFET and planar device (32 nm process technology) at 30 GHz. *Source: From Lee et al. (2019).*

circuitry, thus causing RF performance degradation at both the device and circuit level. The substrate can also couple one FET to another introducing noise into the RF signal. This can be mitigated by adding isolation structures to block the signal injected into the substrate by one FET from reaching others. The active area of a FinFET device lies in the part of the fin surrounded by a gate. This region is fully depleted during the operation. These fins are connected to the silicon substrate via the fin's foot Si regions. This geometry means that the coupling of the source, drain, and channel to the substrate is much smaller than in planar MOSFET of the same effective width. This reduces the effect of substrate bias on the FinFET devices compared to planar FETs both for DC levels as in stacked devices (Fig. 3.11) and for any AC signals in the substrate. The substrate currents from the device see a higher resistance than a planar MOSFET due to the smaller foot region when it passed through the Fin region.

To isolate the active devices from the substrate noise and bias signal, conventional triple well (deep nwell) isolation schemes can be employed. An N-type implant with an energy >500 keV is used to create nwell regions deep in the silicon. Fig. 3.32A shows a schematic of the deep nwell isolation structure, which is used to measure isolation using the S21 parameter measurements (Singh et al., 2018). The space between port 1 and port 2 is kept at 10 μm for both with and without deep nwell isolation structures. Fig. 3.32B compares the measured S21 results from these structures. It can be seen that noise reduction is more

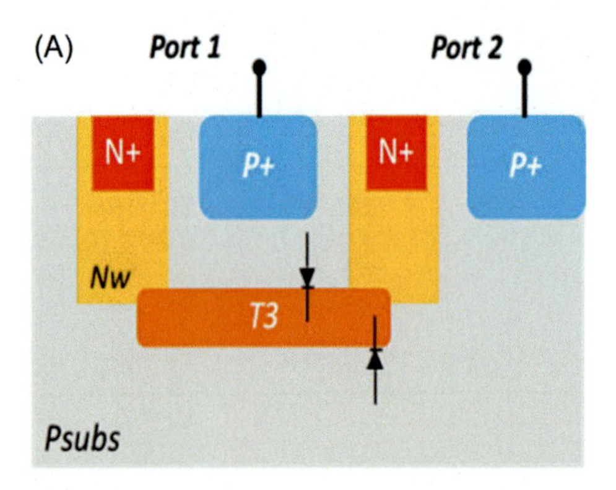


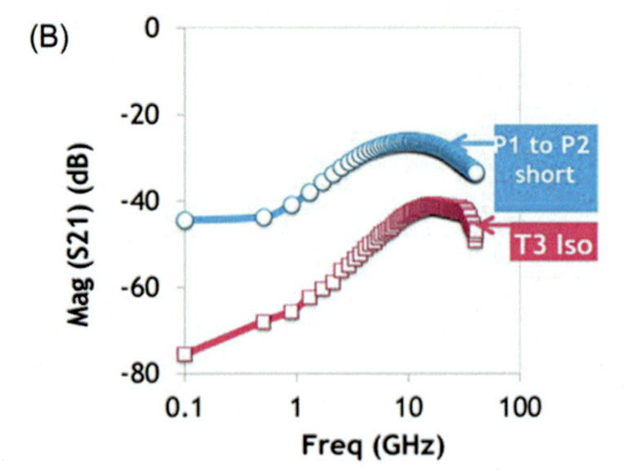


FIGURE 3.32 (A) A cross-section view of the bottom region of deep nwell isolations, (B) substrate coupling (S21) measured silicon results with deep nwell (T3) isolation or without T3 (P1 to P2 short) structures.

effective in the 0.1 to 10 GHz frequency range with deep nwell isolation and achieves a maximum of −75 dB substrate noise reduction at 0.1 GHz. The deep nwell layer can contribute roughly −30 dB reduction substrate noise reduction.

## 3.19 Noise in MOS transistors

The most common noise sources related to RF MOSFETs devices are flicker (*1/f*) and thermal noise.

### 3.19.1 Flicker (*1/f*) noise

The flicker noise is related to random charge carriers trapping and de-trapping by energy states at the silicon-to-gate oxide interface. In the case of FinFET technology, the carrier movement occurs along <110> Si orientation with relatively higher interface trap charges compared to planar <100> Si orientation. The quality of the interface also varies from process to process. Flicker noise is usually modeled as an input-referred gate noise by a voltage source between the gate and source. The voltage spectral density is given by

$$\frac{V_{nf}^2}{\Delta f} = \frac{Kf}{f}\frac{1}{C_{ox}WL} \tag{3.18}$$

where $Kf$ is a device-specific constant, $f$ is frequency, $C_{ox}$ is gate dielectric capacitance per unit area, and $W$ and $L$ is the effective width and length of a MOS device, respectively.

Normalized *1/f* referred as input gate voltage noise spectral densities ($S_{vg}$) of both n and p-type FET from 14 and 28 nm technologies is shown in Fig. 3.33. The FinFET shows a significantly lower $S_{vg}$ which is attributed to the flow of carriers away from the Si and gate oxide interface and concentrates along the centerline of the gate wrap-around fins. A comparison of $S_{vg}$ among two FinFET and one planar technology is also shown in Table 3.6.

### 3.19.2 Thermal noise

Thermal fluctuations in the channel inversion charge will induce noise current in the gate due to capacitive coupling. The gate-induced noise source can be represented after simplification as

$$\frac{v_{ng}^2}{\Delta f} = 4kT\delta R_g \tag{3.19}$$

where $k$ is Boltzmann's constant, $T$ is absolute temperature, $\Delta f$ is bandwidth in Hertz, and $R_g$ is the gate resistance. Fig. 3.34 demonstrates three cases of fins configuration maintaining total device size by modulating the number of gate fingers. The maximum available gain ($G_{max}$) and the minimum noise figure ($NF_{min}$) are adjusted by the number of fins and the delta between $G_{max}$ and $NF_{min}$ is maximized for 4 or 6-fin device at $J_d$ of 0.3 mA/um (Lee et al., 2019; Sell et al., 2018). From the previous section 3.10.2 "FinFET gate resistance", the lowest $R_g$ comes around 4 to 6 fin per transistor. For LNA applications, a low $R_g$ point need

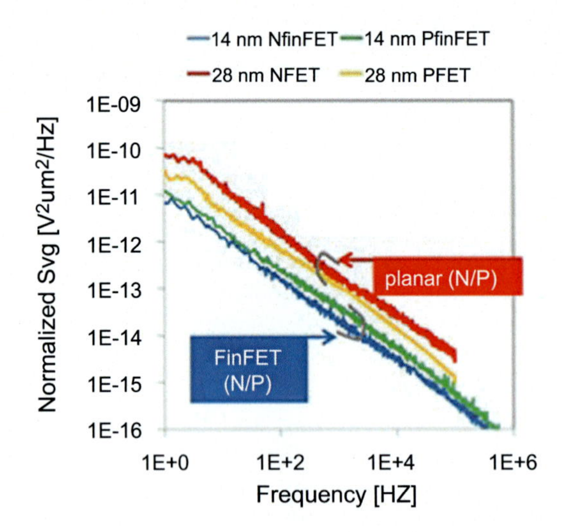


FIGURE 3.33 Normalized input-referred flicker noise (1/f) from 14 nm FinFET and 28 nm planar technologies. *Source: From Singh et al. (2018).*

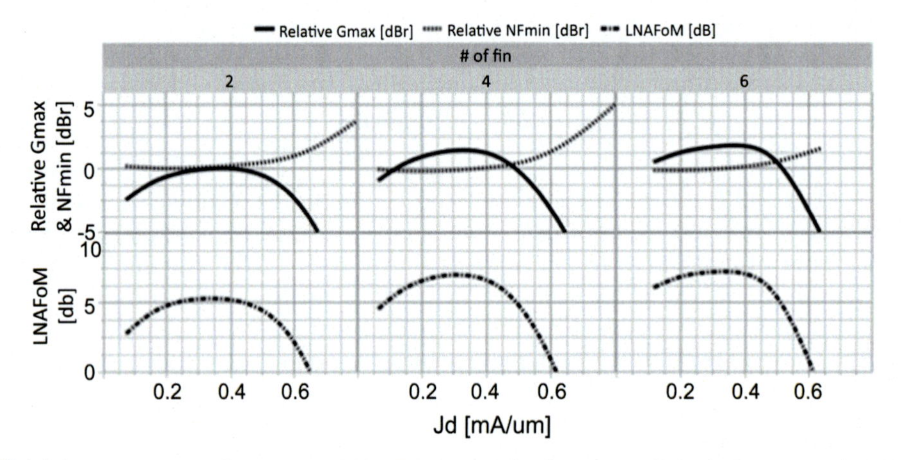


FIGURE 3.34 Device sizing for optimum LNA FoM: 4 or 6-fin devices reach the highest equivalent LNA FoM at ~0.3 mA/um current density. LNA FoM is defined by $G_{max}$[dB]—$NF_{min}$[dB]. $G_{max}$ and $NF_{min}$ are normalized to $G_{max}$ and $NF_{min}$ of 2-fin device. *Source: From Lee et al. (2019).*

to explore, which could vary from one technology to other. A comparison of technologies is shown in Table 3.6. Improved electrostatic behavior of the FinFET technology also contributes to lower thermal noise despite having a higher $R_g$ (Ding et al., 2021).

## 3.20 Summary

FinFET is a higher-density logic technology with a significant drive current performance improvement over the planar 28 nm technologies while maintaining ultra-low leakage current. The RF ($F_t$, $F_{max}$) performance is reaching mid 300 GHz, which meets the device performance requirement of many RF and mmWave circuits. Also, a low noise device can be achieved with a low $R_g$ when $N_{Fin}$ is in the range of 4 to 8. High-voltage FinFET devices also show excellent RF performance. The FinFET technology combines higher performance and ultra-low power FinFET devices. With a lower BEOL stack and relaxed metal pitch, it is also a low-cost technology solution for mobile and RF applications. It is the potential to enable low noise amplifier (LNA), low power PLL, and higher performance PA RF circuits.

## List of symbols

$BV_{dss}$ Drain to source breakdown voltage
$C_{gg}$ Total gate capacitance
$C_{ggi}$ Intrinsic gate capacitance
$C_{gs}$ Gate to source capacitance
$C_{gd}$ Gate to drain capacitance
**dB** Decibel
$F_p$ Fin pitch
$F_{max}$ maximum oscillation frequency when the unilateral power gain becomes unity.
$F_t$ Short Circuit unity current gain cut off frequency
$L_g$ Channel length
$F_{HT}$ Fin Height
$F_W$ Fin Width
$F_{Top}$ Fin top surface dimensions
$G_{ds}$ Output conductance
$G_m$ Transconductance
$I_d$ Drain current
$I_{dlin}$ Linear region MOSFET current
$I_{off}$ Off-state leakage MOSFET current
$I_{dsat}$ Saturation region MOSFET current
$I_{subs}$ Substrate or bulk current
$J_d$ drain current density
**NF** Number of gate fingers
$NF_{50}$ Noise figure with 50Ω load termination
$R_d$ Drain parasitic resistance
$R_g$ Gate resistance
$R_{on}$ MOSFET On state resistance
$R_s$ Source parasitic resistance
$R_{sh}$ sheet resistance in ohm/square
$t_{ox}$ Gate oxide thickness
$t_{si}$ thin silicon layer over a buried oxide
$\mu$ Mobility
$W$ Width of the channel of a planar MOSFET
$W_{eff}$ Effective width of the transistor
$W_{FIN}$ Width of the channel of one fin
$V_d$ Drain voltage
$V_{dd}$ Supply voltage
$V_{dstress}$ HCI stress drain voltage higher than $V_{dd}$
$V_g$ Gate voltage
$V_{gstress}$ Stress gate voltage corresponding to peak $I_{sub}$ for HCI testing
$V_t$ Threshold voltage
$V_{tlin}$ Linear MOSFET operation region threshold voltage
$V_{tsat}$ Saturation MOSFET operation region threshold voltage

# List of acronyms

| | |
|---|---|
| **3D** | three dimensional |
| **BiCMOS** | Bipolar and Complementary Metal Oxide Semiconductor |
| **BJT** | Bipolar junction transistor |
| **$B_{ox}$** | Buried Oxide |
| **CMOS** | Complementary Metal Oxide Semiconductor |
| **COAG** | Contact over active gate |
| **CPP** | Contacted poly pitch |
| **CS** | Common source |
| **$D_{ext}$** | Drain extension region under polygate of the LDMOS |
| **DGC** | Double Gate Contact |
| **DP** | Dummy poly |
| **EoL** | End of Life |
| **FinFET** | fin field-effect or trigate transistor |
| **FoM** | Figure of Merit |
| **GHz** | Gigahertz |
| **HBT** | Heterojunction Bipolar Transistor |
| **HCI** | Hot carrier injection |
| **High-κ** | material with a high dielectric constant (κ, kappa), as compared to silicon dioxide |
| **HiKMG** | High-κ and Metal Gate |
| **HP** | High performance |
| **HV** | High-voltage |
| **IC** | Integrated circuits |
| **LDD** | Lightly doped drain |
| **LDMOS** | Laterally-Diffused Metal Oxide Semiconductor |
| **LNA** | Low noise amplifier |
| **mmWAVE** | Millimeter wave (30–300 GHz band spectrum) |
| **MSG** | The maximum stable gain (MSG) of a device is defined when the maximum available gain is undefined ($K<1$) |
| **nm** | nanometer |
| **PA** | Power amplifier |
| **PC** | Polygate cut |
| **PLL** | Phase lock loop |
| **PPA** | Power performance area |
| **RMG** | Replacement metal gate |
| **RF** | Radio Frequency |
| **S-parameters** | Scattering parameters (the elements of a scattering matrix) |
| **SCE** | Short Channel Effect |
| **SGC** | Single Gate Contact |
| **SiGe** | Silicon Germanium |
| **SNR** | Signal to Noise ratio |
| **SoA** | Safe operating area |
| **SoC** | System on Chip |
| **SOI** | Silicon on Insulator |
| **STI** | Shallow trench isolation |
| **T3** | Deep nwell implant |
| **TCAD** | Technology Computer-Aided Design simulations |
| **TEM** | Transmission electron microscopy |

| | |
|---|---|
| **TEOS** | Tetraethyl orthosilicate oxide |
| **TV** | Television |
| **UTB** | MOSFET Ultra-Thin Body MOSFET |
| **WiFi** | Wireless Fidelity |
| **ZigBee** | standards-based wireless technology developed to enable low-cost, low-power wireless machine-to-machine (M2M) and Internet of Things (IoT) networks |

# References

Ali, K. B., Neve, C. R., Gharsallah, A., & Raskin, J. P. (2014). RF performance of SOI CMOS technology on commercial 200-mm enhanced signal integrity high resistivity SOI substrate. *IEEE Transactions on Electron Devices, 61*(3), 722–728. Available from https://doi.org/10.1109/TED.2014.2302685.

Alvin, J., Jain, V., Ong, S. N, Wolf, R., Lim, S. F., & Singh, J. (2018). Technology positioning for mm wave applications: 130/90nm SiGe BiCMOS vs. 28nm RFCMOS. *2018 IEEE BiCMOS and compound semiconductor integrated circuits and technology symposium (BCICTS)*, 18–21.

Baldwin, G., Ai, J., Benaissa, K., Chen, F., Chidambaram, P.R., Ekbote, S., Ghneim, S., Liu, S., Machala, C., Mehrad, F., Mosher, D., Pollack, G., Tran, T., Williams, B., Yang, J., Yang, S., & Johnson, F.S. (2003). 90nm CMOS RF technology with 9.0 V I/O capability for single-chip radio. In: *Digest of technical papers – Symposium on VLSI technology* (pp. 87–88). United States.

Callender, S., Shin, W., Lee, H. J., Pellerano, S., & Hull, C. (2018). FinFET for mm wave - Technology and circuit design challenges. In: *2018 IEEE BiCMOS and compound semiconductor integrated circuits and technology symposium*, BCICTS 2018 (pp. 168–173). United States: Institute of Electrical and Electronics Engineers Inc. Available from https://doi.org/10.1109/BCICTS.2018.8551125, http://ieeexplore.ieee.org/xpl/mostRecentIssue.jsp?punumber = 8536738.

Carter, R., Mazurier, J., Pirro, L., Sachse, J. U., Baars, P., Faul, J., Grass, C., Grasshoff, G., Javorka, P., Kammler, T., Preusse, A., Nielsen, S., Heller, T., Schmidt, J., Niebojewski, H., Chou, P. Y., Smith, E., Erben, E., Metze, C., … Rice, B. (2017). 22nm FDSOI technology for emerging mobile, Internet-of-Things, and RF applications. In: *Technical digest – International electron devices meeting, IEDM* (pp. 2.2.1–2.2.4). Germany: Institute of Electrical and Electronics Engineers Inc. Available from https://doi.org/10.1109/IEDM.2016.7838029.

Cathelin, A. (2017). Fully depleted silicon on insulator devices CMOS: The 28-nm node is the perfect technology for analog, RF, mmW, and mixed-signal system-on-chip integration. *IEEE Solid-State Circuits Magazine, 9*(4), 18–26. Available from https://doi.org/10.1109/MSSC.2017.2745738, http://ewh.ieee.org/soc/sscs/index.php?option = com_content&task = view&id = 32&Itemid = 22.

Chen, X., Tsai, M. K., Chen, C. H., Lee, R., & Chen, D. C. (2014). Extraction of gate resistance in sub-100-nm MOSFETs with statistical verification. *IEEE Transactions on Electron Devices, 61*(9), 3111–3117. Available from https://doi.org/10.1109/TED.2014.2340871.

Chew, K. W. J., Agshikar, A., Wiatr, M., Wong, J. S., Chow, W. H., Liu, Z., Lee, T. H., Shi, J., Lim, S. F., Sundaram, K., Chan, L. H. K., Cheng, C. H. M., Sassiat, N., Yoo, Y. K., Balijepalli, A., Kumta, A., Nguyen, C. D., Illgen, R., Mathew, A., …Harame, D. (2015). RF performance of 28nm PolySiON and HKMG CMOS devices. In *Digest of papers – IEEE radio frequency integrated circuits symposium* (Vol. 2015-, pp. 43–46). Singapore: Institute of Electrical and Electronics Engineers Inc. Available from https://doi.org/10.1109/RFIC.2015.7337700.

Chiang, C. K., Pai, H., Lin, J. L., Chang, J. K., Lee, M. Y., Hsieh, E. R., Li, K.S., Luo, G. L., Cheng, O., & Chung, S. S. (2021). FinFET Plus: A scalable FinFET architecture with 3D air-gap and air-spacer toward the 3nm generation and beyond. In: *VLSI-TSA 2021 – 2021 International symposium on VLSI technology, systems and applications, proceedings*. Taiwan: Institute of Electrical and Electronics Engineers Inc. Available from https://doi.org/10.1109/VLSI-TSA51926.2021.9440097, http://ieeexplore.ieee.org/xpl/mostRecentIssue.jsp?punumber = 9440054.

Cho, H., & Burk, D. E. (1991). A three-step method for the de-embedding of high-frequency S-parameter measurements. *IEEE Transactions on Electron Devices, 48*(6), 1371–1375. Available from https://doi.org/10.1109/16.81628.

Choi, Y. K., Asano, K., Lindert, N., Subramanian, V., King, T. J., Bokor, J., & Chenming, H. (2000). Ultrathin-body SOI MOSFET for deep-sub-tenth micron era. *IEEE Electron Device Letters, 21*(5), 254–255. Available from https://doi.org/10.1109/55.841313.

Ding, X., Niu, G., Zhang, A., Cai, W., & Imura, K. (2021). Experimental extraction of thermal noise $\gamma$ factors in a 14-nm RF FinFET technology. In: *2021 IEEE 21st topical meeting on silicon monolithic integrated circuits in RF systems, SiRF 2021* (pp. 25–27). United States: Institute of Electrical and Electronics Engineers Inc. Available from https://doi.org/10.1109/SiRF51851.2021.9383331, http://ieeexplore.ieee.org/xpl/mostRecentIssue.jsp?punumber = 9383232.

Erben, E., Hempel, K., & Triyoso, D. (2018). Work function setting in high-k metal gate devices. *IntechOpen*. Available from https://doi.org/10.5772/intechopen.78335.

Jan, C. H., Agostinelli, M., Deshpande, H., El-Tanani, M. A., Hafez, W., Jalan, U., Janbay, L., Kang, M., Lakdawala, H., Lin, J., Lu, Y. L., Mudanai, S., Park, J., Rahman, A., Rizk, J., Shin, W. K., Soumyanath, K., Tashiro, H., Tsai, C., ... Bai, P. (2010). RF CMOS technology scaling in high-k/metal gate era for RF SoC (system-on-chip) applications. In: *Technical digest – International electron devices meeting, IEDM* (p. 27.2.4.). United States. Available from https://doi.org/10.1109/IEDM.2010.5703431.

Jan, C. H., Bhattacharya, U., Brain, R., Choi, S. J., Curello, G., Gupta, G., Hafez, W., Jang, M., Kang, M., Komeyli, K., Leo, T., Nidhi, N., Pan, L., Park, J., Phoa, K., Rahman, A., Staus, C., Tashiro, H., Tsai, C., ... Bai, P. (2012). A 22nm SoC platform technology featuring 3-D tri-gate and high-k/metal gate, optimized for ultra low power, high performance and high density SoC applications. In *Technical digest - International electron devices meeting, IEDM*. United States. Available from https://doi.org/10.1109/IEDM.2012.6478969.

Jeong, E. Y., Song, M., Choi, I., Shin, H., Song, J., Maeng, W., Park, H., Yoon, H., Kim, S., Park, S., You, B. H., Cho, H. J., An, Y. C., Lee, S. K., Kwon, S. D., Jung, S. M. (2017). High performance 14nm FinFET technology for low power mobile RF application. In: *Digest of technical papers – Symposium on VLSI technology* (pp. 142–143). South Korea: Institute of Electrical and Electronics Engineers Inc. Available from https://doi.org/10.23919/VLSIT0.2017.7998155.

Jin, X., Ou, J. J., Chen, C. H., Liu, W., Deen, M. J., Gray, P. R., & Hu, C. (1998). Effective gate resistance model for CMOS RF and noise modeling. In *Technical digest - International electron devices meeting* (pp. 961–964). United States: IEEE.

Jones, S. (2017). Controlling threshold voltage with work-function metals. Semiwiki. Available from https://semiwiki.com/semiconductor-services/ic-knowledge/7259-iedm-2017-controlling-threshold-voltage-with-work-function-metals/.

Joseph, A., Botula, A., Slinkman, J., Wolf, R., Phelps, R., Abou-Khalil, M., Ellis-Monaghan, J., Moss, S., & Jaffe, M. (2013). Power handling capability of an SOI RF switch. In *Digest of papers - IEEE radio frequency integrated circuits symposium* (pp. 385–388). United States. Available from https://doi.org/10.1109/RFIC.2013.6569611.

Kedzierski, J., Ieong, M., Nowak, E., Kanarsky, T. S., Zhang, Y., Roy, R., Boyd, D., Fried, D., & Wong, H. S. P. (2003). Extension and source/drain design for high-performance FinFET devices. *IEEE Transactions on Electron Devices*, *50*(4), 952–958. Available from https://doi.org/10.1109/TED.2003.811412.

Kesapragada, S., Wang, R., Liu, D., Liu, G., Xie, Z., Ge, Z., Yang, H., Lei, Y., Lu, X., Tang, X., Lei, J., Allen, M., Gandikota, S., Moraes, K., Hung, S., Yoshida, N., & Chang, C.P. (2010). High-k/metal gate stacks in gate first and replacement gate schemes. In: *ASMC (advanced semiconductor manufacturing conference) proceedings* (pp. 256–259). United States. Available from https://doi.org/10.1109/ASMC.2010.5551460.

Kissinger, D., Kahmen, G., & Weigel, R. (2021). Millimeter-wave and terahertz transceivers in sige bicmos technologies. *IEEE Transactions on Microwave Theory and Techniques*, *69*(10), 4541–4560. Available from https://doi.org/10.1109/TMTT.2021.3095235, http://ieeexplore.ieee.org/xpl/tocresult.jsp?isYear = 2009&isnumber = 4747395&Submit32 = View + Contents.

Koolen, M. C. A. M., Geelen, J. A. M., & Versleijen, M. P. J. G. (1992). An improved de-embedding technique for on-wafer high-frequency characterization. In: *Proceedings of the 1991 bipolar circuits and technology meeting* (pp. 188–191). IEEE undefined.

Kushwaha, P., Agarwal, H., Lin, Y. K., Kao, M. Y., Duarte, J. P., Chang, H. L., Wong, W., Fan, J., Xiayu, Y. S., Chauhan, S., Salahuddin, C., & Hu. (2018). Modeling of advanced RF bulk FinFETs. *IEEE Electron Device Letters*, *39*(6), 791–794. Available from https://doi.org/10.1109/LED.2018.2825422.

Lakdawala, H., Schaecher, M., Fu, C. T., Limaye, R., Duster, J., Tan, Y., Balankutty, A., Alpman, E., Lee, C. C., Nguyen, K. M., Lee, H. J., Ravi, A., Suzuki, S., Carlton, B. R., Kim, H. S., Verhelst, M., Pellerano, S., Kim, T., Venkatesan, S., ... Soumyanath, K. (2013). A 32 nm SoC with dual core ATOM processor and RF wifi transceiver. *IEEE Journal of Solid-State Circuits*, *48*(1), 91–103. Available from https://doi.org/10.1109/JSSC.2012.2222812.

Lee, H. J., Callender, S., Rami, S., Shin, W., Yu, Q., & Marulanda, J. M. (2020). Intel 22nm low-power FinFET (22FFL) process technology for 5G and beyond. In: *Proceedings of the custom integrated circuits conference*

(Vol. 2020). United States: Institute of Electrical and Electronics Engineers Inc. Available from https://doi.org/10.1109/CICC48029.2020.9075914.

Lee, H. J., Rami, S., Ravikumar, S., Neeli, V., Phoa, K., Sell, B., & Zhang, Y. (2019). Intel 22nm FinFET (22FFL) process technology for RF and mm wave applications and circuit design optimization for FinFET technology. In *Technical digest – International electron devices meeting, IEDM* (Vol. 2018-, pp. 14.1.1–14.1.4). United States: Institute of Electrical and Electronics Engineers Inc. Available from https://doi.org/10.1109/IEDM.2018.8614490.

Lee, J., Kang, B., Joo, S., Lee, S., Lee, J., Kang, S., Jo, L., Ahn, S., Lee, J., Bae, J., Ko, W., Jung, W., Lee, S., Lee, S., Park, E., Lee, S., Woo, J., Lee, J., Lee, Y., ... Kang, L. (2021). 6.1 A low-power and low-cost 14nm FinFET RFIC supporting legacy cellular and 5G FR1. In *Digest of technical papers - IEEE international solid-state circuits conference* (Vol. 64, pp. 90–92). South Korea: Institute of Electrical and Electronics Engineers Inc. Available from https://doi.org/10.1109/ISSCC42613.2021.9365736.

Li, W. K., Chan, W. C., Tsai, T. C., Liu, H. H., Chang, W. M., Lai, C. M., Chiang, T., Lin, C. L., Wu, P. A., Huang, H. W., Yeh, Y. L., Chen, P. N., Hsu, J. L., Chen, S. H., Wang, C. Y., Chang, Y. H., Yang, T. H., Sun, R. B., Hsu, W. H., & Zhan, J. H. C. (2017). A 2 × 2 802.11ac WiFi transceiver supporting per channel 160MHz operation in 28nm CMOS. In *Digest of papers - IEEE radio frequency integrated circuits symposium* (pp. 200–203). Taiwan: Institute of Electrical and Electronics Engineers Inc. Available from https://doi.org/10.1109/RFIC.2017.7969052 9781509046263.

Lu, D.D. (2011). Compact models for future generation CMOS (PhD dissertation). Electrical Engineering and Computer Sciences University of California at Berkeley Technical Report No. UCB/EECS-2011-69 http://www.eecs.berkeley.edu/Pubs/TechRpts/2011/EECS-2011-69.html. May 30, 2011

Mocuta, A., Weckx, P., Demuynck, S., Radisic, D., Oniki, Y., & Ryckaert, J. (2018). Enabling CMOS scaling towards 3nm and beyond. In: *Digest of technical papers – Symposium on VLSI technology* (Vol. 2018-, pp. 147–148). Belgium: Institute of Electrical and Electronics Engineers Inc. Available from https://doi.org/10.1109/VLSIT.2018.8510683.

Morifuji, E., Momose, H. S., Ohguro, T., Yoshitomi, T., Kimijima, H., Matsuoka, F., Kinugawa, M., Katsumata, Y., & Iwai, H. (1999). Future perspective and scaling down roadmap for RF CMOS. In: *Digest of technical papers – Symposium on VLSI technology* (pp. 163–164). Japan: IEEE.

Muralidhar, R., Dennard, R. H., Ando, T., Lauer, I., & Hook, T. (2018). Advanced FDSOI device design: The U-channel device for 7 nm node and beyond. *IEEE Journal of the Electron Devices Society, 6*, 551–556. Available from https://doi.org/10.1109/JEDS.2018.2809587, http://ieeexplore.ieee.org/servlet/opac?punumber = 6245494.

Natarajan, S., Agostinelli, M., Akbar, S., Bost, M., Bowonder, A., Chikarmane, V., Chouksey, S., Dasgupta, A., Fischer, K., Fu, Q., Ghani, T., Giles, M., Govindaraju, S., Grover, R., Han, W., Hanken, D., Haralson, E., Haran, M., Heckscher, M., ... Zhang, K. (2015). A 14nm logic technology featuring 2nd-generation FinFET, air-gapped interconnects, self-aligned double patterning and a 0.0588 μm2 SRAM cell size. In: *Technical digest – International electron devices meeting, IEDM* (Vol. 2015-. February, pp. 3.7.1–3.7.3). United States: Institute of Electrical and Electronics Engineers Inc. Available from https://doi.org/10.1109/IEDM.2014.7046976.

Ong, S. N., Lehmann, S., Chow, W. H., Zhang, C., Schippel, C., Chan, L. H. K., Andee, Y., Hauschildt, M., Tan, K.K. S., Watts, J., Lim, C. K., Divay, A., Wong, J. S., Zhao, Z., Govindarajan, M., Schwan, C., Huschka, A., Bcllaouar, A., Loo, W., ... Harame, D. (2018). A 22nm FDSOI technology optimized for RF/mmWave applications. In: *Digest of papers – IEEE radio frequency integrated circuits symposium* (Vol. 2018-., pp. 72–75). Singapore: Institute of Electrical and Electronics Engineers Inc. Available from https://doi.org/10.1109/RFIC.2018.8429035.

Raskin, J. P., Viviani, A., Flandre, D., & Colinge, J. P. (1997). Substrate crosstalk reduction using SOI technology. *IEEE Transactions on Electron Devices, 44*(12), 2252–2261. Available from https://doi.org/10.1109/16.644646.

Razavi, B. (1999). CMOS technology characterization for analog and RF design. *IEEE Journal of Solid-State Circuits, 34*(3), 268–276. Available from https://doi.org/10.1109/4.748177.

Razavieh, A., Chen, Y., Ethirajan, T., Gu, M., Cimino, S., Shimizu, T., Hassan, M. K., Morshed, T., Singh, J., Zheng, W., Mahajan, V., Wang, H. T., & Lee, T. H. (2021). Extremely-low threshold voltage FinFET for 5G mmWave applications. *IEEE Journal of the Electron Devices Society, 9*, 165–169. Available from https://doi.org/10.1109/JEDS.2020.3046953, http://ieeexplore.ieee.org/servlet/opac?punumber = 6245494.

Razavieh, A., Mahajan, V., Oo, W. L., Cimino, S., Khokale, S. V., Nagahiro, K., Pantisano, L., Ethirajan, T., Lemon, J., Gu, M., Chen, Y., Wang, H. T., & Lee, T. H. (2020). FinFET with contact over active-gate for 5G ultra-wideband applications. In: *Digest of technical papers – Symposium on VLSI technology* (Vol. 2020). United States: Institute of Electrical and Electronics Engineers Inc. Available from https://doi.org/10.1109/VLSITechnology18217.2020.9265095.

Samsung. (2021). Samsung successfully completes 8nm RF solution development to strengthen 5G communications chip solutions. Available from https://news.samsung.com/global/samsung-successfully-completes-8nm-rf-solution-development-to-strengthen-5g-communications-chip-solutions.

Sell, B., Bigwood, B., Cha, S., Chen, Z., Dhage, P., Fan, P., Giraud-Carrier, M., Kar, A., Karl, E., Ku, C. J., Kumar, R., Lajoie, T., Lee, H. J., Liu, G., Liu, S., Ma, Y., Mudanai, S., Nguyen, L., Paulson, L., ... Bai, P. (2018). 22FFL: A high performance and ultra low power FinFET technology for mobile and RF applications. In: *Technical digest – International electron devices meeting*, IEDM (pp. 29.4.1–29.4.4). United States: Institute of Electrical and Electronics Engineers Inc. Available from https://doi.org/10.1109/IEDM.2017.8268475.

Singh, J., Bousquet, A., Ciavatti, J., Sundaram, K., Wong, J.S., Chew, K. W., Bandyopadhyay, A., Li, S., Bellaouar, A., Pandey, S. M., Zhu, B., Martin, A., Kyono, C., Goo, J. S., Yang, H. S., Mehta, A., Zhang, X., Hu, O., Mahajan, S., ... Sohn, D.K. (2017). 14nm FinFET technology for analog and RF applications. In: *Digest of technical papers – Symposium on VLSI technology* (pp. T140–T141). United States: Institute of Electrical and Electronics Engineers Inc. Available from https://doi.org/10.23919/VLSIT.2017.7998154.

Singh, J., Ciavatti, J., Sundaram, K., Wong, J. S., Bandyopadhyay, A., Zhang, X., Li, S., Bellaouar, A., Watts, J., Lee, J. G., & Samavedam, S. B. (2018). 14-nm FinFET technology for analog and RF applications. *IEEE Transactions on Electron Devices*, *65*(1), 31–37. Available from https://doi.org/10.1109/TED.2017.2776838.

TSMC, G. C. (2021). Introducing N6RF: CMOS radio for the 5G Era. Available from https://www.tsmc.com/english/news-events/blog-article-20210603.

Wachnik, R. A., Lee, S., Pan, L. H., Lu, N., Li, H., Bingert, R., Randall, M., Springer, S., & Putnam, C. (2013). Gate stack resistance and limits to CMOS logic performance. In: *Proceedings of the custom integrated circuits conference*. United States: Institute of Electrical and Electronics Engineers Inc. Available from https://doi.org/10.1109/CICC.2013.6658494.

Woerlee, P. H., Knitel, M. J., Van Langevelde, R., Klaassen, D. B. M., Tiemeijer, L. F., Scholten, A. J., & Zegers-Van Duijnhoven, A. T. A. (2001). Netherlands RF-CMOS performance trends. *IEEE Transactions on Electron Devices*, *48*(8), 1776–1782. Available from https://doi.org/10.1109/16.936707, 00189383.

Wu, C. H., Hunter, C., Bae, J., Kim, H., Chang, J., Sharpe, J., Ryu, I., Joo, S., Ha, B., Ko, W., Yim, J., Han, S., Kim, T., Yoon, D., Choi, I., Lee, S., Liu, Q., Kim, M., Lee, J., ... Cho, T.B. (2018). A 28nm CMOS wireless connectivity combo IC with a reconfigurable 2 × 2 MIMO WiFi supporting 80 + 80MHz 256-QAM, and BT 5.0. In: *Digest of papers - IEEE radio frequency integrated circuits symposium* (Vol. 2018-, pp. 300–303). South Korea: Institute of Electrical and Electronics Engineers Inc. Available from https://doi.org/10.1109/RFIC.2018.8428992.

Wu, S. Y., Lin, C. Y., Chiang, M. C., Liaw, J. J., Cheng, J. Y., Yang, S. H., Chang, S. Z., Liang, M., Miyashita, T., Tsai, C. H., Chang, C. H., Chang, V. S., Wu, Y. K., Chen, J. H., Chen, H. F., Chang, S. Y., Pan, K. H., Tsui, R. F., Yao, C. H., ... Ku, Y. (2015). An enhanced 16nm CMOS technology featuring 2nd generation FinFET transistors and advanced Cu/low-k interconnect for low power and high performance applications. In: *Technical digest – International electron devices meeting, IEDM* (Vol. 2015-, pp. 3.1.1–3.1.4.). Taiwan: Institute of Electrical and Electronics Engineers Inc. Available from https://doi.org/10.1109/IEDM.2014.7046970.

Yang, S., Liu, Y., Cai, M., Bao, J., Feng, P., Chen, X., Ge, L., Yuan, J., Choi, J., Liu, P., Suh, Y., Wang, H., Deng, J., Gao, Y., Yang, J., Wang, X. Y., Yang, D., Zhu, J., Penzes, P., ... Chidambaram, P.R. C. (2017). 10nm high performance mobile SoC design and technology co-developed for performance, power, and area scaling. In: *Digest of technical papers – Symposium on VLSI technology* (pp. T70–T71). United States: Institute of Electrical and Electronics Engineers Inc. Available from https://doi.org/10.23919/VLSIT.2017.7998203.

Yeap, G. (2013). Smart mobile SoCs driving the semiconductor industry: Technology trend, challenges and opportunities. In: *Technical digest – International electron devices meeting, IEDM* (pp. 1.3.8). United States. Available from https://doi.org/10.1109/IEDM.2013.6724540.

Yoon, J. S., & Baek, R. H. (2020). Device design guideline of 5-nm-node FinFETs and nanosheet FETs for analog/RF applications. *IEEE Access*, *8*, 189395–189403. Available from https://doi.org/10.1109/ACCESS.2020.3031870, http://ieeexplore.ieee.org/xpl/RecentIssue.jsp?punumber = 6287639.

Zhang, J., Wang, Z., Wang, R., Sun, Z., & Huang, R. (2021). Body bias dependence of bias temperature instability (BTI) in bulk FinFET technology. *Energy and Environmental Materials*. Available from https://doi.org/10.1002/eem2.12232, http://onlinelibrary.wiley.com/journal/25750356.

CHAPTER

# 4

# Gallium nitride technologies for wireless communication

*Nadine Collaert*

**IMEC, Heverlee, Belgium**

## 4.1 Introduction

Gallium nitride (GaN) is a semiconductor material that has garnered significant interest in recent years due to its unique electrical and optical properties. GaN has a wide bandgap, high breakdown voltage, and excellent thermal stability, making it well suited for a variety of applications, including power electronics, light emitting diode (LED) lighting, and wireless communication.

Its first application can be found in LEDs. In the 1960s, researchers at Bell Labs demonstrated that GaN could be used to create efficient blue and ultraviolet LEDs. This was a major milestone, as it paved the way for the development of GaN-based LED technology, which is now widely used in a variety of applications, including lighting, displays, and medical devices. But it wasn't until the 1990s (Asif Khan et al., 1994; Nakamura, 1995) that GaN truly found its success when challenges in epitaxial growth were overcome. Researchers at the University of Tokyo developed a process for growing high-quality GaN crystals using metal-organic chemical vapor deposition (MOCVD) (Nakamura et al., 1991). This process allowed for the mass production of GaN-based devices, such as transistors and power electronics, leading to further advances in the field. Compound semiconductors like GaN and InP (indium phosphide) at that time had to compete with the rising star of complementary metal-oxide-semiconductor (CMOS) technology and were always the next technology to come. The tremendous innovations in CMOS technology, making it more reliable, yielding and more cost-friendly, and thus more suitable for mass market applications, pushed the compound semiconductor technology into niche applications.

In the 2000s, GaN technology began to gain significant traction in the power electronics industry, with the development of high-voltage, high-frequency GaN transistors. These transistors offer several advantages over traditional silicon-based transistors, including higher switching speeds, higher power densities, and improved thermal stability.

*New Materials and Devices Enabling 5G Applications and Beyond*
**DOI: https://doi.org/10.1016/B978-0-12-822823-4.00004-2**

In the 1990s, the most commonly used substrate for the growth of GaN was sapphire ($\alpha$-$Al_2O_3$). This was due to its thermal expansion coefficient being close to that of GaN. Moreover, the substrate is optically transparent and thus ideal to fabricate LEDs.

Since the beginning of the 2000s a lot of effort has been spent to grow GaN transistors, also known as GaN HEMTs, on a silicon substrate using MOCVD. This new process enabled GaN transistors to be produced in the same existing factories as CMOS devices, using almost the same manufacturing processes.

In recent years, GaN technology has continued to advance at a rapid pace, with the development of novel devices and applications. For example, GaN-based photodetectors have been used in the development of high-speed communication systems, and GaN-based materials have been explored for use in next-generation batteries and fuel cells. GaN has also demonstrated the capability to be the displacement technology for silicon semiconductors in power conversion, radio frequency (RF), and analog applications, as well as being a contender to the silicon carbide (SiC) technology. And in general, a bright future is foreseen for GaN technology.

In this chapter, we will delve into the GaN technology, while the performance of GaN circuits will be discussed in Chapter 10. First, we will discuss the benefits of this technology and how a typical GaN transistor works. We will spend some time discussing the different flavors of the technology before discussing the use of these devices in power electronics. The specificities of using this technology for RF applications and its challenges will be discussed afterwards. And the last section of this chapter will give a summary.

## 4.2 Why gallium nitride?

### 4.2.1 Benefits of gallium nitride

Gallium nitride is a compound semiconductor material made up of gallium (group III) and nitrogen (group V). It is known for its wide bandgap of 3.4 eV, which is significantly larger than that of silicon (1.12 eV). This wide bandgap makes it an attractive material for high-power and high-frequency electronic devices.

One of the key properties of GaN is its high electron mobility, which allows it to carry high levels of current with low levels of resistance. This makes it particularly useful for high-power applications, such as power amplifiers and switching devices. The higher electron mobility also allows area scaling, as a GaN device can have a smaller size than a silicon device for a given on-resistance and breakdown voltage.

GaN is also known for its excellent thermal stability and resistance to high temperatures. This makes it well suited for use in aerospace and military applications, where reliable performance in extreme environments is critical. This thermal stability is due to the chemical bond structure between its atoms, where the gallium atoms are covalently bonded to the nitrogen atoms, forming a very strong chemical bond. GaN also has a high melting point of about 1500°C, which means that it can withstand high temperatures without degrading. Additionally, GaN has a smaller thermal expansion coefficient (coefficient of thermal expansion or CTE) than silicon, which means that it does not expand as much when heated, resulting in less strain on the material and reducing the risk of cracking or other types of damage. However, the thermal mismatch with Si or other substrates might introduce problems that need specific tailoring of the growth process when putting GaN on a different substrate.

In addition to its use in electronic devices, GaN is also used in the production of LEDs and laser diodes, which have a wide range of applications in fields such as lighting, communication, and medical treatment.

Despite its many advantages, GaN technology also has several limitations. One key limitation is its poor thermal conductivity, which is lower than that of silicon and even more so when compared to SiC. Specifically, GaN has a thermal conductivity of approximately 1.3 W/cm-K at 300K, while silicon and SiC have thermal conductivities of 1.5 and 4.9 W/cm-K, respectively. This necessitates careful layout planning and packaging strategies to effectively dissipate the heat generated by GaN-based devices.

Another disadvantage of GaN is its high cost of production. GaN-based devices are typically more expensive to manufacture than their silicon-based counterparts due to the specialized fabrication processes and equipment required. This increases the overall cost of production.

Additionally, the quality and performance of GaN-based devices can be affected by defects and damage that occur during the fabrication process. Depending on the process and substrate used, GaN can be highly sensitive to defects, which can negatively impact the performance of the devices. The material is also prone to cracking and other types of damage, which can reduce the yield of GaN-based devices, leading again to higher production costs.

## 4.2.2 Different flavors of gallium nitride

### *4.2.2.1 Introduction*

GaN technology comes in a variety of flavors depending on the starting substrate and the overall build-up of the layer stack. For example, GaN on silicon can offer cost-efficient solutions in power applications, while GaN on sapphire is used in LEDs, and SiC substrates allow the fabrication of high-efficiency RF power amplifiers. Pristine GaN substrates are also being used for fabricating vertical high-power transistors and diodes.

The limited availability and high cost of GaN substrates has led to the use of heteroepitaxy in the industry, which is a technique used to grow high-quality, single crystal films of GaN on non-native substrates. The goal of heteroepitaxy is to minimize defects and achieve a high-quality film on the substrate. However, this process can be complex as the lattice mismatch and thermal expansion coefficient between GaN and commonly used substrates such as sapphire, SiC, and silicon, can be significant. Therefore, a complex structure is often grown to account for these issues and ensure proper adhesion and growth of the GaN film. Additionally, the use of non-native substrates such as sapphire and SiC may require additional processing steps and specialized equipment, which can increase the cost of production.

### *4.2.2.2 Nucleation layer*

In this chapter, we will focus on power and RF devices. LED applications will not be covered. Fig. 4.1 shows a schematic view of a typical GaN device stack. A typical GaN device stack is composed of several layers, starting with the substrate which can be made

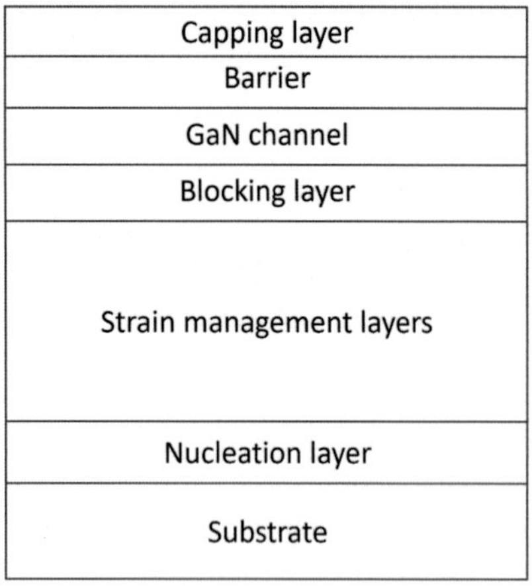

Capping layer
Barrier
GaN channel
Blocking layer
Strain management layers
Nucleation layer
Substrate

**FIGURE 4.1** Schematic view of a typical layer build-up for gallium nitride power or radio frequency devices.

of silicon, SiC, or sapphire. The first layer grown on the substrate is the AlN (aluminum nitride) nucleation layer. This layer serves several purposes, including improving the crystal quality and reducing defects in the GaN layer. AlN has a smaller lattice mismatch with GaN compared to Si or SiC, which can reduce the strain in the GaN layer and improve its structural quality. In addition, the AlN nucleation layer acts as a template for the GaN layer, promoting better crystal growth. This is particularly useful in the case of MOCVD or metal-organic vapor phase epitaxy (MOVPE) techniques, where the layers are grown at high temperatures above 1000°C (Dadgar et al., 2007). In these cases, the Ga can attack the Si substrate through a phenomenon called Ga melt-back, which makes it difficult to directly grow GaN on Si. The AlN layer acts as a barrier, protecting the Si substrate from the Ga-containing molecules and enabling the growth of GaN. Another advantage of the AlN nucleation layer is that it can be grown in situ, which means it can be grown in the same reactor as the GaN layer, which reduces the number of steps required for the fabrication of the device. Other intermediate layers have been explored, but the AlN layer has been found to be the most effective and widely used.

The nucleation layer is a crucial component of the GaN device stack, as the defects that originate in this layer can propagate through the entire stack and negatively impact the performance of the device. These defects, known as threading dislocations (TD), can degrade the electrical and optical properties of the device. Therefore, it is essential to optimize the growth and properties of the nucleation layer to minimize these defects.

However, growing the AlN nucleation layer is not without its challenges. Due to the reactivity and strong bond strength between Al and N, the growth of AlN has low surface mobility and is prone to islanding, where small clusters of material form on the surface. These islands then expand and merge, but the interface between these islands can be a source of defects such as 1D and 2D defects. In MOCVD, the AlN nucleation layer is often grown thick enough (50–150 nm) to ensure the lateral coalescence of the three-dimensional islands.

By increasing the pretreatment temperature and growth temperature, and reducing the V/III ratio, lateral coalescence of the AlN nucleate islands was observed, resulting in improved crystalline quality of the nucleation layer. This was attributed to the removal of the surface contamination and enhanced surface mobility of the Al atoms. In addition, using a high-temperature annealing step can promote the coalescence of the islands and reduce the defects.

Next to that, as we will see later when discussing the impact of the substrate on loss and linearity, the interface between AlN and for example, the Si substrate is of utmost importance. In the case of power applications, the lateral conduction at that interface can lead to a reduction of the breakdown voltage. In the case of RF applications, as mentioned before, this can lead to unwanted losses and harmonic distortion (HD) degrading the linearity.

It should be noted that the optimization and improvement of the interface between the AlN nucleation layer and the substrate is an ongoing research area and that new methods, techniques, and materials are being developed to improve the performance and reliability of the GaN devices.

#### 4.2.2.3 *Buffer layer*

On top of the AlN nucleation layer, a thick buffer layer is grown. The thickness of this layer can range from less than 2 to 10 μm, depending on the starting substrate, the application, and the required breakdown voltage. This buffer layer can be either a graded AlGaN (aluminum gallium nitride) layer or a superlattice (SL) consisting of several cycles of thin AlN/AlGaN layers. The purpose of this layer is to mitigate the difference in lattice constant between the GaN layer and the underlying substrate and to manage the strain in the stack.

As the thermal mismatch between GaN and its typical substrates can be high, the engineering of the stack is crucial to prevent severe wafer bending, cracking, and wafer breakage during and after growth. By designing a buffer layer that has an equal amount of compressive stress, this issue can be solved and flat GaN wafers can be processed. However, this method has some limitations, and the thickness of the buffer layer that can be grown is often limited.

Furthermore, the buffer layer also plays an important role in the electrical performance and reliability of the device. It can be used to adjust the electron concentration and electron mobility, which are essential for power and RF applications. Additionally, the buffer layer can also be used to adjust the polarization charge and the piezoelectric field.

The choice of the buffer layer material and thickness can also depend on the application of the device. For example, for high-voltage applications, a thicker buffer layer with a high electron concentration is typically used to improve the breakdown voltage. On the other hand, for RF applications, a thinner buffer layer with a high electron mobility is typically used as the requirements for breakdown voltage are much lower.

In the case of a graded buffer layer, an AlGaN layer is used where the Al content is reduced from 100% to 0% either continuously or in discrete steps. The layers with a lower Al concentration have a larger lattice constant than the higher Al-content layers at the bottom, which induces compressive stress. This compressive stress helps to counteract the tensile stress induced in the GaN layer due to the thermal mismatch between the GaN and the substrate.

In the case of a SL buffer, AlN interlayers are used. The growth of the GaN layer is thereby continuously interrupted by the AlN layers. The AlN layers are grown at a substantially lower temperature than the rest of the stack (e.g., 700°C) and no build-up of strain happens, thereby relaxing the AlN layer to its lattice constant. The GaN is then grown with standard (high) temperatures and induces strain. This process can be repeated several times, depending on the desired thickness and properties of the buffer layer.

However, the disadvantage of this approach is that the quality of the AlN layers is poor and can give rise to performance degradation of the devices. This is due to the fact that the AlN layers are grown at a lower temperature, which can lead to a higher density of defects and impurities in the AlN layers. Additionally, the SL buffer layer is more complex to fabricate and requires more process steps and time than the graded buffer layer.

An optimized version of the SL buffer is the strained SL buffer, where thin layers of AlN (typically 5 nm) and AlGaN (20–30 nm) are used. The layer thickness is kept below a critical value, thereby avoiding the problem as mentioned above. By changing the ratio of the thicknesses and the amount of cycles or repetitions, the strain in the buffer layer can be precisely controlled.

#### 4.2.2.4 Voltage blocking or confinement layer

On top of the strain management layers, a confinement or voltage blocking layer is grown. This back barrier between the GaN channel and buffer can be a Fe or C-doped GaN layer, an AlGaN layer, or a combination. The purpose of this layer is to prevent electrical breakdown of the device by blocking the flow of current through the device when the voltage exceeds a certain level. This helps to protect the device from damage due to excessive voltage and current, especially in high-power applications.

The confinement or voltage blocking layer is an essential component of the GaN device stack, as it helps to improve the reliability and longevity of the device. The choice of the material and the doping level of this layer can depend on the application of the device. For example, for high-power applications, a high doping level of Fe or C is typically used to improve the blocking voltage of the device. On the other hand, for RF applications, a lower doping level or the use of an AlGaN layer is typically used.

The second purpose of this layer is to provide good confinement of the 2DEG carriers in the GaN channel. The C-GaN layer in fact helps to enhance the gate control. This improves the performance and efficiency of the device by ensuring that carriers are not lost to other layers in the device. And finally, the back barrier also serves to provide carriers to the active layer.

An intentionally C-doped GaN layer is widely used as a back barrier in power electronics. The purpose of this layer is to define a layer with deep acceptor levels by introducing electron traps in the buffer layer. This reduces the leakage current and increases the breakdown voltage, allowing the device to handle higher power levels and improve its efficiency.

By adjusting the C-doping in the GaN back barrier layer, the acceptor levels and resistivity of the layer can be easily changed. This can be achieved through different methods, such as using different precursors or growth conditions during the MOCVD growth process. Alternatively, a dedicated dopant source can be used to introduce carbon into the layer. However, it's important to take into consideration the impact of the C-doping on the device performance when scaling down the devices. As the GaN channel is thinned down to reach higher frequencies, the 2DEG gets weaker due to an increased field and increased ionized C acceptors in the C-doped GaN layer. Therefore, careful optimization of the entire epi stack is necessary to maintain the desired device performance.

An alternative to using C-doped GaN is the use of Fe-doped GaN. However, Fe has a memory effect where Fe atoms segregate through the GaN layers, making it difficult to define abrupt interfaces in the epi stack. Additionally, Fe-doped GaN growth can be more challenging to achieve high crystalline quality as compared to C-doped GaN, leading to poor device uniformity and quality. Furthermore, Fe is a contaminant in silicon and is therefore prohibited in CMOS fabrication and processing, making it incompatible with GaN-Si technologies.

Another option for improving device performance and reliability is the use of AlGaN back barriers. These barriers, which can be created with just a few percent Al, form a high barrier through the band offset and polarization at the GaN/AlGaN interface, effectively confining the 2DEG carriers. This can lead to improvements in current collapse and overall device reliability. However, one drawback of using AlGaN back barriers is their lower thermal conductivity compared to pure GaN. This can result in higher peak temperatures in the device, which can be a concern for device longevity and performance. Additionally, AlGaN can be more difficult to grow than pure GaN and need more sophisticated techniques like MOVPE or MBE, which again increase the cost of the production process.

#### 4.2.2.5 Gallium nitride channel

The GaN channel is grown on top of the back barrier or voltage blocking layers. The typical thickness of the GaN channel is 200–300 nm. However, for RF applications, the gate length may need to be scaled in order to target a specific operating frequency. This gate length scaling should be accompanied by a corresponding scaling of the GaN channel and barrier layer to maintain control over the 2DEG and prevent short channel effects (SCE). Scaling the channel and barrier layers (top and back) appropriately can help to maintain the performance of the device and its reliability. Also, it is important to note that the scaling of the channel and barrier layers should be done in a way that minimizes crystal defects, such as TDs.

#### 4.2.2.6 Top barrier layer

A thin barrier is then grown on top of the GaN channel. The purpose of this layer is to create the 2DEG at the barrier/GaN channel interface. In power applications, AlGaN is used as a barrier material. The thickness is typically between 10 and 30 nm, and the Al content can range from 15% to 30% with a typical value around 25%. This yields a 2DEG between $5 \times 10^{12}/\mathrm{cm}^2$ and $1.5 \times 10^{13}/\mathrm{cm}^2$, and a mobility around 1700–1800 $\mathrm{cm}^2/\mathrm{V\text{-}s}$. Materials like InAlN (indium aluminum nitride) and ScAlN (scandium aluminum nitride) have also been considered. Especially InAlN is of interest for RF applications. This material system has several advantages, such as enhanced scalability and lattice matching to GaN if 17%–18% In is used (Medjdoub et al., 2006; Wang et al., 2010). The InAlN barrier delivers a much higher carrier density ($1.7 \times 10^{13}/\mathrm{cm}^2$), with a mobility around 1758 $\mathrm{cm}^2/\mathrm{V\text{-}s}$, resulting in a much lower sheet resistance $R_{sh} = 210\ \Omega/\mathrm{sq}$ (Peralagu et al., 2019) but with a two times thinner barrier. These layers are also grown at relatively low temperatures, between 700°C and 900°C using MOCVD. In comparison, the growth temperature for AlGaN layers is above 1000°C.

The strong 2DEG created at the barrier/GaN interface is a key benefit when it comes to the performance and scalability of devices. For a given identical thickness, InAlN can provide a two to three times higher 2DEG carrier density when compared to AlGaN. This higher carrier density can lead to improved device performance such as higher breakdown voltage, lower on-resistance, and increased efficiency. Additionally, the use of InAlN can also enable the use of a thinner layer to achieve the same carrier density as AlGaN, which can be beneficial for device scalability and cost-effectiveness (Hardy et al., 2017).

ScAlN has the potential to provide an even higher 2DEG compared to AlGaN. By using 18% Sc, the ScAlN layer can be lattice-matched to GaN, making it an ideal material for the growth of high-performance RF devices. Research has shown that HEMTs (high electron mobility transistors) with 120 nm gate length can achieve a transconductance of over 700 mS/mm and a $f_t$ (cut-off frequency) greater than 70 GHz. In addition, at 30 GHz, these ScAlGaN HEMTs deliver an impressive 5.77 W/mm output power and 47% power added efficiency (PAE) when tuned for maximum power and efficiency (Green et al., 2019, 2020; Kazior et al., 2019). However, growing ScAlN is challenging, due to the lack of a precursor with a vapor pressure high enough to be supplied in industrial MOCVD tools, which is one of the major obstacles to its widespread use.

Aluminum nitride (AlN) has the highest bandgap and carrier density of the III-nitride semiconductors but its lattice mismatch with GaN is high, leading to the need for thin layers to avoid defects. It is commonly used as a spacer layer between the AlGaN or InAlN barrier and the GaN channel. The AlN layer provides electrical insulation and sets the threshold voltage by increasing the effective conduction band offset between AlGaN/GaN, which leads to a larger 2DEG and improves the current. The additional conduction band offset at the interface increases the number of surface states above the Fermi level, leading to a larger 2DEG. This, in turn, improves the mobility of the electrons as the centroid of the charge is shifted away from the interface.

AlN/GaN HEMTs have demonstrated exceptional performance and reliability. In a study by (Kabouche et al., 2020), devices with a 4 nm AlN barrier and 110 nm gate length achieved a PAE of 50% at a drain-source voltage ($V_{DS}$) of 10 V, with a saturated power density ($P_{out}$) of 1.5 W/mm. The PAE slightly decreased to 45% at 20 V with a $P_{out}$ of 2.9 W/mm. Additionally, research by Then et al. (2019, 2021) on GaN MOSFETs utilizing an AlN-only channel region in combination with gate dielectrics yielded outstanding results.

#### *4.2.2.7 Impact of the starting substrate*

Sapphire ($\alpha$-$Al_2O_3$) is a commonly used substrate material for growing GaN, particularly for LEDs, due to its high thermal stability and chemical resistance. It was even utilized in the late 1960s during the pioneering years of GaN epitaxy. However, sapphire is an expensive substrate and growing GaN on it can be challenging due to the large lattice mismatch between the two materials, which is approximately 14%. This mismatch can lead to the formation of defects in the GaN layer, resulting in high defect densities of $\sim 10^{10}$ cm$^{-2}$. Furthermore, sapphire has poor thermal conductivity, making it unsuitable for high-power devices. For electronic applications, SiC and silicon substrates are more commonly used. SiC, in particular, has been widely used for most RF applications due to its relatively low lattice mismatch with GaN, around 4%, and its high thermal conductivity, which makes it suitable for high-power applications that require a large amount of heat dissipation.

SiC substrates can be fabricated using several methods, but one commonly used method is called physical vapor transport (PVT). In PVT, SiC powder is placed in a crucible and heated to a high temperature (above 2300°C) in a vacuum. As the powder melts and evaporates, it condenses on a cooled substrate, forming a single crystal SiC film. This process can be repeated multiple times to grow thicker films. PVT is a costly and difficult process, which is one of the reasons why SiC substrates are still 20 to 50 times more expensive than Si substrates. Additionally, upscaling the wafer size has proven to be challenging. This is why SiC substrates are still mostly limited to small sizes (4-inch, 6-inch) and why there are still quality issues such as high levels of defects and nonuniformities. However, with the success of SiC power devices, there has been a push to upscale to 8-inch wafers. These wafers have been demonstrated, but are not yet commercially available in large quantities. More cost-efficient and ecofriendly alternatives to traditional SiC substrates have also been developed, such as engineered substrates using the SmartCut approach (Olivier et al., 2021). This approach allows for the transfer of thin SiC layers to another substrate, such as silicon, which is more cost-effective and sustainable. By using this approach, only thin layers are used, not the full bulk SiC substrate.

Silicon is a widely used substrate material for semiconductor devices, but its crystal structure differs significantly from that of GaN. This makes it challenging to grow GaN on silicon without introducing defects. The lattice mismatch between Si and GaN is about 17%, so (111) Si substrates are often used for GaN growth. The (111) orientation has a closer lattice constant match with GaN, resulting in fewer defects compared to other orientations such as (100) and (110). Therefore, the (111) orientation is the preferred choice when growing GaN using MOCVD.

In addition to the lattice mismatch, the CTE of silicon is much lower than that of GaN, which can cause issues during device fabrication. The degree of lattice mismatch between GaN and AlN on silicon is similar to that of GaN on sapphire, at 19.2% (AlN to Si) and 17% (GaN to Si) respectively. Therefore, when AlN is used as a seeding layer on silicon, the growth scenario is similar to that of sapphire. This can result in a similar density of misfit dislocations at the interface between the Si substrate and the III-N material, which can degrade the performance and reliability of the device. Therefore, it is important to optimize the nucleation layer in order to minimize these dislocations.

As mentioned before, the mismatch in CTE between the GaN and silicon can lead to high levels of strain in the GaN layers, causing defects and poor device performance. The high strain can cause cracking and dislocations in the epilayer, resulting in decreased electrical performance, reliability, and yield. This can be a major issue for the commercialization of GaN-Si technology. Great progress in mitigating these issues has been made over the years in optimizing the buffer stack to counteract the strain and the use of thick Si substrates (1.15 mm) to provide higher mechanical stability.

Especially in the context of power applications, engineered substrates like QST have shown significant improvement in yield and reduction of wafer breakage. QST is engineered to mitigate stress in the epitaxial layers, enabling the growth of thick layers (tens of micrometers) of high-quality and low-defect-density bulk-like GaN on large diameter wafers with standard thickness. These engineered substrates consist of a ceramic (poly AlN) core that is CTE-matched to GaN. On top of this layer a thin Si (111) is transferred which acts as a seed or template to grow the GaN stack. The technology is designed to be scalable in terms of wafer diameter, with the ability to support growth on 6-inch, 8-inch, and even in the future 12-inch. While these engineered substrates have proven their value for power applications, the high cost still is an issue when it comes to their use for RF (Li et al., 2020).

#### 4.2.2.8 *Different growth techniques*

There are various methods for growing GaN stacks on different substrates, with two of the most popular being MOCVD and molecular beam epitaxy (MBE). The main distinction between these two methods lies in the way the thin films are deposited onto the substrate.

MOCVD utilizes metalorganic precursors that are vaporized and then react with the substrate to form the desired film. This method is known for its high growth rates and ability to produce high-quality layers. It is also the most commonly used in the production of GaN-based electronic devices. It is also considered as the most industrially relevant technique, especially for high-volume manufacturing.

MBE, on the other hand, uses a beam of atoms or molecules that are directed onto the substrate to form the film. This method allows for precise control over the growth of the film and can produce highly uniform and high-quality layers. However, it typically has lower growth rates compared to MOCVD. MBE is commonly used in the production of high-quality GaN-based optoelectronic devices and other specialized applications.

Hydride vapor phase epitaxy (HVPE) is a method used for growing compound semiconductor materials such as gallium arsenide (GaAs), InP, and GaN, and it was the first technique that was used to grow GaN. One of the key advantages of HVPE is its potential to significantly lower the cost of production when compared to the more traditional MOCVD. This cost reduction is achieved through a variety of means, such as reducing the consumption of $NH_3$, using less expensive source materials than in MOCVD, and minimizing capital equipment costs due to the high growth rate of HVPE. HVPE can grow thick layers of high-quality and low-dislocation-density bulk-like GaN on large diameter wafers. As such this technology is not only used to grow epitaxial layers on various substrates including sapphire and SiC, but also to create GaN substrates.

The above discussion shows that the GaN stack is a multilayered structure composed of various materials, each with its own specific function. In this work, we will be focusing on the stacks and device architectures used in power and RF applications, as opposed to those used in LEDs. The next section will delve into the operating principle of the GaN transistor.

### 4.2.3 Operation mode

The typical device architecture used in GaN technology is not the metal oxide semiconductor field effect transistor (MOSFET) but rather the HEMT. The structure of a GaN HEMT is shown in Fig. 4.2. HEMTs and MOSFETs are two of the most commonly used types of transistors for modern electronics. Typically, a HEMT offers much higher performance with lower power consumption and greater efficiency. HEMTs use a two-dimensional electron gas (2DEG) with very high electron mobility in the channel region.

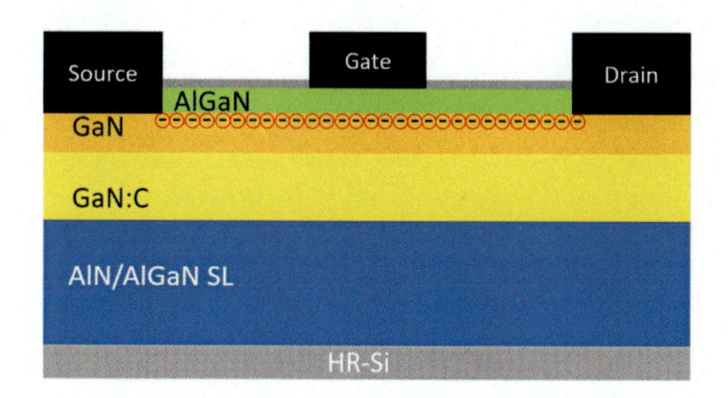


**FIGURE 4.2** Gallium nitride (GaN) high electron mobility transistor structure. Schematic view of a AlGaN/GaN device showing the heterostructure and source, drain and gate areas.

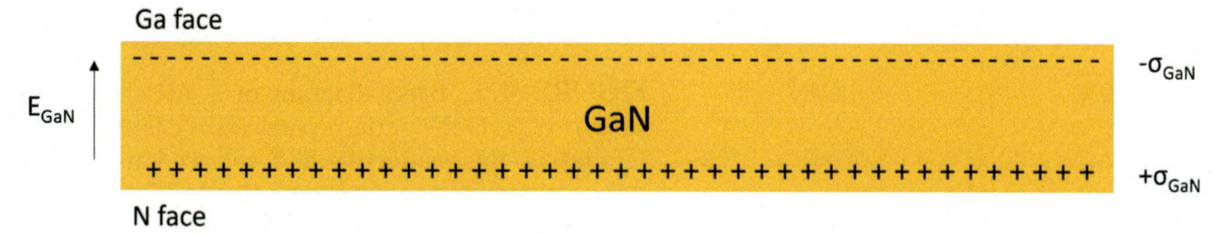


**FIGURE 4.3** Polarization in gallium nitride (GaN). Schematic view of the spontaneous polarization charges in GaN. $E_{GaN}$ is the electric field in the GaN layer.

This is created by sandwiching a channel material, often compound semiconductor like GaAs or GaN, between semiconductor materials with higher bandgap. In the case of GaN this is typically AlGaN or AlN.

A 2DEG, or two-dimensional electron gas, consists of a large number of electrons that are confined to a plane within a low-dimensional structure such as a semiconductor heterostructure. The high mobility of electrons in the 2DEG channel allows it to easily transport a large amount of current and to maintain the current even in the presence of high voltage, making it ideal for use in high-frequency applications.

In the following section, we first consider an AlGaN layer grown on a GaN channel. The two-dimensional electron gas that is formed in a GaN transistor is the result of the piezoelectric and spontaneous polarization that occurs in the AlGaN/GaN heterostructure. GaN shows very strong piezoelectric polarization which aids the accumulation of an enormous amount of carriers at the AlGaN/GaN interface. In GaN, spontaneous polarization arises due to the asymmetry of the wurtzite crystal structure, which is the most stable crystal structure of GaN (Fig. 4.3). It is a property of the material that results from the asymmetry of the atomic structure and the distribution of charges within the material. Specifically, the bond between gallium and nitrogen in the crystal lattice of GaN is ionic, with the nitrogen atom being more electronegative than the gallium atom. As a result, there is a slight positive charge on the gallium atoms and a slight negative charge on the nitrogen atoms, creating a dipole moment within the crystal. This asymmetry results in the presence of a spontaneous polarization electric field within the material. This spontaneous polarization cannot be changed by external fields, and the spontaneous polarization can have a number of important effects on the properties of GaN and the devices made from it. For example, it can affect the dielectric constant of the material, which determines the material's ability to store electrical charge. It can also affect the material's piezoelectric properties, which is the ability of the material to generate an electric charge in response to

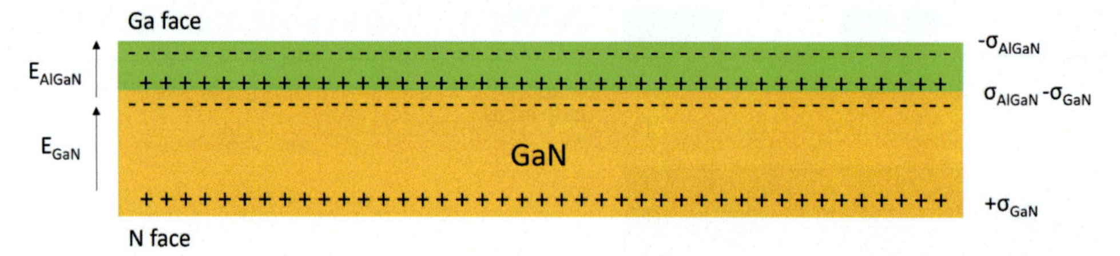


**FIGURE 4.4** Polarization in AlGaN/gallium nitride (GaN) structure. Schematic view of the spontaneous polarization charges in the AlGaN/GaN structure. $E_{GaN}$ is the electric field in the GaN layer and $E_{AlGaN}$ in the AlGaN layer.

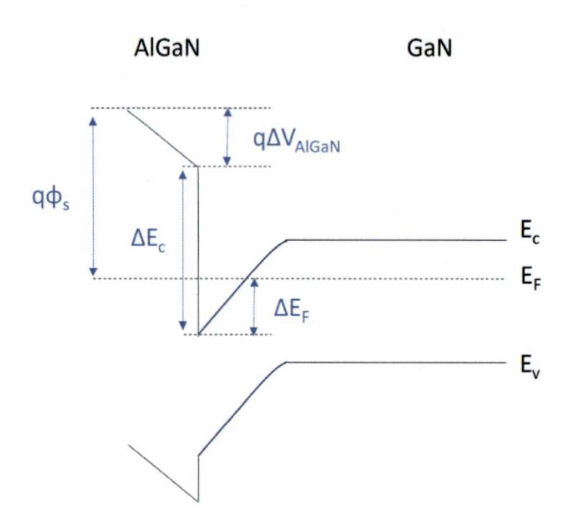


**FIGURE 4.5** Band diagram of a AlGaN/GaN heterostructure; $\varphi_s$ is the surface potential, $E_F$ is the Fermi level, $E_C$ and $E_V$ the conduction and valence band respectively; $\Delta V_{AlGaN}$ the voltage drop over the AlGaN layer; $\Delta E_C$ the conduction band-offset and $\Delta E_F$ the difference between the edge of the conduction band and the Fermi level.

an applied mechanical stress. The piezoelectric effect is caused by the strain in the crystal lattice, which results from the mismatch of atomic sizes between Al, Ga, and N atoms.

Both the AlGaN and GaN layers in a semiconductor heterostructure exhibit polarization, which results in a separation of positive and negative charges as can be seen in Fig. 4.4.

At the heterojunction between AlGaN and GaN, a discontinuity in the polarization of the electrons results in carrier diffusion. Free electrons from the AlGaN are attracted to the GaN channel and create an accumulation of carriers at that interface. At thermodynamic equilibrium, this creates a triangular well as can be seen in Fig. 4.5.

This potential well is a region of lower energy within the material that is surrounded by regions of higher energy. As a result of the potential well, electrons are confined to the interface between the AlGaN and GaN layers and are prevented from moving into the surrounding material due to the higher energy barriers. In the rest of this section, a Ga-face polarity is assumed for the layers when discussing the formation of the 2DEG.

There are three dipoles. Firstly, the AlGaN polarization-induced charge $\sigma_{AlGaN}$; secondly, the GaN polarization-induced charge $\sigma_{AlGaN}$; and thirdly, the 2DEG and the ionized surface charge $\sigma_s$ as is shown in Fig. 4.6. The latter can be accounted for by the unscreened fixed charge that is left by the ionized donors.

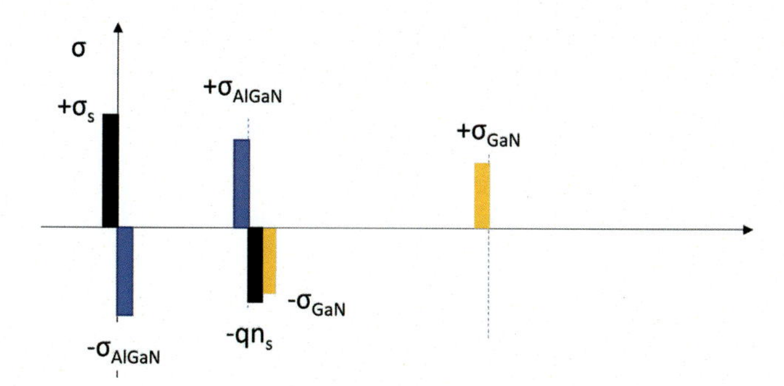


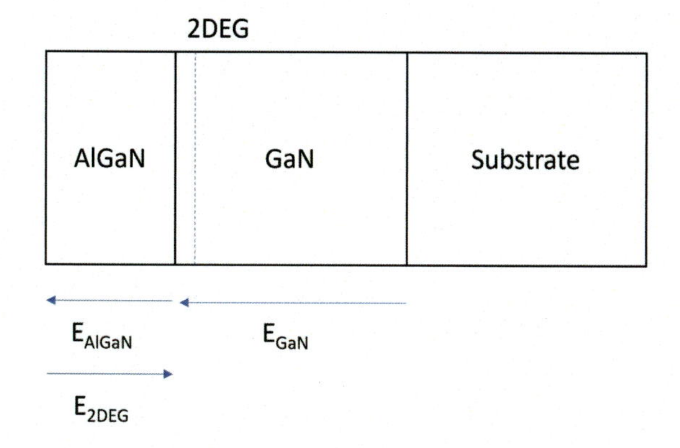


**FIGURE 4.6** The charge distribution. Charge distribution in a typical AlGaN/gallium nitride high electron mobility transistor device.

The polarization effect of an AlGaN layer grown on a GaN buffer generates positive charge at the interface between the two materials and a negative charge at the top of the AlGaN layer. Due to the lattice mismatch, the thinner AlGaN layer is under tensile strain and the total polarization charge at the AlGaN surface $\sigma_{\mathrm{AlGaN}}$ is therefore the sum of the spontaneous polarization $\sigma_{\mathrm{sp-AlGaN}}$ and the piezo-electric polarization $\sigma_{\mathrm{PE-AlGaN}}$.

$$\sigma_{\mathrm{AlGaN}} = \sigma_{\mathrm{sp-AlGaN}} + \sigma_{\mathrm{PE-AlGaN}} \tag{4.1}$$

Also in the GaN part, a uniform electric field is pointing from the bottom of the GaN layer toward the AlGaN/GaN interface.

Electrons will accumulate at the interface and contribute to the 2DEG formation. At the interface the net polarization charge can be defined as the sum of the polarization charge contributed by the AlGaN layer $\sigma_{\mathrm{AlGaN}}$ and the GaN channel layer $\sigma_{\mathrm{GaN}}$.

The polarization field is going to be stronger in the thin AlGaN as compared to the GaN layer and the net charge is therefore positive. This leads to the formation of an electron distribution close to the heterointerface, located at a distance Δd from the heterointerface.

$$\phi_s - \Delta V_{\mathrm{AlGaN}} - \frac{\Delta E_c}{q} + \frac{\Delta E_F}{q} = 0 \tag{4.2}$$

where $\Delta V_{AlGaN}$ is the voltage drop over the AlGaN layer, $\Delta E_C$ the conduction band offset, and $\Delta E_F$ the difference between the edge of the conduction band and the Fermi level. And with

$$\Delta V_{AlGaN} = \frac{(\sigma_{AlGaN} - \sigma_{GaN} - qn_s)d_{AlGaN}}{\varepsilon} \tag{4.3}$$

where $\varepsilon$ is the dielectric constant of the AlGaN. The channel electron density can then be derived as

$$n_s = \frac{\left((\sigma_{AlGaN} - \sigma_{GaN})d_{AlGaN} - \varepsilon\left(\phi_s - \frac{\Delta E_c}{q}\right)\right)}{qd_{AlGaN}} \tag{4.4}$$

An AlN interfacial layer or a GaN cap layer is often added into the AlGaN/GaN HEMT structure to achieve better performance. Following expressions of the 2DEG sheet density $n_s$ for these structures can then be obtained based on (Figs. 4.7 and 4.8).

$$n_s = \frac{\left(\frac{\sigma_{AlGaN}d_{AlGaN}}{\varepsilon_{AlGaN}} + \frac{\sigma_{AlN}d_{AlN}}{\varepsilon_{AlN}} - \phi_s - \frac{E_F}{q} + \frac{\Delta E_{C,AlN/GaN} - \Delta E_{C,AlGaN/AlN}}{q}\right)}{q\left(\frac{d_{AlGaN}}{\varepsilon_{AlGaN}} + \frac{d_{AlN}}{\varepsilon_{AlN}}\right)} \tag{4.5}$$

In a AlGaN/AlN/GaN HEMT, the thicknesses of the AlGaN and AlN layers can significantly affect the properties of the 2DEG that forms at the interface between the AlGaN

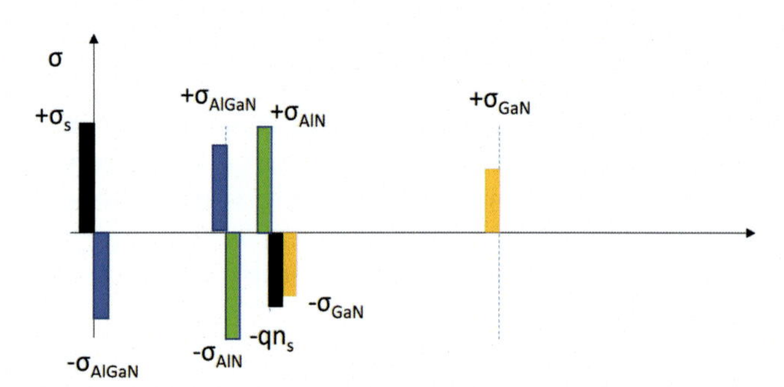


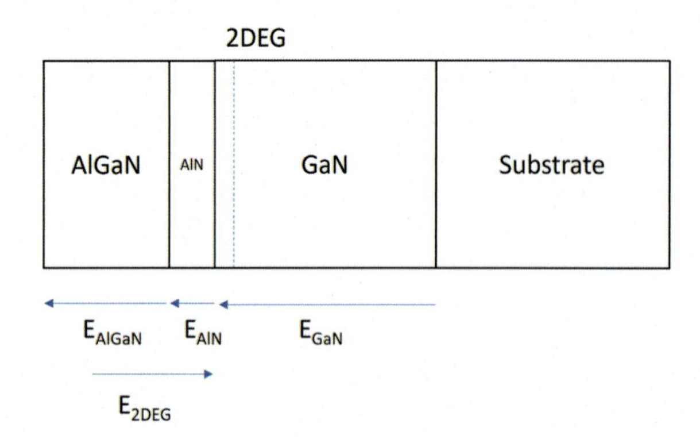


FIGURE 4.7 Charge distribution in a typical AlGaN/AlN/gallium nitride high electron mobility transistor device.

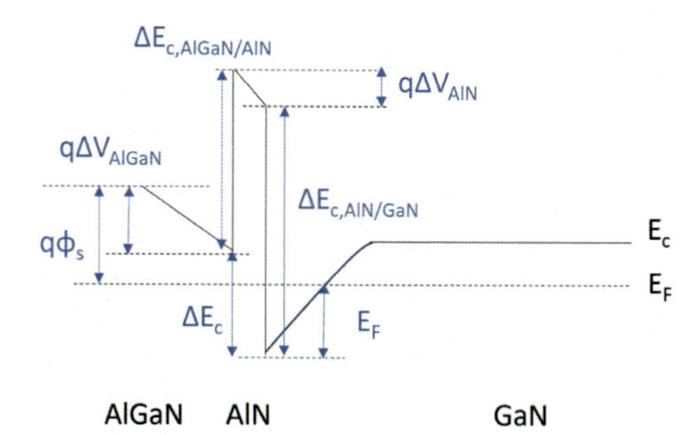


**FIGURE 4.8** Band diagram of a AlGaN/AlN/GaN heterostructure; $E_F$ is the Fermi level, $E_C$ and $E_V$ the conduction and valence band respectively; $\Delta V_{\text{AlGaN}}$ the voltage drop over the AlGaN layer; $\Delta V_{\text{AlN}}$ the voltage drop over the AlN layer; $\Delta E_C$ the effective conduction band-offset, $\Delta E_{C,\text{AlGaN/AlN}}$ the conduction band-offset between the AlGaN and AlN layer, $\Delta E_{C,\text{AlN/GaN}}$ the conduction band-offset between the AlN and GaN layer, and $\Delta E_F$ the difference between the edge of the conduction band and the Fermi level.

and GaN layers. The 2DEG in a HEMT is formed due to the polarization-induced charges that accumulate at the AlGaN/GaN interface. The polarization charges arise due to the different spontaneous polarizations of the AlGaN and GaN layers, which creates a built-in electric field that drives electrons to the interface and forms the 2DEG.

The thickness of the AlN layer in the AlGaN/AlN/GaN HEMT can affect the strength of the polarization field and the density of the 2DEG. A thicker AlN layer can lead to a higher polarization field, which can result in a stronger confinement of the 2DEG and a higher electron mobility. However, if the AlN layer is too thick, it can reduce the density of the 2DEG due to the screening effect of the polarization charges.

The thickness of the AlGaN layer can also affect the properties of the 2DEG. A thicker AlGaN layer can lead to a stronger polarization field, which can increase the density of the 2DEG. However, if the AlGaN layer is too thick, it can reduce the electron mobility due to increased scattering from impurities and defects.

In summary, the thicknesses of the AlGaN and AlN layers in an AlGaN/AlN/GaN HEMT can affect the properties of the 2DEG that forms at the interface between the AlGaN and GaN layers. A suitable balance between the thickness of the AlGaN and AlN layers is required to achieve optimal device performance.

### 4.2.4 Ga-polar versus N-polar

Ga-polar and N-polar refer to the orientation of the crystal lattice in GaN and its alloys such as AlGaN. The difference between these two polarities is the direction in which the crystal lattice is oriented, and this has important implications for the properties and performance of devices that utilize these materials.

Ga-polar GaN refers to a crystal lattice that is oriented such that the gallium atoms form the surface of the material. This is in contrast to N-polar GaN, which refers to a crystal lattice that is oriented such that the nitrogen atoms form the surface of the material. As such, Ga-polar GaN has a positively charged surface, while N-polar GaN has a negatively charged surface.

When it comes to growing Ga-polar and N-polar GaN, there are some differences in the substrates that can be used. Ga-polar GaN can be grown on a variety of substrates, including sapphire, SiC, and silicon. N-polar GaN, on the other hand, is typically grown on sapphire or SiC substrates, as these substrates provide the necessary lattice match for N-polar GaN.

The first N-polar FETs were made using MBE. Later, MOCVD was used to create smooth N-polar films on misoriented sapphire and SiC substrates. This not only reduced the formation of surface hillocks, but also improved the structural and optical properties, leading to the fabrication of MOCVD-grown FETs on both substrates.

Recently, high performing RF devices have been showcased using N-polar GaN, even up to high frequencies. In Liu et al. (2021), the authors discuss the advantages of using GaN-based HEMTs and monolithic microwave integrated circuits (MMICs) for W-band (75–110 GHz) wireless communication and imaging applications due to their superior material properties. While Ga-polar GaN has demonstrated good performance for W-band devices, N-polar GaN has additional benefits such as low ohmic contact resistance, strong back barriers, improved scalability, and the ability to achieve very high RF current density at low knee voltages. This has led to the demonstration of 8 W/mm power density at frequencies from 10 to 94 GHz with an associated PAE of 27.0% at 94 GHz (Romanczyk et al., 2018). Subsequently, Romanczyk et al. (2020) showed that by adding a thin passivation layer of plasma enhanced chemical vapor deposition (PECVD) SiN over the GaN capped access region, they were able to achieve better dispersion control, resulting in an increase in power density to 8.8 W/mm with an associated PAE of 27.0% at 94 GHz. The authors in Liu et al. (2021) pushed the efficiency even to 33% with a 6.3 W/mm output power. To achieve this PAE, the article proposes using an N-polar GaN deep recess MISHEMT with a new atomic layer deposition Ru metallization process to realize T-gates with 48-nm $L_G$ while maintaining the deep recess structure.

Innovations in device design using N-polar GaN have shown that the performance of these devices can be pushed to high frequencies. However, growing a N-polar GaN stack on silicon substrates is not straightforward. There is limited knowledge about the growth of N-polar nitrides on silicon substrates. One of the difficulties in having N-polar GaN on Si substrates is that the polarity of the GaN layer can be affected by the polarity of the underlying Si substrate, which is typically Si(111) and has a polar orientation that is opposite to that of N-polar GaN. This can result in the formation of inversion domains that degrade the performance of N-polar GaN-based devices.

In Keller et al. (2010), the authors conducted a study to investigate the growth of N-polar GaN and GaN/AlGaN/GaN heterostructures on (111) silicon substrates, and found that using misoriented substrates helped to mitigate the formation of hexagonal hillocks and achieve films comparable in quality to Ga-polar material.

In summary, Ga-polar and N-polar refer to the orientation of the crystal lattice in GaN and its alloys, with Ga-polar having a positively charged surface and N-polar having a negatively charged surface. N-polar GaN has shown advantages over Ga-polar for high-frequency RF devices due to its low ohmic contact resistance, strong back barriers, and improved scalability. Innovations in device design using N-polar GaN have demonstrated high performance at high frequencies, but growing N-polar GaN on silicon substrates remains challenging.

## 4.3 Applications

GaN has attracted significant attention in recent years due to its unique electrical and thermal properties. GaN has a higher breakdown field, higher electron mobility, and higher thermal conductivity than traditional semiconductor materials such as silicon, making it an ideal choice for specifically high-power applications.

One of the key applications for GaN technology is in power electronics. GaN-based power electronic devices, such as transistors and diodes, can operate at higher switching frequencies and higher temperatures than traditional silicon-based devices. This makes them ideal for use in high-power converters, inverters, and motor drives, which are commonly found in a variety of industries including renewable energy, automotive, aerospace, and industrial.

Another area where GaN technology has made significant strides is in RF applications. GaN-based RF devices, such as amplifiers and switches, have higher power handling capabilities and higher efficiency compared to silicon-based devices. This makes them suitable for use in high-power RF systems, such as cellular base stations, radar systems, and satellite communications.

Another important property of GaN is its radiation tolerance, which refers to its ability to continue operating after being exposed to ionizing radiation. GaN is generally considered to have good radiation tolerance, with a relatively low single-event upset rate. This means that GaN devices are less likely to suffer from errors or malfunctions after being exposed to radiation, compared to devices made from other materials.

GaN technology has, for a long time, also being explored for use in lighting applications. GaN-based LEDs and lasers have significantly higher luminous efficiency and lifetime compared to traditional lighting technologies, making them an attractive option for a variety of high-power lighting applications including automotive headlights, aviation lighting, and industrial lighting.

In addition to its use in power electronics, RF, and lighting applications, GaN technology is also being explored for use in high-power sensors and imaging systems. GaN-based sensors, such as pyroelectric detectors and thermopiles, have high sensitivity and fast response times, making them suitable for use in high-power sensing applications such as thermal imaging and flame detection.

For long, the adoption of GaN technology in high-power applications has been limited by the high cost of GaN-based devices, as well as the challenges associated with scaling the technology to high-volume production. However, significant progress has been made in recent years in terms of reducing the cost and improving the reliability of GaN-based devices, and it is expected that GaN technology will play a significant role in a variety of high-power applications in the future.

In summary, GaN technology has the potential to revolutionize the way we power and control high-power systems. Its unique electrical and thermal properties make it an ideal choice for use in power electronics, RF, lighting, sensing, and imaging applications. While there are still challenges to be overcome, it is expected that GaN technology will play an increasingly important role in a variety of high-power applications in the coming years. In the next section, we will briefly discuss the use of GaN in power applications, while Section 4.5 will be dedicated to RF GaN.

## 4.4 Gallium nitride for power applications

As the world moves towards more energy-efficient solutions, GaN power devices are becoming increasingly popular. These devices offer many benefits over traditional silicon devices, including higher efficiency, lower heat dissipation, and faster switching speeds. One of the major benefits of GaN power devices is their high efficiency. When compared to silicon devices, GaN power devices can operate at much higher voltages and currents while still maintaining a high level of efficiency. This allows for smaller and more efficient power systems. GaN power devices also have much lower heat dissipation than silicon devices. This is due to their ability to switch faster, which means that less heat is generated during operation. This can lead to cooler operating temperatures and longer device lifetimes. Finally, GaN power devices have much faster switching speeds than silicon devices. This means that they can respond more quickly to changes in load, resulting in a more efficient overall system. The advantages listed above make them a great choice for use in a variety of applications where efficiency and performance are critical.

Typical power applications include data centers, audio visual equipment, industrial equipment, reusable energy, home energy management systems, white goods, LED lighting, USB-PD (USB power delivery), electric vehicles and quick chargers, and contactless power transfer.

Wireless charging technology is one of GaN's most inventive applications; GaN's high efficiency minimizes power losses by delivering more energy to the receiving devices. These GaN-based systems typically consist of an RF receiver and a power amplifier that operate at a frequency of 6.78 or 13.56 MHz. GaN transistors achieve a solution with a very compact size as compared to conventional Si-based devices, which is essential for wireless charging applications. In drones, where the available space is limited, one such use is charging the drone while it is hovering over the charger from a close distance.

In the past, the main challenges for GaN were reliability and cost. With commercial devices operating at a junction temperature exceeding 200°C, the initial reliability concerns for power applications have been fully resolved, and these days a mean time to failure of over 1 million hours is achieved.

GaN also presents significant prospects for the data center industry, which is currently looking to achieve both higher performance and lower cost. Voltage converters are frequently used in data centers, where cloud servers run continuously, with typical values of 48 V, 12 V, and even lower voltages for powering the multiprocessor system cores. Power-conversion efficiency has emerged as a crucial consideration for businesses looking to achieve net zero—especially those running data centers and providing cloud computing services. This is due to the constantly rising worldwide electricity generation.

With the price difference between GaN and traditional silicon devices decreasing, GaN presents significant prospects for the data center industry. The price difference has dramatically decreased from initial GaN manufacturing on 2-inch, 4-inch, and 6-inch wafers and, most recently, 8-inch (200 mm) wafers. It is believed that GaN device manufacturing costs will continue to further decline thanks to recent innovations and continued process advancements, which will drive down their price even further.

## 4.5 Gallium nitride for wireless communication

### 4.5.1 Introduction

Previous paragraph demonstrates that GaN-Si has been widely adopted by the industry for power applications. This is, to this date, not yet the case for RF and specifically wireless communication. GaN-SiC is the most widely used technology and particularly for infrastructure. Next to that it is being used for defense and military purposes, and in satellite communication. An even more expensive and not widely available flavor GaN-diamond is expected to gain interest in applications where extreme high power needs to be generated. The benefit of diamond is its high thermal conductivity of around 2000 W/m-K, which is much higher than that of most traditional materials used for heat sinks such as aluminum (about 200 W/m-K) and copper (about 400 W/m-K).

Besides the use of GaN for infrastructure applications (e.g., base stations), the question of whether GaN-Si can be effectively used in handset applications remains open. Over the last few years, improvements in speed, efficiency and output power have been shown even for low voltages. Reliability and thermal aspects on the other hand remain challenges to overcome. In the next sections, we will give an overview of the fabrication of typical GaN-Si RF devices, the challenges related to reliability, thermal dissipation, and the impact of the Si substrate. Next, device modeling and the quest for enabling an E-mode device will be addressed, before comparing GaN to GaAs RF devices.

## 4.5.2 From lab devices to CMOS-compatible transistors

A typical Au-free GaN flow starts with the MOCVD growth of the buffer and the active layers (AlGaN/AlN or InAlN/AlN) as is shown in Fig. 4.9 and described in Parvais et al. (2020) and Peralagu et al. (2019). The starting substrate is a 200 mm high resistivity (HR) (111) Si substrate. This can also include the growth of a high temperature nitride layer to protect the top surface. It acts as passivation layer, etch stop, and can even be used as gate dielectric in MOSFET and MISHEMT/MOSHEMT devices. The benefit of this in-situ nitride cap has been discussed in Yu et al. (2021). Low surface state densities (DSS) varying between

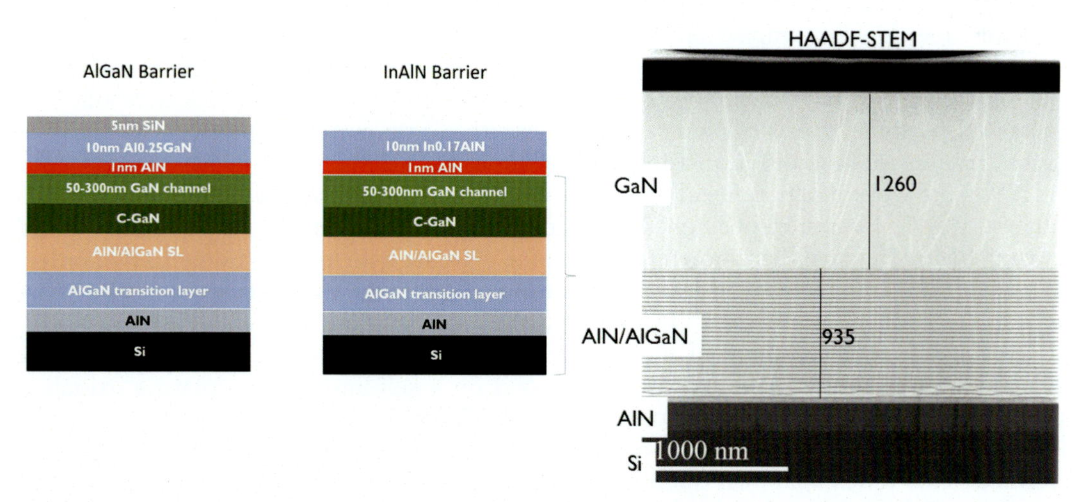


**FIGURE 4.9** Schematic picture of a typical AlGaN/AlN/GaN device stack (left) and InAlN/AlN/GaN device stack (middle); on the right a high-angle annular dark-field scanning transmission electron microscopy (HAADF-STEM) figure showing the ~2.5 μm buffer.

$1.4 \times 10^{12}$ and $4.7 \times 10^{12}$ $/eV - cm^2$ have been demonstrated which proves that successful AlGaN passivation with the in-situ SiN can be done. Another option can be to grow a thin GaN capping layer as passivation.

After that an oxide layer is deposited and nitrogen implantation is done to define the field areas. This implant effectively destroys the 2DEG in the field areas.

The study in Yu, Putcha et al. (2022) describes the impact of this N implant on the device performance and proposes a novel method to estimate the net active defect density caused by this implant. The implant is a three-step implant, and the net active defect densities that have been extracted in the GaN and AlN layers are $\sim 2 \times 10^{19}$ and $\sim 2 \times 10^{18}\,cm^{-3}$, respectively. The analysis of isolated heterostructures with varied AlGaN or AlN thicknesses indicates common electron leakage paths at the surface of GaN. The electrostatics of the leakage path is determined by an interplay between the high densities of defects created by the implants, the net sheet polarization charges between the III-nitrides, and the AlGaN surface states. Therefore, in general, careful optimization of this isolation implant is needed.

Then the flow moves into the gate module, starting with opening the dielectric on top of the GaN stack and defining the stem of the T-gate, doing a pre-clean, depositing an optional gate dielectric (in the case of a MOSHEMT/MISHEMT or MOSFET) before the gate metallization and patterning to form the T-gate. A low gate resistance ($R_g$) is critical for obtaining a high $f_{max}$ in RF devices. A thick Au layer can be utilized in a lift-off technique to readily achieve a low $R_g$; however, this method is incompatible with CMOS-like VLSI processing. Instead, a TiN/Ti/AlCu/Ti/TiN stack is used that is patterned using the subtractive litho/etch method as described before. This is a gate first flow. The benefit of a gate first flow is that one can prevent any potential ohmic contact degradation that is typical of a gate last flow, as this gate first technique can accommodate the possibility of MOSHEMT/MISHEMT high temperature anneals and passives integration. However, in a gate first flow the gate stack does see the thermal budget of the ohmic contact module, and this does have an impact on the gate resistance. The Ti in the gate stack mixes with the AlCu to create TiAl, which has a higher sheet resistance ($R_s$) than AlCu. By taking Ti out of the gate metal stack, the $R_s$ can drop from 0.55 to 0.2 $\Omega$/sq. As a result, $f_{max}$ increased by more than 30%, indicating the fact that gate metal optimization is a crucial factor for non-Au based RF transistors (Elkashlan et al., 2020).

TiN is the work function metal used in the gate stack. The Schottky gate leakage is high when a physical vapor deposition (PVD) TiN layer is used. The use of an ionized metal plasma (IMP) TiN layer has shown to reduce the off-state gate leakage by up to two orders of magnitude. The decrease in leakage is attributable to a 0.2 eV increase in the IMP TiN layer's work function over the PVD layer (Hu et al., 2016).

Local thinning or removal of the active layers (e.g., AlGaN) in the gate area can be achieved by an atomic layer etch method based on a self-limiting $O_2$- plasma oxidation and $BCl_3$-plasma etch, prior to depositing the gate dielectric and metallization.

After the gate module, another oxide layer is deposited and the ohmic contacts are defined. This relies again on opening areas in the dielectric, deposition of the contact layers and patterning of those layers. The ohmic contacts are defined by depositing Si/Ti/Al/Ti/TiN stack over a recessed barrier followed by alloying at a temperature of 565°C. By careful optimization of the barrier etch and post etch cleaning process contact resistance values as low as 0.15 Ω/mm with excellent uniformity have been shown (Parvais et al., 2020).

Scaling the front end of line (FEOL), and particularly the gate length $L_g$ and also the gate-to-source distance $L_{gs}$ and gate-to-drain distance $L_{gd}$, is required to push the device's performance to mm-wave frequencies.

A litho/etch technique is employed for the ohmic and gate modules in the FEOL used in the mm-wave flow, which is comparable to the flow reported in Peralagu et al. (2019). The flow presented in this publication had $L_g$ down to 200 nm. Taking a rule of thumb that the $f_t/f_{max}$ needs to be about 3 times to 5 times larger than the operating frequency, we see that for sub-6GHz $L_g$ down to 200–250 nm are required, while for mm-wave frequencies, preferably the gate length needs to be scaled down to 70 nm. Therefore, the lithography needs to be enhanced from I-Line/DUV (deep UV) to DUV/193 nm to get more scaled devices. Using 193 nm for the gate module, a minimum $L_g$ of 110 nm can be achieved as in Parvais et al. (2020). Further reduction of the gate length can be done by either trimming or using an internal spacer that is formed after trench etch in the dielectric and before gate stack deposition.

After the ohmic module, the flow moves into the back end of line (BEOL). The flow defined in Peralagu et al. (2019) uses a relaxed Al BEOL. Another approach is to use a Cu BEOL as shown in Parvais et al. (2020) which diverges from the typical Al-based approach. By using chemical mechanical polishing, an interlayer dielectric layer made up of a thick SiO layer is created and made planar.

While typically Schottky gate HEMT devices are used for RF applications addressing the power amplifiers, also other device architectures like MISHEMT/MOSHEMT and MOSFETs have been made and their performance compared (Peralagu et al., 2019). The IV characteristics for devices with an $L_g = 300$ nm are shown in Fig. 4.10. The HEMT has the lowest $I_{on}/I_{off}$ ratio of $10^3$ and improves to $10^4$ in the absence of AlN spacer, albeit at the expense of on-state performance. To lower the gate leakage $I_g$, gate metals with a higher work function are required like the use of IMP TiN or other materials like Ni or Ru (Liu et al., 2021). The MOSFET has an $I_{on}/I_{off}$ ratio of $10^8$, which is the highest of all the device architectures. After processing, the MISHEMT SiN cap in the gate stack is only 2 nm thick (this is the nitride layer in-situ grown after the buffer and active layers using MOCVD), which explains why the $I_g$ is significantly higher.

High temperatures are often used in the fabrication of AlGaN/AlN/GaN and even InAlN/AlN/GaN devices. During growth of the epitaxial stack, temperatures can rise well above 1000°C. This can lead to Al diffusion into the underlying GaN layer and can increase alloy disorder scattering (ADS) of the 2DEG, and degrade the 2DEG mobility (Yu, Parvais et al., 2022; Yu, Putcha et al., 2022). ADS is a significant mobility degradation mechanism and it is found that the scattering rate increases with the 2DEG density and the Al diffusion length. The observed trends in the literature that the 2DEG mobility and sheet resistances of GaN heterostructures degrade with increased thermal budgets, especially after >900°C processing, could be explained by the model presented in (Yu, Parvais et al., 2022; Yu, Putcha et al., 2022), and calls for lowering the temperature growth and processing where possible.

Fig. 4.11 shows the PAE versus width-normalized output power $P_{out}$, extracted from device-level load-pull measurements, comparing GaN-Si with GaN-SiC devices at

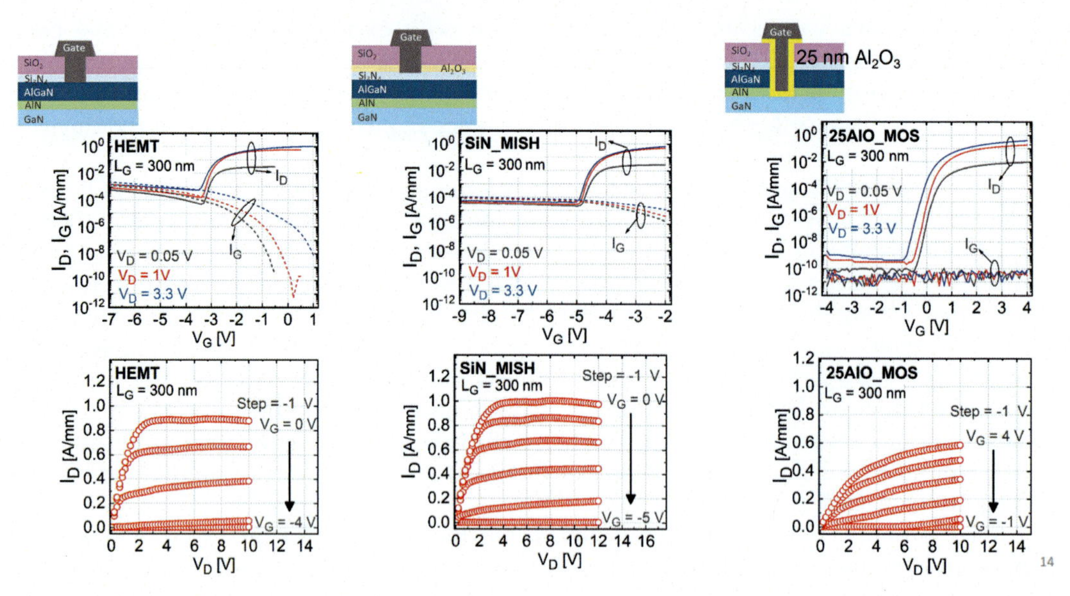


**FIGURE 4.10** Transfer and output characteristics of a (left) high electron mobility transistor, (middle) MISHEMT, and (right) MOSFET using AlGaN barriers, with $L_g/L_{gs}/L_{gs} = 0.3/1.5/1.75\ \mu m$. *Source: From Peralagu, U., De Jaeger, B., Fleetwood, D. M., Wambacq, P., Zhao, M., Parvais, B., Waldron, N., Collaert, N., Alian, A., Putcha, V., Khaled, A., Rodriguez, R., Sibaja-Hernandez, A., Chang, S., Simoen, E., & Zhao, S. E. (2019). CMOS-compatible GaN-based devices on 200mm-Si for RF applications: Integration and performance. In* Technical digest - International electron devices meeting, IEDM *(Vols. 2019-). Institute of Electrical and Electronics Engineers Inc. https://doi.org/10.1109/IEDM19573.2019.8993582.*

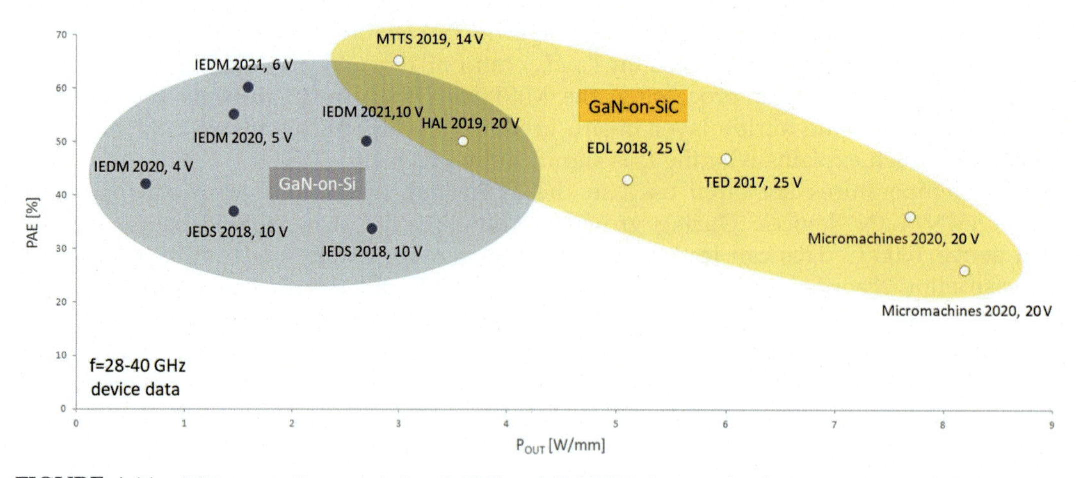


**FIGURE 4.11** PAE versus $P_{out}$ comparing GaN-Si and GaN-SiC devices; the data are extracted from load-pull measurements at 28–40 GHz at different $V_{DD}$.

28–40 GHz. The efficiencies of GaN-Si are becoming quite comparable to those of GaN-SiC, even at reduced bias conditions ($V_{DD} < 15$ V), which makes them interesting candidates for battery-powered user equipment applications.

However, there are still a number of challenges that need to be addressed before GaN-Si will be widely adopted in RF applications. Some of these challenges will be addressed in the next sections.

## 4.5.3 Challenges

### *4.5.3.1 Reliability*

Defects in GaN can occur due to a variety of factors, including impurities in the material, defects in the crystal structure, and damage caused by processing or handling. These defects can affect the electrical and optical properties of the material, and can lead to a range of performance issues in GaN devices. Surface defects resulting from a nonoptimized passivation can also have an impact on the device performance.

Some examples of the impact of defects on GaN devices include:

1. Decreased electrical conductivity: Defects in GaN can disrupt the flow of electrons through the material, leading to decreased electrical conductivity and reduced device performance. These defects can act as scattering centers that increase the resistance of GaN devices, leading to increased power dissipation and reduced efficiency.
2. Decreased breakdown voltage: Defects in GaN can reduce the breakdown voltage of the material, leading to decreased device reliability and performance.
3. Decreased lifetime: Defects in GaN devices can reduce their lifetime and increase their susceptibility to failure.

Overall, defects in GaN devices can significantly degrade their performance and reliability, and it is important to minimize or eliminate defects in order to produce high-quality GaN devices.

In Putcha et al. (2021), four kind of defects have been identified:

1. C/O/H impurities in the energy range of 0.43–0.49 eV.
2. Si dopants in the range of 0.37–0.4 and 0.59 eV that behave as acceptor traps, or gallium interstitials/nitrogen vacancies at ~0.4 eV that behave as donor traps.
3. Surface defects at the AlGaN/SiN interface where the SiN or any other material is used as passivation layer.
4. Deep defects such as gallium vacancy–oxygen complex (VGa–ON) at ~0.6 eV and nitrogen vacancies at ~0.61 eV.

Next to that, further investigation is still required to understand the physical origins of the defects with activation energy $E_A$ in the range of ~0.21 to 0.29 eV (shallow acceptors) and ~0.7 eV (deep acceptors).

As a result, current collapse, knee voltage walkout, drain and gate lag are the typical problems that can be associated with the defects in GaN devices.

There are several methods that can be used to characterize charge trapping phenomena in GaN devices. Some common methods include:

Deep level transient spectroscopy: This technique involves applying a voltage pulse to the device and measuring the resulting current response. The current response can be used to identify the presence of deep-level traps in the material, which are defects that can trap charge carriers.

Capacitance-voltage measurements: In this method, the capacitance of the device is measured as a function of applied voltage. Charge trapping can be detected by observing changes in the capacitance as the voltage is varied.

Photoluminescence measurements: In this method, the device is illuminated with light, and the resulting emission of light is measured. Charge trapping can be detected by observing changes in the intensity and spectral properties of the emitted light.

Thermally stimulated current measurements: In this method, the device is heated and the resulting current response is measured. Charge trapping can be detected by observing changes in the current as the device is heated.

Electron spin resonance measurements: In this method, the device is exposed to an oscillating magnetic field, and the resulting resonance signal is measured. Charge trapping can be detected by observing changes in the resonance signal as the magnetic field is varied.

These defects have as mentioned before an impact on the device performance, manifesting itself as degradation of the dynamic-$R_{on}$, current collapse or knee voltage walkout.

Dynamic-$R_{on}$ refers to the on-resistance of the device during switching operation, which includes both the steady-state on-resistance and the transient resistance during the turn-on and turn-off transitions. It is an important parameter to consider for power switching applications as it affects the power dissipation and efficiency of the device. A low dynamic-$R_{on}$ is desirable.

Current collapse refers to a sudden and dramatic decrease in the current flowing through the transistor when the device is operated at high power levels. Current collapse can have significant negative impacts on the performance of RF devices, which operate at high frequencies and typically require high levels of power. In RF devices, current collapse can lead to a sudden and dramatic decrease in the output power, which can severely degrade the device's performance. Current collapse can also affect the efficiency of RF devices, as it can result in the dissipation of large amounts of power in the transistor. This can lead to increased power consumption and reduced battery life in portable devices, as well as higher operating temperatures and increased thermal management requirements in larger systems.

To prevent current collapse in RF devices, it is important to carefully design the devices to minimize the likelihood of current collapse occurring. This can involve optimizing the device structure and fabrication process, as well as carefully controlling the operating conditions of the transistor. In addition, it may be necessary to use protective measures such as limiting the power levels of the device or using circuit-level techniques to prevent current collapse from occurring.

Knee voltage walkout relates to an increase in knee voltage of the device. The term knee voltage refers to a specific voltage point on a current–voltage curve of a transistor. The knee voltage is where the device transitions from a low-conduction state to a high-conduction state or vice versa.

All phenomena discussed above can be related to some extent to the existence of surface and buffer defects.

Surface traps can be controlled by field plates, but bulk traps originate from the deep level dopants in the buffer introduced intentionally to control leakage and SCEs. In Putcha et al. (2021) the complex role of C/O/H impurities and the acceptor/donor energy levels associated with other buffer defects in the DC-RF dispersion, observed for TLM devices subjected to buffer stress bias conditions, was reported. In this work, a new, more robust optimization method for defect spectroscopy was developed to study the charge-trapping kinetics of defects in GaN-Si buffers. Various energy levels that cause dynamic$-R_{on}$, with an activation energy $E_A$ ranging from 0.2 to 0.7 eV, are extracted from the time constant spectra and indicate the presence of both acceptor and donor type defects. The heat maps and time constant spectra were also used to study the impact of the buffer stack configuration and GaN channel layer thickness on dynamic$-R_{on}$. On the other hand, dynamic$-R_{on}$ was observed to increase for thinner GaN channels. It was attributed to the increasing proximity between defects in the buffer layers (plausibly GaN:C layer) and the 2DEG channel.

The charge-trapping kinetics associated with surface/interface defects have been previously described using a time-dependent "virtual-gate" extension model (Kaushik, 2012). This model describes the charge injection from the gate metal into surface defects via the Schottky-Read-Hall theory (Schroder, 2006) and the charge emission from surface defects through non-radiative multi-phonon theory (Grasser, 2014). The interaction of the 2DEG with defects in barrier and bulk/buffer layers depends on both the Fermi energy level ($E_F$) position and the temperature under stress conditions. Therefore, it is essential to accurately estimate the 2D potential profile in order to extract an accurate energy distribution of barrier defects under the required $V_{GS}$-$V_{DS}$ operating bias conditions. This potential profile can change considerably with distance from the gate edge in the drain side access region (Parvais et al., 2021).

The back barrier is a layer of material that is typically inserted between the GaN channel layer and the substrate. This layer serves several purposes as mentioned before. A typical back barrier is the C-doped GaN layer, but it can also be made of a different material than the GaN channel, such as AlGaN. In Yu, Parvais, Peralagu et al. (2022), it was observed that by increasing the 2DEG density in the GaN channel the $R_{on}$ dispersion can be reduced. In fact, a 50% increase in charge density results in a 30% reduction in $R_{on}$. Additionally, inserting an intrinsic AlGaN back barrier of 100 nm can reduce the dispersion even further.

While solutions at the device and material level can reduce the nonidealities like dispersion and memory effects, also at system level techniques exist to account for this nonideal behavior of the GaN devices.

Digital predistortion (DPD) is a technique used to compensate for the nonlinearities of power amplifiers (PAs) in wireless communication systems. It works by applying a predistortion function to the input signal before it is amplified, in order to cancel out the nonlinearities that will be introduced by the PA. This can improve the performance of the PA, particularly in terms of its linearity, and can help to reduce the amount of intermodulation distortion and error vector magnitude (EVM).

DPD is often implemented using digital signal processing algorithms, which can be implemented in software or hardware. It is a flexible and widely used technique that can be applied to a variety of PA architectures and can be adapted to changing system conditions. It is particularly useful in high-speed wireless systems, where the nonlinearities of the PA can significantly degrade the performance of the overall system.

#### 4.5.3.2 *Self-heating*

Self-heating refers to the phenomenon in which an object or system generates heat as a result of its own internal processes or activities. This can occur in a variety of contexts, and can be caused by a number of different factors.

One example of self-heating is the generation of heat in an electronic device due to the flow of current through its components. As current flows through a conductor, it can generate heat as a result of electrical resistance. This can lead to self-heating in electronic devices, particularly if they are operating at high power levels or have high resistance components.

Self-heating can also occur as a result of chemical reactions, such as those that occur during the operation of a battery or fuel cell. In these cases, the heat is generated as a byproduct of the chemical reaction taking place within the device.

Self-heating can be a concern in many different types of systems, as it can affect the performance and reliability of the system. For example, in electronic devices, self-heating can lead to thermal runaway, in which the temperature of the device increases rapidly and can cause damage or failure. In other systems, self-heating can lead to changes in the properties of materials or the operation of the system, and may need to be carefully managed or controlled.

Predicting the impact of self-heating on a device and determining the primary factors for enhancing its performance at the device level is a challenging task. Additionally, translating this behavior from the device level to the macro scale including the circuit, system, and packaging, poses another significant challenge.

In Vermeersch et al. (2022), an in-house Monte Carlo simulator that solves the Boltzmann transport equation (BTE) governing the microscopic heat carriers with first-principles phonon/electron dispersions and scattering rates was developed. This tool brings together phonon thermal transport in semiconductors, electron thermal transport in metals, and heat diffusion in amorphous materials into a unified particle-based simulation. Complementing its use in CMOS logic applications, this methodology is also highly suited for the thermal modeling of RF devices on two conceptual levels: in a hybrid way where effective thin film thermal conductivities obtained from 1D BTE analysis serve as input in conventional diffusive models, and through full 2D and 3D Monte Carlo thermal transport simulations of detailed device geometries.

This platform allows not only to see the impact of material choices (thickness, impact of starting substrate, etc) and GaN stack build-up but also the impact of layout (number of gate fingers, length of the gate fingers, pitch, etc.) on the self-heating of GaN devices. The tool serves as a means to predict and optimize these devices. In a next step, electrothermal simulations are envisioned where trade-offs in thermal dissipation and electrical performance can be explored and evaluated. This will allow for a deeper understanding of the interplay between self-heating and device performance, and will enable the development of more efficient and reliable electronic devices. Moreover, the tool can also aid in the design and optimization of thermal management strategies, ensuring that devices remain within safe operating temperatures while maintaining optimal performance.

#### 4.5.3.3 *Impact of the Si substrate on loss and linearity*

The finite resistivity of the Si substrate can degrade the performance of switches and power amplifiers, and passives. Therefore the GaN stack is typically grown on HR Si substrates. The work on RF-silicon on insulator (SOI) has shown that even with the use of HR-Si degradation of linearity and effective substrate resistivity $\rho_{\mathrm{eff}}$ can be seen coming from the substrate manufacturing process and subsequent device processing (Kerr et al., 2008; Lederer & Raskin, 2005; Neve & Raskin, 2012).

For RF front-end modules, $\rho_{\mathrm{eff}} > 3\ \mathrm{k}\Omega \cdot \mathrm{cm}$ is required to achieve quasi lossless substrates (Neve & Raskin, 2012).

Despite using a HR-Si substrate, it was shown that the interface loss is dominated by the parasitic conductive channel formed in the Si substrate near the interface between the epitaxial layer (AlN nucleation layer) and Si substrate. This occurs because thermal diffusion of Al and Ga into the Si substrate from the epitaxial layer creates a parasitic conductive channel through which current can flow. This is still one of the major challenges for GaN-Si devices.

This conductive nature of Si leads to increased substrate crosstalk, RF losses, and nonlinearities. The top buffer layers can be doped with a high concentration of Fe or C atoms to increase their resistivity. However, these dopants create deep traps throughout the buffer and lead to issues with reliability and performance loss as explained in the section on reliability.

Next to that, in Cardinael et al. (2022) stress/relaxation sequences were used to characterize the transient behavior of RF losses in a GaN-Si HEMT stack. It was shown that relaxation times in reaction to a chuck bias step can exceed 1000 seconds, during which $\rho_{\mathrm{eff}}$ can vary by a factor of 2. The temperature dependence of nonmonotonous features was linked to the redistribution of charge trapped in the deep levels of the C-doped GaN buffer. Finally, properly considering these transient phenomena was shown to be necessary for accurate substrate nonlinearities characterization and modeling in GaN-Si substrates.

### 4.5.4 D-mode versus E-mode

Traditional AlGaN/GaN HEMTs are depletion mode (D-mode) devices as a high-density 2DEG forms in the AlGaN/GaN interface.

A D-mode transistor is a type of transistor that is normally in the on-state. This means that the transistor is normally allowing current to flow between the source and drain terminals, and, for an n-type transistor, a negative voltage needs to be applied to the gate terminal, to turn the transistor off, blocking the current flow between the source and drain terminals. The challenge lies in developing a HEMT that is normally off. This would reduce overhead at circuit level to enable the negative voltages needed to switch off the device.

These enhancement mode devices (E-mode) typically have lower performance than D-mode devices. Especially for InAlN devices, designing a device with positive threshold voltage $V_T$ is more challenging than in AlGaN/GaN devices. Typically, the stronger 2DEG in these devices leads to more negative $V_T$.

There are several ways to create a E-mode device, either by engineering the epi-stack or the gate stack. The former involves the use of a p-type buffer and/or channel, thinning the GaN channel and introducing a back barrier. The latter relates to the use of a gate dielectrics in combination with barrier thinning, and/or the use of work function tuning. Especially the use of gate dielectrics has shown to reduce the gate leakage current significantly and as well as increase the breakdown voltage. One of the key challenges in fabricating GaN MISHEMTs/MOSHEMTs and MOSFETs is finding an appropriate gate dielectric material. Typically combinations of $Al_3O_2$ and $HfO_2$ are used. In Ye et al. (2013) and Then et al. (2019, 2021), GaN transistors with a high-k gate dielectric with good performance have been shown. In the former work, $ZrO_2$ was used as gate dielectric, and while still negative threshold voltages were shown, reduction in gate leakage and increase in gate voltage swing were shown.

The latter work by Intel on 300 mm Si shows E-mode devices with $V_T$ around 0.5 V using a composite high-k layer.

In power devices, p-GaN HEMT is used as an E-mode device. This requires the deposition of a p-type (Mg-doped) GaN or AlGaN layer on top of the barrier, either blanket, after the growth of the buffer and active layers, or selectively in the channel area.

Gate stack reliability is an important consideration in the design and fabrication of GaN-based devices, as the gate stack plays a crucial role in determining the performance and reliability of the device. One of the main issues related to gate stack reliability in GaN devices is the high electric field strength that can be present in the device. This high electric field strength can lead to a phenomenon known as "hot electron injection" in which high-energy electrons are injected into the gate dielectric layer and become trapped there, leading to the formation of defects and degrading the reliability of the device.

Another issue that can affect gate stack reliability in GaN devices is the high temperature and high current density that can be present in the device during operation. These conditions can lead to the formation of defects in the gate stack and degrade its reliability.

There are several methods that can be used to characterize the reliability of the gate stack in GaN devices, including electrical measurements such as current-voltage (I-V) and capacitance-voltage (C-V) measurements, as well as microscopic techniques such as transmission electron microscopy and atomic force microscopy.

There are several approaches that can be used to improve the reliability of the gate stack in GaN devices. These include the use of novel gate dielectric materials with lower defect densities, the use of passivation layers to prevent the injection of hot electrons into the gate dielectric layer, and the optimization of the device design (e.g., use of a field plate) to reduce the temperature and current density during operation.

In conclusion, while traditional AlGaN/GaN HEMTs are depletion mode devices, the development of enhancement mode devices is critical for reducing overhead at the circuit level and enabling good switches. Designing an HEMT that is normally off is a challenging task, but various approaches can be taken, including engineering the epi-stack or the gate stack. Gate stack reliability is a critical consideration in the design and fabrication of GaN-based devices, as it plays a crucial role in determining device performance and reliability. Characterization techniques and approaches to improving gate stack reliability are available and continue to be studied to develop more efficient and reliable GaN devices.

### 4.5.5 Device modeling

To fully understand the trade-offs in device design, and see the impact at circuit level, we need physics-based models to account for nonidealities. These nonidealities come from the increased defect density, reduced thermal conductivity in the case of GaN-Si, as compared to GaN-SiC, and substrate losses as we have seen in the sections above. In Parvais et al. (2021), the model requirements to establish a Design-Technology Co-Optimization (DTCO) loop for GaN devices are reviewed that give insight if nonidealities can be dealt with at circuit and system level by using techniques such as DPD, or if innovations at material and device level are needed.

DTCO is a methodology first introduced in CMOS scaling. It is a process used to develop new semiconductor process nodes. It combines both design and technology considerations to optimize the performance, power, and area of the process node. DTCO encompasses all aspects of the semiconductor process, including logic, static random access memory (SRAM), analog, IO, and other components. It is used to optimize library cell PPA (power performance area) estimates for gate-all-around devices and new transistor architectures. DTCO analysis is also used to ensure the manufacturability of the semiconductor process. A schematic view of the DTCO loop is shown in Fig. 4.12.

Different from logic applications, establishing a DTCO loop for RF applications is more challenging. The appropriate figures-of-merit (FOMs) need to be considered, next to the availability of accurate models for example, when considering power amplifiers a good model for PAE and output power needs to be there, but until now large-signal modeling remains a challenge. Next to that, transistor-level modeling requirements need to be understood. This allows for instance to study the impact at circuit level of any technology booster like the use of compound semiconductors, being introduced into an RF application.

As the RF-FEM is also becoming more and more heterogeneous, co-optimization and co-design of antenna, package, and ICs is a must and calls for multidomain EDA tools and hybrid Process Design Kitts.

Technology Computer Aided Design (TCAD) and compact models are at the center of the DTCO loop. Having good models than can accurately represent the device behavior, both the intrinsic and extrinsic but also the impact of all nonidealities is imperative.

For CMOS, BSIM is typically used. BSIM is short for Berkeley Short-Channel IGFET Model, which is a compact model for predicting the electrical behavior of MOSFETs.

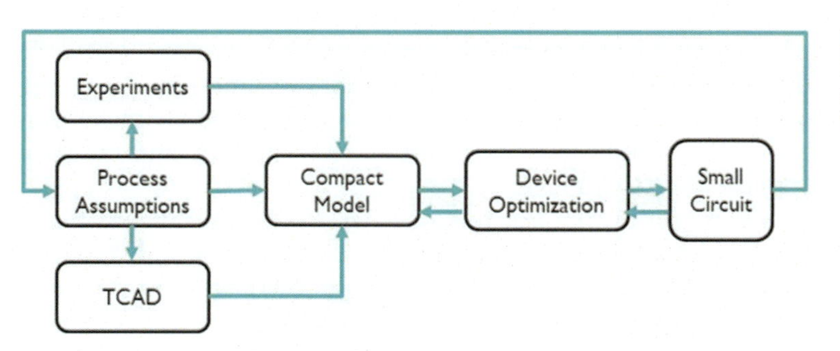


**FIGURE 4.12** The DTCO loop concept: from device modeling and exploration to benchmark circuits.

It was developed at the University of California, Berkeley and is widely used in the semiconductor industry to simulate and design MOSFETs and other types of integrated circuits. The BSIM model is based on physical principles and takes into account various factors that affect the performance of MOSFETs, such as gate length, channel width, and doping levels. It is an important tool for optimizing the performance of MOSFETs and other devices in the design of integrated circuits.

There are a number of similar models that have been derived for GaN HEMTs.

The ASM-HEMT model (advanced spice model for high-electron-mobility transistors) is a physical model that is used to predict the electrical and thermal behavior of GaN HEMTs.

The ASM-HEMT model is based on a set of equations that describe the transport of carriers (electrons and holes) in the GaN material, the generation and recombination of carriers, and the interactions between carriers and phonons (lattice vibrations). It takes into account the multi-region nature of the GaN HEMT, including the channel, barrier, and buffer regions, and it is able to predict the device's electrical and thermal behavior as a function of temperature, voltage, and current.

The ASM-HEMT model is a powerful tool for predicting the performance of GaN HEMTs and optimizing their design for specific applications. It has been widely used in the design of GaN-based power electronic circuits, such as inverters and motor drives, and in the prediction of the performance of GaN-based RF devices, such as amplifiers and switches.

There are alternative models like the Massachusetts Institute of Technology (MIT) virtual source GaNFET-high voltage (MVSG-HV) model. This model was developed by researchers at the MIT. One of the advantages of the MVSG-HV model is that it is able to accurately predict the electrical behavior of GaN HEMTs at high voltage and high frequency, which is important for RF devices that operate in these regimes. The model is also able to capture the nonlinear behavior of the device.

One of the challenges of using the MVSG-HV model for RF devices is that it requires a detailed understanding of the device's physical properties, which can be challenging to obtain in some cases. Additionally, the model is computationally intensive, which can make it challenging to solve the equations in real time.

Regarding the FOMs, when assuming a power amplifier, the cut-off frequency $f_t$, the unilateral power gain $f_{max}$, the PAE and the output power $P_{out}$ of the transistor are the most important metrics.

Next to that, the nonlinearity of the devices and circuits needs to be modeled, not only the compression points (P1dBs) but also the HD as well as AM/PM. These parameters must be known to predict the circuit EVM. Compound semiconductor devices also suffer from higher defectivity than group IV materials, which means they are more prone to failure. The trap levels in the stack induce a time dependence of the device characteristics and together with thermal effects reliability must be carefully characterized and modeled to assess the real impact at device and circuit level.

The advancements in metrology within the last few decades have made it possible to check nonlinear models across a wide range of frequencies. Nonlinear vector network analyzer (VNA) can be employed for extracting direct FOMs such as AM/AM and AM/PM. Both AM/AM and AM/PM are important metrics in characterizing the linearity of a device or system. AM/AM stands for amplitude modulation to amplitude modulation and refers to the change in output amplitude of a device or system as a function of input amplitude. AM/PM stands for amplitude modulation to phase modulation and refers to the change in output phase of a device or system as a function of input amplitude.

These methods are also useful during model calibration and validation stages. As soon as compact models take into account all relevant effects mentioned above, they become ideal tools to assess the circuit FOM in correlation with technology elements. This provides helpful feedback from the technology level towards the design phase while allowing feed-forward from the circuitry development stage.

When grown on a Si platform, substrate modeling brings another challenge since lossy substrates affect the quality factor of passive components, but also the PA performance (Yamaguchi et al., 2016). The small- and large-signal characteristics of substrates are studied from coplanar transmission lines. The direct relationship between RF losses and distortion observed in SOI substrates does not hold in the case of GaN-Si stacks comprising a multi-layer semiconducting buffer, even if competing performance has been reported (Yadav et al., 2020). The strong hysteresis observed in Fig. 4.13 and described in Cardinael et al. (2022) is attributed to the long emission time constants of traps located inside the GaN buffer layers (Cardinael et al., 2022). Therefore, the prediction of substrate nonlinearity as a function of DC bias is more challenging and requires precise modeling of the buffer dynamic effects. Although the substrate nonlinearity can be modeled using normal TCAD (Rack & Raskin, 2017), the approach is only applicable to transverse electromagnetic mode propagation since there isn't a simulator that combines semiconductor and Maxwell's equations as of yet.

The large signal characterization of a RF device relies on measuring the performance of the device under high power and large signal conditions. Large signal characterization is important because it allows the device to be tested and evaluated under conditions that are similar to its normal operating conditions.

There are several methods that can be used to perform large signal characterization of an RF device. One method is to use a VNA. A VNA is a type of test equipment that measures the transmission and reflection characteristics of an RF device over a wide range of frequencies. The VNA sends a swept frequency signal through the device and measures the transmission and reflection characteristics at each frequency. This allows the device to be characterized over a wide range of frequencies and under different load conditions.

Another method for large signal characterization is to use a load-pull system. A load-pull system is a specialized test setup that is used to measure the performance of RF devices under different load conditions. The load-pull system consists of a signal

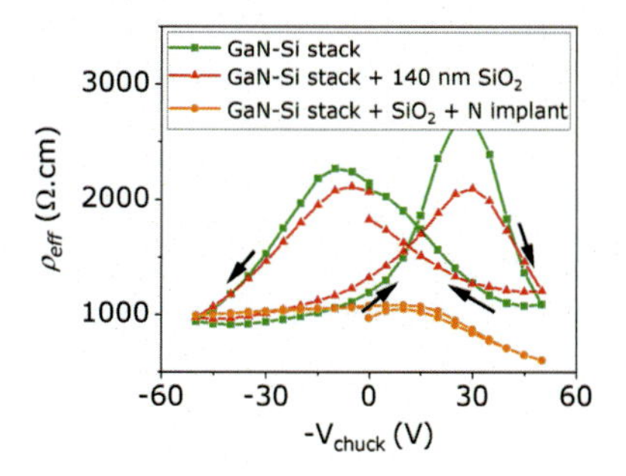


**FIGURE 4.13** The effective substrate losses measured on CPW lines with 50 μm separation versus chuck bias comparing various GaN-Si stacks, after growth, growth and oxide deposition, and after oxide deposition and N implantation. *Source: From Parvais, B., Elkashlan, R., Yu, H., Sibaja-Hernandez, A., Vermeersch, B., Putcha, V., Cardinael, P., Rodriguez, R., Khaled, A., Alian, A., Peralagu, U., Zhao, M., Yadav, S., Gramegna, G., Driessche, J. V., & Collaert, N. (2021). Transistor modelling for mm-Wave technology pathfinding. In* International conference on simulation of semiconductor processes and devices, SISPAD *(Vols. 2021-, pp. 247–250). Institute of Electrical and Electronics Engineers Inc. https://doi.org/10.1109/SISPAD54002.2021.9592530.*

generator, an RF device, and a variable load. The signal generator is used to generate a large signal that is applied to the input of the RF device. The RF device amplifies the signal and delivers it to the load. The performance of the RF device is then measured by measuring the output power and the efficiency of the device under different load conditions.

Next to measuring the large signal performance, enabling good models that can describe and predict the large signal behavior of a device are crucial, but this remains still one of the major challenges for high-power RF devices.

### 4.5.6 Gallium nitride versus gallium arsenides

One of the main competitors to GaN HEMTs is the GaAs HEMT, which is made from GaAs, a semiconductor material with similar properties to GaN.

GaAs is a semiconductor material that is commonly used in the production of high-speed and high-frequency electronics, such as wireless communication devices. GaAs has several properties that make it well suited for these applications, including a high electron mobility, a high breakdown voltage, and a direct bandgap, which allows for efficient emission of light.

Both GaAs heterojunction bipolar transistors (HBTs) and HEMTs are used in high-speed applications. HBTs are typically used for the lower frequencies (below 6 GHz) and HEMTs for the higher frequencies (above 6 GHz).

There are two main types of GaAs HEMTs: pseudomorphic HEMTs (pHEMT) and metamorphic HEMTs (mHEMT).

In pHEMT, the channel is typically made of InGaAs. The indium content in the InGaAs channel layer is usually adjusted to match the lattice constant of the underlying GaAs substrate, allowing for strain to be induced in the channel layer, which in turn enhances the carrier mobility. The InGaAs channel is capped with a AlGaAs layer leading to a heterostructure and the formation of a high-density 2DEG at the interface between the InGaAs and AlGaAs layers. In addition to AlGaAs, other materials that can be used in a pseudomorphic GaAs HEMT include InAlAs.

mHEMT uses a strain relaxed buffer layer (SRB) to accommodate the lattice mismatch between the substrate and the III-V channel material which is typically also a InGaAs layer. This allows for the use of InGaAs channels with a much larger lattice constant than the substrate, or in other words a larger In content in the InGaAs channel, leading to higher mobility and thus improved device performance.

The materials used in GaAs mHEMTs typically include a channel layer like InGaAs and barrier layer such as AlGaAs or InAlAs. The SRB is made of InGaAs or InAlAs, and is designed to gradually relax the strain induced by the lattice mismatch between the substrate and the channel layer.

mHEMTs offer several benefits over pHEMTs. They can be designed with wider range of bandgap materials and heterostructures, allowing for improved performance at higher frequencies and higher power levels. The use of metamorphic buffers in mHEMTs also reduces the density of traps at the interface between the channel and barrier layers, improving the device performance and reliability. However, there are also some drawbacks to using mHEMTs, like the higher fabrication complexity. The use of a metamorphic buffer layer adds complexity to the device fabrication process, making it more difficult and expensive to manufacture.

Overall, the choice between mHEMTs and pHEMTs depends on the specific requirements of the application, including frequency range, power levels, and device size. GaAs HEMTs have been widely used in the past few decades for their excellent high-frequency performance, but they face limitations in terms of power density due to the lower breakdown voltage and thermal conductivity of GaAs.

One of the key advantages of GaN HEMTs over GaAs HEMTs is their higher breakdown voltage, which allows for operation at higher voltages and higher output powers. This is especially important for high-power applications.

Another advantage of GaN HEMTs is their high thermal conductivity, especially when grown on SiC substrates, which allows for better heat dissipation and improved reliability.

However, there are also some drawbacks to GaN HEMTs. The epitaxial growth of GaN is more challenging compared to GaAs, which makes it more difficult and expensive to fabricate GaN-based devices. In addition, GaN HEMTs require more complex fabrication processes compared to GaAs HEMTs, which increases the cost and reduces the yield. Therefore, being able to process GaN-Si in a CMOS-compatible fab allows to reduce cost and improve yield. Finally, GaN-based devices are still relatively new compared to GaAs, which means that there is less mature infrastructure and less established design knowledge for GaN-based devices.

In summary, while GaAs HEMTs have been the standard for high-frequency applications for many years, GaN-based HEMTs are emerging as a promising alternative for high-power and high-frequency applications due to their high breakdown voltage, high electron mobility, and high thermal conductivity. However, the higher cost, more complex fabrication processes, and less mature infrastructure and design knowledge for GaN-based devices are still challenges that need to be addressed.

## 4.6 Summary

In this chapter, we explored the various applications of GaN technology and the benefits and challenges associated with its use. GaN is a compound semiconductor material that has a wide bandgap and high breakdown voltage, making it ideal for use in high-power and high-frequency electronic devices. Some of the key applications of GaN include RF power amplifiers, high-voltage power electronics, and high-brightness LEDs.

There are several different flavors of GaN, depending on the starting substrate used. These include GaN on silicon, GaN on sapphire, and GaN on SiC. Each type of substrate has its own unique properties and limitations, and the choice of substrate can have a significant impact on the performance of the final device. In particular, the use of GaN on silicon substrates has opened up the possibility of fabricating GaN devices in a CMOS fab, which can make them more yielding and cost-friendly. Optimization of the GaN devices also requires the development of new models and dedicated multidomain EDA (electronic design automation) tools to accurately predict the performance of GaN devices and assess their impact at circuit level and even system level.

While GaN has many advantages, there are still many challenges. One of the main challenges of using GaN is related to reliability. GaN materials can contain defects that can affect their performance and lifetime. In addition, self-heating can be a concern, which can lead to thermal runaway if not properly managed. Another challenge is the need to enable a normally off device with the same performance as the typical Schottky gate HEMTs standardly being used in GaN technology.

Finally we have also compared the performance of GaN devices to GaAs devices, which can be seen as the major contender to GaN in the mm-wave regime.

Overall, GaN technology offers many benefits for RF applications, including high-power density and high-frequency operation. However, it also presents some challenges, such as self-heating and the need for specialized processing techniques. Despite these challenges, the unique properties of GaN make it an attractive choice for a wide range of electronic applications, and it is likely to continue to be an important technology in the future.

## References

Asif Khan, M., Kuznia, J., Olson, D., & Bhattarai, A. (1994). Wide bandgap Ai x Ga 1x N material system for visible and ultraviolet optoelectronic devices. *Leos 1993 Summer topical meeting digest on optical microwave interactions/visible semiconductor lasers/impact of fiber nonlinearities on lightwave systems/hybrid optoelectronic integration and packaging/gigabit networks, LEOSST 1993* (pp. 21–22). United States: Institute of Electrical and Electronics Engineers Inc. Available from https://doi.org/10.1109/LEOSST.1993.696829, http://ieeexplore.ieee.org/xpl/mostRecentIssue.jsp?punumber = 700.

Cardinael, P., Yadav, S., Zhao, M., Rack, M., Lederer, D., Collaert, N., Parvais, B., & Raskin, J. P. (2022). Time dependence of RF losses in GaN-on-Si substrates. *IEEE Microwave and Wireless Components Letters*, *32*(6), 688–691. Available from https://doi.org/10.1109/LMWC.2022.3162028, Institute of Electrical and Electronics Engineers Inc Belgium, https://ieeexplore.ieee.org/servlet/opac?punumber = 7260.

Dadgar, A., Veit, P., Schulze, F., Bläsing, J., Krtschil, A., Witte, H., Diez, A., Hempel, T., Christen, J., Clos, R., & Krost, A. (2007). MOVPE growth of GaN on Si - Substrates and strain. Thin solid films. Germany, 515(10), 4356–4361. Available from https://doi.org/10.1016/j.tsf.2006.07.100.

Elkashlan, R. Y., Rodriguez, R., Yadav, S., Khaled, A., Peralagu, U., Alian, A., Waldron, N., Zhao, M., Wambacq, P., Parvais, B., & Collaert, N. (2020). Analysis of gate-metal resistance in CMOS-compatible RF GaN HEMTs. *IEEE Transactions on Electron Devices*, *67*(11), 4592–4596. Available from https://doi.org/10.1109/TED.2020.3017467, Institute of Electrical and Electronics Engineers Inc., Belgium, https://ieeexplore.ieee.org/xpl/mostRecentIssue.jsp?punumber = 16.

Grasser, T. (2014). The capture/emission time map approach to the bias temperature instability. *Bias temperature instability for devices and circuits* (pp. 447–481). New York, Austria: Springer. Available from https://doi.org/10.1007/978-1-4614-7909-3_17, http://doi.org/10.1007/978-1-4614-7909-3.

Green, A. J., Gillespie, J. K., Fitch, R. C., Walker, D. E., Lindquist, M., Crespo, A., Brooks, D., Beam, E., Xie, A., Kumar, V., Jimenez, J., Lee, C., Cao, Y., Chabak, K. D., & Jessen, G. H. (2019). ScAlN/GaN high-electron-mobility transistors with 2.4-A/mm current density and 0.67-S/mm transconductance. *IEEE Electron Device Letters*, *40*(7), 1056–1059. Available from https://doi.org/10.1109/LED.2019.2915555, Institute of Electrical and Electronics Engineers Inc United States, https://ieeexplore.ieee.org/servlet/opac?punumber = 55.

Green, A. J., Moser, N., Miller, N. C., Liddy, K. J., Lindquist, M., Elliot, M., Gillespie, J. K., Fitch, R. C., Gilbert, R., Walker, D. E., Werner, E., Crespo, A., Beam, E., Xie, A., Lee, C., Cao, Y., & Chabak, K. D. (2020). RF power performance of Sc(Al,Ga)N/GaN HEMTs at Ka-Band. IEEE electron device letters. United States, 41(8), 1181–1184. Available from http://doi.org/10.1109/LED.2020.3006035, https://ieeexplore.ieee.org/servlet/opac?punumber = 55.

Hardy, M. T., Downey, B. P., Nepal, N., Storm, D. F., Katzer, D. S., & Meyer, D. J. (2017). ScAlN: A novel barrier material for high power GaN-based RF transistors. *ECS Transactions*, *80*(7), 161–168. Available from https://doi.org/10.1149/08007.0161ecst, Electrochemical Society Inc, United States, http://ecst.ecsdl.org/.

Hu, J., Stoffels, S., Lenci, S., Bakeroot, B., De Jaeger, B., Van Hove, M., Ronchi, N., Venegas, R., Liang, H., Zhao, M., Groeseneken, G., & Decoutere, S. (2016). Performance optimization of Au-free lateral AlGaN/GaN schottky barrier diode with gated edge termination on 200-mm silicon substrate. *IEEE Transactions on Electron Devices*, *63*(3), 997–1004. Available from https://doi.org/10.1109/TED.2016.2515566.

Kabouche, R., Harrouche, K., Okada, E., & Medjdoub, F. (2020). Short-term reliability of high performance Q-band AlN/GaN HEMTs. In *IEEE international reliability physics symposium proceedings* (2020). France:

Institute of Electrical and Electronics Engineers Inc, Available from https://doi.org/10.1109/IRPS45951.2020.9129322, http://ieeexplore.ieee.org/xpl/conhome.jsp?punumber = 1000627.

Kaushik, J. (2012). Semicond science and techn.

Kazior, T. E., Chumbes, E. M., Schultz, B., Logan, J., Meyer, D. J., & Hardy, M. T. (2019). High power density ScAlN-based heterostructure FETs for mm-wave applications. *IEEE MTT-S international microwave symposium digest* (2019, pp. 1136–1139). United States: Institute of Electrical and Electronics Engineers Inc. Available from https://doi.org/10.1109/mwsym.2019.8701055.

Keller, S., Dora, Y., Wu, F., Chen, X., Chowdury, S., DenBaars, S. P., Speck, J. S., & Mishra, U. K. (2010). Properties of N-polar GaN films and AlGaN/GaN heterostructures grown on (111) silicon by metal organic chemical vapor deposition. *Applied Physics Letters*, *97*(14), 142109. Available from https://doi.org/10.1063/1.3499428.

Kerr, D. C., Gering, J. M., McKay, T. G., Carroll, M. S., Neve, C. R., & Raskin, J. P. (2008). Identification of RF harmonic distortion on Si substrates and its reduction using a trap-rich layer. In *IEEE topical meeting on silicon monolithic integrated circuits in RF systms - Digest of papers* (pp. 151–154). United States: SiRF. Available from https://doi.org/10.1109/SMIC.2008.44.

Lederer, D., & Raskin, J. P. (2005). Effective resistivity of fully-processed SOI substrates. Solid-State Electronics. Belgium, 49(3), 491–496. Available from https://doi.org/10.1016/j.sse.2004.12.003.

Li, X., Geens, K., Wellekens, D., Zhao, M., Magnani, A., Amirifar, N., Bakeroot, B., You, S., Fahle, D., Hahn, H., Heuken, M., Odnoblyudov, V., Aktas, O., Basceri, C., Marcon, D., Groeseneken, G., & Decoutere, S. (2020). Integration of 650 v GaN Power ICs on 200 mm engineered substrates. *IEEE Transactions on Semiconductor Manufacturing*, *33*(4), 534–538. Available from https://doi.org/10.1109/TSM.2020.3017703, Institute of Electrical and Electronics Engineers Inc. Belgium, https://ieeexplore.ieee.org/servlet/opac?punumber = 66.

Liu, W., Romanczyk, B., Guidry, M., Hatui, N., Wurm, C., Li, W., Shrestha, P., Zheng, X., Keller, S., & Mishra, U. K. (2021). 6.2 W/Mm and record 33.8% PAE at 94 GHz from N-polar GaN deep recess MIS-HEMTs with ALD Ru gates. *IEEE Microwave and Wireless Components Letters*, *31*(6), 748–751. Available from https://doi.org/10.1109/LMWC.2021.3067228, Institute of Electrical and Electronics Engineers Inc., United States, https://ieeexplore.ieee.org/servlet/opac?punumber = 7260.

Medjdoub, F., Carlin, J. F., Gonschorek, M., Feltin, E., Py, M. A., Ducatteau, D., Gaquière, C., Grandjean, N., & Kohn, E. (2006). Can InAlN/GaN be an alternative to high power/high temperature AlGaN/GaN devices? In *Technical digest - International electron devices meeting*, IEDM. Germany, Available from https://doi.org/10.1109/IEDM.2006.346935.

Nakamura, S. (1995). InGaN/AlGaN blue-light-emitting diodes. Journal of Vacuum Science & Technology A: Vacuum. *Surfaces, and Films*, *13*(3), 705–710. Available from https://doi.org/10.1116/1.579811.

Nakamura, S., Harada, Y., & Seno, M. (1991). Novel metalorganic chemical vapor deposition system for GaN growth. Applied Physics Letters. Japan, 58(18), 2021–2023. Available from https://doi.org/10.1063/1.105239.

Neve, C. R., & Raskin, J. P. (2012). RF harmonic distortion of CPW lines on HR-Si and trap-rich HR-Si substrates. IEEE Transactions on Electron Devices. Belgium, 59(4), 924–932. Available from https://doi.org/10.1109/TED.2012.2183598.

Olivier, B., Eric, G., Walter, S., & Gonzalo, P. (2021). A greener SiC wafer with \nSmart Cut technology. *Compound Semiconductor*.

Parvais, B., Alian, A., Peralagu, U., Rodriguez, R., Yadav, S., Khaled, A., Elkashlan, R. Y., Putcha, V., Sibaja-Hernandez, A., Zhao, M., Wambacq, P., Collaert, N., & Waldron, N. (2020). GaN-on-Si mm-wave RF devices integrated in a 200mm CMOS Compatible 3-Level Cu BEOL. *Technical digest - International electron devices meeting, IEDM* (2020, pp. 8.1.1–8.1.4). Belgium: Institute of Electrical and Electronics Engineers Inc. Available from https://doi.org/10.1109/IEDM13553.2020.9372056.

Parvais, B., Elkashlan, R., Yu, H., Sibaja-Hernandez, A., Vermeersch, B., Putcha, V., Cardinael, P., Rodriguez, R., Khaled, A., Alian, A., Peralagu, U., Zhao, M., Yadav, S., Gramegna, G., Driessche, J. V., & Collaert, N. (2021). Transistor modelling for mm-Wave technology pathfinding. *International conference on simulation of semiconductor processes and devices, SISPAD* (2021, pp. 247–250). Belgium: Institute of Electrical and Electronics Engineers Inc. Available from https://doi.org/10.1109/SISPAD54002.2021.9592530.

Peralagu, U., De Jaeger, B., Fleetwood, D. M., Wambacq, P., Zhao, M., Parvais, B., Waldron, N., Collaert, N., Alian, A., Putcha, V., Khaled, A., Rodriguez, R., Sibaja-Hernandez, A., Chang, S., Simoen, E., & Zhao, S. E. (2019). CMOS-compatible GaN-based devices on 200 mm-Si for RF applications: Integration and Performance. *Technical digest-International electron devices meeting, IEDM* (2019). Belgium: Institute of Electrical and Electronics Engineers Inc. Available from https://doi.org/10.1109/IEDM19573.2019.8993582.

Putcha, V., Cheng, L., Alian, A., Zhao, M., Lu, H., Parvais, B., Waldron, N., Linten, D., & Collaert, N. (2021). On the impact of buffer and GaN-channel thickness on current dispersion for GaN-on-Si RF/mmWave devices. In *IEEE international reliability physics symposium proceedings*. Belgium: Institute of Electrical and Electronics Engineers Inc. Available from https://doi.org/10.1109/IRPS46558.2021.9405139, http://ieeexplore.ieee.org/xpl/conhome.jsp?punumber = 1000627.

Rack, M., & Raskin, J. P. (2017). RF harmonic distortion modeling in silicon-based substrates including non-equilibrium carrier dynamics. *IEEE MTT-S international microwave symposium digest* (pp. 91–94). Belgium: Institute of Electrical and Electronics Engineers Inc. Available from https://doi.org/10.1109/MWSYM.2017.8058737.

Romanczyk, B., Wienecke, S., Guidry, M., Li, H., Ahmadi, E., Zheng, X., Keller, S., & Mishra, U. K. (2018). Demonstration of constant 8 W/mm power density at 10, 30, and 94 GHz in state-of-the-art millimeter-wave N-polar GaN MISHEMTs. *IEEE Transactions on Electron Devices*, *65*(1), 45–50. Available from https://doi.org/10.1109/TED.2017.2770087, Institute of Electrical and Electronics Engineers Inc., United States.

Romanczyk, B., Mishra, U. K., Zheng, X., Guidry, M., Li, H., Hatui, N., Wurm, C., Krishna, A., Ahmadi, E., & Keller, S. (2020). W-band power performance of SiN-passivated N-polar GaN deep recess HEMTs. *IEEE Electron Device Letters*, *41*(3), 349–352. Available from https://doi.org/10.1109/LED.2020.2967034, Institute of Electrical and Electronics Engineers Inc. United States, https://ieeexplore.ieee.org/servlet/opac?punumber = 55.

Schroder, D. K. (2006). Semiconductor material and device characterization (3rd ed.). Wiley.

Then, H. W., Huang, C. Y., Krist, B., Jun, K., Lin, K., Nidhi, N., Michaelos, T., Mueller, B., Paul, R., Peck, J., Rachmady, W., Dasgupta, S., Staines, D., Talukdar, T., Thomas, N., Tronic, T., Fischer, P., Hafez, W., Radosavljevic, M., ... Holybee, B. (2019). 3D heterogeneous integration of high performance high-K metal gate GaN NMOS and Si PMOS transistors on 300mm high-resistivity Si substrate for energy-efficient and compact power delivery, RF (5G and beyond) and SoC applications. *Technical digest - International electron devices meeting, IEDM* (2019). United States: Institute of Electrical and Electronics Engineers Inc. Available from https://doi.org/10.1109/IEDM19573.2019.8993583.

Then, H. W., Radosavljevic, M., Koirala, P., Thomas, N., Nair, N., Ban, I., Talukdar, T., Nordeen, P., Ghosh, S., Bader, S., Hoff, T., Michaelos, T., Nahm, R., Beumer, M., Desai, N., Wallace, P., Hadagali, V., Vora, H., Oni, A., Weng, X., Joshi, K., Meric, I., Nieva, C., Rami, S., & Fischer, P. (2021). Advanced scaling of enhancement mode high-K gallium nitride-on-300mm-Si(111) transistor and 3D layer transfer GaN-silicon finfet CMOS integration. *Technical digest - International electron devices meeting, IEDM* (2021, pp. 11.1.1–11.1.4). United States: Institute of Electrical and Electronics Engineers Inc. Available from https://doi.org/10.1109/IEDM19574.2021.9720710.

Vermeersch, B., Rodriguez, R., Sibaja-Hernandez, A., Vais, A., Yadav, S., Parvais, B., & Collaert, N. (2022). Thermal modelling of GaN & InP RF devices with intrinsic account for nanoscale transport effects. *IEDM*.

Wang, R., Saunier, P., Xing, X., Lian, C., Gao, X., Guo, S., Snider, G., Fay, P., Jena, D., & Xing, H. (2010). Gate-recessed enhancement-mode InAlN/AlN/GaN HEMTs with 1.9-A/mm drain current density and 800-ms/mm transconductance. IEEE Electron Device Letters. United States, 31(12), 1383–1385. Available from https://doi.org/10.1109/LED.2010.2072771.

Yadav, S., Cardinael, P., Zhao, M., Vondkar, K., Khaled, A., Rodriguez, R., Vermeersch, B., Makovejev, S., Ekoga, E., Pottrain, A., Waldron, N., Raskin, J. P., Parvais, B., & Collaert, N. (2020). Substrate RF losses and non-linearities in GaN-on-Si HEMT technology. *Technical digest - International electron devices meeting, IEDM* (2020, pp. 8.2.1–8.2.4). Belgium: Institute of Electrical and Electronics Engineers Inc. Available from https://doi.org/10.1109/IEDM13553.2020.9371893.

Yamaguchi, Y., Kamioka, J., Shinjo, S., Yamanaka, K., & Oishi, T. (2016). Physical model of RF leakage in GaN HEMTs on Si substrates based on atomic diffusion analysis at buffer/substrate interface. *Technical digest - IEEE compound semiconductor integrated circuit symposium, CSIC* (2016). Japan: Institute of Electrical and Electronics Engineers Inc. Available from https://doi.org/10.1109/CSICS.2016.7751058.

Ye, G., Wang, H., Arulkumaran, S., Ng, G. I., Hofstetter, R., Li, Y., Anand, M. J., Ang, K. S., Maung, Y. K. T., & Foo, S. C. (2013). AlGaN/GaN MISHEMTs on silicon using atomic layer deposited $ZrO_2$ as gate dielectrics. In *Device research conference - Conference digest, DRC* (pp. 71–72). Singapore. Available from https://doi.org/10.1109/DRC.2013.6633798.

Yu, H., Alian, A., Peralagu, U., Zhao, M., Waldron, N., Parvais, B., & Collaert, N. (2021). Surface state spectrum of AlGaN/AlN/GaN extracted from static equilibrium electrostatics. *IEEE Transactions on Electron Devices*, *68*(11), 5559–5564. Available from https://doi.org/10.1109/TED.2021.3115086, Institute of Electrical and Electronics Engineers Inc. Belgium, https://ieeexplore.ieee.org/xpl/mostRecentIssue.jsp?punumber = 16.

Yu, H., Putcha, V., Peralagu, U., Zhao, M., Yadav, S., Alian, A., Parvais, B., & Collaert, N. (2022). Leakage mechanism in ion implantation isolated AlGaN/GaN heterostructures. *Journal of Applied Physics, 131*(3). Available from https://doi.org/10.1063/5.0076243, American Institute of Physics Inc., Belgium, http://scitation.aip.org/content/aip/journal/jap.

Yu, H., Parvais, B., Zhao, M., Rodriguez, R., Peralagu, U., Alian, A., & Collaert, N. (2022). Thermal budget increased alloy disorder scattering of 2DEG in III−N heterostructures. *Applied Physics Letters, 120*(21), 213504. Available from https://doi.org/10.1063/5.0093839, AIP Publishing.

Yu, H., Parvais, B., Peralagu, U., Elkashlan, R. Y., Rodriguez, R., Khaled, A., Yadav, S., Alian, A., Zhao, M., De Almeida Braga, N., Cobb, J., Fang, J., Cardinael, P., Sibaja-Hernandez, A., & Collaert, N. (2022). Back barrier trapping induced resistance dispersion in GaN HEMT: Mechanism, modeling, and solutions. *Technical digest - International electron devices meeting, IEDM* (2022, pp. 3061−3064). Belgium: Institute of Electrical and Electronics Engineers Inc. Available from https://doi.org/10.1109/IEDM45625.2022.10019489.

CHAPTER

# 5

# Heterojunction bipolar transistors for sub-THz applications

*Nadine Collaert*

**IMEC, Heverlee, Belgium**

## 5.1 Introduction

Sub-THz frequencies are electromagnetic radiation that have a frequency in the range of 0.1 to 0.3 THz, even to 1 THz (depending on the source of information). This range of frequencies is often referred to as the sub-THz band, and it is located between the microwave and infrared regions of the electromagnetic spectrum.

Sub-THz frequencies have a number of unique properties that make them useful for a variety of applications. They have relatively short wavelengths, which allows them to penetrate through many materials that are opaque to other types of electromagnetic radiation. This makes them useful for imaging and sensing applications, as they can be used to see inside objects or to detect hidden objects.

In addition, sub-THz frequencies have a relatively low energy, which means that they are nonionizing and do not have the same risks associated with ionizing radiation. This makes them safe for use in a wide range of applications, including medical imaging and wireless communication.

Sub-THz frequencies are an important and increasingly useful part of the electromagnetic spectrum, and they have the potential to revolutionize a variety of industries and technologies. As mentioned before, there is an increasing interest in using these frequencies for wireless communication, especially in the context of 6G. The D-band around 140 GHz is the first band of interest.

Power amplifiers (PAs) working at these frequencies require devices that have high speed, can deliver high output power and efficiency. As detailed out in Chapter 1, widely used complementary metal-oxide-semiconductor (CMOS) technologies might not be the right choice for these circuits working at D-band. Bipolar junction transistors (BJTs) that in general have much higher breakdown and thus can deliver more power can be interesting device architectures for these PAs working at frequencies above 100 GHz.

*New Materials and Devices Enabling 5G Applications and Beyond*
**DOI: https://doi.org/10.1016/B978-0-12-822823-4.00005-4**

In this chapter, we will first give an overview of the different bipolar transistors that exist, we will review the operation of a bipolar transistor and its typical optimization metrics and trade-offs. Then we will dive into two particular implementations that are being considered for sub-THz transceivers: the indium phosphide (InP) and the silicon-germanium (SiGe) heterojunction bipolar transistor (HBT). We will discuss several ways of upscaling InP HBTs before diving a bit deeper into the device scaling and nonidealities. Finally, we will conclude the chapter with a short summary.

## 5.2 Bipolar transistors

### 5.2.1 Introduction

Bipolar transistors have played a crucial role in the development of modern electronics from the start. They were first developed in the late 1940s and early 1950s, and they quickly became an essential component in a wide range of electronic circuits.

The first bipolar transistor was developed at Bell Labs in 1947 by a team of researchers led by John Bardeen, Walter H. Brattain, and William Shockley. These researchers were working on the development of the point-contact transistor, which was an early version of the bipolar transistor. The device they created was made from a slice of germanium with two electrodes, or "fingers," placed on either side. When a small current was applied to one of the electrodes, it caused a larger current to flow through the germanium and out the other electrode. This amplification of the electrical signal made the bipolar transistor a key component in the development of modern electronics. They were awarded the Nobel Prize in Physics in 1956 for their work on the transistor.

In the years following the invention of the bipolar transistor, researchers continued to improve upon the design and develop new technologies based on the device. One of the major improvements was the development of the p-n junction, which allowed for more precise control of the current flowing through the transistor. This paved the way for the development of the integrated circuit, in which multiple transistors and other electronic components could be combined on a single chip.

The bipolar transistor also played a crucial role in the development of the computer. In the early days of computers, they were large and expensive, and they used vacuum tubes as their electronic components. These tubes were prone to failure and required a lot of power to operate. The development of the bipolar transistor allowed for the creation of smaller, more reliable, and more energy-efficient computers.

In the 1960s and 1970s, the bipolar transistor was widely used in a variety of electronic devices, including radios, televisions, and computers. However, as technology continued to advance, other types of transistors, such as the field-effect transistor (FET), began to gain popularity. FETs offered several advantages over bipolar transistors, including faster switching speeds and lower power consumption.

While it might look like the BJT has been replaced by other device architectures in many applications, this is far from the truth. Typically they are used for very specific functions which can not be done by metal-oxide semiconductor field effect transistor (MOSFETs). The amount of devices in these applications is far from the high density required for digital applications, but the few BJTs need to excel in speed, power delivery, and efficiency.

Today, bipolar transistors are being used in a variety of applications, including radio frequency (RF) and microwave communications, power amplification, and microwave integrated circuits. However, with the ongoing device research and development, new and emerging applications are being explored. These emerging applications include 5G and beyond. The high frequency and high gain characteristics of BJTs make them ideal for use in 5G and future wireless communication systems. Also, BJTs are being studied for use in power electronics applications such as high-frequency switching and power conversion, and millimeter-wave and terahertz technologies (imaging, sensing, and communication). In automotive applications, such as power inverters for electric vehicles, as well as other applications requiring their high-power and high-frequency capabilities are of high interest. BJTs are also being studied for use in Internet of Things applications, where they can be used to improve the performance of wireless sensor networks and other low-power devices. And finally in quantum computing, BJTs are considered to improve the performance of qubits.

## 5.2.2 Different types of bipolar transistors

Bipolar transistors are semiconductor devices that are used to amplify signals and control the flow of electrical current. They are called "bipolar" because they use both positive and negative charge carriers (holes and electrons) to operate. While there is a perception that bipolar transistors and MOSFETs are fundamentally different, this is actually not the case as presented in (Johnson, 1973). They use the same basic physical principles to operate. Moreover, the main difference between a bipolar transistor and a MOSFET that is often used is that a bipolar transistor is a current-controlled device while a MOSFET is a voltage-controlled device. This is not entirely true as we will see later. In practice currents are used in bipolar transistors because it is easier to control a current than a voltage. Because of their exponential dependence on the forward base-emitter bias, injecting a current and setting the forward bias using that current is much easier and better controlled. But the device is essentially, just like a MOSFET, operated by lowering a barrier by forward biasing the base-emitter junction.

The three important areas in a MOSFET are the source, drain, and gate. In a MOSFET, the gate terminal or voltage controls the current flowing through the source-drain channel. The three layers of a bipolar transistor are called the emitter, the base, and the collector. The emitter is the layer that emits charge carriers, the base is the layer that controls the flow of charge carriers, and the collector is the layer that collects the charge carriers. The operation of a bipolar transistor is based on the fact that a small current flowing through the base layer can control a much larger current flowing between the emitter and collector layers. This allows a bipolar transistor, just like a MOSFET, to function as an amplifier, switching device, or a combination of both. A schematic view and comparison of both devices is shown in Fig. 5.1.

There are several different types of bipolar transistors, each with their own set of unique characteristics and applications.

One of the most common types of bipolar transistors is the BJT. A BJT is a three-layer semiconductor device in which the emitter, base, and collector are made of the same semiconductor material. In most cases this is silicon, but it can also be made of compound semiconductors (III–V).

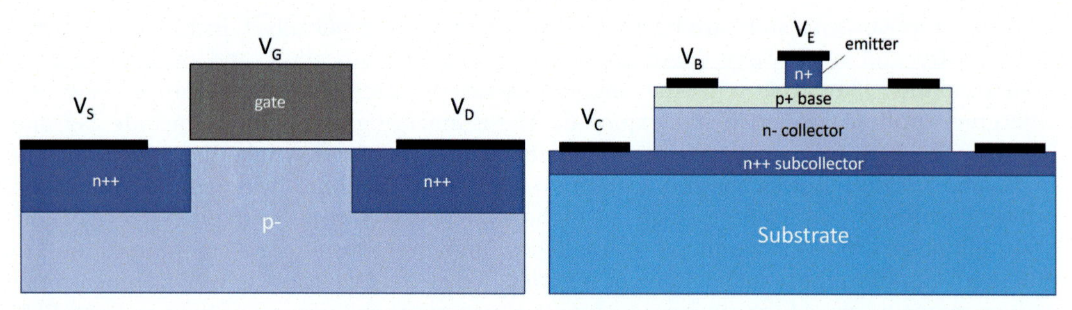


FIGURE 5.1 Metal-oxide semiconductor field effect transistor (MOSFET) versus bipolar junction transistor (BJT). Simple schematic view of (left) a nMOSFET and (right) BJT comparing the two different device architectures; a BJT is essentially a vertical device where the different layers of the stack are typically epitaxially grown and contacts to the different levels need to be fabricated.

There are two main types of bipolar transistors: NPN and PNP. An NPN transistor has a layer of p-type semiconductor material between two layers of n-type semiconductor material, while a PNP transistor has a layer of n-type semiconductor material between two layers of p-type semiconductor material. In other words, in a NPN transistor the base-emitter junction is a p-n junction, and the base-collector is also a p-n junction, while in a PNP transistor the emitter and collector are both p-type and base n-type. The type of bipolar transistor used in a particular application depends on the desired characteristics and the circuit in which it is being used.

Another type of bipolar transistor is the HBT. HBTs are similar to BJTs in that they have the three layers of semiconductor material, but the layers are made of different materials. This type of transistor is often made from a combination of two or more semiconductor materials with different bandgaps. The engineered heterostructure in HBTs allows for improved electron transport and higher frequency operation than traditional (silicon-based) bipolar transistors. These HBTs can be made using either IV-based materials like Si and SiGe, or using III–V semiconductors like GaAs, InGaAs, InGaP, InP, etc., and the structure can be quite complex going from the introduction of a single heterojunction to double heterojunction to even a superlattice made of compound semiconductor materials.

In SiGe HBTs, the base layer is made of a SiGe alloy, which has a higher electron mobility than pure silicon. The emitter and collector layers are typically made of silicon. In III–V HBTs, the collector and base layers can be made of a semiconductor material such as gallium arsenide (GaAs), while the emitter layer is made of a different material, such as indium gallium phosphide (InGaP). Depending on the required performance and frequency different material combinations are possible and this will be the focus of a later section in this chapter. HBTs are commonly used in high-frequency and microwave applications, such as in wireless communication systems because they can operate at higher frequencies and have better current and power handling than homojunction transistors.

### 5.2.3 Operating principle

As mentioned before, there are two main types of bipolar transistors: NPN and PNP. The main difference between the two is the direction of current flow. In a NPN transistor, current flows from the collector to the emitter when a voltage is applied to the base. In a PNP transistor, current flows from the emitter to the collector when a voltage is applied to the base.

A bipolar transistor consists of three layers of semiconductor material, each of which has a different type of doping. The inner layer is called the base, and it is lightly doped. The outer layers are called the collector and the emitter, and they are heavily doped with either n-type or p-type impurities depending on the fact if it is a NPN or PNP transistor. The collector typically consists of a subcollector and collector whereby the subcollector is very heavily doped to make good contact and the collector itself is lightly doped, even lighter than the base region. A schematic view of a NPN transistor and band structure are shown in Fig. 5.2. $V_{bi}$ is in this case the built-in voltage of the p-n diode at the base-emitter side and $E_v$, $E_c$, and $E_F$ represent the valence band, conduction band, and Fermi level, respectively.

The bipolar transistor is just like the MOSFET in essence a barrier-controlled device. The biasing scheme for a NPN and PNP transistor is shown in Fig. 5.3 where $I_B$, $I_E$, and $I_C$ represent the base, emitter, and collector currents, and $V_{XY} = V_X\text{-}V_Y$ the bias between the terminals $X$ and $Y$ where $X$ and $Y$ can be the base (B), the emitter (E), and the collector (C). In a NPN transistor, the base-emitter junction is biased in forward when operating in forward regime.

This means a positive voltage will be applied to the base with respect to the emitter ($V_{BE} > 0$ V). This will pull down the barrier between the emitter and the base. However, between the base and the collector a reverse bias will be applied ($V_{BC} < 0$ V or $V_{CB} > 0$ V). This will pull the energy in the collector down. This is similar to the saturation regime in a MOSFET.

The lowering of the base/emitter barrier allows for electrons from the emitter to be injected into the base. They diffuse across the p-type base and are swept by the high electric field to the collector contact.

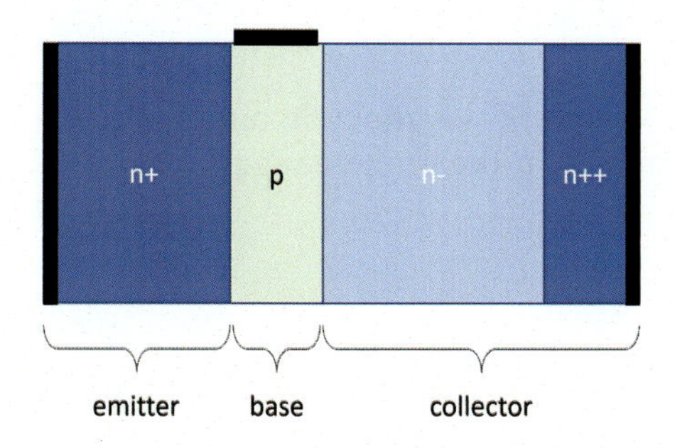


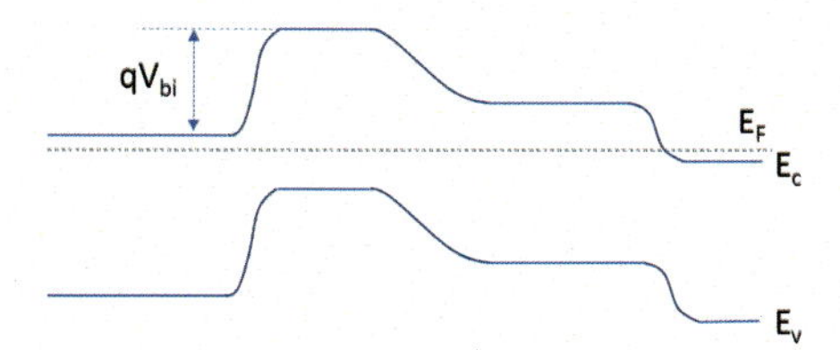


**FIGURE 5.2** Schematic presentation of an NPN transistor indicating the different layers and accompanying band structure.

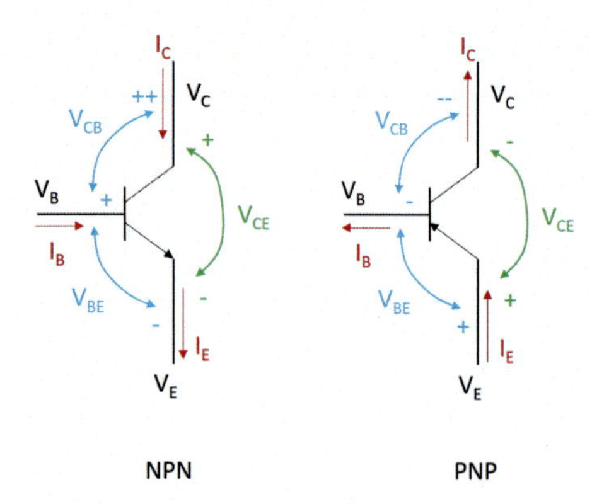


**FIGURE 5.3** Biasing. Overview of the biasing and current flow in (left) a NPN bipolar transistor and (right) PNP bipolar transistor.

The collector current that flows can be defined as follows:

$$I_C = A_E J_n = qA_E \frac{D_n}{W_B} \frac{n^2{}_{iB}}{N_{AB}} \left( e^{\frac{qV_{BE}}{kT}} - 1 \right) \tag{5.1}$$

with $q$ being the elementary charge, $A_E$ the area of the emitter-base junction where the carriers are injected, $J_n$ the electron current density, $W_B$ the width of the base, $D_n$ is the electron diffusion coefficient or diffusivity, $n_{iB}$ the intrinsic carrier concentration in the base region, $N_{AB}$ the acceptor doping level in the base, $V_{BE}$ the bias applied between base and emitter, $k$ the Boltzmann's constant, and $T$ the temperature.

Next to that, holes are being injected from the base into the emitter. The holes need to be replenished by a current flowing in the base contact $I_B$. The base current is given by the following expression:

$$I_B = A_E J_p = qA_E \frac{D_p}{W_E} \frac{n^2{}_{iE}}{N_{DE}} \left( e^{\frac{qV_{BE}}{kT}} - 1 \right) \tag{5.2}$$

with $J_p$ the hole current density, $W_E$ the width of the emitter, $D_p$ is the hole diffusion coefficient or diffusivity, $n_{iE}$ the intrinsic carrier concentration in the emitter region, and $N_{DE}$ the donor doping level in the emitter.

The emitter current $I_E$ is then finally defined as

$$I_E = I_C + I_B \tag{5.3}$$

The base current needs to be as small as possible, usually a few percent of the emitter or collector current. It is used to control the base-emitter forward bias, but it is essentially a parasitic current.

The gain of the transistor is an important figure of merit (FOM) for the forward operation, and is defined as the ratio between the collector current and the base current. Using the previous expressions for these currents, the gain of the bipolar transistor can be written as

$$\beta = \frac{I_C}{I_B} = \frac{D_n}{D_p} \frac{W_E}{W_B} \frac{N_{DE}}{N_{AB}} \left( \frac{n_{iB}}{n_{iE}} \right)^2 \tag{5.4}$$

Typical output characteristics $I_C$-$V_{CE}$ of a device look very similar to the output characteristics of a MOSFET with main difference that $I_B$ is used to step up the current rather than the $V_{GS}$ bias in a MOSFET. This is shown in Fig. 5.4. This figure shows the different regimes of operation: saturation, active, and breakdown.

In the saturation regime, the base-emitter junction is forward biased, and the collector-base junction is reverse biased. The collector current is almost constant, and the collector-emitter voltage is close to zero. The transistor is said to be "saturated" because the current flowing through it cannot increase any further. In this region, the transistor operates like a switch.

In the active regime, the base-emitter junction is forward-biased, and the collector-base junction is also forward biased. The collector current is controlled by the base current, and the collector-emitter voltage is not zero. The transistor is said to be "active" because it is amplifying the input signal.

In the breakdown regime, the collector-base junction is reverse biased and the voltage applied to it exceeds the breakdown voltage. The transistor is said to be in "breakdown" because it can no longer control the current flowing through it. This can cause damage to the device.

And just like in a MOSFET, the transfer characteristics can be plotted as shown in Fig. 5.5.

In the transfer characteristics, we can distinguish a subthreshold regime with a subthreshold swing of 60 mV/dec at 300K. A saturation-like regime or leveling off of the current is typically seen at large $V_{BE}$. This is not the same as the saturation that occurs in MOSFETs. The reduction in current (as compared to its ideal 60 mV/dec current) is related to the impact of the series resistance.

In a circuit, there are three possible ways to connect a bipolar transistor:

Common emitter configuration: In this configuration, the emitter terminal of the transistor is grounded, the collector is connected to a positive voltage supply, and the base is the input.

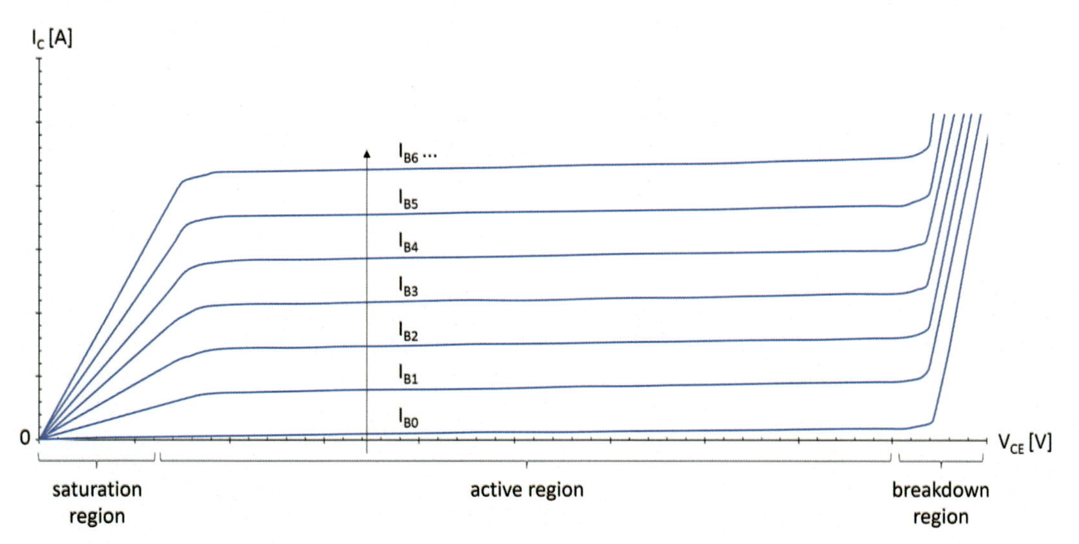


**FIGURE 5.4** Ideal $I_C$-$V_{CE}$ characteristics for a bipolar transistor showing the saturation region, active region, and breakdown region. In this device the base current $I_B$ is used to step up the current.

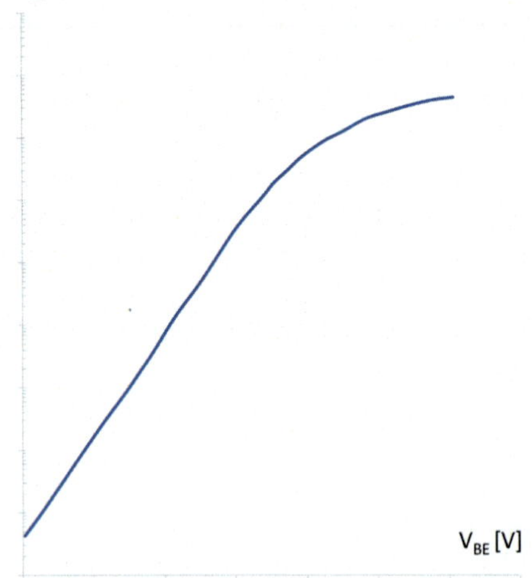


**FIGURE 5.5** Collector current $I_C$ versus $V_{BE}$.

This configuration provides amplification of both voltage and current, and is commonly used in low-frequency amplifier circuits.

Common base configuration: In this configuration, the base terminal of the transistor is grounded, the emitter is the output, and the collector is the input. This configuration provides high voltage gain and low current gain, and is commonly used in RF amplifier circuits.

Common collector configuration (also known as emitter follower): In this configuration, the collector terminal is grounded, the emitter is the output, and the base is the input. This configuration provides high current gain and low voltage gain, and is commonly used as a buffer amplifier to isolate high impedance circuits from low impedance circuits.

Other analog/RF FOMs which are important are the transconductance $g_m$ of the transistor and related to that is the cut-off frequency $f_t$. The $g_m$ represents the change in collector current ($I_C$) with respect to the change in base-emitter voltage ($V_{BE}$) when the collector-emitter voltage ($V_{CE}$) is held constant.

$$g_m = \frac{I_c}{kT/q} \tag{5.5}$$

Different from a MOSFET, this device typically exhibits much larger transconductance (factor 10–20 larger than the MOSFET) and as such higher drive currents. However, this does not necessarily translate into a significantly larger cut-off frequency. As we can see below the cut-off frequency depends on $g_m$, but also the total capacitance and there the MOSFET has an advantage, having lower capacitance than a typical bipolar transistor.

$$f_t = \frac{g_m}{2\pi C_{tot}} = \frac{1}{2\pi\tau} \tag{5.6}$$

where $\tau$ is the delay time.

For a bipolar transistor the delay time can be written as

$$\tau = \frac{C_{eb+}C_{\text{par}}}{g_m} = t_t + \frac{C_{\text{par}}kT/q}{I_c} \tag{5.7}$$

where $C_{\text{eb}}$ is the emitter-base junction capacitance, $C_{\text{par}}$ is the parasitic capacitance, and $t_t$ is the transit time or the time it takes for an electron to go from the emitter, across the base to the collector.

The impact of $C_{\text{par}}$ can be made negligible when the collector current is high. Therefore, for high speed applications bipolar transistors are biased in a regime where the collector current is high.

In a PNP transistor, the operation is essentially the same, but the direction of the current flow is reversed.

A bipolar transistor can be used in two main modes: as an amplifier or as a switch.

When used as an amplifier, a bipolar transistor amplifies the input signal applied to the base terminal. As we have seen before, the current flowing through the base terminal controls the current flowing through the collector terminal, which in turn amplifies the input signal. The amount of amplification is determined by the transistor's current gain as defined above, which is known as the $h_{\text{FE}}$ or beta (β) of the transistor.

When used as a switch, a bipolar transistor is used to turn on or off the current flowing through the collector-emitter terminals.

A HBT is a modified version of the BJT. If we look at the equation for the gain of the transistor (Eq. 10.4), we know that for a homojunction BJT where the base and the emitter are made of the same material $n_{\text{iB}} = n_{\text{iE}}$ and that simplifies the equation to

$$\beta = \frac{D_n}{D_p}\frac{W_E}{W_B}\frac{N_{\text{DE}}}{N_{\text{AB}}} \tag{5.8}$$

This means that high gain is typically achieved by doping the emitter a lot higher than the base ($N_{\text{DE}} >> N_{\text{AB}}$). However, to reduce the base resistance we want to have high doping in the base region. Moreover, for high speed the thickness of the base needs to be scale so the transit time can be reduced. However, in HBTs this issue of having to increase the doping of the emitter much higher than the doping of the base can be circumvented by bandgap engineering. By using a different material for the emitter (e.g., wide bandgap material), we can optimize the gain of the transistor as follows. Knowing that the intrinsic carrier concentration can be written as

$$n_i = N_C N_v e^{\frac{-E_G}{kT}} \tag{5.9}$$

with $N_C$ effective density of states in the conduction band, $N_V$ the effective density of states in the valence band, and $E_G$ the bandgap of the material. We can then approximate the gain as follows (assuming similar $N_C$ and $N_V$).

$$\beta \sim \frac{D_n}{D_p}\frac{W_E}{W_B}\frac{N_{\text{DE}}}{N_{\text{AB}}} e^{\frac{\Delta E_G}{kT}} \tag{5.10}$$

where $\Delta E_G = E_{\text{GE}}\text{-}E_{\text{GB}}$ is the difference in bandgap between the emitter ($E_{\text{GE}}$) and base material ($E_{\text{GB}}$).

To have achieve a high gain, it is not necessary to play with doping levels of the emitter and base, but engineering the bandgap is an extra parameter that can be used to optimize the gain, whereby it is important to have $E_{GE} > E_{GB}$.

Often a double heterojunction bipolar transistor (DHBT) is used where the same material as used in the emitter is also used in the collector, because it is a more symmetric device, it reduces the collector off-set voltage and can achieve a higher breakdown (because of the wider bandgap in the collector).

If we go back to Fig. 5.4, we see that the current in this case goes through the origin but in reality this is typically not the case and that there is a certain off-set, collector-emitter off-set. This is undesirable as it leads to higher power consumption. This is related to the asymmetrical structure of the bipolar transistor. Even the DHBT is never exactly symmetrical.

Secondly, thermal effects are quite important as the devices are often run at high current densities, as we will discuss later.

$F_{max}$ values higher than 1 THz have been shown for these. In general, one can say that the HBTs can offer similar $f_t$ and $f_{max}$ as III–V high electron mobility transistors (HEMTs), but they can deliver higher power and densities than III–V HEMTs.

## 5.2.4 InP heterojunction bipolar transistor versus silicon-germanium heterojunction bipolar transistor

SiGe HBTs and InP HBTs are both types of bipolar transistors that are used in high-frequency electronic circuits and devices. However, there are some key differences between these two types of transistors:

Material: SiGe HBTs are made from silicon and germanium where the SiGe layer is used in the base region of the transistor, while III–V HBTs are made from InP or GaAs. These materials have different electrical and physical properties, which can affect the performance of the transistor.

Bandgap: The bandgap of a material is the energy required to excite an electron from the valence band to the conduction band. SiGe HBTs have a smaller bandgap than InP HBTs.

Voltage handling: InP HBTs are able to handle higher voltage levels than SiGe HBTs, which makes them suitable for use in high-voltage circuits.

Cost: InP HBTs are typically more expensive to manufacture than SiGe HBTs, due to the higher cost of the InP material.

A typical III–V DHBT has a wide bandgap InP emitter, a InGaAs base, and a wide bandgap InP collector. Often a graded layer can be found between the InGaAs base and the InP collector to ensure that there is no discontinuity ($E_c$ band off-set) at the base–collector interface that could trap the electrons in the base.

The fabrication process is much simpler than Si CMOS as can be found in Rodwell et al. (2008). After growing the HBT stack on the InP substrates, using either metal-organic chemical vapor deposition (MOCVD) or molecular beam epitaxy (MBE), the emitter metal is deposited and then the emitter is etched (metal and emitter III–V stack) stopping on the base. Then the base metal contact is formed self-aligned to the emitter,

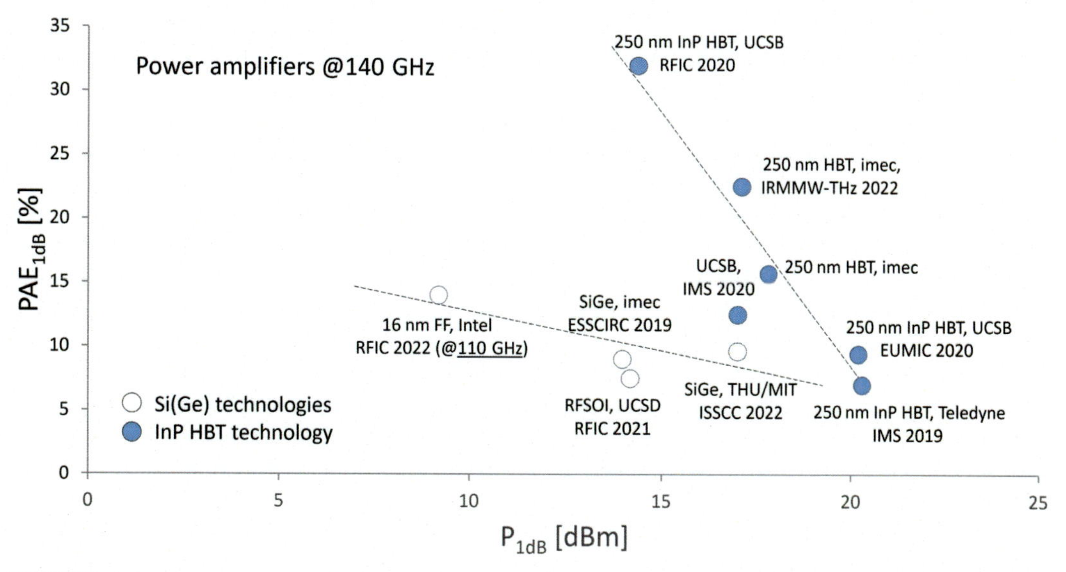


**FIGURE 5.6** $PAE_{1db}$ versus $P_{1db}$ comparing PAs designed in SiGe BiCMOS, FinFET, RFSOI, and InP heterojunction bipolar transistor at 110–140 GHz. *Source: From Collaert, N., Alian, A., Banerjee, A., Boccardi, G., Cardinael, P., Chauhan, V., Desset, C., Elkashlan, R., Khaled, A., Ingels, M., Kunert, B., Mols, Y., O'Sullivan, B., Peralagu, U., Pinho, N., Rodriguez, R., Sibaja-Hernandez, A., Sinha, S., Sun, X., ... Peeters, M. (2022). III-V/III-N technologies for next generation high-capacity wireless communication. In:* Technical digest – International electron devices meeting, IEDM *(Vols. 2022-, pp. 1151–1154). Institute of Electrical and Electronics Engineers Inc. https://doi.org/10.1109/IEDM45625.2022.10019555.*

and the base is patterned and the collector is exposed, followed by depositing the collector contacts.

Overall, the choice between a SiGe HBT and an InP HBT will depend on the specific requirements of the application and the trade-offs that are acceptable. Both types of transistors have their own advantages and disadvantages, and the best choice will depend on the specific needs of the circuit or device. In Fig. 5.6, a comparison is shown between PAs designed in InP HBT and Si(Ge) technologies at 110–140 GHz. Overall, we see higher efficiencies and output power for InP-based PAs.

In the next sections, we'll dive a bit deeper in the fabrication and device design of SiGe and III–V HBTs.

## 5.3 Silicon-germanium heterojunction bipolar transistor

SiGe BICMOS (silicon-germanium bipolar complementary metal-oxide-semiconductor) combines the benefits of both silicon and germanium (SiGe) materials with those of bipolar and complementary metal-oxide-semiconductor technologies. It has a long history, with the development of SiGe materials and devices dating back to the 1970s.

The first SiGe materials were developed in the 1970s as a way to overcome the limitations of silicon, which had been the dominant material used in the microelectronic industry up to that point. Silicon has several desirable properties, such as a high melting point and good thermal stability, but it also has some limitations, such as a lower mobility compared to other materials. Germanium, on the other hand, has a higher electron and hole mobility but is not as thermally stable as silicon.

Until the 1980s, the bipolar transistors were fully made out of silicon, and optimization of the BJT was not straightforward. By playing with the doping of the layers where the doping of the emitter is higher than the doping of the base which is again higher than the doping of the collector, one could improve the gain of the transistor, but often at the expense of other metrics like $f_t$ and $f_{max}$. Polysilicon was commonly used as the material for the base electrode (to reduce the base current) in bipolar transistors during the early days of the transistor technology.

In the 1980s, SiGe HBTs were developed, which utilized the high mobility of germanium to improve the performance of bipolar transistors, next to the ability to make an heterostructure that allowed more degrees of freedom to optimize the transistor for both gain and speed. These transistors were used in a variety of applications, such as high-speed analog and RF circuits.

In the 1990s, SiGe technology began to be used in the development of BICMOS devices, which combine bipolar and CMOS technologies on a single wafer. CMOS technology, which is based on the use of MOSFETs, is widely used in microelectronic devices due to its low power consumption and high integration density. However, CMOS technology is not well suited for certain applications, such as high-speed analog and RF circuits, due to its relatively low bandwidth.

The combination of SiGe materials with bipolar and CMOS technologies in BICMOS devices allows for the creation of devices with improved performance and versatility. SiGe BICMOS technology has been used in a variety of applications, including high-speed analog and RF circuits, mixed-signal circuits, and high-resolution imaging sensors. It has also been used in the development of microelectromechanical systems and other microelectronic devices.

Over the years, SiGe BICMOS technology has undergone significant development and improvement. One of the major challenges in the development of SiGe BICMOS technology has been the integration of SiGe materials with CMOS technology, which required the development of new fabrication processes and materials. In recent years, significant progress has been made in this area, leading to the development of advanced SiGe BICMOS devices with improved performance and capabilities.

SiGe BICMOS technology has had a significant impact on the microelectronic industry and has enabled the development of a wide range of devices with improved performance and capabilities. It is likely that this technology will continue to play a significant role in the development of new microelectronic devices in the future.

State-of-the-art SiGe HBTs have achieved impressive performance levels, with maximum oscillation frequencies approaching 500 GHz and even going up to 720 GHz with $f_t$ of 500 GHz and breakdown voltage $BV_{CEO}$ of 1.6 V as shown in Heinemann et al. (2017). These exceptional speed improvement was attributed to several key factors: an optimized vertical profile of the device, which led to enhanced carrier transport efficiency and reduced parasitic resistance. The base and emitter resistance were reduced by using a combination of millisecond annealing and low-temperature backend processing. This resulted

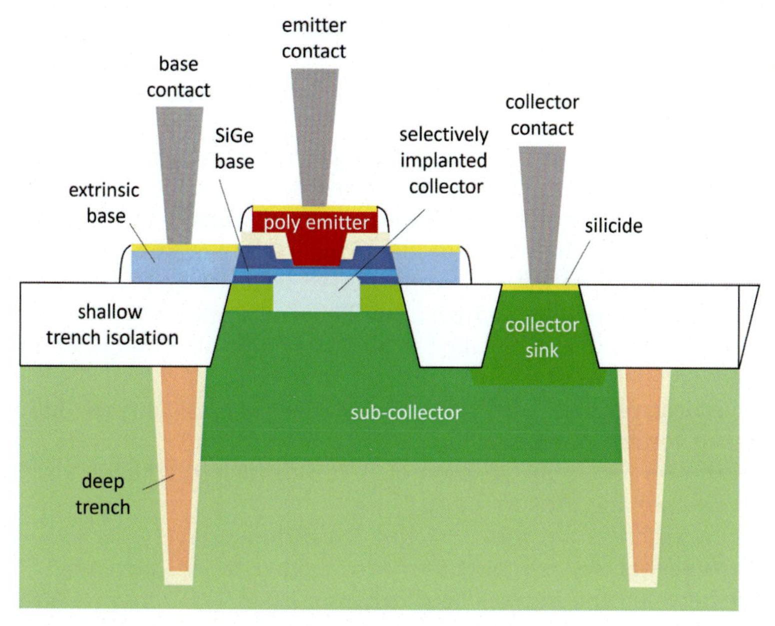


**FIGURE 5.7** Schematic view of an advanced silicon-germanium HBT.

in improved carrier injection and extraction, leading to faster switching times. This was all done in combination with lateral device scaling techniques to reduce the device dimensions and further improve the speed performance.

The ability to achieve these remarkable improvements in SiGe HBT performance can be attributed to the fact that SiGe HBTs largely take use of processes and toolsets that have been developed over decades for CMOS technology. In that sense, SiGe HBTs have an advantage over InP HBTs.

A schematic view of an advanced SiGe HBT is shown in Fig. 5.7 (Pekarik, et al., 2021; Rucker & Heinemann, 2018). In Pekarik et al. (2021), the integration of SiGe HBTs in a 45 nm partially depleted silicon-on-insulator process is described. In the next section, we will briefly address some of the challenges in fabricating SiGe HBTs.

### 5.3.1 Fabrication of silicon-germanium heterojunction bipolar transistor

The fabrication flow of a SiGe HBT is a multi-step process that involves the following steps: the epitaxial growth of the Si/SiGe layers, dielectric deposition, emitter mesa fabrication, collector area lithography, emitter-base lithography, polysilicon deposition for the emitter, n-type doping and anneal of the emitter, polysilicon patterning, contacts, and metallization and passivation.

The most critical and key processing steps are listed here below:

SiGe epitaxy: The growth of the SiGe layer with the desired composition and thickness is critical for controlling the base width of the SiGe HBT.

Base implantation: The doping profile and activation of the base region can significantly affect the performance of the SiGe HBT.

Collector epitaxy: The growth of the Si collector layer is critical for ensuring a high gain and low noise in the SiGe HBT.

Spacer formation: The formation of the spacer surrounding the SiGe HBT is crucial for isolating it from other devices in the BICMOS process.

Contact formation: The formation of low-resistance contacts to the different layers of the SiGe HBT is critical for achieving high performance and minimizing parasitic effects.

Passivation: The quality of the passivation layer is important for protecting the SiGe HBT from external contamination and ensuring long-term reliability.

Challenges are described in Chevalier et al. (2015) and also described below. The following challenges can be seen:

Controlling vertical profiles: The reduction of the process thermal budget allows for controlling steeper vertical profiles, which is important for increasing the $f_t$ of SiGe HBTs. However, this can also result in reduction of Boron diffusion, which is detrimental to the extrinsic base resistance and thus to the $f_{max}$.

Developing new architectures for contacting the different layers: To address the issue of the extrinsic base resistance, new architectures needed to be developed that provide a lower resistance between the intrinsic and extrinsic base. This often requires forming the extrinsic base after the deposition of the intrinsic base.

Sensitivity of the CMOS devices: CMOS devices, especially the more scaled CMOS nodes, are becoming more sensitive to any modification of the process thermal budget. This limits the maximum allowed thermal budget added by the SiGe HBT integration, and makes it difficult to modify the source/drain anneal to fit SiGe HBT needs without having to tune the implants of the MOS devices. In general, high-k/metal gate technology poses a challenge for integrating BICMOS. The thermal budget of SiGe HBT processing can impact the crystallinity of the high-k material and complicate gate patterning. This problem becomes even more complex in fully depleted silicon on insulator (FDSOI) technology, where the gate stack determines the threshold voltages of long MOS transistors, and there is no pocket and channel implant to adjust the $V_T$ in response to the added thermal budget from the SiGe HBT fabrication. Therefore, it is crucial to determine the maximum thermal budget that can be added to the CMOS core process, depending on the bipolar integration schemes. A study conducted on 28 nm FDSOI through the evaluation of a 18 Mb SRAM yield (Chevalier et al., 2015) indicates that the location of the thermal budget is more critical than its value.

Improved patterning: The use of 193 nm deep UV lithography and thin resists adapted to the low topography of MOS devices has improved patterning. However, the height of bipolar transistors is usually much higher than that of CMOS gates, which may require modification of CMOS patterning steps to protect bipolar devices from gate etching.

Pre-metal dielectric (PMD) thickness: The height of the bipolar transistor is a concern with respect to the PMD thickness, which shrinks continuously with CMOS scaling. Thickening of the PMD in BICMOS could circumvent this limitation, but there is little margin for error since augmenting the contact height/width aspect ratio could lead to open contacts issues. Therefore, a dramatic reduction of the bipolar transistor height is mandatory for next BICMOS generations. In the next sections, we will focus on the growth of the SiGe/Si layers and the challenges related to doping and activating the dopants in the layers.

#### 5.3.1.1 *Growth of silicon-germanium layers*

Growing SiGe layers on Si for bipolar transistors can be challenging due to a few factors.

One of the main challenges is the lattice mismatch between Si and SiGe. SiGe is a compound semiconductor made of silicon and germanium, and the lattice constant of SiGe is larger than that of Si. The mismatch is heavily dependent on the germanium concentration. The critical thickness is the thickness at which the strain in the SiGe layer becomes large enough to cause the formation of defects, such as dislocations or stacking faults. This can occur when the lattice mismatch between the Si substrate and the SiGe layer exceeds a certain threshold. The critical thickness is also related to the composition of the SiGe layer, with a higher Ge content requiring a thinner critical thickness. The critical thickness can be influenced by the growth conditions, such as temperature and pressure, as well as the surface preparation of the substrate. Typically, the critical thickness is in the range of 5–10 nm for SiGe layers grown on Si substrates.

Precise control of the SiGe alloy composition is critical to achieving desired device characteristics. However, achieving uniform and precise alloy composition can be challenging due to variations in material properties and growth conditions.

Next to that, the quality of the interface between the Si and SiGe layers is critical to device performance. Precise transitions between the layers are required. Any defects or impurities at the interface can affect carrier transport and the device reliability.

The growth of Si and SiGe layers can result in surface roughness and morphology issues, which can affect the quality of subsequent layers and device performance.

Until the beginning of 1980s, MBE was used to grow the layers. In this process, high-energy beams of Si and Ge atoms are directed at the substrate, causing them to react and deposit the material. It allows to precisely control the thickness and composition of the layers, but is a less commercially suitable process. Today, epitaxial growth using CVD is widely used to deposit the SiGe and Si layers. In this process, a gas mixture containing the precursor gases for SiGe or Si is passed over the substrate, which reacts and deposits the desired material on the substrate.

#### 5.3.1.2 *Doping of silicon-germanium layers*

Doping the SiGe layer and activation of the dopants in the layer are crucial steps in the fabrication of SiGe HBTs, as they control the electrical properties of the base layer and ultimately determine the performance of the transistor. Typically in a NPN transistor, dopants like boron (B) are used to achieve P-type doping of the base layer in SiGe HBTs and dopants like phosphorus (P) or arsenic (As) are used to achieve N-type doping of the device. Carbon doping is also used to improve the performance of SiGe HBTs (Bouhouche et al., 2009; Gauthier et al., 2018). In this case, a small amount of carbon is added to the base region of the SiGe HBT during the growth process. The carbon atoms substitute for some of the germanium atoms in the SiGe crystal lattice, creating carbon-germanium (C-Ge) bonds. The use of carbon doping in SiGe HBTs has several advantages. First, it can significantly increase the base doping concentration, which can improve the transistor's current gain. Secondly, it can improve the thermal stability of the transistor by reducing the diffusion of dopants from the base into the emitter and collector regions. Finally, it can reduce the recombination rate in the base region, which can improve the transistor's high-frequency performance.

There are several ways to dope the layers:

Ion implantation: Ion implantation is a widely used technique for doping SiGe and Si layers in SiGe HBTs. In this process, impurity atoms are implanted into the substrate using an ion beam. This technique allows for precise control over the doping concentration and depth profile.

Diffusion: Diffusion is a process where impurity atoms are introduced into the SiGe or Si layer by heating the substrate in the presence of a dopant source, typically a layer that is put on top of the target layers, containing a large amount of dopants. This technique is less precise than ion implantation but can be used for low doping concentrations.

Epitaxial growth: Epitaxial growth can also be used to dope SiGe and Si layers. In this process, impurity atoms are incorporated into the SiGe or Si layer during growth. This technique can be used to achieve high doping concentrations and precise doping profiles. Doping of the SiGe layer in SiGe HBTs is typically achieved through in-situ doping during the growth of the SiGe layer. The two main methods used for in-situ doping are MBE and MOCVD during the growth of the layers itself. In MBE, dopant atoms such as boron or phosphorus are introduced into the growth chamber and deposited onto the substrate along with the Si and Ge atoms. The dopant atoms are incorporated into the growing SiGe layer as it is formed, resulting in a highly uniform and precise doping profile.

In MOCVD, dopant atoms are introduced into the growth chamber as a gas, and they react with the Si and Ge precursors to form the SiGe layer. The doping concentration can be precisely controlled by adjusting the flow rate of the dopant gas.

Next to doping the Si/SiGe layers, the collector region is selectively implanted to achieve high doping concentrations in this region. This is achieved using a technique called selectively implanted collector. In this process, the collector region is implanted with dopant atoms after the base and emitter regions have been defined. This allows for higher doping concentrations in the collector region without affecting the doping profiles in the base and emitter regions.

However, there are several issues that arise when doping the layers in SiGe HBTs:

Doping nonuniformity: The doping process is not always uniform across the entire SiGe layer, resulting in variations in electrical properties and performance. This can lead to device-to-device variability and lower overall performance.

Doping depth: The depth of the doped layer is difficult to control and can lead to nonuniformity in the electrical properties of the base layer.

Doping concentration: The doping concentration is a critical factor in determining the electrical properties of the base layer. However, it is difficult to control the doping concentration precisely, which can lead to variations in performance.

Interdiffusion: Interdiffusion of the dopant atoms with the SiGe layer can occur during high-temperature processing, leading to changes in the electrical properties of the base layer.

Finally, even if the layers are in-situ doped during the growth, annealing techniques often complement the process to further increase the active dopant level. Advanced annealing techniques like dynamic surface annealing (DSA) (Gauthier et al., 2017) and laser annealing (Lorito et al., 2008) are being considered for implementation in BICMOS flows.

DSA is a technique that uses a fast-moving strip heater to anneal the surface of the silicon substrate. The strip heater is moved across the surface of the substrate at a high speed, creating a transient temperature rise that anneals the surface layer of the silicon. This technique allows for the creation of high-quality SiGe layers with controlled doping profiles, improved interface properties, and reduced defects.

Laser annealing, on the other hand, uses a high-energy laser beam to anneal the SiGe layer. The laser beam heats the SiGe layer to a high temperature, causing the atoms to rearrange and forming a single-crystal structure. This technique is particularly useful for creating shallow junctions, reducing the sheet resistance, and improving the mobility of carriers.

The use of these advanced annealing techniques in SiGe BICMOS technology offers several benefits. For instance, it can improve the overall performance of the device by reducing defects, improving dopant profiles, and creating high-quality SiGe layers. Additionally, these techniques can enable the creation of advanced structures such as ultra-shallow junctions and SiGe channels with improved mobility.

However, there are also some disadvantages associated with the use of DSA and laser annealing. For example, they can be expensive and time-consuming, as they require specialized equipment and extensive process optimization. Moreover, they can also increase the risk of damaging the device (both HBT and CMOS) during the annealing process, leading to reduced yield and reliability.

### 5.3.2 Silicon-germanium heterojunction bipolar transistor versus SiGe BICMOS

As we have seen before, SiGe BICMOS is a combination of bipolar transistors and CMOS transistors on the same substrate. The BICMOS process allows for the inclusion of high-performance SiGe HBTs and low-power CMOS transistors on the same substrate. This technology can provide the high-frequency performance of HBTs and the low-power consumption of CMOS, making it suitable for applications that require both high-frequency and digital signal processing.

The performance of a standalone SiGe HBT and an HBT in a SiGe BICMOS process may differ due to the integration of the HBT with CMOS transistors. The SiGe HBT in BICMOS process goes through additional processing steps compared to a standalone SiGe HBT, which can affect the performance of the HBT. For example, the SiGe HBT in BICMOS process may be subjected to additional thermal cycles during the CMOS processing, which can affect the device's electrical characteristics such as the breakdown voltage, current gain, and noise figure. Additionally, the integration of the HBT with CMOS transistors can also introduce additional parasitics, which can also affect the HBT's performance. However, the performance difference can be small or significant depending on the specific process and the design of the circuit.

## 5.4 InP heterojunction bipolar transistor

### 5.4.1 Introduction

The history of III–V HBTs can be traced back to the 1970s, when researchers first began experimenting with using different semiconductor materials to improve the performance of bipolar transistors. One of the first III–V HBTs to be developed was the InP HBT, which was made using InP as the base and emitter material. InP HBTs had much higher electron mobility than silicon bipolar transistors, making them well suited for high-frequency applications such as microwave and millimeter-wave communications.

Next to InP HBTs, GaAs HBTs have been quite successful and used in a variety of applications. GaAs HBTs are made using a combination of GaAs (base and collector) and indium gallium phosphide GaA or InGaP to create the heterojunction (emitter). GaAs HBTs have a high breakdown voltage and high thermal stability, which made them particularly well suited for high-power, high-frequency applications, but they have lower mobility and thus lower $f_t/f_{max}$ than InP.

InP and GaAs HBTs both have advantages in their own respective application areas. InP HBTs have high electron mobility and fast switching speeds, making them well suited for high-frequency applications such as microwave and millimeter-wave communications, L-band and C-band amplifiers. GaAs HBTs have higher breakdown voltages, and they are commonly used in high-power and high-frequency applications such as satellite communication systems and cellular base stations. Moreover, these devices are used in the RF-FEM of high-end smartphones. GaAs HBTs are typically used for sub-6GHz applications, while InP HBTs target frequencies above 100 GHz.

Apart from the typical InP HBT using a p-type doped InGaAs layer as base, alternative structures have been proposed for example, in Bolognesi et al., 1998. In this work, a InP-GaAsSb-InP structure is described. One of the advantages using GaAsSB is related to the staggered band alignment, whereby the GaAsSb conduction band sits above the InP conduction band. This staggered band line-up at InP-GaAsSb interfaces eliminates any possibility of current blocking at the base-collector junction, allowing for ballistic electron injection and high saturated drift velocity with low rates of impact ionization. The InP-GaAsSb-InP DHBTs can be implemented without compositional grading and with nominally abrupt interfaces, making turn-on and offset voltages determined by band line-ups, doping, and junction areas rather than effectiveness of compositional grading. The active doping level of C-doped MOCVD-grown GaAsSb bases can reach values up to $10^{20}$ $cm^{-3}$ without H-passivation effects. Low p-type Schottky barrier heights on antimony compounds lead to low resistance base ohmic contacts.

Devices with $f_{max}$ up to 1.2 THz have been shown (Arabhavi et al., 2022), and PAs with saturated output power of 14.5 dBm were shown at 94 GHz (Hamzeloui et al., 2022).

## 5.4.2 Upscaling the InP heterojunction bipolar transistor

One of the challenges with using InP as a substrate for electronic devices is that it is more expensive and less widely available than silicon, which is the dominant material used in the semiconductor industry. InP is also more difficult to process than silicon, which can add to its cost. In addition, InP has a smaller substrate size (up to 6-inch) compared to silicon, which can be a limitation for some applications.

InP also has a lower thermal conductivity compared to silicon, which can be an issue for high-power devices that generate a lot of heat. InP is generally considered to be brittle, and it is more prone to cracking and breaking when subjected to mechanical stress than some other semiconductor materials. This can make it difficult to achieve the desired level of device performance and yield. There are a few factors that contribute to the brittleness of InP. First, InP has a high melting temperature (1635°C) and is therefore sensitive to thermal expansion mismatch with other materials Coefficient of Thermal Expansion (CTE). When InP is subjected to temperature changes, it expands and contracts more than materials with lower melting temperatures, which can cause it to become stressed and brittle.

Secondly, InP is a compound semiconductor that is made up of two elements (indium and phosphorus) with very different atomic sizes. This causes InP to have a high degree of internal strain.

Finally, InP is sensitive to moisture and can corrode in the presence of water. This can further contribute to its brittleness, as the corrosion can weaken the material and make it more prone to breaking.

While for many applications up to now, the use of InP substrates has been acceptable, it hampers the uptake of this technology for more high volume, mass market applications for example, and the use of InP technology in future 6G applications. Availability, maturity, yield, and cost need to be tackled. Currently, several ways of upscaling InP are under investigation and they are summarized in Fig. 5.8.

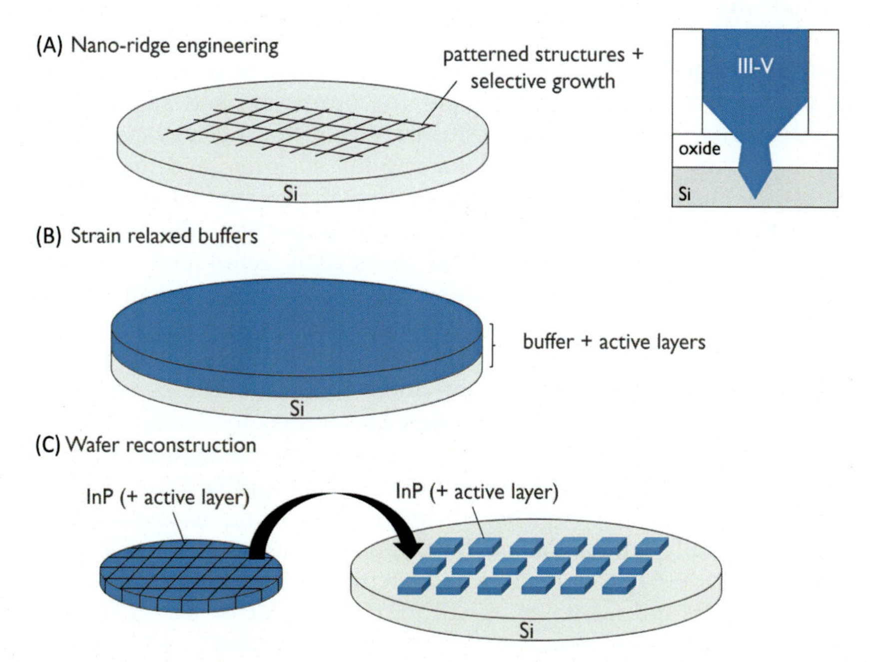


**FIGURE 5.8** Schematic presentation of the different InP technologies under investigation; (A) nano-ridge engineering, (B) blanket strain relaxed buffer, and (C) wafer reconstruction. *Source: From Collaert, N., Alian, A., Banerjee, A., Boccardi, G., Cardinael, P., Chauhan, V., Desset, C., Elkashlan, R., Khaled, A., Ingels, M., Kunert, B., Mols, Y., O'Sullivan, B., Peralagu, U., Pinho, N., Rodriguez, R., Sibaja-Hernandez, A., Sinha, S., Sun, X., … Peeters, M. (2022). III-V/III-N technologies for next generation high-capacity wireless communication. In:* Technical digest – International electron devices meeting, IEDM *(Vols. 2022-, pp. 1151–1154). Institute of Electrical and Electronics Engineers Inc. https://doi.org/10.1109/IEDM45625.2022.10019555.*

The first approach involves direct growth of the III–V material on Si, with both blanket and selective area growth methods being investigated for upscaling. Nano-ridge engineering (NRE), which combines defect trapping and epitaxial lateral overgrowth, has been shown to be the most effective way of reducing defectivity and increasing the active area. NRE has been used for the integration of lasers and GaAs HBT devices (Vais et al., 2019, 2022).

The other approach involves transferring InP tiles, with or without the active layers, onto a large-size and cost-effective wafer using die-to-wafer bonding, also known as wafer reconstruction technology. One advantage of this approach is that it allows for starting with high crystal quality. In the next sections, we will go a bit deeper into several of these integration schemes. We will include microtransfer printing as well, as a version of wafer reconstruction.

#### 5.4.2.1 *Microtransfer printing*

Microtransfer printing is a technique used to transfer microscale materials or devices from one substrate to another with high precision and accuracy. This technique is at this moment used for photonics applications, particularly for transferring InP lasers and photodetectors (Roelkens et al., 2023). By using microtransfer printing, one can fabricate high-performance InP devices on a separate substrate and then transfer them onto the final application substrate.

Microtransfer printing is a complex multi-step process that involves several steps. The process requires careful control of the temperature, pressure, and adhesion properties of the materials involved to ensure a successful transfer of the InP devices. The InP devices can be fully or partially processed on the starting InP substrate and then transferred onto the target substrate that can be a (pre-processed) Si wafer or another substrate.

The steps that are involved are described below. In a first step, the InP devices are fabricated on a donor substrate using standard InP fabrication techniques. Then a thin layer of sacrificial material, such as polyimide or photoresist, is deposited onto the donor substrate. This stamp allows the pickup of several coupons or InP devices/circuits simultaneously. The transfer or target substrate, such as glass or silicon, is coated with an adhesive layer that can bond to the sacrificial layer on the donor substrate. The donor and target substrates are brought into contact with each other and heated to a temperature that creates a strong bond between the two substrates. The donor substrate is then separated from the target substrate, leaving the InP devices bonded to the adhesive layer on the target substrate. Further processing of the transferred devices is possible using standard fabrication techniques. This typically entails the processing of additional metal layers to enable the proper connections with the other devices and circuits on the target substrate.

In Carter et al., 2019 and Carter et al. (2019), the transfer of InP HBTs onto a several kinds of substrates has been demonstrated using microtransfer printing, showing the potential of this technique to enable heterogeneous InP/CMOS circuits.

#### 5.4.2.2 *Nanoridge engineering*

NRE is a technique used to grow III–V materials, such as (In)GaAs and InP, onto Si substrates. The technique of growing III–V materials onto Si substrates has a long history, dating back to the 1970s when it was first demonstrated that GaAs could be grown on Si using metalorganic vapor phase epitaxy. However, the growth of high-quality III–V materials on Si substrates continues to be a challenge due to the lattice mismatch between the two materials, which can cause defects and reduce the quality of the material.

In recent years, NRE has emerged as a promising approach for overcoming this challenge. In this technique, a pattern of small trenches, or "nano-ridges," is created on the surface of the Si substrate using techniques such as standard photolithography or electron beam lithography. The III–V material is then grown selectively in the nano-ridges. The high aspect ratio trench allows to trap most threading dislocations and planar defects in the trench. In a second step, the III–V material is grown out of the trench. That material has significantly lower defectivity and can be used to create the devices.

This technique has been demonstrated for GaAs HBTs (Vais et al., 2019, 2022) and for the integration of lasers and photodetectors in photonic applications (Ozdemir et al., 2021; Shi et al., 2019). A cross-section Transmission electron microscopy (TEM) and topview scanning elecron microscopy (SEM) of a GaAs HBT device is shown in Fig. 5.9, highlighting not only the good crystal quality and the ability to fabricate a complex heterostructure using this technique, but as well the unique layout of a device built up using these nano-ridges.

There are several advantages to using NRE for growing III–V materials onto Si substrates. One advantage is that it allows for the growth of high-quality III–V materials with

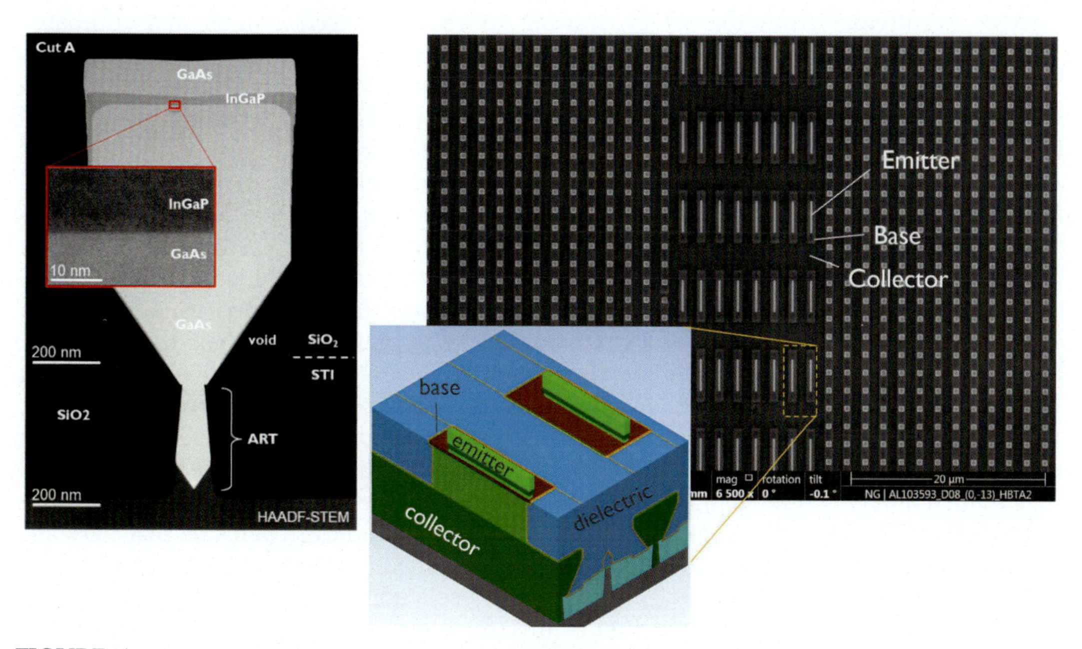


**FIGURE 5.9** (Left) TEM of a grown nano-ridge structure showing the InGaP/GaAs heterostructure, (right) topview SEM of a GaAs device on 300 mm Si after emitter patterning showing the unique layout of the nano-ridge HBT. *Source: From Collaert, N., Alian, A., Banerjee, A., Boccardi, G., Cardinael, P., Chauhan, V., Desset, C., Elkashlan, R., Khaled, A., Ingels, M., Kunert, B., Mols, Y., O'Sullivan, B., Peralagu, U., Pinho, N., Rodriguez, R., Sibaja-Hernandez, A., Sinha, S., Sun, X., ... Peeters, M. (2022). III-V/III-N technologies for next generation high-capacity wireless communication. In:* Technical digest – International electron devices meeting, IEDM *(Vols. 2022-, pp. 1151–1154). Institute of Electrical and Electronics Engineers Inc. https://doi.org/10.1109/IEDM45625.2022.10019555.*

reduced defects and improved performance. As compared to blanket layer strain relaxed buffer (SRB) approaches (see next section), the advantage of this technique lies that much less III–V material is needed to reduce defectivity. Another advantage is that it is compatible with existing Si-based technologies, enabling the integration of III–V materials with these technologies in the same substrate. This could potentially lead to the development of new and improved devices that are faster, more energy efficient, and more reliable.

There are also some disadvantages to using NRE. One disadvantage is that it requires precise control of the size and shape of the nano-ridges, which can be challenging to achieve. Another disadvantage is that it can be difficult to grow III–V materials with high uniformity along the nano-ridges, which can impact the performance of the material.

Despite these challenges, NRE is a promising approach for growing III–V materials on Si substrates and has the potential to enable the upscaling of these materials for a wider range of applications. Research in this area is ongoing, and it is likely that advances in the technology will lead to further improvements in the quality and performance of III–V materials grown using this technique.

#### *5.4.2.3 Strain relaxed buffers*

SRBs are used to alleviate the lattice mismatch between III–V semiconductors and Si substrates or in principal any other target substrate. The lattice constant of III–V semiconductors is different from that of Si, leading to strain-induced defects and ultimately degraded device performance. The SRB allows the epitaxial growth of high-quality III–V materials on Si substrates by accommodating the lattice mismatch and relaxing the strain. Different from NRE, this is a blanket process where the III–V layer is grown over the entire substrate without the need for advanced, scaled features.

In general, there are two main types of SRBs: graded and compositionally step-graded. Graded SRBs have a continuously varying lattice constant from the Si substrate to the III–V layer, while compositionally step-graded SRBs have abrupt changes in lattice constant due to the introduction of different III–V alloys with varying lattice constants.

For GaAs-based devices, a graded SRB is typically used. The SRB is usually made of AlAs or AlGaAs, which have a lattice constant closer to that of GaAs than Si. The graded SRB allows for gradual accommodation of the lattice mismatch between the GaAs and Si, resulting in a lower defect density and improved device performance. However, the thickness of the SRB needs to be optimized to balance the tradeoff between defect density and strain relaxation. If the SRB is too thin, it may not be able to accommodate the lattice mismatch adequately, resulting in a high defect density in the epitaxial layer. On the other hand, if the SRB is too thick, it can introduce strain relaxation mechanisms that may also increase the density of threading dislocations in the epitaxial layer.

InP-based devices, on the other hand, typically use a compositionally step-graded SRB. This approach involves the introduction of InGaAs layers with varying In composition to gradually accommodate the lattice mismatch between InP and Si. Also InAlAs can be used for the SRBs. InAlAs has also a lattice constant between that of InP and Si, making it suitable for accommodating the lattice mismatch between InP-based materials and Si substrates.

Similar to AlAs or AlGaAs-based SRBs for GaAs-based devices, InAlAs-based SRBs can also be graded or compositionally step graded. The choice of the SRB design depends on the specific device requirements and growth conditions. However, it's worth noting that the growth of InAlAs-based SRBs can be more challenging than that of AlAs or AlGaAs-based SRBs due to the formation of indium-rich interfacial layers, which can result in increased defect density. Therefore, the growth conditions need to be carefully controlled to optimize the quality of the InAlAs-based SRB and the resulting III–V material stack grown on top.

Step-graded SRBs can provide a lower defect density compared to graded SRBs, as the abrupt changes in lattice constant can effectively block the propagation of dislocations. However, the growth conditions need to be carefully controlled to prevent the formation of undesirable interfacial layers.

One of the main advantages of using SRBs is the ability to integrate high-performance III–V materials with existing Si-based technology. This can lead to improved device performance and reduced fabrication costs for the III–V devices as, being grown on a Si substrate, they can use the toolsets and processes developed for CMOS devices. Additionally, the use of SRBs can enable the integration of new functionalities, such as III–V quantum wells or quantum dots, with Si-based electronics. Complex buffers made of GaAs, InP, and InAlAs have been shown with total thickness less than 1 um in (Huang et al., 2015, 2016) and InGaAs devices fabricated using 300 mm processes.

However, there are also some challenges associated with SRBs. The growth of SRBs requires precise control of the growth conditions, including temperature, growth rate, and composition. The introduction of dislocations at the SRB interface can also limit the quality of the III–V material grown on top. In addition, the use of SRBs can result in increased processing complexity and cost due to the need for additional growth steps. From a cost perspective, while the processing of the devices using typical CMOS toolsets can be beneficial for cost, the growth of thick III–V SRBs brings a significant additional cost with it. Cost modeling shows that if some form of reusability can be used, this would benefit the overall cost of fabricating the substrate. Reusability can be achieved for example, by using SmartCut (Sollier et al., 2015) after the growth of the SRB, whereby the top layer of the SRB is transferred to the target substrate. The SRB on the donor substrate as such can be used a few times, making the cost of the SRB more acceptable.

In SmartCut, originally used to fabricate silicon-on-insulator (SOI) wafers, a substrate is first hydrogen-implanted to create a thin layer of hydrogen in the material. This hydrogen layer acts as a separation layer between the donor substrate and the active layer. Next a target wafer is bonded to the hydrogen-implanted substrate using a process called wafer bonding. This results in a bonded wafer structure, combining the target and donor substrate. Then the bonded wafer is heated to a high temperature, which causes the hydrogen to diffuse and creates a weak point between the donor substrate and the active layer. At this stage, a technique called temperature-induced cleavage is applied, where the wafer is cooled rapidly, causing the active layer to separate from the substrate along the weak point created by the hydrogen implant. The active layer is then removed from the donor substrate and becomes part of the target substrate. In this way, thin slices (or active layer) of the donor substrate can be transferred to several target substrates.

#### 5.4.2.4 *Reconstructed wafers*

Reconstructed wafers are wafers that are composed of multiple layers or dies of different materials, which have been bonded together to form a single reconstructed wafer.

It involves transferring InP tiles, with or without the active layers, onto a large, inexpensive, and robust wafer (often silicon, but could in principle be extended to other substrates) using a process called die-to-wafer bonding. This method allows for the use of high-quality InP crystals as a starting point, which is essential for the high-performance devices required in many applications. This is different from the chiplet approach as the transferred dies typically have not seen any processing prior besides the possible epitaxial growth of the active layers.

Despite its potential advantages, wafer reconstruction also presents several challenges. One key challenge is the efficient transfer of the materials, which requires careful attention to the bonding process and the selection of suitable adhesion layers. Another challenge is the removal of the thick InP substrate, which can be accomplished using techniques such as grinding, laser lift-off, or SmartCut (Strinati et al., 2022), each with its own advantages and disadvantages.

In addition to these challenges, wafer reconstruction also requires careful control of the CTE mismatch between the different materials in the reconstructed assembly, which can cause warpage, delamination, and planarization issues. These challenges must be addressed through careful design and optimization of the bonding and processing steps to achieve a successful and reliable wafer reconstruction process.

Despite these challenges, wafer reconstruction technology has the potential to significantly reduce the cost of producing InP-based devices and enable their use in a wider range of applications, while potentially maintaining good crystal quality. Furthermore, this technology is also being investigated for other compound semiconductor devices for applications in micro-LEDs, highlighting its versatility and potential impact on various fields of research and industry.

### 5.4.3 Heterojunction bipolar transistor versus high electron mobility transistor

InP HEMT and InP HBT are both types of transistors that are used in wireless communication systems.

An InP HEMT has a unique structure that includes a buried layer of (In)GaAs on top of the InP substrate capped with either a InAlAs or InP layers, forming a complex heterostructure. This structure allows for a high electron mobility, which makes the device well suited for high-frequency applications such as microwave and millimeter-wave communications.

The main difference between the two types of transistors is the way they operate. HEMT is a type of FET, while HBT is a BJT.

The HEMT has some advantages over the HBT. These include lower input capacitance, high electron mobility, high-frequency response, and lower noise figure, important for low noise amplifiers (LNA), while the HBT advantages lie in the high current gain and breakdown voltage that can be achieved, next to the high reliability, stability, and higher power and current densities that are typically found for HBTs.

In wireless communications, HEMT devices are often used in the front-end amplifier stages of the receive paths, while HBT devices can be used in the PA stages.

Let's dive a bit deeper in the operation of a HEMT device and compare it to the operating principle of a HBT as we have seen before.

An HEMT device is a barrier-controlled transistor. The very first III–V transistors were GaAs metal semiconductor field effect transistors or MESFETs. These devices had no gate insulator because of the difficulties to fabricate well-controlled and reliable gate dielectrics on GaAs. The structure is shown in Fig. 5.10. It has a lightly n-type doped channel and highly doped source and drain areas. The gate is a Schottky gate. It is a normally on device or depletion mode device (D-mode). By putting a reverse bias on the gate, the depletion region is pushed down, preventing the current to flow and switching the device off. This structure, however, was not ideal as relatively high doping of the channel was needed to get enough charge or carriers into the device, thereby decreasing the overall mobility. Moreover, the limitation of the Schottky barrier gate is that the amount of bias that can be applied is limited.

The concept of modulation doping, introduced end of the 1970s (Dingle et al., 1978), was interesting. It allowed to increase the amount of carriers in the channel without doping the channel region and thus degrading the mobility. When a highly doped wide bandgap material (e.g., AlGaAs) is brought into contact with a undoped small bandgap material (GaAs or InGaAs), electrons will flow from the highly doped wide bandgap material to the undoped small bandgap material. The electrons in the undoped layer do not experience scattering because of the ionized impurities and as such we have a high concentration of carriers with high mobility in this small bandgap material. The band structure is shown in Fig. 5.11. At the heterojunction interface between the different materials a

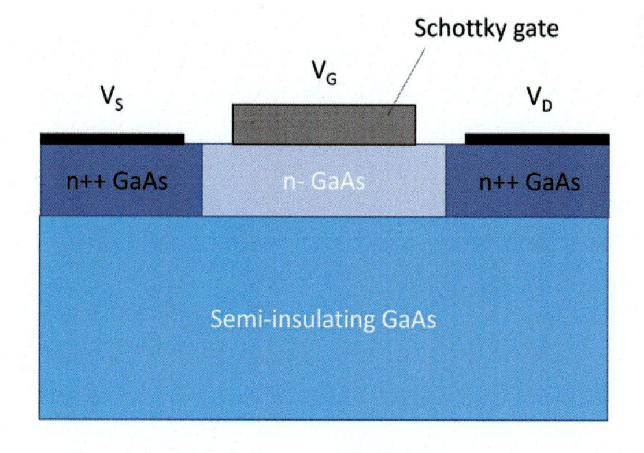


**FIGURE 5.10** Schematic presentation of a GaAs MESFET.

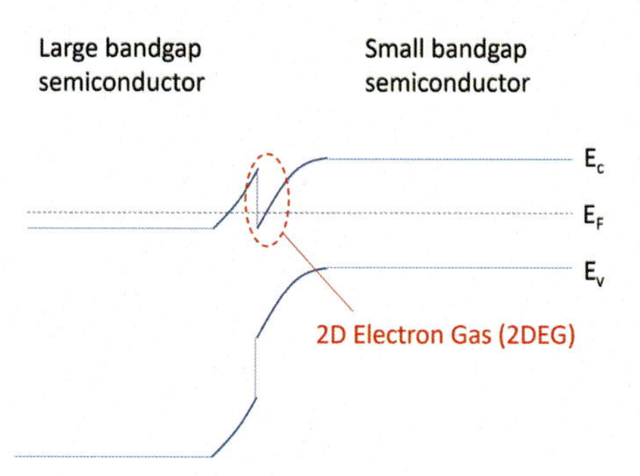


**FIGURE 5.11** The band structure of a highly doped wide bandgap material in contact with a intrinsic small bandgap material forming the 2DEG, where $E_C$ represents the conduction band energy, $E_V$ the valence band energy, and $E_F$ the Fermi level.

two dimensional electron gas (2DEG) is created, very similar to the inversion layer in a MOSFET that is also created by the potential well at the oxide-semiconductor interface.

A more realistic band structure is shown in Fig. 5.12 taken into account the Schottky gate as well. In this figure only the conduction band is shown. When designing this HEMT, we need to make sure that no parasitic conducting layer is formed in the wide bandgap, coming from an nondepleted region in this layer. These carriers have a low mobility and degrade the performance of the device. Therefore, the sum of the two depletion layers $W_1$ and $W_2$, coming from the gate metal and the semiconductor on one hand and coming from the heterojunction between the wide and small bandgap materials, needs to be equal to the thickness of the wide bandgap material, in this case defined as $t_{ins}$. This top wide bandgap material can be seen as an insulator layer, similar to the gate dielectric in a MOSFET, with main difference that the bandgap of this layer is not as wide as the $SiO_2$ layer in a MOSFET and much lower voltage can be applied to this layer.

Often an undoped wide bandgap material is inserted between the doped wide bandgap semiconductor and the low bandgap semiconductor to reduce the remote impurity scattering. The dopants in the wide bandgap material are then moved away from the electrons in the 2DEG. The thickness needs to be optimized such that it is not too thick whereby it inhibits the transfer for electrons to the small bandgap material and that it is not too thin such that remote impurity scattering affects the mobility of the carriers in the 2DEG.

Instead of using a highly doped wide bandgap material, more advanced structures these days will use a delta doping in this layer. This delta doping is an atomic layer of dopants inserted in the undoped wide bandgap material. The dopants are located in a well-defined position in the material, moved away from the small bandgap material. There are a numbers of benefits linked to this delta doping. A higher amount of carriers can be introduced. The delta doping induces an electric field that can increase the breakdown voltage of the device. The gate electrode can be put closer to the channel thereby improving the transconductance $g_m$ and thus the $f_t$ and $f_{max}$ of the device.

While in the early days AlGaAs/GaAs combinations were used, today InAlAs/InGaAs and InP/InGaAs are used.

A typical device has either a InP or GaAs substrate with a buffer layer grown by MBE or MOCVD on top of this substrate, typically a InAlAs buffer. On top of the buffer the

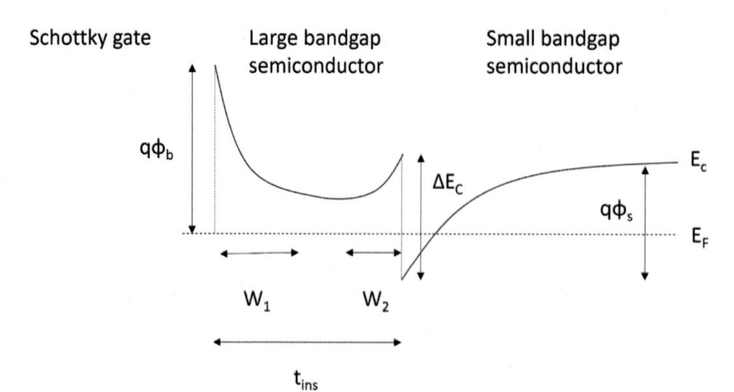


FIGURE 5.12 Band structure for the Schottky gate/wide bandgap/small bandgap structure of the high electron mobility transistor; $E_C$ represents the conduction band energy, $E_F$ the Fermi level, $\varphi_b$ the Schottky barrier, $\Delta E_C$ the conduction band off-set, $\varphi_s$the surface potential, and $W_1$ and $W_2$ the depletion regions in the wide bandgap material.

small bandgap material is grown. This is often an InGaAs layer. Depending on the starting substrate and buffer, this layer can contain low In concentrations (<53% for GaAs substrate) or high concentrations (>53% when starting from an InP substrate). The higher the In concentration, the higher the mobility and the lower the bandgap. The extreme case is when InAs is used (Kharche et al., 2009). But because of the lattice mismatch the thickness of this layer can only be a few nanometers.

On top of the channel layer, the insulating layer or wide bandgap material is grown with the delta doping. This is typically an InAlAs layer.

To make good ohmic contacts lower bandgap materials will be put on top of the InAlAs layer in the source and drain areas. The metal contacts itself are typically using Au.

The gate has a typical T-shape, which is optimized for high speed performance. The stem of the gate (bottom part) is typically below 50 nm, while the top of the gate is larger allowing to reduce the gate resistance and therefore optimize the $f_{max}$. The structure is schematically shown in Fig. 5.13. This transistor also operates very close to the ballistic limits (Del Alamo et al., 2021; Kim & Del Alamo, 2011).

When comparing the III–V HEMTs (and especially the InP HEMT) with other technologies like CMOS, SiGe BICMOS, GaN HEMT, and InP HBT, we see that these devices can achieve very high $f_t$ and $f_{max}$, even higher than InP HBT as shown in Fig. 5.14. However, they have still have a lower breakdown voltage and can therefore deliver less power than InP HBTs. On the other hand, they exhibit excellent low noise figures (NF) making them very suitable for LNAs.

Over the last few decades, with the revival of III–V for digital applications, the work on III–V MOSFETs has also progressed. The key challenge for III-V MOSFETs, however, is still enabling a high-quality gate dielectric that has low interface states and in general excellent reliability performance (e.g., bias temperature instability or BTI) to be used across several applications. Currently, the most widely studied dielectric materials for III–V MOSFETs are still $Al_2O_3$, $HfO_2$, and $ZrO_2$.

In summary, in this section we have looked at the operation principle of III–V HEMTs as an alternative to III–V HBTs. Just like MOSFETs these transistors are barrier-controlled

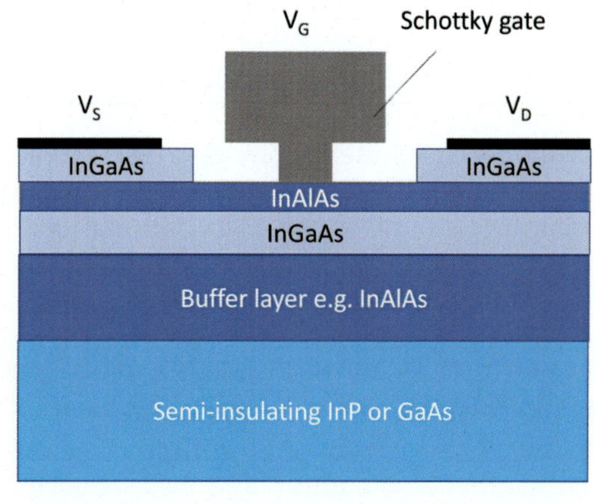


**FIGURE 5.13** A schematic view of a typical HEMT showing the layer stack and source and drain designs.

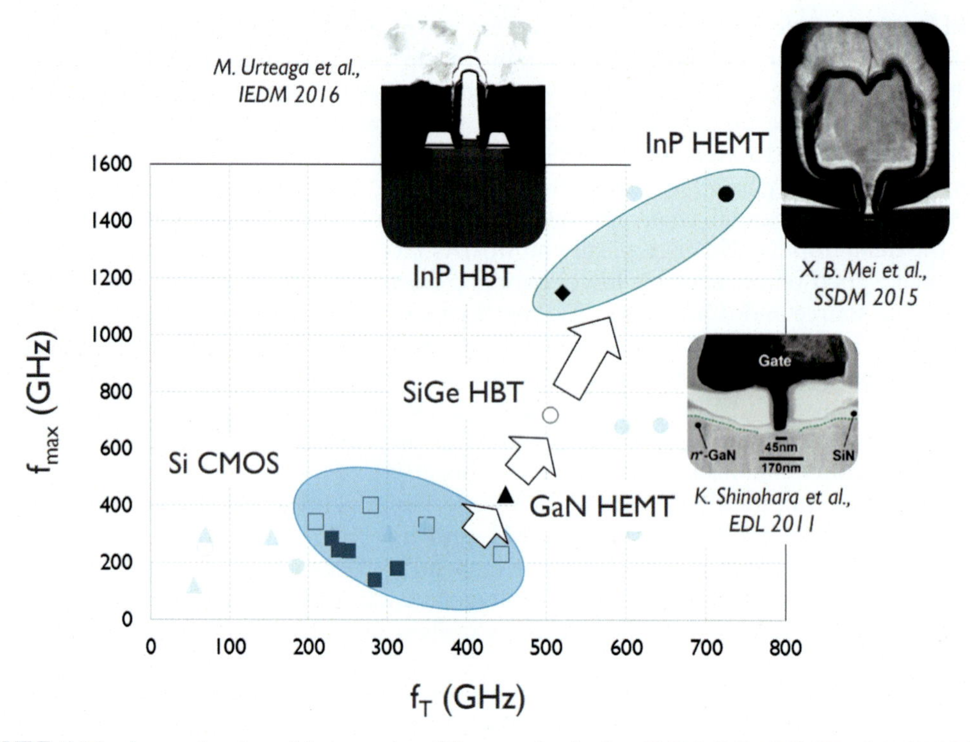


**FIGURE 5.14** $f_{max}$ as function of $f_t$ comparing different technologies: CMOS, SiGe BiCMOS, GaN HEMT, InP HBT, and InP HEMT.

devices where the benefits of III–V materials are explored when it comes to high mobility and building advanced heterostructures. Just like HBTs they are used in high speed RF applications and can achieve operation in the THz regime. Finally, these III–V HEMTs operate in the quasi-ballistic regime.

## 5.5 Device modeling

There are several commonly used device models for bipolar transistors, including the Ebers-Moll model, the Gummel-Poon model, and the high current model (HiCUM). These models have different levels of accuracy and computational complexity, and choosing the appropriate model for a given application requires careful consideration of the accuracy and computational requirements of the specific circuit or system being analyzed.

One of the earliest models of bipolar transistors was the Gummel-Poon model, developed in the 1960s by Gummel & Poon, 1970; Gummel & Poon, 1970. This model is a two-port model that describes the behavior of the transistor in terms of its small-signal equivalent circuit. It considers the emitter, base, and collector regions of the transistor and includes parameters such as the base-emitter and base-collector junction capacitances, as well as the base transit time and the emitter and collector resistances. Using these parameters, the Gummel-Poon model can be used to derive a set of nonlinear equations that describe the behavior of the transistor. These equations can be solved numerically to obtain the transistor's input and output characteristics. The Gummel-Poon model is a more complex model than the Ebers-Moll model, but it is also more computationally intensive to use.

Another important device model for bipolar transistors is the Ebers-Moll model, which was developed in the 1950s by J. Ebers and J. Moll (1954). The Ebers-Moll model is a simple, but useful model for analyzing bipolar transistors. It represents the transistor as two back-to-back diodes, with one diode representing the base-emitter junction and the other representing the base-collector junction. The model assumes that the base-emitter junction is forward biased, while the base-collector junction is reverse biased. The Ebers-Moll model describes the transistor's behavior in terms of the following parameters: the forward current gain, the reverse current gain, the emitter injection efficiency, and the base transport factor. The Ebers-Moll model can be used to analyze the behavior of the transistor in different operating regions. Overall, the Ebers-Moll model is a simple, yet an effective way to analyze the behavior of bipolar transistors, and is still widely used today for both educational and practical purposes. However, it should be noted that the model is less accurate than more sophisticated models, such as the Gummel-Poon model, which take into account additional physical effects that are not captured by the Ebers-Moll model.

The HiCUM is a more advanced model for bipolar transistors compared to the Ebers-Moll and Gummel-Poon models (Schroter & Chakravorty, 2010). It is a physics-based model that incorporates more physical effects and parameters, allowing for more accurate and detailed simulation of the transistor's behavior. The HiCUM was developed to improve the accuracy of bipolar transistor models for use in high-performance analog and mixed-signal circuit design. It was designed to account for a variety of physical effects that were not included in previous bipolar transistor models, such as the non-quasi-static effect and the high injection effect. These effects become significant at high currents and high frequencies, and can have a significant impact on the performance of bipolar transistor circuits. The HiCUM also includes temperature-dependent parameters, allowing for accurate simulation of the transistor's behavior over a wide range of temperatures.

Since its introduction, the HiCUM has become widely used in the semiconductor industry for designing and optimizing bipolar transistor circuits. It has also been adopted as a standard model by various industry organizations, such as the Compact Model Council and the Semiconductor Industry Association.

Specifically for InP HBTs, the Keysight HBT model can be used. This compact model is based on the Gummel-Poon model. The Keysight HBT model includes several components, including the intrinsic base resistance, the collector-emitter saturation voltage, and the transit time delay. It also includes an emitter-crowding effect, which is important for accurately simulating the behavior of InP HBTs.

One advantage of the Keysight HBT model is that it has been validated against a wide range of InP HBT devices, including devices with varying geometries and doping profiles. This makes the model suitable for simulating a wide range of InP HBTs for various applications, including high-speed digital circuits, microwave amplifiers, and more.

In general, today there are many different models available for bipolar transistors, ranging from simple empirical models to complex physics-based models. These models are used in the design and analysis of a wide range of bipolar transistor circuits, from low-frequency amplifiers to high-speed digital circuits and power electronics. The development of these models has been essential to the continued success and evolution of bipolar transistor technology over the past several decades.

## 5.6 Optimizing the bipolar transistor

Bipolar transistors are important devices in modern electronics, and optimizing their performance requires an understanding of several key metrics and parameters. The most important device metrics of a bipolar transistor include the current gain, voltage gain, forward current transfer ratio, breakdown voltage, and capacitances. These metrics are essential for evaluating the performance of the transistor in different circuit configurations.

Optimizing the performance of a bipolar device requires a careful balance of several parameters. One of the most critical parameters is the base width, which determines the current density in the base region. A smaller base width allows for higher current densities and output powers, but it also results in a higher base resistance and lower $f_{max}$. Similarly, the emitter width and collector width also play critical roles in determining the current density and output power of the device. However, increasing these widths also increases the resistance and reduces the $f_{max}$. Therefore, finding the right balance between these parameters is crucial for optimizing the performance of the device.

Doping levels in the base, emitter, and collector regions are also critical parameters to consider. Higher doping levels result in lower resistance and better performance, but they also increase the base-collector capacitance, which again reduces the $f_{max}$. The epitaxial layer thickness in collector and base is also an essential parameter, as a thinner active layer allows for better high-frequency response but reduces for example, the breakdown voltage, thereby reducing the output power.

Additionally, the quality of the interface between the layers must be carefully considered, as a good quality interface with low defects and low doping gradients is necessary for achieving good performance, particularly at high frequencies.

Key challenges for III–V HBTs are in general related to reducing the contact resistivities to operate the transistor at high current densities for high $f_t$ and $f_{max}$. These high current densities lead to strong self-heating of the transistor. Thermal resistance is therefore an important parameter to consider, as high thermal resistance can increase thermal noise and degrade the performance at high frequencies. Parasitic capacitances are also performance detractors. Finally, as the size of the device is scaled down, perimeter effects due to mesa etching start to play an increasingly important role in maintaining high current gain.

However, as stated in Rodwell et al. (2008), the current gain and power gain cutoff frequencies are sometimes of limited value in predicting the bandwidth of mixed-signal integrated circuits (ICs). Transistors designed exclusively for high $f_t$ at the expense of other parameters will perform poorly in circuits. For reactively tuned amplifiers used in radio transceivers, $f_{max}$ defines the highest frequency at which power gain can be obtained and is of primary importance. In HBT mixed-signal ICs, gate delay dominates over $f_t$ and $f_{max}$ due to large-signal operation, fan-in, and fan-out. Minimizing $R_{ex}$ (emitter resistance) and having low $(C_{je} + C_{cb})/I_C$ (with $I_C$ the collector current, $C_{je}$ the emitter depletion capacitance, $C_{cb}$ the collector-base capacitance) is crucial.

In the next section, we will briefly have a look at the scaling scenarios for III–V HBTs.

### 5.6.1 Scaling the heterojunction bipolar transistor

Just like the MOSFET, several scaling scenarios have been proposed for the HBT, which to a large extent have been detailed out in Rodwell et al. (2008). Depending on the trade-offs that are allowed, different scaling scenarios can be considered. In the remainder of the paragraph, we will use the scaling parameters and dimensions defined in Fig. 5.15. The critical dimensions are: $W_E$ the emitter junction width, $W_C$ the collector junction width, $W_{\text{undercut}}$ the width of the undercut of the collector junction under the base, $W_{\text{CB}}$ the width of the base contact, $W_{\text{BE}}$ the distance between emitter and base contact, $t_B$ the thickness of the base, and $t_C$ the thickness of the collector. The dimension that is not shown in this figure, but is still important is $L_E$, the length of the emitter perpendicular to the figure.

The goal of the scaling is to improve (or increase) the bandwidth of a circuit with a factor $\alpha$. To do this, one needs to reduce the capacitances and the transit delays by a factor $\alpha$, while keeping the parasitic resistances, the voltages, transconductance $g_m$, and operating current $I_C$ constant.

The scaling involves reducing the collector and base transit times, $\tau_C$ and $\tau_B$, respectively, which can be achieved by reducing the collector thickness $t_C$ by a factor $\alpha$ and the base thickness $t_B$ by slightly more than $\sqrt{\alpha}$ given the following dependencies.

$$\tau_C \sim \frac{t_C}{2v_{\text{eff}}} \tag{5.11}$$

$$\tau_B \sim \frac{{t_B}^2}{2D_n} + \frac{t_B}{v_{\text{eff}}} \tag{5.12}$$

where $v_{\text{eff}}$ is the collector effective high-field velocity and $D_n$ the electron diffusivity in the base region.

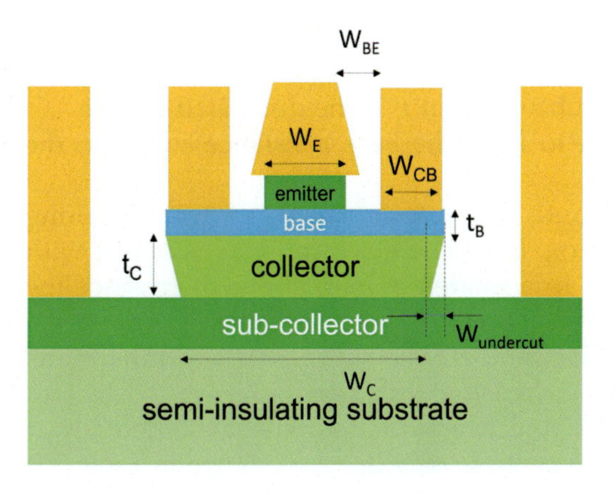


**FIGURE 5.15** Schematic cross-section of a heterojunction bipolar transistor device indicating the critical dimensions.

However, reducing $t_C$ would increase the collector-base capacitance $C_{CB}$ if the junction areas were held constant. As such, the emitter and junction areas must be reduced in proportion with a factor $\alpha^2$ as well to achieve the desired reduction in $C_{CB}$. As the emitter area reduces by this factor and given the fact that the collector current $I_C$ stays constant, the emitter current density increases by $\alpha^2$. The Kirk effect, however, puts some boundaries on the increase that is possible. The Kirk effect is a phenomenon that occurs in bipolar transistors, where an increase in the current density in the emitter region leads to an increase in the transit time of carriers through the base region. This is due to the increased electric field in the base region, which slows down the transport of carriers and reduces the efficiency of the transistor. The Kirk effect is a limiting factor in the performance of high-speed bipolar transistors, and it imposes a limit on how small the emitter region can be scaled down while maintaining a high level of performance. Thinning of the base-emitter depletion layer needs to happen to avoid series resistance effects and space-charge storage effects.

The emitter resistance $R_E$ needs to stay constant. In a first scaling strategy the width and length of the transistor is scaled. As the emitter area $A_E = W_E L_E$ scales with a factor $\alpha^2$, the emitter resistivity $\rho_E$ (bulk resistivity and contact resistivity) needs to scale accordingly with a factor $\alpha^2$.

Regarding the base resistance for a HBT with negligible base-collector undercut $W_{under} << W_{CB}$. This undercut is typically used in the collector of an InP HBT to reduce the parasitic capacitance. The undercut refers to the etching of a portion of the collector layer beneath the base layer, leaving a narrow, suspended collector pedestal that supports the base and emitter layers as shown in Fig. 5.15.

$$R_B \sim \frac{\rho_{B,v}}{2L_E W_{CB}} + \frac{\rho_s}{2}\frac{W_E}{L_E}\left(\frac{W_{CB}}{3W_E} + \frac{W_{BE}}{W_E} + \frac{1}{6}\right) \tag{5.13}$$

where $\rho_s$ is base sheet resistance and $\rho_{B,v}$ the vertical base contact resistivity.

With all width ratios staying the same, the scaling law for the base contact resistivity requires the need to scale the base contact resistivity by a factor $\alpha^2$.

This scaling strategy is unfortunately not viable as it results in a rapid increase in temperature with scaling. The HBT junction temperature rise can be approximated as

$$\Delta T \sim \frac{P}{\pi K L_E}\left(\ln\left(\frac{L_E}{W_E}\right) + 1\right) \tag{5.14}$$

where $K$ is the thermal conductivity of the InP substrate and $P$ is the dissipated power.

When both $L_E$ and $W_E$ scale with a factor $\alpha$ to keep the base resistance constant, the temperature rises with a factor $\alpha$.

The second strategy scales the width and length of the transistor differently, resulting in a moderate increase in temperature with scaling. In this case typically $W_E$ is scaled with a factor $\alpha^2$ and $L_E$ is kept constant. This results in a moderate logarithmic increase of the junction temperature of an isolated HBT.

The third strategy maintains a constant junction temperature by varying the dimensions as follows: scaling of $W_E$ with a factor $\alpha^3$ and $L_E$ increasing with a factor $\alpha$.

The intermediate strategy of scaling the width and length of the transistor differently is selected for enabling a roadmap extending all the way up to 480-GHz as is shown in Table 5.1.

**TABLE 5.1** Heterojunction bipolar transistor scaling roadmap.

| Parameter | Scaling law | 500 nm | 250 nm | 130 nm | 64 nm | 32 nm | 16 nm |
|---|---|---|---|---|---|---|---|
| $f_t$ [THz] | $\alpha$ | 0.37 | 0.52 | 0.73 | 1 | 1.4 | 2 |
| $f_{max}$ [THz] | $\alpha$ | 0.49 | 0.85 | 1.3 | 2 | 2.8 | 4 |
| $BV_{CEO}$ [V] | | 4.5 | 4 | 3.3 | 2.75 | | |
| Current density $J_C$ [mA/μm²] | $\alpha^2$ | 4.5 | 9 | 18 | 36 | 72 | 140 |
| $\Delta T$ [K] | | 39 | 50 | 61 | 72 | | |
| $I_E/L_E$ [mA/μm] | 1 | 2.3 | 2.3 | 2.3 | 2.3 | 2.3 | 2.3 |
| Emitter width $W_E$ [nm] | $\alpha^{-2}$ | 500 | 250 | 130 | 64 | 32 | 16 |
| Emitter resistivity [Ω μm²] | $\alpha^{-2}$ | 16 | 8 | 4 | 2 | 1 | 0.5 |
| Base thickness $t_B$ [nm] | $\alpha^{-1/2}$ | 30 | 25 | 21 | 18 | 15 | 13 |
| Doping [cm$^{-3}$] | 1 | 7e19 | 7e19 | 7e19 | 7e19 | 7e19 | 7e19 |
| Sheet resistance [Ω] | $\alpha^{1/2}$ | 500 | 600 | 708 | 830 | 990 | |
| Contact resistivity [Ω μm²] | $\alpha^{-2}$ | 20 | 10 | 5 | 2.5 | 1.25 | 0.63 |
| Base contact width $W_{CB}$ [nm] | $\alpha^{-2}$ | 300 | 175 | 120 | 60 | 30 | 15 |
| Collector width $W_C$ [nm] | $\alpha^{-2}$ | 1200 | 600 | 360 | 180 | 90 | 45 |
| Collector thickness $t_C$ [nm] | $\alpha^{-1}$ | 150 | 106 | 75 | 53 | 37.5 | 26 |
| $Area_{collector}/area_{emitter}$ | 1 | 2.4 | 2.4 | 2.9 | 2.8 | 2.8 | 2.8 |

Regarding the discussion above, the following topics have been identified as critical and enablers for HBT scaling: emitter and base resistivity reduction, self-heating related to increased current density, and collector-base width scaling. Contact and thermal resistivities are the major bottlenecks to scaling InP HBTs, as they lead to increased power consumption and reduced device performance.

Several approaches have been proposed in the scientific literature to reduce contact resistivity in InP HBTs. One such approach involves the use of metal-semiconductor alloy contacts, which have been shown to reduce contact resistivity and improve device performance. For example, a study by Mehari et al. (2012) demonstrated a significant reduction in base contact resistivity by using a Ni-InGaAs alloy as an ohmic contact to p-type InGaAs. The specific contact resistivity between Ni-InGaAs layer and p-type InGaAs obtained from the structure was 6.2 Ω/μm$^{-2}$. It is smaller than the contact resistivity of Pt contacts to the p-type InGaAs layer which is typically in the order of 50–80 Ω/μm$^{-2}$. However, careful control of the Ni thickness is needed. Jain et al. (2011) report the use of a Mo/W/TiW refractory emitter metal contact, which allows for biasing the transistors at high emitter current densities without problems of electromigration or contact diffusion under electrical stress.

In order to decouple the doping level in the base from the doping level in the base contact, base regrowth processes have been proposed (Ida et al., 1996). Increasing the base doping is a good way to lower the contact resistance. However, it has a number of problems, such as reduced minority carrier lifetime, dopant out-diffusion, and reduced reliability. A solution to this is having the heavily doped layers only in the extrinsic-base region.

In Rode et al. (2015), a novel two-step deposition process has been developed to self-align the metallization of sub-20-nm bases, using a metal stack with an ultrathin layer of platinum for low contact resistance and a thick refractory diffusion barrier for stable operation at high current densities and elevated temperatures. This technology enables the deposition of low-sheet-resistivity base electrodes, further improving the overall base access resistance and the $f_{max}$ bandwidth.

In conclusion, in this section we have looked at the scaling scenarios for the InP HBT and challenges related to that. In general, we see that the devices can achieve remarkable speed with $f_t$ and $f_{max}$ well above 1 THz. But it comes with a number of challenges. As the size of the device is scaled down, the amount of heat generated per unit area increases. This can cause self-heating, which can lead to reduced device performance and reliability. This will be addressed in the next sections. As device dimensions are scaled down and as such also the contact area, the contact resistance between the metal and the semiconductor becomes a more significant factor in device performance. High contact resistance can reduce the overall efficiency of the device. Several approaches have been investigated. As device dimensions are scaled down, the fabrication process becomes more complex and challenging, which can lead to reduced yields and increased costs. And finally, as device dimensions are scaled down, the quality of the materials used in the device becomes more critical, and defects in the materials can have a more significant impact on device performance.

### 5.6.2 Reliability

Reliability is a critical factor in the design and performance of HBT devices made from InP. Their reliability can be compromised by various factors, including device geometry, process variations, and material defects. In this section, we will briefly discuss some of the major reliability concerns in InP HBT devices and the solutions that have been proposed to mitigate these issues.

Hot carrier degradation (HCD) is a phenomenon that occurs in HBTs due to the generation of hot carriers (i.e., high-energy carriers) in the base region of the transistor. In InP HBTs, HCD is a major reliability concern that can significantly impact the device performance and lifetime (Wang & Tian, 2004).

HCD occurs when the hot carriers gain enough energy to cause damage to the lattice structure of the semiconductor material. This damage can lead to changes in the transistor's electrical properties, such as a decrease in current gain and an increase in base resistance, which can ultimately cause device failure. HCD is more severe in InP HBTs compared to other semiconductor materials due to the high carrier mobility and the narrow bandgap of InP, which makes it more susceptible to hot carrier effects.

HCD mainly occurs in the base region of the transistor, where the carriers gain energy due to the high electric field and carrier density. The electrons and holes can then collide with the lattice atoms, leading to the creation of defects and traps in the semiconductor material. This can cause a decrease in carrier mobility and an increase in recombination rate, which can ultimately degrade the device performance.

There are several solutions to mitigate HCD in InP HBTs, such as optimizing the device design and fabrication process to reduce the electric field and carrier density in the base region. Another approach is to use passivation techniques to reduce the surface recombination rate and minimize the generation of hot carriers. The surface recombination occurs when electrons are injected directly from the emitter sidewall into the extrinsic base. The passivation ledge structure was proposed as a feature to solve this (Kashio et al., 2008, 2010). This is also shown in Fig. 5.16. The idea behind the ledge structure is to reduce the electric field and carrier density in the base region by introducing a recessed region (i.e., a "ledge") between the emitter and base layers of the transistor, where the carriers gain the most energy and are most likely to cause hot carrier degradation. The recessed region creates a gradual transition in the doping profile.

In addition to reducing HCD, the ledge structure can also improve the transistor's high-frequency performance by reducing the parasitic capacitance and improving the device's thermal stability. Devices with high current gain of 60 and excellent reliability have been shown in Kashio et al. (2008, 2010). An extrapolated mean time to failure of over 108 hours at a junction temperature of 125°C was shown.

However, the implementation of the ledge structure can be challenging as it requires precise control of the etching process to create the recessed region. Additionally, the ledge structure can increase the base resistance and reduce the current gain of the transistor.

Another reliability concern in InP HBT devices is aging, which occurs over time due to the gradual accumulation of defects in the material structure. This can lead to a reduction in the performance of the device, as well as a decrease in its lifetime. To mitigate aging, various solutions have been proposed, such as the use of passivation layers to reduce the rate of defect accumulation, and the use of compensation techniques to reduce the effect of aging on device performance.

Device-to-device and process-to-process variations are also a concern in InP HBT devices, as they can lead to significant variations in the device performance. This can cause significant difficulties in the design and testing of high-performance HBT devices. To mitigate these variations, various solutions have been proposed, such as the use of

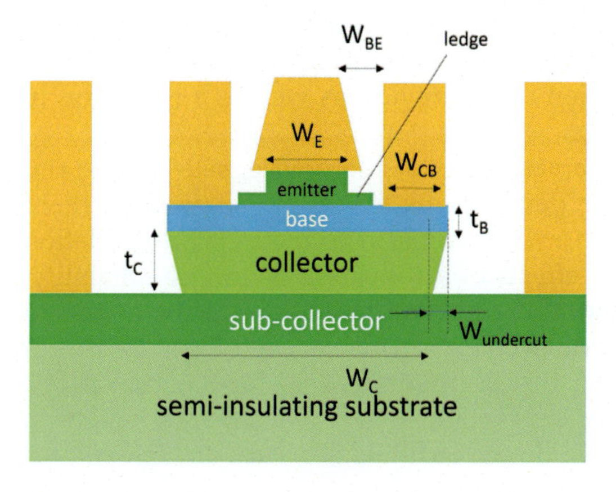


**FIGURE 5.16** Schematic cross-section of a heterojunction bipolar transistor device indicating the ledge structure. In this case the emitter can consist of a highly doped InGaAs contact layer and a undoped InP interface/passivation layer to the base.

process control techniques to reduce the impact of process variations, and the use of statistical methods to optimize device performance.

Finally, material defects are another major concern in InP HBT devices, as they can cause significant variations in the device performance and reliability. To mitigate the impact of material defects, various solutions have been proposed, such as the use of advanced fabrication techniques to reduce the formation of defects, and the use of process control techniques to detect and correct defects during the manufacturing process.

In conclusion, reliability is a critical factor in the design and performance of InP HBT devices. The reliability of HBT devices can be compromised by various factors, including device degradation due to hot carrier injection, aging, process and device-to-device variations, and material defects. To mitigate these reliability concerns, various solutions have been proposed, including the use of process control techniques, passivation layers, compensation techniques, statistical methods, and advanced fabrication techniques. The continued development of these solutions is critical to ensure the reliability and performance of InP HBT devices in high-speed and high-frequency applications.

### 5.6.3 Self-heating

Self-heating refers to the process by which the operation of a device generates heat, which in turn can affect the device's performance. Self-heating can be a significant factor in the performance of bipolar transistors, particularly at high power levels, and it is important to consider the effects of self-heating when designing and analyzing circuits that include these devices.

One of the main ways in which self-heating can impact the performance of bipolar transistors is through its effect on the device's electrical characteristics. Many of the electrical characteristics of bipolar transistors, such as the collector-emitter current, are strongly dependent on temperature. As the device heats up due to self-heating, the electrical characteristics of the device will change, which can affect the overall performance of the circuit in which it is used.

For example, consider a bipolar transistor operating in the active region, where the collector current is controlled by the base current. As the device heats up due to self-heating, the collector current may increase, leading to increased power dissipation and possibly even device failure. Similarly, the base current may increase, leading to increased power dissipation and possibly even device failure. In general, self-heating can lead to a reduction in the maximum allowable power dissipation of the device, as well as a reduction in the maximum allowable collector-emitter voltage.

In addition to its effects on the electrical characteristics of the device, self-heating can also impact the reliability of the device. As the device heats up, it may experience more stress due to thermal expansion and contraction, which can lead to increased degradation and failure over time. In order to minimize the effects of self-heating on device reliability, it is important to design the device and its packaging to dissipate heat effectively and to minimize the temperature gradients within the device.

There are several ways in which self-heating can be minimized in bipolar transistors. One approach is to use devices with low power dissipation, which will generate less heat. This can be achieved through the use of advanced device technologies, such as high-voltage bipolar transistors or lateral bipolar transistors, which have lower power dissipation than traditional vertical bipolar transistors.

Another approach is to use devices with improved thermal packaging, such as devices with improved heat sinking or devices with improved thermal management. These devices are designed to dissipate heat more effectively, which can help to minimize the effects of self-heating on device performance and reliability.

One approach to reducing thermal resistivity involves the use of high thermal conductivity materials for the device layers and substrates. In Carter et al. (2019), researchers have proposed to place InP HBTs on a metallic subcollector ohmic contact deposited on a high-thermal conductivity SiC substrate using microtransfer printing. They achieve an increase in RF power density (up to 4x) compared to conventional mesa-HBTs with a similar device footprint, while maintaining RF performance. Another study by Nosaeva et al. (2015) demonstrated a significant reduction in thermal resistance by using a diamond substrate for an InP HBT.

Also microcooling has been proposed. For example, a study by Khalkhali and Kurabayashi (2004) demonstrated a significant reduction in thermal resistance by incorporating a microcooling system in an InP HBT circuit.

Finally, self-heating can be minimized through careful circuit design, including the use of proper bias conditions and the use of proper thermal management techniques. For example, using a low collector current and a low collector-emitter voltage can help to reduce the power dissipation of the device and minimize the effects of self-heating.

In conclusion, self-heating is an important factor to consider in the design and analysis of circuits that include bipolar transistors. It can have significant impacts on the performance and reliability of these devices, and it is important to consider these impacts when choosing devices and designing circuits that will operate at high power levels. There are several ways in which self-heating can be minimized, including the use of low-power devices, improved thermal packaging, and careful circuit design.

## 5.7 Summary

In this chapter, we have provided an overview of bipolar transistors, including their operating principle, different types, and a comparison of InP HBT and SiGe HBT. We briefly focused on the fabrication and growth of SiGe layers in SiGe HBT, as well as a comparison of SiGe HBT and SiGe BICMOS. Next, we discussed the challenges and innovations in upscaling InP HBT needed to make this a cost-effective, reliable, and mature technology that can be used for next generation wireless communication. Techniques such as microtransfer printing, NRE, and SRBs have been highlighted as possible ways to do this. In this chapter we also compared HBT and HEMT devices, the chapter concluded with a discussion on device modeling and optimizing bipolar transistors, including scaling, reliability, and self-heating.

## References

Arabhavi, A. M., Ciabattini, F., Hamzeloui, S., Fluckiger, R., Saranovac, T., Han, D., Marti, D., Bonomo, G., Chaudhary, R., Ostinelli, O., & Bolognesi, C. R. (2022). InP/GaAsSb double heterojunction bipolar transistor emitter-fin technology with f MAX = 1.2 THz. *IEEE Transactions on Electron Devices*, *69*(4), 2122–2129. Available from https://doi.org/10.1109/ted.2021.3138379.

Bolognesi, C.R., Matine, N., Xu, X., Dvorak, M. W., Watkins, S.P., & Thewalt, M.L.W. (1998). Low-offset NpN InP-GaAsSb-InP double heterojunction bipolar transistors with abrupt interfaces and ballistically launched collector electrons. In: *56th annual device research conference digest (Cat. No.98TH8373)* (pp. 30–31). Charlottesville, VA, USA. Available from https://doi.org/10.1109/DRC.1998.731108.

Bouhouche, M., Latreche, S., & Gontrand, C. (2009). Effect of implantation defects and carbon incorporation on Si/SiGe bipolar characteristics. In: *2009 International conference on computer and electrical engineering, ICCEE 2009* (Vol. 2, pp. 201–204). Algeria. Available from https://doi.org/10.1109/ICCEE.2009.180.

Carter, A. D., Urteaga, M. E., Griffith, Z. M., Lee, K. J., Roderick, J., Rowell, P., Bergman, J., Hong, S., Patti, R., Petteway, C., Fountain, G., Ghosel, K., & Bower, C. A. (2019). Si/InP heterogeneous integration techniques from the wafer-scale (hybrid wafer bonding) to the discrete transistor (micro-transfer printing). In: *2018 IEEE SOI-3D-subthreshold microelectronics technology unified conference, S3S 2018*. United States: Institute of Electrical and Electronics Engineers Inc. Available from https://doi.org/10.1109/S3S.2018.8640196, http://ieeexplore.ieee.org/xpl/mostRecentIssue.jsp?punumber = 8636845.

Carter A.D., Urteaga M.E., Rowell P., Bergman J., & Arias A. (2019). Microtransfer-printed InGaAs/InP HBTs utilizing a vertical metal sub-collector contact. In: *2019 device research conference (DRC)* (pp. 45–46). Ann Arbor, MI, USA. Available from https://doi.org/10.1109/DRC46940.2019.9046426.

Chevalier, P., Avenier, G., Canderle, E., Montagné, A., Ribes, G., & Vu, V. T. (2015). Nanoscale SiGe BiCMOS technologies: From 55 nm reality to 14 nm opportunities and challenges. In: *Proceedings of the IEEE bipolar/BiCMOS circuits and technology meeting* (Vol. 2015-, pp. 80–87). France: Institute of Electrical and Electronics Engineers Inc. Available from https://doi.org/10.1109/BCTM.2015.7340556.

Del Alamo, J. A., Cai, X., Zhao, X., Vardi, A., & Grajal, J. G. (2021). Nanoscale InGaAs FinFETs: Band-to-band tunneling and ballistic transport. In: *European solid-state device research conference* (Vol. 2021-, pp. 203–206). Editions Frontiers United States. Available from https://doi.org/10.1109/ESSDERC53440.2021.9631768, http://ieeexplore.ieee.org/xpl/conferences.jsp.

Dingle, R., Störmer, H. L., Gossard, A. C., & Wiegmann, W. (1978). Electron mobilities in modulation-doped semiconductor heterojunction superlattices. *Applied Physics Letters*, *33*(7), 665–667. Available from https://doi.org/10.1063/1.90457.

Ebers, J. J., & Moll, J. L. (1954). Large-signal behavior of junction transistors. *Proceedings of the IRE*, *42*(12), 1761–1772. Available from https://doi.org/10.1109/JRPROC.1954.274797.

Gauthier, A., Borrel, J., Chevalier, P., Avenier, G., Montagne, A., Juhel, M., Duru, R., Clement, L. R., Borowiak, C., Buczko, M., & Gaquiere, C. (2018). 450 GHz f-T SiGe:C HBT featuring an implanted collector in a 55-nm CMOS node. In: *2018 IEEE BiCMOS and compound semiconductor integrated circuits and technology symposium, BCICTS 2018* (pp. 72–75). France: Institute of Electrical and Electronics Engineers Inc. Available from https://doi.org/10.1109/BCICTS.2018.8551057, http://ieeexplore.ieee.org/xpl/mostRecentIssue.jsp?punumber = 8536738.

Gauthier, A., Chevalier, P., Avenier, G., Ribes, G., Rellier, M. L., Campidelli, Y., Beneyton, R., Celi, D., Haury, G., & Gaquiere, C. (2017). SiGe HBT / CMOS process thermal budget co-optimization in a 55-nm CMOS node. In: *Proceedings of the IEEE bipolar/BiCMOS circuits and technology meeting* (Vol. 2017-, pp. 58–61). France: Institute of Electrical and Electronics Engineers Inc. Available from https://doi.org/10.1109/BCTM.2017.8112911.

Gummel, H. K., & Poon, H. C. (1970). An integral charge control model of bipolar transistors. *Bell System Technical Journal*, *49*(5), 827–852. Available from https://doi.org/10.1002/j.1538-7305.1970.tb01803.x.

Gummel, H., & Poon, H. (1970). A compact bipolar transistor model. In: *1970 IEEE international solid-state circuits conference. Digest of technical papers* (pp. 78–79), Philadelphia, PA, USA. Available from https://doi.org/10.1109/ISSCC.1970.1154800.

Hamzeloui, S., Arabhavi, A. M., Ciabattini, F., Fluckiger, R., Marti, D., Ebrahimi, M., Ostinelli, O., & Bolognesi, C. R. (2022). High power InP/Ga(In)AsSb DHBTs for millimeter-wave PAs: 14.5 dBm output power and 10.4 mw/μm 2 power density at 94 GHz. *IEEE Journal of Microwaves*, *2*(4), 660–668. Available from https://doi.org/10.1109/jmw.2022.3202854.

Heinemann, B., Rucker, H., Barth, R., Barwolf, F., Drews, J., Fischer, G. G., Fox, A., Fursenko, O., Grabolla, T., Herzel, F., Katzer, J., Korn, J., Kruger, A., Kulse, P., Lenke, T., Lisker, M., Marschmeyer, S., Scheit, A., Schmidt, D., ... Wolansky, D. (2017). SiGe HBT with fx/fmax of 505 GHz/720 GHz. In: *Technical digest – International electron devices meeting, IEDM* (pp. 3.1.1–3.1.4). Germany: Institute of Electrical and Electronics Engineers Inc. Available from https://doi.org/10.1109/IEDM.2016.7838335.

Huang, M. L., Chang, S. W., Chen, M. K., Fan, C. H., Lin, H. T., Lin, C. H., Chu, R. L., Lee, K. Y., Khaderbad, M. A., Chen, Z. C., Lin, C. H., Chen, C. H., Lin, L. T., Lin, H. J., Chang, H. C., Yang, C. L., Leung, Y. K., Yeo, Y. C., Jang, S. M., ... Diaz, C. H. (2015). In0.53Ga0.47As MOSFETs with high channel mobility and gate stack quality fabricated on 300 mm Si substrate. In: *Digest of technical papers – Symposium on VLSI technology* (Vol. 2015-, pp. T204–T205). Taiwan: Institute of Electrical and Electronics Engineers Inc. Available from https://doi.org/10.1109/VLSIT.2015.7223675.

Huang, M. L., Chang, S. W., Chen, M. K., Oniki, Y., Chen, H. C., Lin, C. H., Lee, W. C., Lin, C. H., Khaderbad, M.A., Lee, K. Y., Chen. Z. C., Tsai. P. Y., Lin, L. T., Tsai,M. H., Hung, C. L., Huang, T. C., Lin, Y. C., Yeo, Y. C., Jang, S. M., ... Diaz, C. H. (2016). High performance In0.53Ga0.47As FinFETs fabricated on 300 mm Si substrate. In: *Digest of technical papers – Symposium on VLSI technology* (Vol. 2016-). Taiwan: Institute of Electrical and Electronics Engineers Inc. Available from https://doi.org/10.1109/VLSIT.2016.7573361.

Ida, M., Yamahata, S., Kurishima, K., Ito, H., Kobayashi, T., & Matsuoka, Y. (1996). Enhancement of fmax in InP/InGaAs HBT's by selective MOCVD growth of heavily-doped extrinsic base regions. *IEEE Transactions on Electron Devices*, *43*(11), 1812–1817. Available from https://doi.org/10.1109/16.543012.

Jain, V., Rode, J. C., Chiang, H. W., Baraskar, A., Lobisser, E., Thibeault, B. J., Rodwell, M., Urteaga, M., Loubychev, D., Snyder, A., Wu, Y., Fastenau, J. M., & Liu, W. K. (2011). 1.0 THz fmax InP DHBTs in a refractory emitter and self-aligned base process for reduced base access resistance. In: *Device research conference – Conference digest, DRC* (pp. 271–272). United States. Available from https://doi.org/10.1109/DRC.2011.5994528.

Johnson, E. O. (1973). Insulated-gate field-effect transistor – A bipolar transistor in disguise. *R.C.A. Review*, *34*(1), 80–94.

Kashio, N., Kurishima, K., Fukai, Y. K., Ida, M., & Yamahata, S. (2010). High-speed and high-reliability InP-based HBTs with a novel emitter. *IEEE Transactions on Electron Devices*, *57*(2), 373–379. Available from https://doi.org/10.1109/TED.2009.2037461.

Kashio, N., Kurishima, K., Fukai, Y. K., & Yamahata, S. (2008). Highly reliable submicron inp-based hbts with over 300-GHz ft. *IEICE Transactions on Electronics*, *E91-C*(7), 1084–1090. Available from https://doi.org/10.1093/ietele/e91-c.7.1084, http://www.jstage.jst.go.jp/article/transele/E91.C/7/1084/-pdf.

Khalkhali, H., & Kurabayashi, K. (2004). Micromachined thermosyphon for on-chip cooling of high-power InP HBT circuits. In: *The ninth intersociety conference on thermal and thermomechanical phenomena in electronic systems* (IEEE Cat. No.04CH37543). Available from https://doi.org/10.1109/ITHERM.2004.1318382.

Kharche, N., Klimeck, G., Kim, D. H., Del Alamo, J. A., & Luisier, M. (2009). Performance analysis of ultra-scaled InAs HEMTs. In: *Technical digest – International electron devices meeting, IEDM* (p. 20.3.4). United States. Available from https://doi.org/10.1109/IEDM.2009.5424315.

Kim, T. W., & Del Alamo, J. A. (2011). Injection velocity in thin-channel InAs HEMTs. In: *Conference proceedings – International conference on indium phosphide and related materials* (pp. 10928669). United States.

Lorito, G., Gonda, V., Scholtes, T. L. M., & Nanver, L. K. (2008). SiGe HBTs implemented with implanted laser-annealed emitters to completely eliminate the transient enhanced diffusion. In: *2008 26th International conference on microelectronics, proceedings, MIEL 2008* (pp. 291–294). The Netherlands. Available from https://doi.org/10.1109/ICMEL.2008.4559281.

Mehari, S., Gavrilov, A., Cohen, S., & Ritter, D. (2012). Study of the Ni-InGaAs alloy as an ohmic contact to the p-type base of InP/InGaAs HBTs. In: *Conference proceedings – International conference on indium phosphide and related materials* (pp. 200–203). Israel. Available from https://doi.org/10.1109/ICIPRM.2012.6403357.

Nosaeva, K., Weimann, N., Rudolph, M., John, W., Krueger, O., & Heinrich, W. (2015). Improved thermal management of InP transistors in transferred-substrate technology with diamond heat-spreading layer. *Electronics Letters*, *51*(13), 1010–1012. Available from https://doi.org/10.1049/el.2015.1135, http://scitation.aip.org/dbt/dbt.jsp?KEY = ELLEAK.

Ozdemir, C. I., De Koninck, Y., Yudistira, D., Kuznetsova, N., Baryshnikova, M., Van Thourhout, D., Kunert, B., Pantouvaki, M., & Van Campenhout, J. (2021). Low dark current and high responsivity 1020nm InGaAs/GaAs nano-ridge waveguide photodetector monolithically integrated on a 300-mm Si wafer. *Journal of Lightwave Technology*, *39*(16), 5263–5269. Available from https://doi.org/10.1109/JLT.2021.3084324, https://ieeexplore.ieee.org/xpl/mostRecentIssue.jsp?punumber = 50.

Pekarik, J., Jain, V., Kenney, C., Holt, J., Khokale, S., Saroop, S., Johnson, J. B., Stein, K., Ontalus, V., Durcan, C., Nafari, M., Nesheiwat, T., Saudari, S., Yarmoghaddam, E., Chaurasia, S., & Joseph, A. (2021). SiGe HBTs with

$f_T/f_{max}\sim$375/510GHz integrated in 45nm PDSOI CMOS. In: *2021 IEEE BiCMOS and compound semiconductor integrated circuits and technology symposium, BCICTS 2021*. United States: Institute of Electrical and Electronics Engineers Inc. Available from https://doi.org/10.1109/BCICTS50416.2021.9682454, http://ieeexplore.ieee.org/xpl/mostRecentIssue.jsp?punumber = 9682201.

Rode, J. C., Chiang, H. W., Choudhary, P., Jain, V., Thibeault, B. J., Mitchell, W. J., Rodwell, M. J. W., Urteaga, M., Loubychev, D., Snyder, A., Wu, Y., Fastenau, J. M., & Liu, A. W. K. (2015). Indium phosphide heterobipolar transistor technology beyond 1-THz bandwidth. *IEEE Transactions on Electron Devices, 62*(9), 2779–2785. Available from https://doi.org/10.1109/TED.2015.2455231.

Rodwell, M. J. W., Le, M., & Brar, B. (2008). InP bipolar ICs: Scaling roadmaps, frequency limits, manufacturable technologies. *Proceedings of the IEEE, 96*(2) 271–286 Available from http://ieeexplore.ieee.org/xpl/RecentIssue.jsp?punumber = 5. https://doi.org/10.1109/JPROC.2007.911058.

Roelkens, G., Zhang, J., Bogaert, L., Billet, M., Wang, D., Pan, B., Kruckel, C. J., Soltanian, E., Maes, D., Vanackere, T., Vandekerckhove, T., Cuyvers, S., De Witte, J., Lufungula, I. L., Guo, X., Li, H., Qin, S., Muliuk, G., Uvin, S., ... Baets, R. (2023). Micro-transfer printing for heterogeneous Si photonic integrated circuits. *IEEE Journal of Selected Topics in Quantum Electronics*, 29(3). Available from https://doi.org/10.1109/JSTQE.2022.3222686, http://ieeexplore.ieee.org/xpl/RecentIssue.jsp?punumber = 2944.

Rucker, H., & Heinemann, B. (2018). *Silicon-germanium heterojunction bipolar transistors for Mm-wave systems technology, modeling and circuit, applications.* Edited by N. Rinaldi & M. Schröter. River Publishers. ISBN 9788793519619.

Schroter, M., & Chakravorty, A. (2010). *Compact hierarchical modeling of bipolar transistors with HICUM*. World Scientific.

Shi, Y., Baryshnikova, M., Mols, Y., Pantouvaki, M., Van Campenhout, J., Kunert, B., & Van Thourhout, D. (2019). Loss-coupled DFB nano-ridge laser monolithically grown on a standard 300-mm si wafer. In: *2019 conference on lasers and electro-optics Europe & European quantum electronics conference (CLEO/Europe-EQEC)* (pp. 1–1). Munich, Germany. Available from https://doi.org/10.1109/CLEOE-EQEC.2019.8873310.

Sollier, S., Widiez, J., Gaudin, G., Mazen, F., Baron, T., Martin, M., Roure M.C., Besson, P., Morales, C., Beche, E., Fournel, F., Favier, S., Salaun, A., Gergaud, P., Cordeau, M., Veytizou, C., Ecarnot, L., Delprat, D., Radu, I., & Signamarcheix, T. (2015). 300 mm InGaAsOI substrate fabrication using the Smart Cut™ technology. In: *2015 IEEE SOI-3D-subthreshold microelectronics technology unified conference, S3S 2015*. France: Institute of Electrical and Electronics Engineers Inc. Available from https://doi.org/10.1109/S3S.2015.7333495.

Strinati, E. C., Peeters, M., Neve C. R., Gomony, M. D., Cathelin, A., Boldi, M. R., Ingels, M., Banerjee, A., Chevalier, P., Kozicki, B., & Belot, D. (2022). The hardware foundation of 6G: The NEW-6G approach. In: *2022 Joint European conference on networks and communications and 6 G summit, EuCNC/6 G summit 2022* (pp. 423-428). France: Institute of Electrical and Electronics Engineers Inc. Available from https://doi.org/10.1109/EuCNC/6GSummit54941.2022.9815700, http://ieeexplore.ieee.org/xpl/mostRecentIssue.jsp?punumber = 9815454.

Vais, A., Alcotte, R., Ingels, M., Wambacq, P., Parvais, B., Langer, R., Kunert, B., Waldron, N., Collaert, N., Witters, L., Mols, Y., Hernandez, A. S., Walke, A., Yu, H., Baryshnikova, M., Mannaert, G., & Deshpande, V. (2019). First demonstration of III-V HBTs on 300 mm Si substrates using nano-ridge engineering. In: *Technical digest – International electron devices meeting, IEDM* (Vol. 2019-). Belgium: Institute of Electrical and Electronics Engineers Inc. Available from https://doi.org/10.1109/IEDM19573.2019.8993539.

Vais A., Yadav S., Mols Y., Vermeersch B., Kodandarama K. V., Baryshnikova M., Mannaert G., Alcotte R., Boccardi G., Wambacq P., Parvais B., Langer R., Kunert B., & Collaert N. (2022). III-V HBTs on 300 mm Si substrates using merged nano-ridges and its application in the study of impact of defects on DC and RF performance. In: *European solid-state device research conference* (Vol. 2022-, pp. 261–264). Editions Frontiers Belgium. Available from https://doi.org/10.1109/ESSDERC55479.2022.9947124, http://ieeexplore.ieee.org/xpl/conferences.jsp.

Wang, H., & Tian, Y. (2004). Physically based analysis of hot carrier induced degradation in InP/InGaAs double heterojunction bipolar transistors. In: *Conference proceedings – International conference on indium phosphide and related materials* (pp. 765–767). Singapore.

CHAPTER

# 6

# InP-based monolithic microwave integrated circuit technologies for 5G and beyond

*Hiroshi Hamada*

NTT Device Technology Labs, NTT Corporation, Atsugi-shi, Kanagawa-ken, Japan

## 6.1 InP devices for millimeter-wave/terahertz wireless communications toward beyond 5G

The data rate of wireless communications in beyond 5G (B5G) is considered 100 Gb/s to 1 Tb/s (David & Berndt, 2018). To achieve such an extremely high data rate, the use of millimeter-wave (MMW) and terahertz (THz) is effective due to their wide bandwidth and frequency availability. The frequency range above 275 GHz is not allocated to wireless communications, so there are possibilities for use in future B5G. In fact, the information about the usage of frequencies of 275–450 GHz for land, mobile, and fixed service applications was newly added in the Final Acts of ITU-R World Radiocommunication Conference 2019 (WRC-19) (ITU-R ITU-R World Radiocommunication Conference 2019, 2020).

The 300-GHz band (275–320 GHz in this case) is known to have medium atmospheric attenuation (<10 dB/km) and wide bandwidth (45 GHz) in the frequency range of 275–450 GHz. Therefore, it is considered suitable for high-speed wireless communications such as B5G. Many studies on 300-GHz-band transceivers (TRXs) have been recently reported. Fig. 6.1 shows the data rate and link distance of reported TRXs with complementary metal-oxide-semiconductor (Abdo et al., 2020, 2021; Hara et al., 2017; Lee et al., 2019), silicon-germanium (SiGe) (Rodríguez-Vázquez et al., 2018a, 2018b, 2019, 2020; Sarmah et al., 2016), and indium phosphide (InP) (Boes et al., 2014; Dan et al., 2020; Hamada et al., 2018; Hamada, Sugiyama et al., 2019; Hamada, Tsutsumi, Matsuzaki, et al., 2020; Kallfass et al., 2011; Kallfass, Boes, et al., 2015; Kallfass, Dan, et al., 2015; Song et al., 2016) technologies in the vicinity of 300 GHz. As can be seen from this figure, InP-based TRXs tend to have both high data rates and link distances. One of the major reasons of this high performance is the outstanding

*New Materials and Devices Enabling 5G Applications and Beyond*
DOI: https://doi.org/10.1016/B978-0-12-822823-4.00006-6

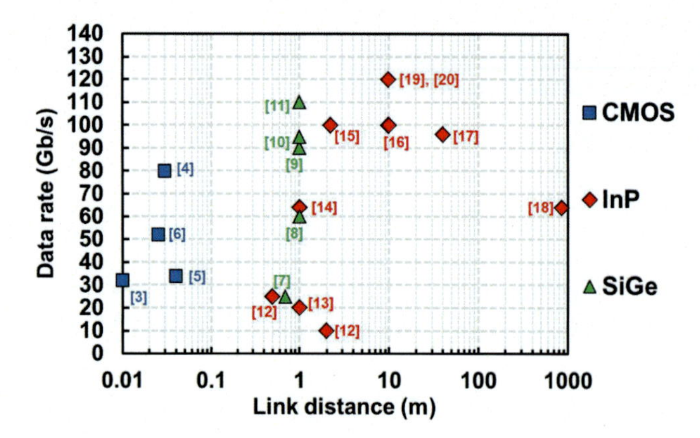


FIGURE 6.1 Data rate and link distance of reported CMOS, SiGe, and InP TRXs in vicinity of 300 GHz.

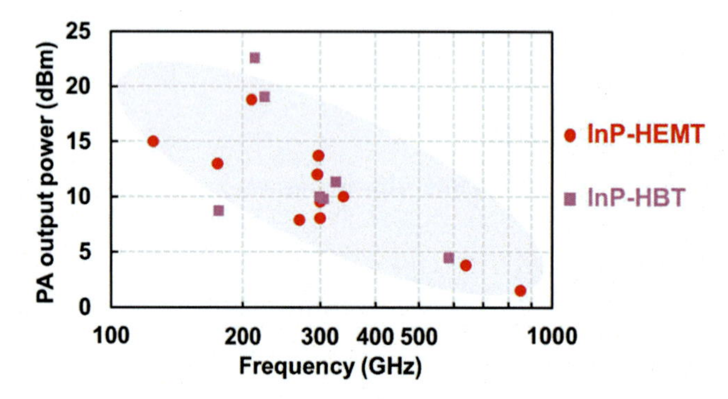


FIGURE 6.2 Reported output power of amplifiers with InP devices.

characteristics of InP amplifiers with the aid of extremely high-speed InP transistors, that is, InP heterojunction bipolar transistors (InP-HBTs) and InP high electron mobility transistors (InP-HEMTs). There are two widely used figures-of-merit of high-speed transistors, cut-off frequency ($f_T$) and maximum oscillation frequency ($f_{MAX}$). The $f_T$ and $f_{MAX}$ are respectively defined as frequencies at which the current gain and power gain of the transistor is unity. Considering amplifier operation, $f_{MAX}$ is the key index for the implementation of the MMW/THz analog front end, rather than $f_T$. The reported maximum $f_{MAX}$ of InP-HBTs and InP-HEMTs are 1.15 THz (Urteaga et al., 2017) and 1.5 THz (Deal et al., 2016), respectively. By using the high $f_{MAX}$ characteristic, many high-power MMW/THz InP amplifiers have been reported. Fig. 6.2 shows the trend in output power versus operation frequency of reported MMW/THz amplifiers with InP transistors. In 300-GHz band, high output power of around 10 dBm is available with these transistors. High-performance 300-GHz-band TRXs can be implemented using these high-power amplifiers (PAs) in a radio frequency (RF) front end.

In this chapter, InP device technologies and 300-GHz monolithic microwave integrated circuits (MMICs) for a 300-GHz high data rate ($>100$ Gb/s) TRX are described. Wireless data transmission experimental results with this TRX are also explained to show the feasibility of InP-based TRXs for B5G. These technologies are mainly being developed at NTT Device Technology Laboratories, NTT Corporation.

## 6.2 InP device technologies

There are two mandatory technologies for the implementation of 300-GHz InP MMICs: the transistor itself and passive technologies.

### 6.2.1 InP-heterojunction bipolar transistors

Let us start with InP-based transistors. The advantage of InP-based transistors is derived from the $In_xGa_{1-x}As$ ($0 < x < 1$) mixed crystal, which can be epitaxially grown on an InP substrate by choosing the ratio of In ($x$) as 0.53. The electron mobility of $In_{0.53}Ga_{0.47}As$ is around 10,000 $cm^2/V$. This is twice as high as the electron mobility of GaAs, which is also known as a high-speed compound semiconductor. In addition, the saturated electron velocity in $In_{0.53}Ga_{0.47}As$ is also high, $2.7 \times 10^7$ cm/s. In-based transistors (InP-HEMTs and InP-HBTs) use this high-electron-mobility InGaAs as an electron transport layer to achieve high-speed characteristics. Furthermore, the mobility and saturated electron velocity in $In_xGa_{1-x}As$ can be enhanced by increasing $x$. By using this characteristic, thin-film In-rich InGaAs, such as $In_{0.8}Ga_{0.2}As$, are often used with lattice-matched $In_{0.53}Ga_{0.47}As$ to further improve the speed of the transistors (Jo et al., 2019). This technique is called composite channel.

Fig. 6.3A shows a cross-section of the NTT's in-house 80-nm InP-HEMT. It uses the composite channel of $In_{0.8}Ga_{0.2}As$ and $In_{0.53}Ga_{0.47}As$ (Sugiyama et al., 2010), as mentioned above. The gate length is shrunk to 80 nm to shorten the carrier transit time from the source to drain, which enables high $f_T$. To reduce the parasitic capacitances between the gate and

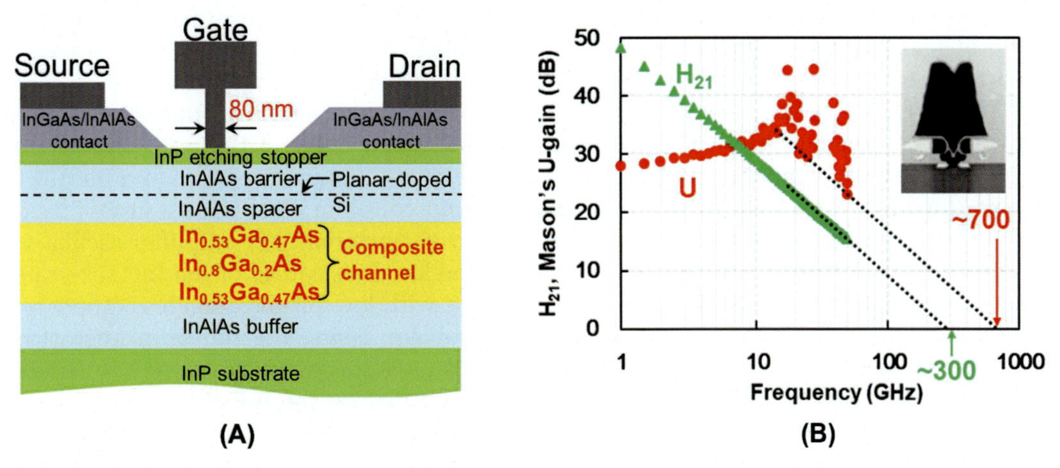


**FIGURE 6.3** (A) Schematic and (B) measured characteristics (H21 and Mason's unilateral power gain) of NTT's in-house 80-nm InP-HEMT. *Source: From Hamada, H., Tsutsumi, T., Matsuzaki, H., Fujimura, T., Abdo, I., Shirane, A., Okada, K., Itami, G., Song, H.-J., Sugiyama, H., & Nosaka, H. (2020). 300-GHz-Band 120-Gb/s Wireless front-end based on InP-HEMT PAs and mixers. IEEE Journal of Solid-State Circuits, 55(9), 2316–2335. https://doi.org/10.1109/JSSC.2020.3005818.*

drain ($C_{DG}$) and gate and source ($C_{GS}$), the recess etching technique is used in the vicinity of the gate metal. The RF characteristics, that is, current amplification factor ($H_{21}$) and Mason's unilateral power gain (U-gain) of this HEMT, are shown in Fig. 6.3B, which are values of the intrinsic HEMT structure obtained from the S-parameter measurement of the HEMT with de-embedding of the parasitic RF pad structure by using the open-short method (Ito & Masu, 2008). The $f_T$ and $f_{MAX}$ are 300 and 700 GHz, respectively. These values are reasonable for the design of 300-GHz MMICs.

### 6.2.2 Substrate-mode-reduction technique

Because the frequency of 300 GHz is quite high, there are special technologies related to the passive components for 300-GHz MMICs. It is well known that III-V compound semiconductor substrates, such as InP and GaAs, have low-loss characteristics due to their semi-insulated nature. Therefore, in InP MMICs, the low-loss coplanar waveguide (CPW) and microstrip line (MSL) are often used as RF transmission lines (TLs). There is one notable difference from low-frequency MMICs when using these TLs in the 300-GHz-band, the need for considering "substrate modes" (Schmückle et al., 2011). Substrate modes are the electromagnetic (EM) modes that propagate in the substrate. The propagation principle of substrate modes is similar to transverse electric (TE) and transverse magnetic (TM) modes in rectangular waveguides (WGs). An EM wave can be guided in the substrate with confinement from two metal layers of the front and back sides of the InP substrate. The same as rectangular WGs, the substrate modes will be guided when the electrical size of the guide structure (substrate in this case) is comparable to or larger than the wavelength of the EM wave. Considering the permittivity of InP (12.3), the wavelength of a 300-GHz EM wave in the InP substrate is around 290 μm. Therefore, a substrate mode can propagate in commercially available 600-μm-thick InP substrate. The problem with substrate modes is explained as follows. Fig. 6.4A shows an example of a one-stage common-source (CS) amplifier using an InP-HEMT. The layout schematic of this CS amplifier implemented using 600-μm-thick InP substrate is illustrated in Fig. 6.4B. In this case, there is a risk that an unintentional coupling path between the CS amplifier input and output is formed due to the substrate modes. This input-output coupling often causes instability of the amplifier operation, such as ripples on frequency characteristics or amplifier oscillation (Tsutsumi et al., 2019). Unfortunately, because of the low-loss nature of the InP substrate, these substrate modes can be guided in the substrate with low propagation loss. There are several ways to circumvent the substrate-mode problem. Using the thin film microstrip line (TFMSL) as an RF TL is one option. The TFMSL uses the two metal layers separated by a thin-film dielectric, such as benzocyclobutene, formed on the InP substrate as signal and ground. Therefore, the InP substrate is isolated from the TFMSL propagation mode by the ground metal layer. However, this approach is affected by the high-loss nature of the TFMSL; in other words, it sacrifices the advantage of the low-loss nature of the semi-insulated InP substrate. Using substrate thinning and through-substrate vias (TSVs) (Hamada, Sugiyama et al., 2019; Tsutsumi et al., 2019) is a technique to cut out substrate modes without impacting the low-loss InP TLs, as illustrated in Fig. 6.4C. The vertical and horizontal spaces of the InP substrate, which an EM wave can propagate, are reduced by

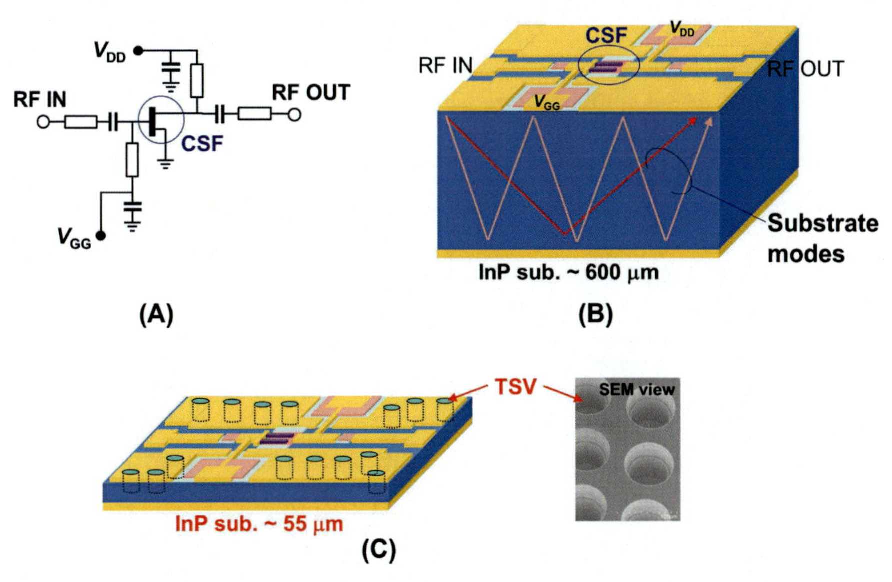


**FIGURE 6.4** (A) Circuit diagram of one-stage CS HEMT amplifier, (B) its layout schematic on 600-μm-thick InP substrate with substrate modes, and (C) its layout schematic when using thinned 55-μm-thick InP substrate with dense TSV formation for rejection of substrate modes.

the thinned substrate (55 μm) and TSV (diameter: 50 μm) formation. By forming sufficiently dense TSVs, substrate-mode propagation can be rejected. Fig. 6.5A and B shows the EM simulation results of substrate-mode propagation in a 55-μm-thick InP substrate when 50-μm-diameter TSVs are arranged with pitches (TSV edge-to-edge distance) of 100 and 50 μm. For both cases, the most dominant mode ($TE_{10}$ mode) propagation is simulated by inputting the mode from the edge of the chip (incident port in Fig. 6.5A). From these figures, arranging TSVs with 50-μm pitch is good to cut out the substrate mode, whereas 100-μm pitch is clearly not sufficient. Fig. 6.5C shows the calculated transmittance of the $TE_{10}$ mode for TSV pitches of 150, 100, and 50 μm. By using the 50-μm TSV pitch, the cut-off frequency of the $TE_{10}$ mode becomes much higher than 300 GHz, and the 300-GHz substrate mode cannot propagate, as also illustrated in Fig. 6.5B. Therefore, a TSV pitch of around 50 μm is used for 300-GHz MMIC design.

## 6.2.3 MMIC-to-WG transition

The most frequently used media of EM waves in MMW/THz region is a rectangular WG. Therefore, a low-loss MMIC packaging technique in the WG module is also an important passive technique to implement practical TRXs. The essential component for the low-loss packaging is the RF transition between the MMIC and WG. The important characteristics of the transition are loss and bandwidth. Several transitions have been reported in the 300-GHz band (Kosugi et al., 2014; Tessmann et al., 2013; Zamora et al., 2014). The dipole coupler (Zamora et al., 2014) is one of the most commonly used transitions. The dipole coupler integrated in an InP chip with around 1-dB WG coupling loss was reported

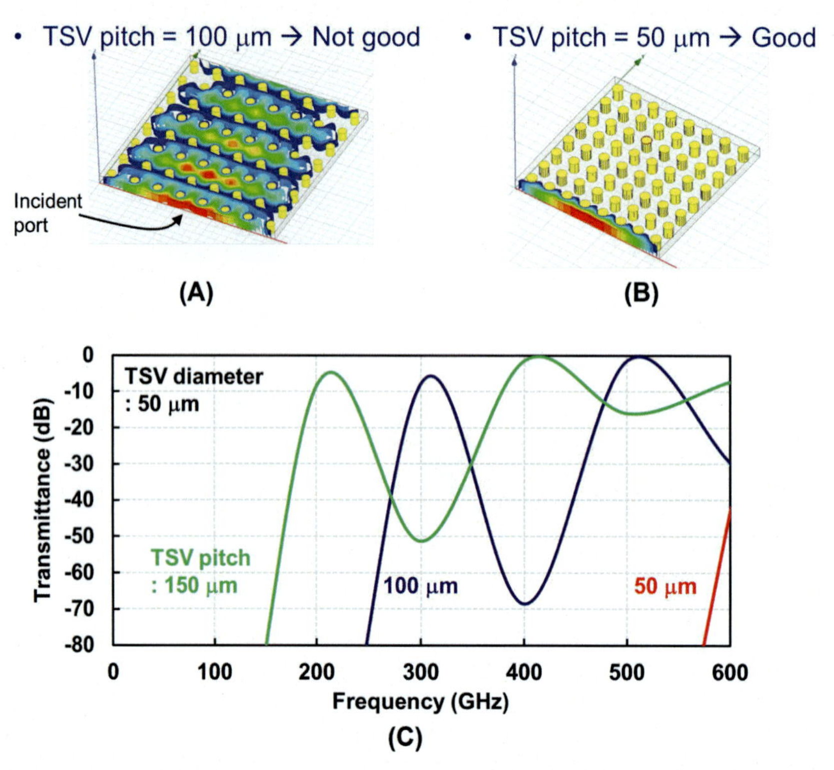


**FIGURE 6.5** Substrate mode (TE10 mode) propagation in 50-μm-thick InP substrate when TSV pitches are (A) 100 μm and (B) 50 μm. (C) Simulated transmittance of TE10 mode transmittance. *Source: From Hamada, H., Tsutsumi, T., Sugiyama, H., Matsuzaki, H., Song, H.-J., Itami, G., Fujimura, T., Abdo, I., Okada, K., & Nosaka, H. (2019). Millimeter-wave InP device technologies for ultra-high speed wireless communications toward beyond 5G. 2019 IEEE International Electron Devices Meeting (IEDM), 9.2 0.1–9.2 0.4. https://doi.org/10.1109/IEDM19573.2019.8993540.*

at 200–300 GHz (Zamora et al., 2014). One minor difficult point of this type of integrated transition is that the on-wafer ground-signal-ground (GSG) probe measurement, which is the most basic method of MMIC evaluation, cannot be applied to MMICs with these couplers. If the transition supports the coupling between the normal GSG pad (i.e., CPW) and WG, the same MMIC chip can be measured using the on-wafer probe and mounted on the module with the transition. In previous NTT studies, a ridge coupler (Hamada, Tsutsumi, Matsuzaki, et al., 2020; Kosugi et al., 2014) was investigated on the basis of this consideration. It couples the WG and normal CPW, as illustrated in Fig. 6.6A. This transition uses the metal ridge inserted in the center of the WG. The ridge transforms the EM mode between the WG and CPW. The ridge and signal line of the CPW are electrically connected by a bonding wire, as shown in Fig. 6.6A. Two metal blocks called earth posts (EPs) are used to form the connection of ground between the WG and CPW. The principle of mode transformation is illustrated in Fig. 6.6B. The $TE_{10}$ mode in the WG gradually transforms its mode profile and impedance by the tapered ridge and is converted to microstrip-line (MSL) mode at the top of the ridge. The gap of the ridge top and WG is designed for MSL mode to have its impedance of 50 Ω. This 50-Ω impedance is also kept for the MSL mode

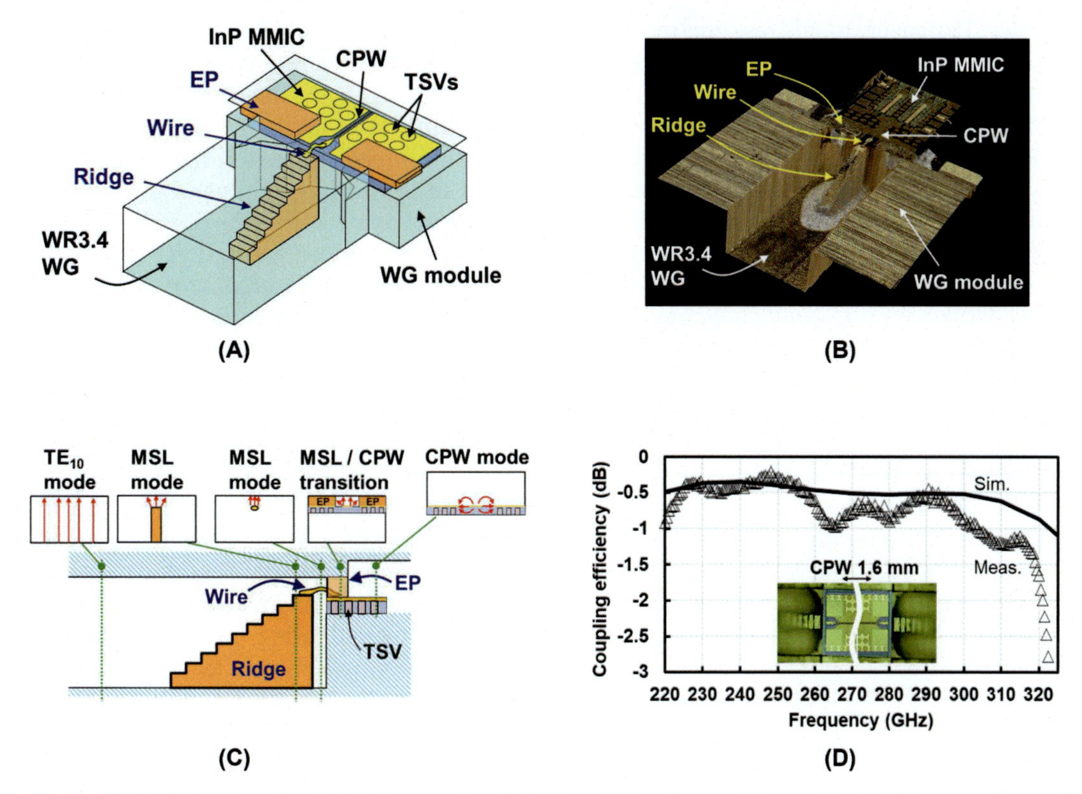


**FIGURE 6.6** (A) Schematic, (B) 3-D photograph, (C) CPW-to-WG mode translation profile, and (D) transmittance of ridge coupler. *Source: From Hamada, H., Tsutsumi, T., Matsuzaki, H., Fujimura, T., Abdo, I., Shirane, A., Okada, K., Itami, G., Song, H.-J., Sugiyama, H., & Nosaka, H. (2020). 300-GHz-Band 120-Gb/s wireless front-end based on InP-HEMT PAs and mixers. IEEE Journal of Solid-State Circuits, 55(9), 2316–2335. https://doi.org/10.1109/JSSC.2020.3005818.*

guided in the bonding wire. The ground of MSL and CPW are connected by the EPs. A 3D photograph of the WG package using a ridge coupler is shown in Fig. 6.6C. The coupling loss of the ridge coupler is small because both the conductive and dielectric losses are small. The simulated and measured transmittance of the ridge coupler is shown in Fig. 6.6D. The simulation and measurement well matched, and very broadband low loss transmittances was obtained. Measured coupling loss is below 1 dB from 220 to 305 GHz.

## 6.3 InP MMICs for 300-GHz-band transceiver

In the initial 300-GHz TRX studies, a simple modulation scheme such as amplitude shift keying (ASK) (Hamada et al., 2015; Song et al., 2016) or quadrature phase shift keying (QPSK) (Song et al., 2014) is used. Song et al. (2016) reported 20-Gb/s wireless transmission with a link distance of 1 m. These studies are suitable for real-time data transfer such as for a kiosk download system (Hamada et al., 2016) because the off-line demodulation system of a complicated modulation scheme (which causes delay) was not used in these studies. However, to address the required data rate for B5G of 100 Gb/s–1 Tb/s in a limited bandwidth, it is mandatory to use a higher spectral efficiency modulation scheme, such as quadrature amplitude modulation (QAM). By using the bandwidth of 25 GHz with 16 QAM modulation, 100 Gb/s transmission (Hamada et al., 2018) can be achieved. There are two TRX architectures for QAM modulation, direct conversion and heterodyne. A direct conversion TRX has many advantages due to its simple architecture. In a direct conversion TRX, the baseband data signal can be directly connected to the TRX by generating an I-Q modulated signal in the RF band. Many high-performance direct conversion TRXs are reported in MMW range (Okada et al., 2011; Shahramian et al., 2013; Wu et al., 2014). However, it is quite challenging to implement this architecture in the 300-GHz band due to the difficulty of generating a 300-GHz I-Q signal with accurate 90-degree phase difference due to its extremely high frequency. It is also difficult to make compensation circuits such as local oscillator (LO) feed-through cancellation circuit (Abdo et al., 2021; Wu et al., 2014), in 300-GHz InP MMICs. Therefore, a heterodyne architecture is preferable as a first step in 300-GHz TRX development.

Fig. 6.7 shows the ideal schematic of the 300-GHz heterodyne TRX considered in this chapter. In this configuration, the fundamental mixer is used to avoid performance degradation of the TRX caused by the LO harmonics of the harmonic mixer. Therefore, both the RF and LO frequency range is the 300-GHz band in Fig. 6.7. The red parts in this figure are 300-GHz RF building blocks, which are implemented using InP-HEMT MMICs. The necessary 300-GHz building blocks are an RF PA, low noise amplifier (LNA), mixer, and LO PA. In this section, these building blocks, especially the RF PA and mixer, are described.

### 6.3.1 300-GHz power amplifier

Let us start with the RF PA. To amplify the QAM signal with a high signal-to-noise ratio (SNR), the important parameter of the PA is linearity. The QAM signal has a multi-amplitude level, for example, three levels for 16QAM. When linearity is not sufficient, the gain of a higher amplitude symbol will be saturated, and the signal constellation will be distorted. This will degrade the SNR of the QAM signal. Therefore, the PA should be designed to have high linear output power, that is, high output power of 1-dB compression point (OP1dB).

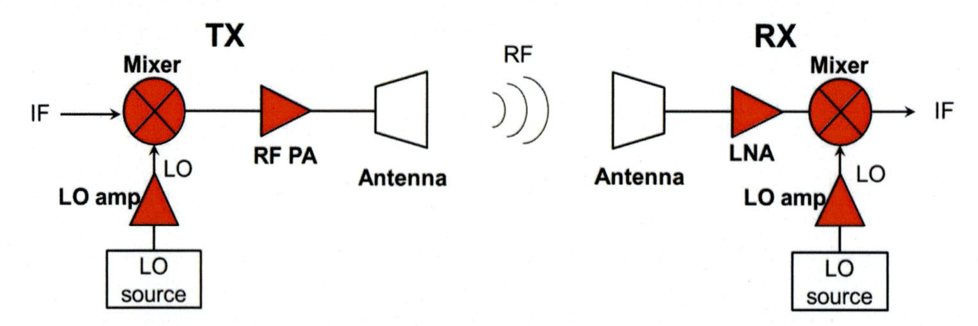


FIGURE 6.7 Schematic of 300-GHz heterodyne TRX using fundamental mixers.

The schematic of the PA studied in previous studies (Hamada, Itami et al., 2019; Hamada, Tsutsumi, Matsuzaki, et al., 2020; Hamada, Pander et al., 2020; Hamada, Sugiyama et al., 2019;) is shown in Fig. 6.8A. This PA is composed of unit PAs (UPAs), which consist of six-stage CS amplifiers, as illustrated in Fig. 6.8B. Each UPA is designed to have input/output impedances of 50 Ω. The PA has 8-parallelled UPAs in its output stage with a power combiner to sum up each output power of the UPAs and achieve high output power. A UPA is designed with two- and four-finger HEMT CS amplifiers to increase its output power and enhance the OP1dB of the PA. The four-finger HEMT has higher output power than the two-finger one because of its large current capacity. However, it has lower $f_{MAX}$ than the two-finger one due to relatively large parasitic capacitances and inductances accompanied with the complex finger metal layout. Therefore, the first three stages are designed as a gain-boosting stage with high-speed two-finger HEMTs. The latter three stages are implemented as a power-boosting stage with four-finger HEMTs. The simulated small-signal gain of a UPA is around 10 dB at 300 GHz. In the high-frequency region, the loss of RF components (TLs, matching network, combiners) is much higher than the low-frequency region due to the large conductive loss ( ∝ square root of frequency), dielectric loss ( ∝ frequency), and radiation loss. This PA was designed considering a loss reduction to enhance the gain and output power as described below.

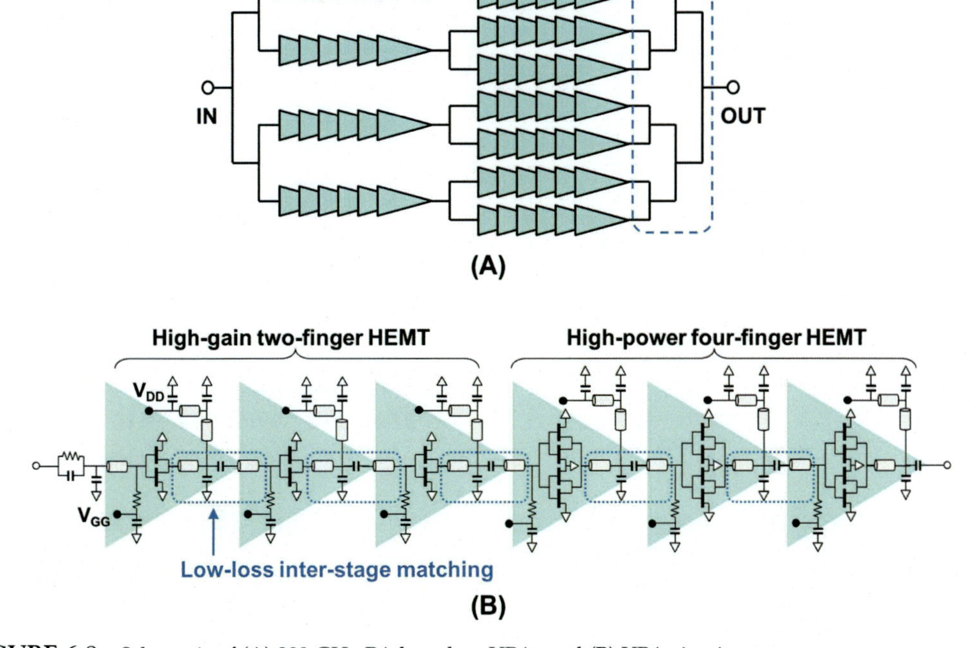


FIGURE 6.8 Schematic of (A) 300-GHz PA based on UPAs and (B) UPA circuitry.

In a cascaded UPA design, as used in this PA, the design of the inter-stage matching (ISM) circuits without degrading the gain and bandwidth is important. Therefore, the ISM circuits should have low-loss, broadband characteristics. In the UPA in Fig. 6.8B, the ISMs are designed in consideration of the input and output impedances ($Z_G$ and $Z_D$) of HEMTs. The $Z_G$ and $Z_D$ are generally very high at low frequencies; however, in the 300-GHz band, because of its extremely high frequency, these values are quite low due to the parasitic capacitances of the HEMT, as shown in Fig. 6.9A. Therefore, $Z_G$ and $Z_D$ are very low, 7.5−j4.5 and 10−j10 Ω. In such a case, the ISM circuits should also be designed with low matching impedances. In the UPA, the ISM is carried out in the vicinity of 10 Ω. By using such low impedance matching, the matching loss can be decreased and UPA gain can be increased. Fig. 6.9B shows the small-signal gain ($S_{21}$) of the UPA with conventional 50 and 10-Ω ISM design. The gain of the UPA using the 10-Ω ISM is more than 5 dB larger than that with the 50-Ω ISM.

The low-loss design of the output combiner is also important to achieve high output power. The eight-way combiner is needed to implement the PA shown in Fig. 6.8A. In the lower frequency range, a Wilkinson coupler comprised quarter-wavelength (λ/4) TLs and isolation resistance is often used due to its superior port-to-port isolation (Kim & Nguyen, 2018; Law & Pham, 2010). In such a paralleled UPA fashion, like Fig. 6.8A, the isolation of each UPA is necessary to avoid bias voltage fluctuations by the output signal from the adjacent UPA. If the UPA bias voltage is changed, the UPA gain and output power will decrease, which will degraded the PA gain and output power. However, in the 300-GHz band, the isolation between UPAs is not so important because the output signal of each

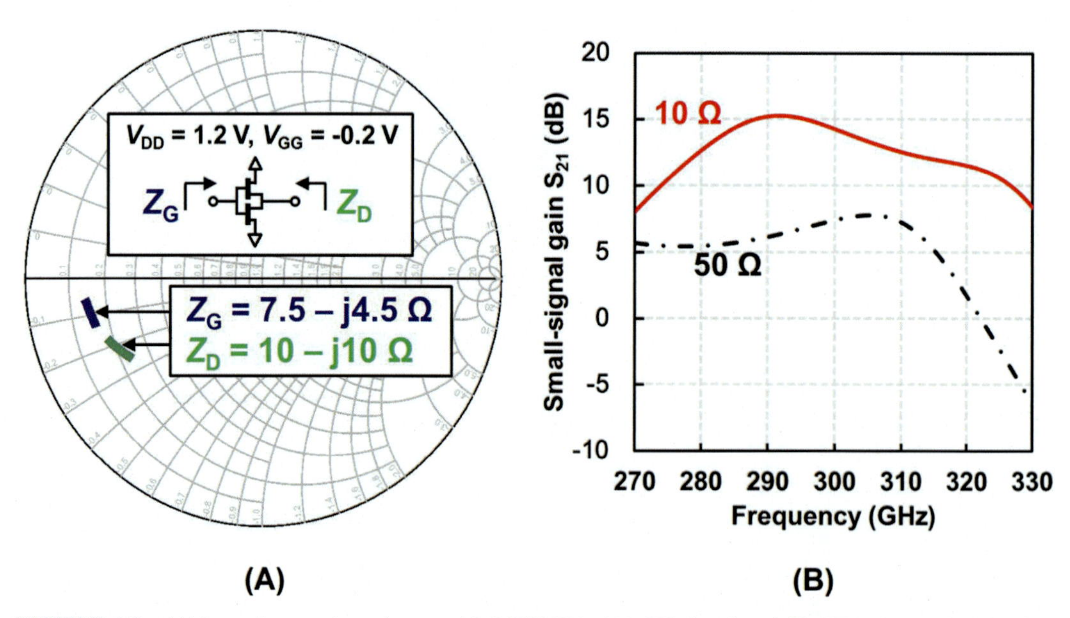


FIGURE 6.9 (A) Input/output impedances of InP-HEMT in 300-GHz-band and (B) UPA gain with 50 and 10-Ω ISMs. *Source: From Hamada, H., Tsutsumi, T., Matsuzaki, H., Fujimura, T., Abdo, I., Shirane, A., Okada, K., Itami, G., Song, H.-J., Sugiyama, H., & Nosaka, H. (2020). 300-GHz-Band 120-Gb/s wireless front-end based on InP-HEMT PAs and mixers. IEEE Journal of Solid-State Circuits, 55(9), 2316–2335. https://doi.org/10.1109/JSSC.2020.3005818.*

UPA is not so high (below 5 dBm in typical case). For example, when the UPA output is 5 dBm and the isolation between the adjacent UPA is zero (worst case), the drain-voltage swing fluctuation by the adjacent UPA is around 0.5 V peak to peak taking into account the drain output impedance of around 10 Ω (Fig. 6.9A). This voltage fluctuation is not such a large problem considering the amplifier's drain bias voltage (~1.2 V) and HEMT knee voltage (~0.5 V). The drain bias point will not move to the low-gain triode region (below knee voltage) and will stay in the high-gain saturation region even with this fluctuation. Therefore, for the PA in Fig. 6.8A, a Wilkinson coupler is not used and a simpler power combiner is used to decrease the combining loss at the cost of isolation, as described below. The schematics of the eight-way power combiner with Wilkinson couplers and a simple combiner are shown in Fig. 6.10A and B, respectively. In Fig. 6.10A, the output signal of each UPA should pass the three (=$\log_2 8$) $\lambda/4$ TLs of the Wilkinson

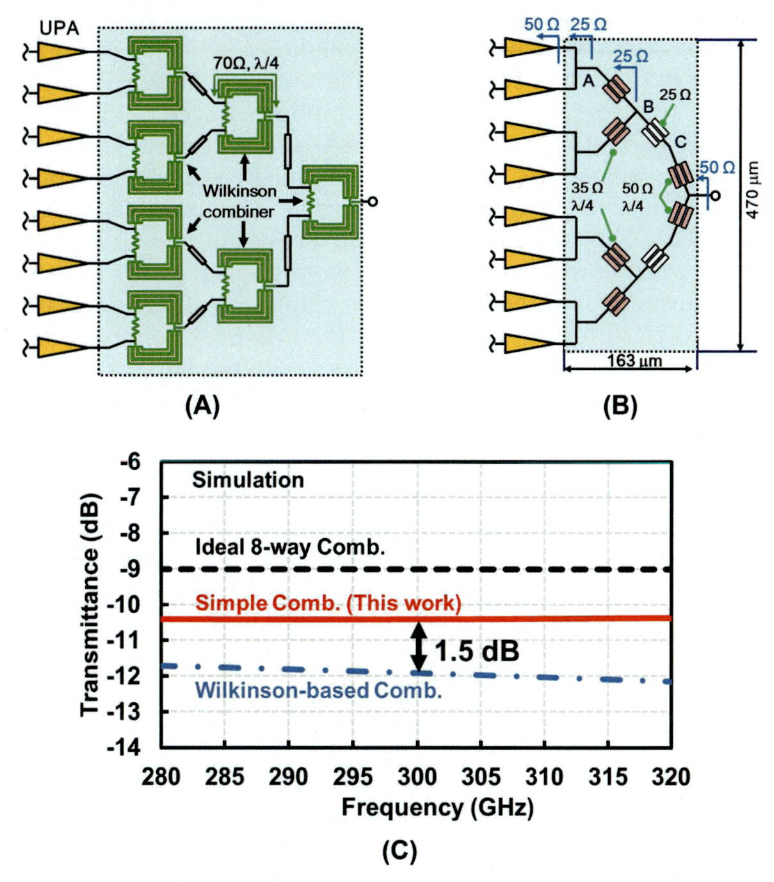


**FIGURE 6.10** Schematic of eight-way combiners with (A) Wilkinson couplers and (B) simple combiner in this study. Simulated transmittance comparison of these eight-way combiners. *Source: From Hamada, H., Tsutsumi, T., Matsuzaki, H., Fujimura, T., Abdo, I., Shirane, A., Okada, K., Itami, G., Song, H.-J., Sugiyama, H., & Nosaka, H. (2020). 300-GHz-Band 120-Gb/s wireless front-end based on InP-HEMT PAs and mixers. IEEE Journal of Solid-State Circuits, 55 (9), 2316–2335. https://doi.org/10.1109/JSSC.2020.3005818.*

couplers before arriving at the PA output port. These $\lambda/4$ TLs have high transmission losses ($\sim$1 dB) and will cause large power combining loss, reducing the PA output power. In the simple combiner in Fig. 6.10B, the adjacent UPAs are directly connected and the impedance transformation network is designed with 35 and 50-$\Omega$ $\lambda/4$ TLs. In this case, the number of $\lambda/4$ TLs for each UPA output signal is reduced to 2, which should result in a smaller power combining loss than in Fig. 6.10A. The calculated power combining losses for these two combiners are shown in Fig. 6.10C. The simple combiner has 1.5-dB smaller loss than the Wilkinson-based one. This means that by using the simple combiner, the output power of the PA can improve by 1.5 dB from the design using a conventional Wilkinson coupler.

The last loss reduction technique is the special process technique called backside DC line (BDCL). BDCL is especially effective for a PA with large current consumption. As mentioned above, because of the lack of signal power available from each UPA in the 300-GHz band, the UPAs should be arranged in parallel. This means the current consumption will proportionally increase with the number of paralleled UPAs. In addition, each UPA is designed with cascaded amplifier stages due to the small available gain of the transistor in the 300-GHz band. Therefore, the current consumption of each UPA is also large. As a result, the total drain current of the PA in Fig. 6.8A is quite large at 1.85 A. In such a case, the difficulty in PA layout occurs. The amplifiers and RF combiners should be laid out with wide-width DC lines to support the large current. One possible layout of such a PA is shown in Fig. 6.11A. The wide DC line (width $\sim$ 400 μm) for UPAs intersects the RF TLs. From the view point of RF signal, this DC-RF crossing TL means the addition of extra TL length, which causes an undesirable loss. The chip size will also increase due to these DC-RF crossing TLs. This problem is serious for the PAs because the length of a crossing TL (width of wide DC line) proportionally increases with the number of paralleled UPAs. This means that to obtain large power by increasing the number of paralleled UPAs, the loss will also increase and as a result, output power and gain of PAs will not increase so much. To overcome this issue, the interference of DC lines to RF TLs should be eliminated. BDCL enables the layout of bias lines in the backside of the InP chip. By using this technique, the RF components (UPAs, RF TLs, combiners) are released from the crossing of DC lines. Fig. 6.11B shows an image of a PA with BDCL. The wide-width DC lines are formed on the back side and DC connection of back and front sides are carried out using the TSVs described in Section 2. By introducing BDCL, the chip length of the PA is reduced by 400 m and the gain of the PA is enhanced by 3 dB, as shown in Fig. 6.11C.

A photograph of the fabricated PA chip and module is shown in Fig. 6.12A and B. The chip size is 1 $\times$ 2 mm. To reduce the RF loss at the PA output stage, the eight-way combiner is laid out as close as possible to the PA output signal pad. The BDCL is used nearly in the center of the PA chip. The PA chip is fixed on the WG module by using the silver paste except for BDCL section. A nonconductive resin paste is used for the BDCL section to isolate the BDCL from the ground. The input/output RF connection between the PA chip and WG is carried out using the ridge couplers, as described in Section 2. The module size is 15 $\times$ 22 $\times$ 25 mm. The small-signal characteristics (S-parameters) of this PA module are measured using the vector network analyzer (Keysight PNA-X) and frequency extension module (VDI WR3.4 VNAX). In the measurement, the input and output power for the PA module are controlled to be $-30$ dBm to avoid the nonlinear response of the PA.

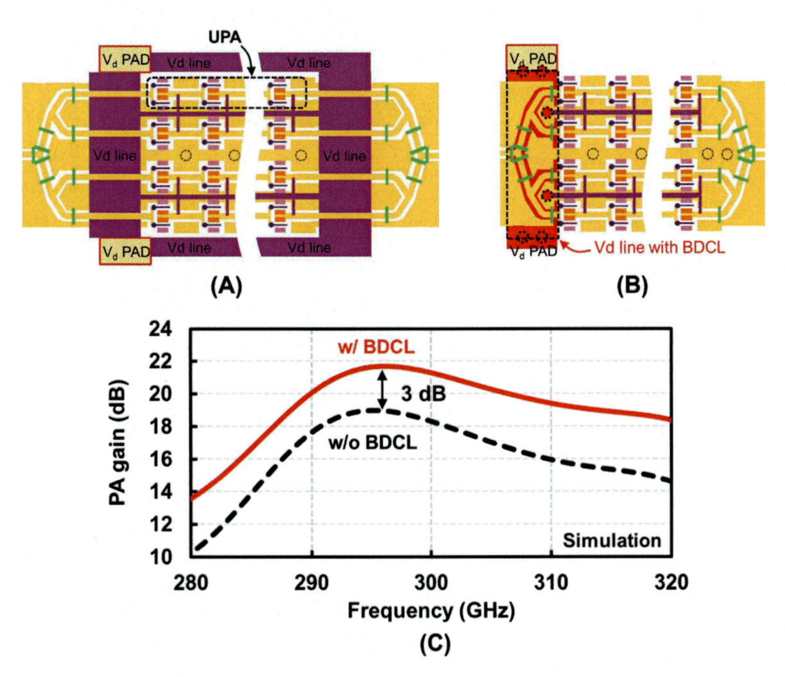


**FIGURE 6.11** Schematic of PA layout with (A) conventional DC line and (B) BDCL in this work. PA gain enhancement by loss reduction with BDCL (C). *Source: From Hamada, H., Tsutsumi, T., Matsuzaki, H., Fujimura, T., Abdo, I., Shirane, A., Okada, K., Itami, G., Song, H.-J., Sugiyama, H., & Nosaka, H. (2020). 300-GHz-Band 120-Gb/s wireless front-end based on InP-HEMT PAs and mixers. IEEE Journal of Solid-State Circuits, 55(9), 2316–2335. https://doi.org/10.1109/JSSC.2020.3005818.*

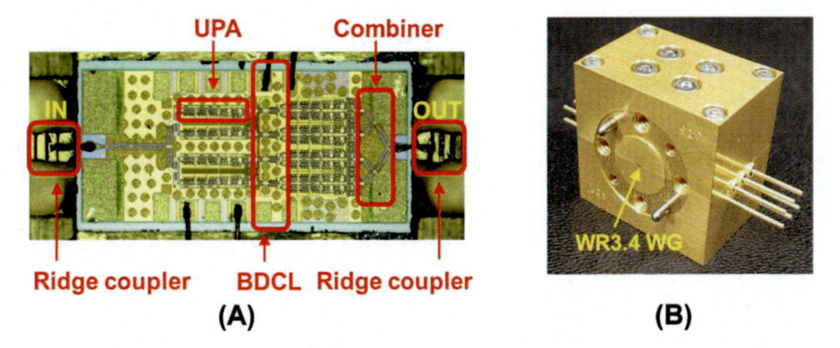


**FIGURE 6.12** Photograph of (A) PA chip and (B) PA module. *Source: From Hamada, H., Tsutsumi, T., Matsuzaki, H., Fujimura, T., Abdo, I., Shirane, A., Okada, K., Itami, G., Song, H.-J., Sugiyama, H., & Nosaka, H. (2020). 300-GHz-Band 120-Gb/s wireless front-end based on InP-HEMT PAs and mixers. IEEE Journal of Solid-State Circuits, 55 (9), 2316–2335. https://doi.org/10.1109/JSSC.2020.3005818.*

The measured and simulated S-parameters of the PA module are shown in Fig. 6.13A. The measured maximum gain ($S_{21}$) is 21 dB at 295 GHz, and the 3-dB bandwidth is 24 GHz. The simulation and measurement matches except for the frequencies above 295 GHz. The reverse coupling ($S_{12}$) is quite low in full WR 3.4 band, less than −40 dB. This means the stability of the PA module is high. This is because the substrate modes are removed by

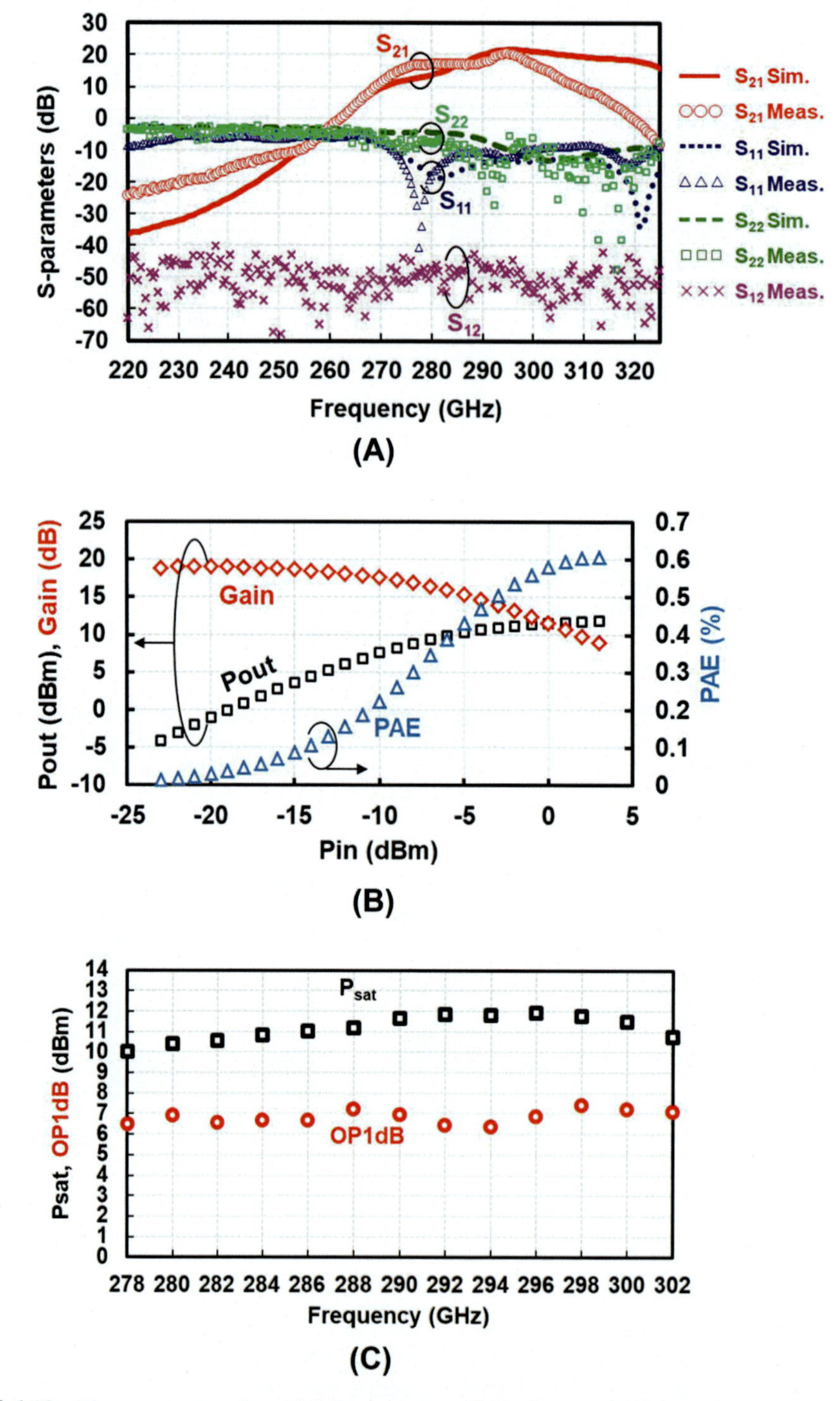


FIGURE 6.13 Measurement results of (A) S-parameters, (B) Pin-Pout, and (C) frequency response of Psat and OP1dB of PA module. *Source: From Hamada, H., Tsutsumi, T., Matsuzaki, H., Fujimura, T., Abdo, I., Shirane, A., Okada, K., Itami, G., Song, H.-J., Sugiyama, H., & Nosaka, H. (2020). 300-GHz-Band 120-Gb/s Wireless Front-End Based on InP-HEMT PAs and Mixers. IEEE Journal of Solid-State Circuits, 55(9), 2316–2335. https://doi.org/10.1109/JSSC.2020.3005818.*

the densely formed TSVs in the 55-μm-thick InP chip as described in Section 2. Next, the large-signal characteristics are measured using a commercially available ×18 frequency multiplier as a 300-GHz signal source. To measure the input-output characteristics (Pin-Pout), the input signal level of the PA is varied using the variable attenuator (VATT) inserted between the ×18 frequency multiplier and PA module. The measured Pin-Pout at 296 GHz is shown in Fig. 6.13B. High linearity is obtained. The saturated output power (Psat) and OP1dB are 12 and 6.8 dBm, respectively. The measured maximum power-added efficiency (PAE) is 0.61% at the output power of 12 dBm. The frequency response of Psat and OP1dB is also investigated as shown in Fig. 6.13C. The Psat and OP1dB are larger than 10 and 6 dBm for 278 to 302 GHz, respectively. This broadband high-linearity PA characteristic is suitable for use in high-speed 300-GHz TRXs.

The noise figure (NF) of this PA module is also measured using the NF measurement system (Hamada, Tsutsumi, Matsuzaki et al., 2020). The measured NF is 10–18 dB for 280–300 GHz as shown in Fig. 6.14. This value is not very bad considering that the typical NF of the LNA in the 300-GHz band is 7–10 dB (Samoska et al., 2016; Tessmann et al., 2017). Therefore, this PA is used as the LNA in the 300-GHz receiver (RX), as described in Section 4.

### 6.3.2 300-GHz mixer

Differential mixers, such as Gilbert-cell mixer, are widely used in microwave and MMW TRXs (Shahramian et al., 2013; Wu et al., 2014) due to their high conversion gain (CG) and LO leakage cancellation property. In the 300-GHz band, the implementation of the differential mixer is challenging since the generation of LO signals with perfectly 180-degree difference, which is important for differential mixer operation, is difficult in this high-frequency band. Furthermore, the necessary LO power for the differential mixer is generally higher than the single-ended one. This also makes it difficult to use differential mixers in the THz region due to the lack of high-power LO sources. Therefore, in the 300-GHz band, single-ended mixers are commonly used (Dan et al., 2019; Hamada et al., 2018). The resistive mixer (Maas, 1987) is a well-known passive-type FET mixer which has broadband and high-linearity characteristics. This mixer can operate for both up- and down-frequency conversion. Therefore, the resistive mixer can be used for both the transmitter (TX) and RX. The circuit schematic of the 300-GHz resistive mixer designed with

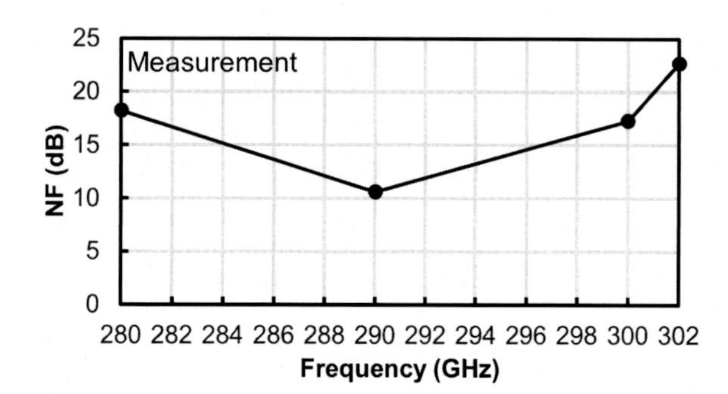


**FIGURE 6.14** Measured NF of PA module. *Source: From Hamada, H., Tsutsumi, T., Matsuzaki, H., Fujimura, T., Abdo, I., Shirane, A., Okada, K., Itami, G., Song, H.-J., Sugiyama, H., & Nosaka, H. (2020). 300-GHz-Band 120-Gb/s Wireless Front-End Based on InP-HEMT PAs and Mixers. IEEE Journal of Solid-State Circuits, 55(9), 2316–2335. https://doi.org/10.1109/JSSC.2020.3005818.*

NTT's in-house InP-HEMT technology is shown in Fig. 6.15A. As a manner of resistive topology, the LO signal is input to the gate and intermediate frequency (IF) and RF signals are connected to the drain of the common-sourced HEMT. The drain bias is zero and gate bias is set to the threshold voltage of the HEMT. The mixing-operation principle of the resistive mixer is the change in the impedance from the view point of the IF and RF signals connected to the drain caused by the drain-to-source resistance change ($\Delta R_{DS}$) due to the LO injection to the gate. Therefore, to obtain high CG, the impedance change of the drain port with LO injection should be designed as high as possible. In down-conversion operation, from the view point of RF signal, the impedance change ($\Delta Z_{DS}$) decreases from $\Delta R_{DS}$ due to the IF matching circuit, which is parasitically connected in parallel with $R_{DS}$. This $\Delta Z_{DS}$ decrease occurs in the IF signal due to the RF matching circuit for up-conversion operation. In the mixer in Fig. 6.15A, to avoid $\Delta Z_{DS}$ decrease, both the IF and RF matching circuits are designed to have very high impedance from the view point of the RF and IF signals. Two quarter-wavelength TLs in the IF matching circuit and small series capacitance in the RF matching circuit are used for this purpose. The detailed calculation and design of this point is described in a previous study (Hamada, Tsutsumi, Matsuzaki et al., 2020). For the mixer implementation, a 40-μm gate-width HEMT (two-fingered 20-μm HEMT) is used as its drain impedance is close to 50 Ω when the gate is biased at the threshold voltage. The fabricated mixer chip is shown in Fig. 6.15B. The chip size is $1 \times 1$ mm. This chip is mounted on the WG module using a ridge coupler as well as the PA chip. A photograph of the mixer module is shown in Fig. 6.15C. A V-connector is used as the IF signal interface. The mixer module characteristics are evaluated using the mixer measurement option of Keysight PNA-X. The LO signal is generated using the WR3.4 frequency extension module and amplified to a sufficiently large value to saturate the mixer operation by NTT's in-house InP-HEMT PA (Hamada et al., 2015). The measured CG of the mixer in up-conversion mode is shown in Fig. 6.16A. The LO frequency and power are set to 270 GHz and 5 dBm. The IF signal power is set to −30 dBm to avoid the nonlinear response of the mixer. The CG is around −15 dB for a RF frequency from 270 to 300 GHz. This broadband characteristic is based on the resistive IF/RF impedances of resistive

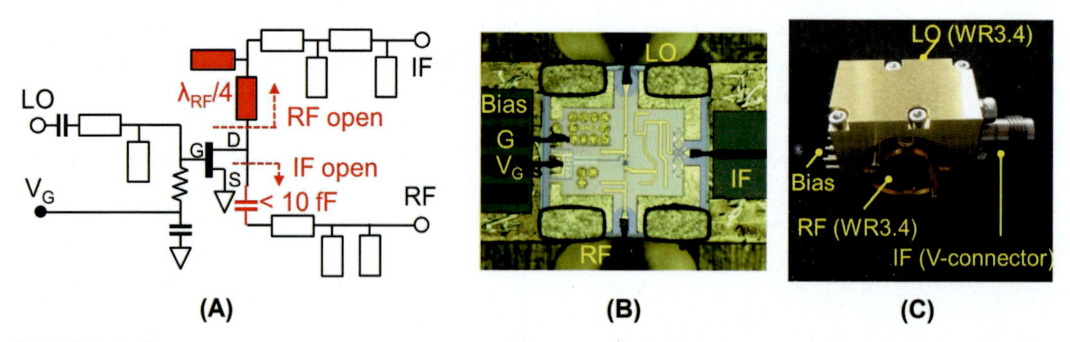


**FIGURE 6.15** (A) Schematic of 300-GHz resistive mixer circuit. Photographs of (B) fabricated mixer chip and (C) module. *Source: From Hamada, H., Tsutsumi, T., Matsuzaki, H., Fujimura, T., Abdo, I., Shirane, A., Okada, K., Itami, G., Song, H.-J., Sugiyama, H., & Nosaka, H. (2020). 300-GHz-Band 120-Gb/s Wireless Front-End Based on InP-HEMT PAs and Mixers. IEEE Journal of Solid-State Circuits, 55(9), 2316–2335. https://doi.org/10.1109/JSSC.2020.3005818.*

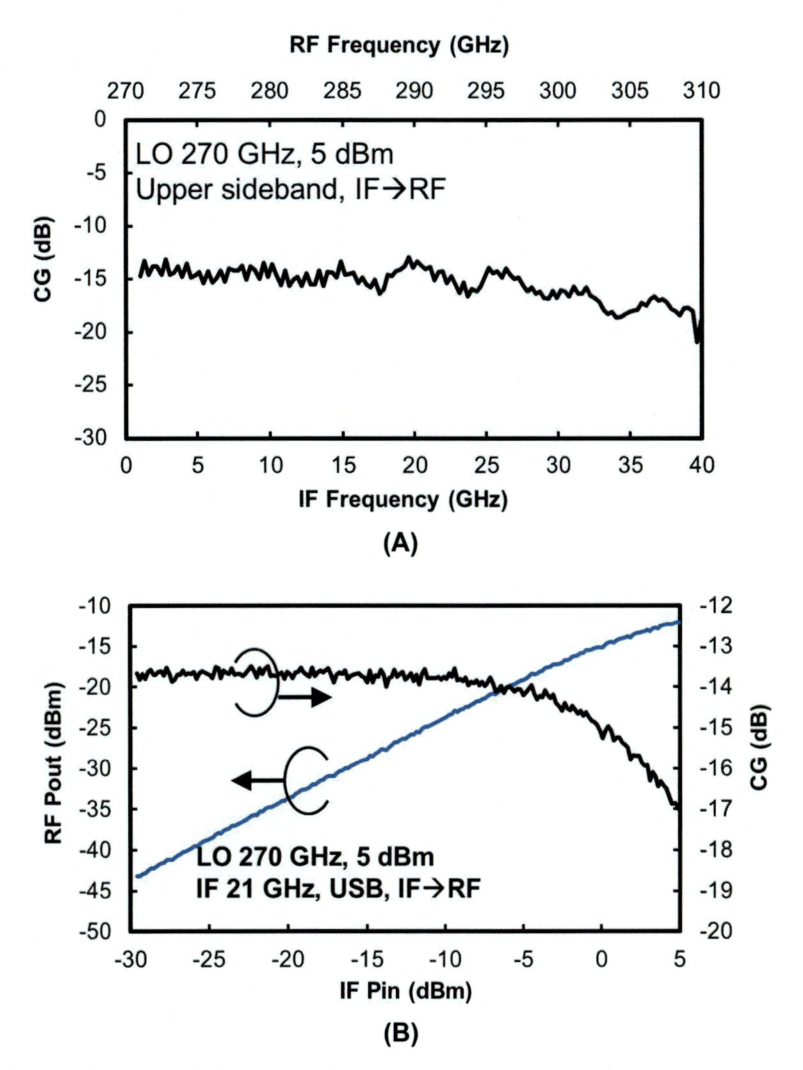


**FIGURE 6.16** Measurement results of (A) CG versus frequency and (B) Pin-Pout of mixer module. *Source: From Hamada, H., Fujimura, T., Abdo, I., Okada, K., Song, H.-J., Sugiyama, H., Matsuzaki, H., & Nosaka, H. (2018). 300-GHz, 100-Gb/s InP-HEMT wireless transceiver using a 300-GHz fundamental mixer. In* 2018 IEEE/MTT-S international microwave symposium - IMS, *(pp. 1480–1483). https://doi.org/10.1109/MWSYM.2018.8439850.*

topology and the choice of appropriate HEMT gate width as described above. The linearity of the mixer module is also evaluated with the same measurement setup. Fig. 6.16B shows the measured Pin-Pout characteristic of the mixer in up-conversion mode. The IF frequency is set to 21 GHz. The measured input P1dB (IP1dB) and OP1dB is −1.5 and −16.5 dB, respectively. The LO leakage to the RF port is −16.4 dBm at 270 GHz, as described in a previous study (Hamada, Tsutsumi, Matsuzaki et al., 2020).

## 6.4 300-GHz-band InP transceiver and 120 Gb/s wireless data transmission

The 300-GHz-band heterodyne TRX is built using the PAs and mixers, as shown in Fig. 6.17A. The configuration of the fabricated TRX is almost the same as simplified TRX illustrated in Fig. 6.7. The forms of all the building blocks in Fig. 6.17A are WR3.4 WG modules, as shown in Fig. 6.17B. The LO and IF frequencies are set to 270 and 20 GHz, respectively. The upper side band (USB) is used as an RF signal. Therefore, in the TX, the high-pass filter (HPF) with cut-off frequency of 271 GHz is inserted between the mixer and RF PA to reject the lower side band signal output from the TX. This HPF also acts as the LO leakage rejection filter to protect the RF PA from the intermodulation of LO leakage and RF signal. The band-pass filters are used between the mixer and LO PA for both the TX and RX to cut out unwanted harmonic signals from the LO source ($\times 18$ multiplier). The same resistive mixers are used for both the TX and RX with the aid of the commutative property of this mixer. The same RF PA is used as the LNA in the RX as the NF of RF PA is not high (10–15 dB).

First, the CG and linearity of the TX and RX are evaluated in the same manner as the mixer measurement discussed in Section 3. The measured frequency dependence of the TX CG is shown in Fig. 6.18A. The CG is around 3 dB over 290–295 GHz. The 3-dB bandwidth is 17 GHz. This CG frequency dependence reflects those of PA and mixer in Figs. 6.13A and 6.16A. Considering the broadband characteristics of the mixer CG, the bandwidth of the TX CG is mainly determined by the PA. The Pin-Pout characteristic of the TX is shown in Fig. 6.18B. The Psat of the RF output power is 5 dBm and OP1dB is

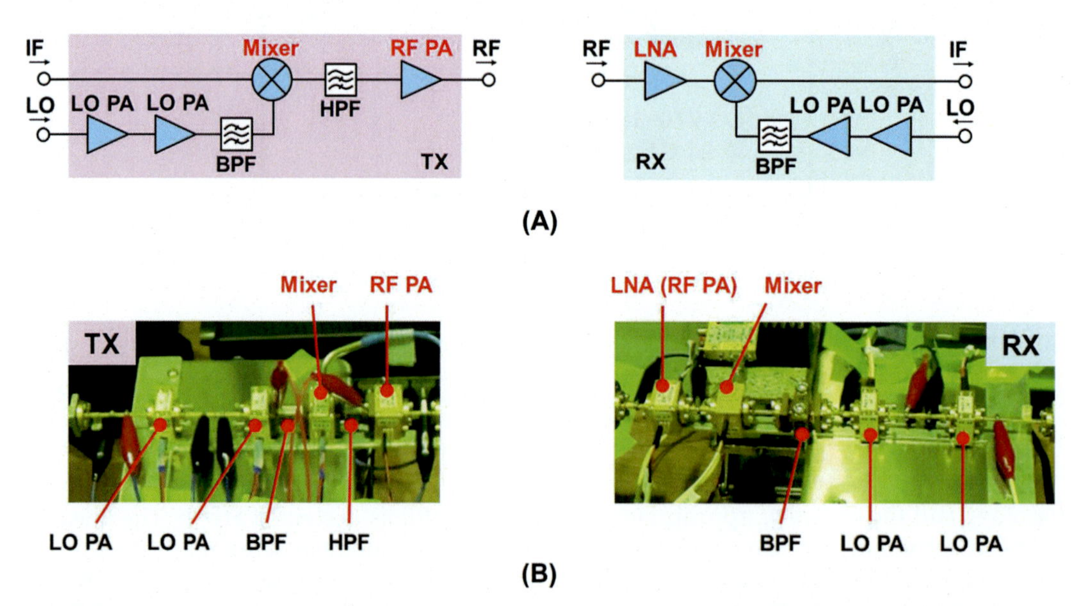


FIGURE 6.17 (A) Schematic and (B) photograph of InP-HEMT 300-GHz TRX. *Source: From Hamada, H., Tsutsumi, T., Matsuzaki, H., Fujimura, T., Abdo, I., Shirane, A., Okada, K., Itami, G., Song, H.-J., Sugiyama, H., & Nosaka, H. (2020). 300-GHz-Band 120-Gb/s wireless front-end based on InP-HEMT PAs and mixers. IEEE Journal of Solid-State Circuits, 55(9), 2316–2335. https://doi.org/10.1109/JSSC.2020.3005818.*

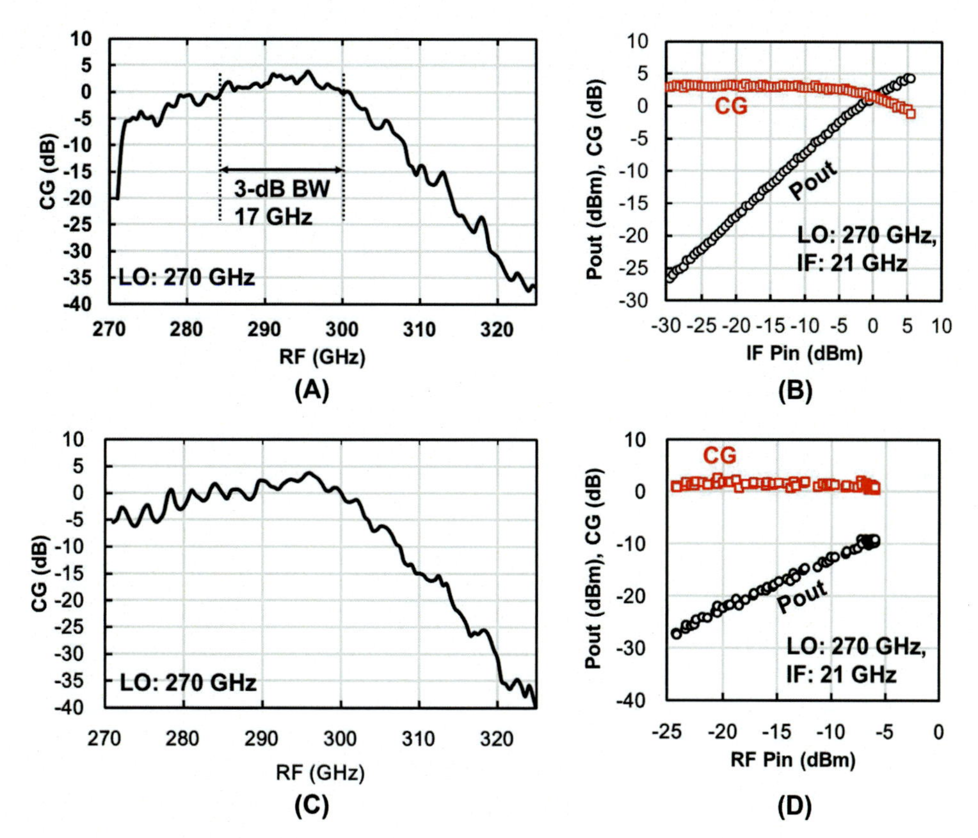


**FIGURE 6.18** (A) CG and (B) Pin-Pout of TX, (C) CG and Pin-Pout of RX. *Source: From Hamada, H., Tsutsumi, T., Matsuzaki, H., Fujimura, T., Abdo, I., Shirane, A., Okada, K., Itami, G., Song, H.-J., Sugiyama, H., & Nosaka, H. (2020). 300-GHz-Band 120-Gb/s wireless front-end based on InP-HEMT PAs and mixers. IEEE Journal of Solid-State Circuits, 55(9), 2316–2335. https://doi.org/10.1109/JSSC.2020.3005818.*

1.5 dBm. The OP1dB matches the theoretical value of 1.6 dBm which is calculated using the formula (Pozar, 2012) in the cascaded form of the mixer (OP1dB = −16.5 dBm) and PA (OP1dB = 6.8 dBm) shown in Fig. 6.16B and Fig. 6.13B. The RX CG and Pin-Pout are also measured as shown in Fig. 6.18C and D. The maximum CG and 3-dB bandwidth are 3 dB and 17 GHz. These values are similar to those of the TX. The Pin-Pout shows the linear response of the RX up to its input power of −7 dBm.

By using the above TX and RX, the data transmission experiments were carried out. Before wireless data transmission, back-to-back data transmission was carried out to clarify the maximum data rate of this TRX. In this configuration, the TX and RX are connected through the WG ATT, as shown in Fig. 6.19A. The role of this ATT is to avoid large-power input to the RX because a large input will cause nonlinearity to the RX, which degrades the SNR. The IF data signal was generated from an arbitrary waveform generator (AWG, Keysight M8196A) and detected with a digital storage oscilloscope (DSO, Keysight Z634A). In the RX, the IF amplifier (SHF S807B) was used to amplify the received IF signal to supply sufficient voltage

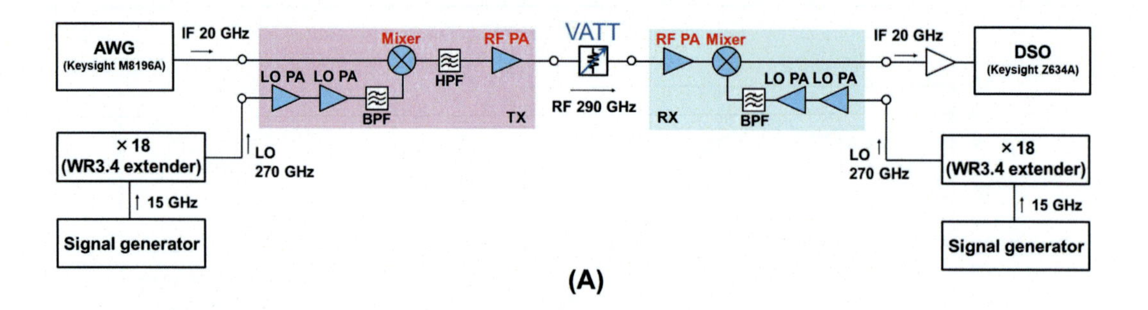


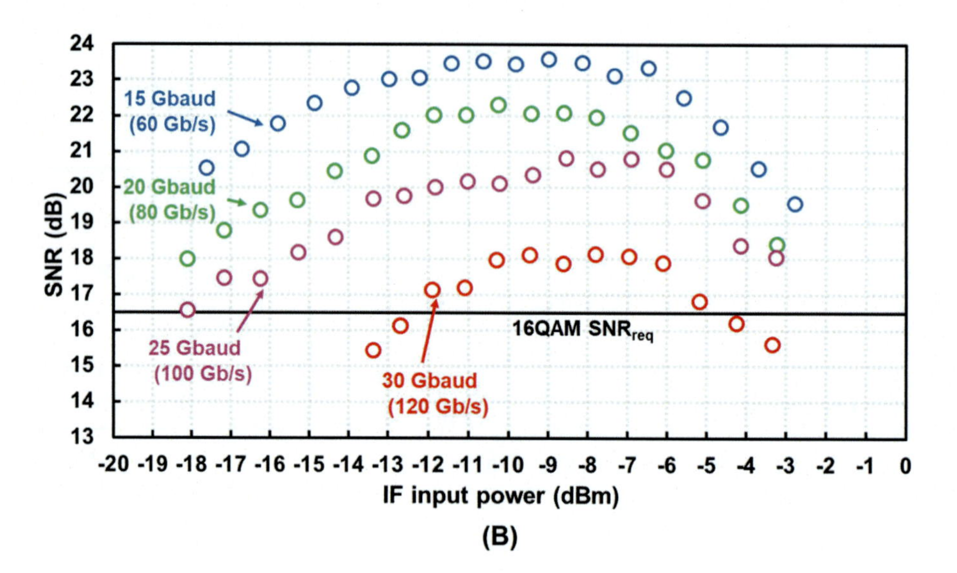


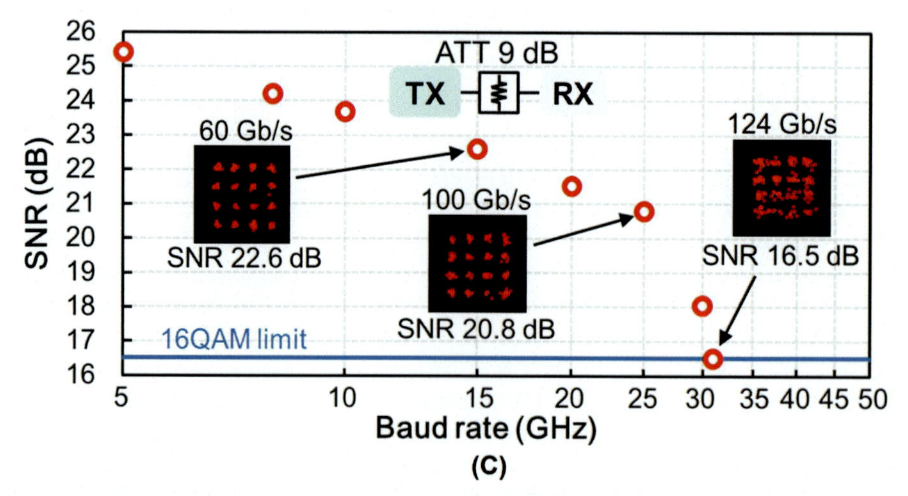


FIGURE 6.19 (A) Schematic of back-to-back data transmission setup, measured (B) SNR versus IF input power, and (C) SNR versus baud rate. *Source: From Hamada, H., Tsutsumi, T., Matsuzaki, H., Fujimura, T., Abdo, I., Shirane, A., Okada, K., Itami, G., Song, H.-J., Sugiyama, H., & Nosaka, H. (2020). 300-GHz-Band 120-Gb/s wireless front-end based on InP-HEMT PAs and mixers. IEEE Journal of Solid-State Circuits, 55(9), 2316–2335. https://doi.org/10.1109/JSSC.2020.3005818.*

amplitude to the DSO. The IF center frequency was set to 20 GHz and 16QAM was used as the modulation format. A pseudo-random-bit sequence of the modulation data was set to $2^{11}-1$. Signal equalization was applied to the received IF signal by using Keysight VSA software. The constellation and SNR of the received IF signal were measured to evaluate the quality of the data transmission. The data transmission was judged successful when the SNR of the received IF signal was larger than the required SNR ($SNR_{req}$) of the modulation format. The required SNR is determined as the SNR that is equivalent to the bit-error rate (BER) of $10^{-3}$. The required SNR of the 16QAM signal is 16.5 dB. The dependence of the SNR and TX IF input power was measured to find the optimum IF power for maximizing the SNR. While the TX IF input power is small, the SNR is mainly determined from the noise level. In this case, the measured SNR will increase proportionally with the TX IF input power. When the IF input power is sufficiently large, the nonlinearity of the TRX occurs and the SNR will degrade with increasing IF input power. Therefore, there is an optimum IF input power. Fig. 6.19B shows the measurement results of the SNR versus IF input power for the baud rates of 15 Gbaud (60 Gb/s), 20 Gbaud (80 Gb/s), 25 Gbaud (100 Gb/s), and 30 Gbaud (120 Gb/s). The measured SNRs are maximum at the IF input power of around $-9$ dBm. Therefore, in the following measurement, the IF input power was set to $-9$ dBm. The maximum baud rate of the TRX was also estimated with the back-to-back configuration. Fig. 6.19C shows the measured baud-rate dependence of the SNR and the constellations of the 16QAM IF signal obtained at the RX side. The maximum baud rate at which the SNR is equal to 16QAM $SNR_{req}$ of 16.5 dB is 31 Gbaud (124 Gb/s). In the low baud rate region, the measured SNR is inversely proportional to the baud rate due to the noise-power increase proportional to the signal bandwidth. At a baud rate larger than 25 Gbaud, the SNR decreases more rapidly. This is because of the bandwidth limitation of the TRX as shown in Fig. 6.18A and C. The CGs of the TX and RX decrease rapidly above 300 GHz and this will degrade the SNR. Therefore, the maximum data rate can be increased by improving the frequency responses of the TX and RX. The main bandwidth limitation block in both TX and RX are the PAs in this case. Recently, the wideband 300-GHz InP PAs are reported. Hamada, Tsutsumi, Pander, et al. (2020) and John et al. (2020) use the neutralization technique (Bameri & Momeni, 2017) to both increase the gain and bandwidth. Using these PAs are good option to enhance the TRX data rate.

Next, wireless data transmission was carried out using the TRX in an anechoic chamber. The measurement setup was same as the back-to-back data transmission except for the use of antennas. Lensed horn antennas with an antenna gain of 50 dBi were used for both the TX and RX. A photograph of the measurement system is shown in Fig. 6.20A. The link distance (antenna-to-antenna distance) was fixed to 9.8 m. This link distance is virtually changed by adjusting the attenuation value of the VATT in the TX, as shown in Fig. 6.20A. The VATT equivalently increases the propagation loss between the TX and RX. By using the relationship between the link distance and propagation loss

$$\text{linkdistance} \propto \sqrt{\text{propagationloss}}$$

the SNR dependence of the link distance can be estimated. The measurement results of 16QAM data transmission are shown in Fig. 6.20B. Wireless data transmission was successfully demonstrated up to the baud rate of 30 Gbaud (120 Gb/s). The maximum link

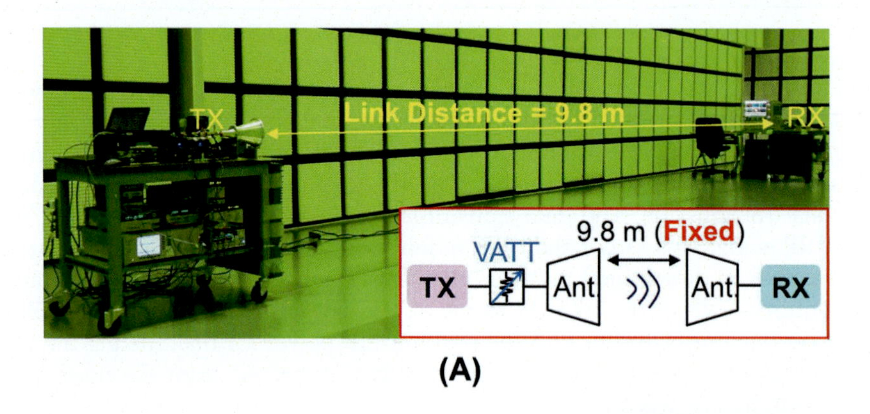


**(A)**

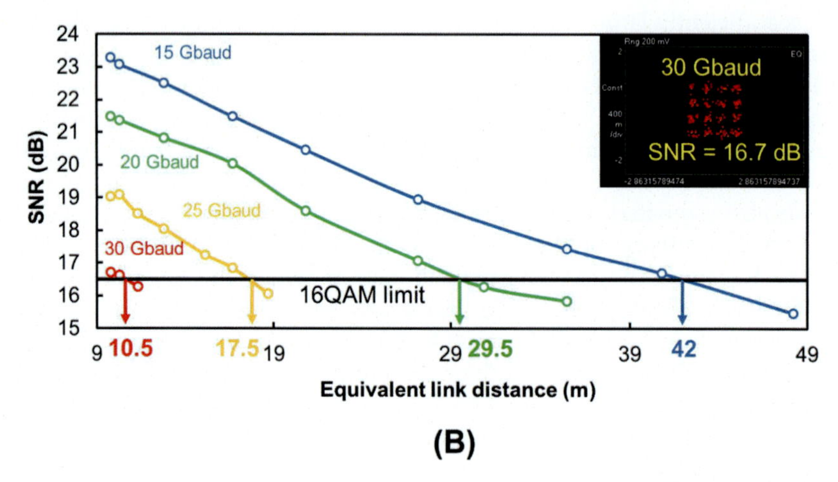


**(B)**

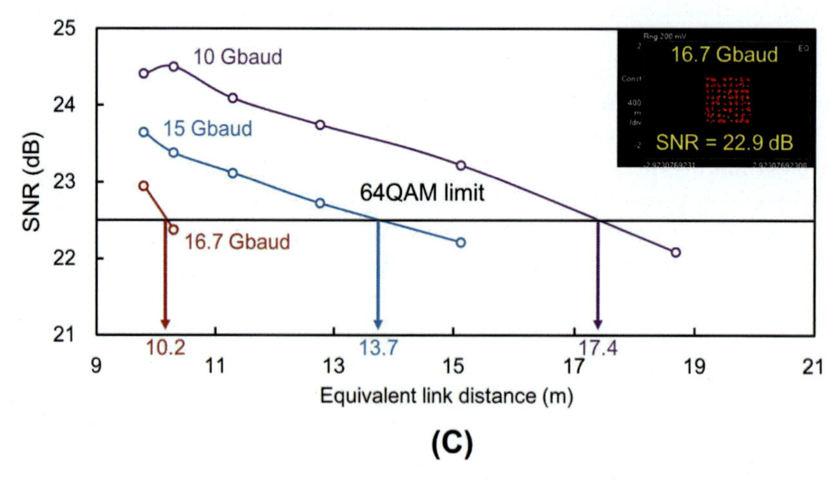


**(C)**

**FIGURE 6.20** (A) Photograph and setup of 300 GHz wireless data transmission, measured SNR versus equivalent link distance of (B) 16QAM and (C) 64QAM wireless data transmissions. *Source: From Hamada, H., Tsutsumi, T., Matsuzaki, H., Fujimura, T., Abdo, I., Shirane, A., Okada, K., Itami, G., Song, H.-J., Sugiyama, H., & Nosaka, H. (2020). 300-GHz-Band 120-Gb/s Wireless Front-End Based on InP-HEMT PAs and Mixers. IEEE Journal of Solid-State Circuits, 55(9), 2316–2335. https://doi.org/10.1109/JSSC.2020.3005818.*

distances where the measured SNR is equal to 16.5 dB ($SNR_{req}$ of 16QAM) are 10.5, 17.5, 29.5, and 42 for the baud rates of 30 (120 Gb/s), 25 (100 Gb/s), 20 (80 Gb/s), and 15 (60 Gb/s), respectively. Because the measured SNRs were so high, we propagated a more SNR-demanding higher-order modulation format of 64QAM ($SNR_{req}$: 22.5 dB). The measured SNRs of the 64QAM transmission are shown in Fig. 6.20C. The maximum baud rate is 16.7 Gbaud (100.2 Gb/s), and the maximum equivalent link distances are 10.2, 13.7, and 17.4 for the baud rates of 16.7 (100.2 Gb/s), 15 (90 Gb/s), and 10 (60 Gb/s), respectively. Although the integration level of the TRX is not high (combination of the individual WG modules of PAs and mixers), these results indicate that the InP-based MMIC technology is a good candidate to support the high data rate 300-GHz-band wireless systems for B5G.

## 6.5 Conclusion

In this chapter, several InP technologies for wireless TRXs toward beyond 5G were described from device level to wireless front-end level. NTT's in-house 80-nm InP-HEMT with $f_T$ and $f_{MAX}$ of 300 and 700 GHz is used as a transistor for 300-GHz-band MMICs. The substrate thinning and dense TSV formation technique is introduced to avoid the negative impact of the substrate modes. By using these back-end techniques, 300-GHz InP MMICs are thinned to 55 μm with 50-μm-pitch TSVs formation. The fabricated InP MMIC chips are mounted on WR3.4 WG modules with a ridge coupler as the low-loss (~1 dB) IC-to-WG transition. By using these device techniques, the MMICs for 300-GHz-band TRX, that is, PAs and mixers are designed and fabricated. The PA is designed to achieve high output power by reducing the RF loss with a simple eight-way output power combiner and BDCL technique. The small-signal gain of the PA is 21 dB at 295 GHz, and the 3-dB bandwidth is 24 GHz. The Psat and OP1dB of the PA are 12 and 6 dBm, respectively. The mixer is designed using resistive topology to achieve broad bandwidth and high linearity. The CG of the mixer is −15 dB, and the 3-dB bandwidth in the USB operation is 32 GHz with its LO frequency of 270 GHz. The 300-GHz TRX was built on the basis of the individual WG modules of these PAs and mixers. It achieved back-to-back 16QAM data transmission up to 124 Gb/s. Wireless data transmission was also carried out with high-gain lensed horn antennas. The TRX achieved the 120 Gb/s wireless data transmission with a link distance of 9.8 m. Although the integration level of the InP TRX is not high, these results indicate the potential of InP-based 300-GHz-band TRXs for B5G.

## Acknowledgements

The authors would like to thank Takuya Tsutsumi, Hideaki Matsuzaki, Hideyuki Nosaka, Go Itami, Adam Pander, Munehiko Nagatani, Hiroki Sugiyama, Hiroshi Yamazaki at NTT Device Technology Laboratories, Kanagawa, Japan, and Kenichi Okada, Atsushi Shirane, Ibrahim Abdo, Takuya Fujimura at Tokyo Institute of Technology, Tokyo, Japan for their fruitful discussions. The authors also would like to thank the continuous assistance of Yukio Yago and Ryukichi Kamada as well as the support with the measurement setup from Katsumi Fujii, and Akifumi Kasamatsu at National Institute of Information and Communications Technology (NICT).

## References

Abdo, I., Da Gomez, C., Wang, C., Hatano, K., Li, Q., Liu, C., Yanagisawa, K., Aviat Fadila, A., Pang, J., Hamada, H., Nosaka, H., Shirane, A., & Okada, K. (2021). A 300GHz-band phased-array transceiver using bi-directional outphasing and hartley architecture in 65nm CMOS. In *2021 IEEE international solid-state circuits conference (ISSCC)*.

Abdo, I., Fujimura, T., Miura, T., Tokgoz, K. K., Hamada, H., Nosaka, H., Shirane, A., & Okada, K. (2020). A 300GHz wireless transceiver in 65nm CMOS for IEEE802.150.3d using push-push subharmonic mixer, 2020*IEEE MTT-S International Microwave Symposium (IMS)*, 623–626. Available from https://doi.org/10.1109/IMS30576.2020.9224033.

Bameri, H., & Momeni, O. (2017). A high-gain mm-wave amplifier design: An analytical approach to power gain boosting. *IEEE Journal of Solid-State Circuits*, *52*(2), 357–370. Available from https://doi.org/10.1109/JSSC.2016.2626340.

Boes, F., Messinger, T., Antes, J., Meier, D., Tessmann, A., Inam, A., & Kallfass, I. (2014). Ultra-broadband MMIC-based wireless link at 240GHz enabled by 64 GS/s DAC, 2014*39th International Conference on Infrared, Millimeter, and Terahertz waves (IRMMW-THz)*. Available from https://doi.org/10.1109/IRMMW-THz.2014.6956202.

Dan, I., Grötsch, C. M., Schoch, B., Wagner, S., John, L., Tessmann, A., & Kallfass, I. (2019). A 300GHz quadrature down-converter S-MMIC for future terahertz communication, 2019*IEEE International Conference on Microwaves, Antennas, Communications and Electronic Systems (COMCAS)*. Available from https://doi.org/10.1109/COMCAS44984.2019.8958300.

Dan, I., Szriftgiser, P., Peytavit, E., Lampin, J. F., Zegaoui, M., Zaknoune, M., Ducournau, G., & Kallfass, I. (2020). A 300-GHz wireless link employing a photonic transmitter and an active electronic receiver with a transmission bandwidth of 54 GHz. *IEEE Transactions on Terahertz Science and Technology*, *10*(3), 271–281. Available from https://doi.org/10.1109/TTHZ.2020.2977331.

David, K., & Berndt, H. (2018). 6G vision and requirements: Is there any need for beyond 5G? *IEEE Vehicular Technology Magazine*, *13*(3), 72–80. Available from https://doi.org/10.1109/MVT.2018.2848498.

Deal, W. R., Leong, K., Yoshida, W., Zamora, A., & Mei, X. B. (2016). InP HEMT integrated circuits operating above 1,000 GHz, 2016*IEEE International Electron Devices Meeting (IEDM)*. Available from https://doi.org/10.1109/IEDM.2016.7838502.

Hamada, H., Fujimura, T., Abdo, I., Okada, K., Song, H. J., Sugiyama, H., Matsuzaki, H., & Nosaka, H. (2018). 300-GHz, 100-Gb/s InP-HEMT wireless transceiver using a 300-GHz fundamental mixer, 2018*IEEE/MTT-S International Microwave Symposium - IMS*, 1480–1483. Available from https://doi.org/10.1109/MWSYM.2018.8439850.

Hamada, H., Kosugi, T., Song, H. J., Matsuzaki, H., El Moutaouakil, A., Sugiyama, H., Yaita, M., Tajima, T., Nosaka, H., Kagami, O., Kawano, Y., Takahashi, T., Nakasha, Y., Hara, N., Fujii, K., Watanabe, I., & Kasamatsu, A. (2016). 20-Gbit/s ASK wireless system in 300-GHz-band and front-ends with InP MMICs, 2016*URSI Asia-Pacific Radio Science Conference (URSI AP-RASC)*. Available from https://doi.org/10.1109/URSIAP-RASC.2016.7601303.

Hamada, H., Kosugi, T., Song, H. J., Yaita, M., El Moutaouakil, A., Matsuzaki, H., & Hirata, A. (2015). 300-GHz Band 20-Gbps ASK transmitter module based on InP-HEMT MMICs, 2015*IEEE Compound Semiconductor Integrated Circuit Symposium (CSICS)*. Available from https://doi.org/10.1109/CSICS.2015.7314461.

Hamada, H., Tsutsumi, T., Itami, G., Sugiyama, H., Matsuzaki, H., Okada, K., & Nosaka, H. (2019). 300-GHz 120-Gb/s wireless transceiver with high-output-power and high-gain power amplifier based on 80-nm InP-HEMT technology, 2019*IEEE BiCMOS and Compound semiconductor Integrated Circuits and Technology Symposium (BCICTS)*. Available from https://doi.org/10.1109/BCICTS45179.2019.8972756.

Hamada, H., Tsutsumi, T., Sugiyama, H., Matsuzaki, H., Song, H.-J., Itami, G., Fujimura, T., Abdo, I., Okada, K., & Nosaka, H. (2019). Millimeter-wave InP device technologies for ultra-high speed wireless communications toward beyond 5 G, 2019*IEEE International Electron Devices Meeting (IEDM)*, 9.2.1–9.2.4. Available from https://doi.org/10.1109/IEDM19573.2019.8993540.

Hamada, H., Tsutsumi, T., Matsuzaki, H., Fujimura, T., Abdo, I., Shirane, A., Okada, K., Itami, G., Song, H. J., Sugiyama, H., & Nosaka, H. (2020). 300-GHz-Band 120-Gb/s wireless front-end based on InP-HEMT PAs and mixers. *IEEE Journal of Solid-State Circuits*, *55*(9), 2316–2335. Available from https://doi.org/10.1109/JSSC.2020.3005818.

Hamada, H., Tsutsumi, T., Pander, A., Nakamura, M., Itami, G., Matsuzaki, H., Sugiyama, H., & Nosaka, H. (2020). 230–305 GHz, > 10-dBm-output-power wideband power amplifier using low-Q neutralization technique in 60-nm InP-HEMT technology. IEEE BiCMOS Compound Semiconductor Integr. *Circuits Technol. Symp.* (BCICTS).

Hara, S., Katayama, K., Takano, K., Dong, R., Watanabe, I., Sekine, N., Kasamatsu, A., Yoshida, T., Amakawa, S., & Fujishima, M. (2017). A 32Gbit/s 16QAM CMOS receiver in 300GHz band, 2017*IEEE MTT-S International Microwave Symposium (IMS)*, 1703–1706. Available from https://doi.org/10.1109/MWSYM.2017.8058969.

Ito, H., & Masu, K. (2008). A simple through-only de-embedding method for on-wafer S-parameter measurements up to 110 GHz, 2008*IEEE MTT-S International Microwave Symposium Digest*, 383–386. Available from https://doi.org/10.1109/MWSYM.2008.4633183.

ITU-R ITU-R World Radiocommunication Conference 2019 (WRC-19) Final Acts 2019. (2020). http://www.itu.int > opb > act > R-ACT-WRC.14–2019-PDF-E.pdf

John, L., Tessmann, A., Leuther, A., Neininger, P., Merkle, T., & Zwick, T. (2020). Broadband 300-GHz power amplifier MMICs in InGaAs mHEMT technology. *IEEE Transactions on Terahertz Science and Technology*, *10*(3), 309–320. Available from https://doi.org/10.1109/TTH10Z.2020.2965808.

Jo, H. B., Yun, D. Y., Baek, J. M., Lee, J. H., Kim, T. W., Kim, D. H., Tsutsumi, T., Sugiyama, H., & Matsuzaki, H. (2019). Lg = 25 nm InGaAs/InAlAs high-electron mobility transistors with both fT and fmax in excess of 700 GHz. *Applied Physics Express*, *12*(5), 054006-1–054006-4.

Kallfass, I., Antes, J., Schneider, T., Kurz, F., Lopez-Diaz, D., Diebold, S., Massler, H., Leuther, A., & Tessmann, A. (2011). All active MMIC-based wireless communication at 220 GHz. *IEEE Transactions on Terahertz Science and Technology*, *1*(2), 477–487. Available from https://doi.org/10.1109/TTHZ.2011.2160021.

Kallfass, I., Boes, F., Messinger, T., Antes, J., Inam, A., Lewark, U., Tessmann, A., & Henneberger, R. (2015). 64 Gbit/s transmission over 850m fixed wireless link at 240GHz carrier frequency. *Journal of Infrared, Millimeter, and Terahertz Waves*, *36*, 221–233. Available from https://doi.org/10.1007/s10762-014-0140-6.

Kallfass, I., Dan, I., Rey, S., Harati, P., Antes, J., Tessmann, A., Wagner, S., Kuri, M., Weber, R., Massler, H., Leuther, A., Merkle, T., & Kürner, T. (2015). Towards MMIC-based 300GHz indoorwireless communication systems. *IEICE Transactions on Electronics*, *98*(12), 1081–1090. Available from https://doi.org/10.1587/transele.E98.C.1081.

Kim, K., & Nguyen, C. (2018). A V-band power amplifier with integrated wilkinson power dividers-combiners and transformers in 0.18-μm SiGe BiCMOS. *IEEE Transactions on Circuits and Systems—II*, *66*(33), 337–341. Available from https://doi.org/10.1109/TCSII.2018.2850899.

Kosugi, T., Hamada, H., Takahashi, H., Song, H.J., Hirata, A., Matsuzaki, H., & Nosaka, H. (2014). 250–300GHz waveguide module with ridge-coupler and InP-HEMT IC. In *2014 Asia-Pacific microwave conference* (pp. 1133–1135).

Law, C. Y., & Pham, A. V. (2010). A high-gain 60GHz power amplifier with 20dBm output power in 90nm CMOS, 2010*IEEE International Solid-State Circuits Conference - (ISSCC)*. Available from https://doi.org/10.1109/ISSCC.2010.5433882.

Lee, S., Hara, S., Yoshida, T., Amakawa, S., Dong, R., Kasamatsu, A., Sato, J., & Fujishima, M. (2019). An 80Gb/s 300GHz-band single-chip CMOS transceiver. *IEEE Journal of Solid-State Circuits*, *54*(12), 3577–3588. Available from https://doi.org/10.1109/JSSC.2019.2944855.

Maas, S. A. (1987). A GaAs MESFET mixer with very low intermodulation. *IEEE Transactions on Microwave Theory and Techniques*, *35*(4), 425–429. Available from https://doi.org/10.1109/TMTT.1987.1133665.

Okada, K., Li, N., Matsushita, K., Bunsen, K., Murakami, R., Musa, A., Sato, T., Asada, H., Takayama, N., Ito, S., Chaivipas, W., Minami, R., Yamaguchi, T., Takeuchi, Y., Yamagishi, H., Noda, M., & Matsuzawa, A. (2011). A 60-GHz 16QAM/8PSK/QPSK/BPSK direct-conversion transceiver for IEEE802.15.3c. *IEEE Journal of Solid-State Circuits*, *46*(12), 2988–3004. Available from https://doi.org/10.1109/JSSC.2011.2166184.

Pozar, D. M. (2012). *Microwave engineering*. Hoboken, NJ, USA: Wiley, 4.

Rodríguez-Vázquez, P., Grzyb, J., Heinemann, B., & Pfeiffer, U. R. (2019). A 16-QAM 100-Gb/s 1-M wireless link with an EVM of 17% at 230GHz in an SiGe technology. *IEEE Microwave and Wireless Components Letters*, *29*(4), 297–299. Available from https://doi.org/10.1109/LMWC.2019.2899487.

Rodríguez-Vázquez, P., Grzyb, J., Heinemann, B., & Pfeiffer, U. R. (2020). A QPSK 110-Gb/s polarization-diversity MIMO wireless link with a 220–255GHz tunable LO in a SiGe HBT technology. *IEEE Transactions on Microwave Theory and Techniques*, *68*(9), 3834–3851. Available from https://doi.org/10.1109/TMTT.2020.2986196.

Rodríguez-Vázquez, P., Grzyb, J., Sarmah, N., Heinemann, B., & Pfeiffer, U. R. (2018a). Towards 100 Gbps: A fully electronic 90 Gbps one meter wireless link at 230GHz, 2018*48th European Microwave Conference (EuMC)*, 1389–1392. Available from https://doi.org/10.23919/EuMC.2018.8541410.

Rodríguez-Vázquez, P., Grzyb, J., Sarmah, N., Heinemann, B., & Pfeiffer, U. R. (2018b). A 65 Gbps QPSK one meter wireless link operating at a 225–255GHz tunable carrier in a SiGe HBT technology, 2018*IEEE Radio and Wireless Symposium (RWS)*, 146–149. Available from https://doi.org/10.1109/RWS.2018.8304970.

Samoska, L., Fung, A., Varonen, M., Lin, R., Peralta, A., Soria, M., Lee, C., Padmanabhan, S., Sarkozy, S., & Lai, R. (2016). Miniature packaging concept for LNAs in the 200–300GHz range, 2016*IEEE MTT-S International Microwave Symposium (IMS)*. Available from https://doi.org/10.1109/MWSYM.2016.7540162.

Sarmah, N., Vazquez, P. R., Grzyb, J., Foerster, W., Heinemann, B., & Pfeiffer, U. R. (2016). A wideband fully integrated SiGe chipset for high data rate communication at 240GHz, 2016*11th European Microwave Integrated Circuits Conference (EuMIC)*, 181–184. Available from https://doi.org/10.1109/EuMIC.2016.7777520.

Schmückle, F.J., Doerner, R., Phung, G.N., Heinrich, W., Williams, D., & Arz, U. (2011). Radiation, multimode propagation, and substrate modes in W-band CPW calibrations. In *2011 41st European microwave conference* (pp. 297–300). Available from https://doi.org/10.23919/EuMC.2011.6101804.

Shahramian, S., Baeyens, Y., Kaneda, N., & Chen, Y. K. (2013). A 70–100GHz direct-conversion transmitter and receiver phased array chipset demonstrating 10 Gb/s wireless link. *IEEE Journal of Solid-State Circuits*, *48*(5), 1113–1125. Available from https://doi.org/10.1109/JSSC.2013.2254536.

Song, H. J., Kim, J. Y., Ajito, K., Kukutsu, N., & Yaita, M. (2014). 50-Gb/s direct conversion QPSK modulator and demodulator MMICs for terahertz communications at 300GHz. *IEEE Transactions on Microwave Theory and Techniques*, *62*(3), 600–609. Available from https://doi.org/10.1109/TMTT.2014.2300844.

Song, H. J., Kosugi, T., Hamada, H., Tajima, T., El Moutaouakil, A., Matsuzaki, H., Kawano, Y., Takahashi, T., Nakasha, Y., Hara, N., Fujii, K., Watanabe, I., Kasamatsu, A., & Yaita, M. (2016). Demonstration of 20-Gbps wireless data transmission at 300GHz for KIOSK instant data downloading applications with InP MMICs, 2016*IEEE MTT-S International Microwave Symposium (IMS)*. Available from https://doi.org/10.1109/MWSYM.2016.7540141.

Sugiyama, H., Matsuzaki, H., Yokoyama, H., & Enoki, T. (2010). High-electron-mobility In0.53Ga0.47As / In0.8Ga0.2As composite-channel modulation-doped structures grown by metal-organic vapor-phase epitaxy. In *2010 22nd International conference on indium phosphide and related materials (IPRM)* (pp. 477–480). Available from https://doi.org/10.1109/ICIPRM.2010.5516265.

Tessmann, A., Hurm, V., Leuther, A., Massler, H., Weber, R., Kuri, M., Riessle, M., Stulz, H.P., Zink, M., Schlechtweg, M., Ambacher, O., & Närhi, T. (2013). A 243 GHz low-noise amplifier module for use in next-generation direct detection radiometers. In *2013 European microwave integrated circuit conference* (pp. 220–223).

Tessmann, A., Leuther, A., Wagner, S., Massler, H., Kuri, M., Stulz, H.-P., Zink, M., Riessle, M., & Merkle, T. (2017). A 300GHz low-noise amplifier S-MMIC for use in next-generation imaging and communication applications, 2017*IEEE MTT-S International Microwave Symposium (IMS)*. Available from https://doi.org/10.1109/MWSYM.2017.8058687.

Tsutsumi, T., Hamada, H., Sano, K., Ida, M., & Matsuzaki, H. (2019). Feasibility study of wafer-level backside process for InP-based ICs. *IEEE Transactions on Electron Devices*, *66*(9), 3771–3776. Available from https://doi.org/10.1109/TED.2019.2928849.

Urteaga, M., Griffith, Z., Seo, M., Hacker, J., & Rodwell, M. J. W. (2017). InP HBT technologies for THz integrated circuits. *Proceedings of the IEEE*, *105*(6), 1051–1067. Available from https://doi.org/10.1109/JPROC.2017.2692178.

Wu, P. Y., Gupta, A. K., & Buckwalter, J. F. (2014). A dual-band millimeter-wave direct-conversion transmitter with quadrature error correction. *IEEE Transactions on Microwave Theory and Techniques*, *62*(12), 3118–3130. Available from https://doi.org/10.1109/TMTT.2014.2362123.

Zamora, A., Leong, K. M. K. H., Reck, T., Chattopadhyay, G., & Deal, W. (2014). A 170–280 GHz InP HEMT low noise amplifier, 2014*39th International Conference on Infrared, Millimeter, and Terahertz waves (IRMMW-THz)*. Available from https://doi.org/10.1109/IRMMW-THz.2014.6956402.

CHAPTER

# 7

# RF-MEMS for 5G: high performance switches and reconfigurable passive networks

*Jacopo Iannacci*
**Fondazione Bruno Kessler (FBK), Trento, Italy**

The scope of this chapter is that of sketching the potentialities of RF-MEMS technology, that is microelectromechanical systems (MEMS) for radio frequency (RF) applications, in realizing high-performance and wide-reconfigurable passive components suitable to meet the challenging requirements of 5G emerging applications. This is done by introducing at first the MEMS/RF-MEMS background since the early days of discussion within the research international community. Furthermore, the 5G scenario is depicted, starting from high-level specifications, and then stepping down to specifications and requirements that will be posed on RF passive components. Subsequently, practical examples of RF-MEMS devices fabricated in a dedicated technology platform will be reviewed, including comparative discussion on their measured and simulated performance.

## 7.1 A recap of RF-MEMS across two decades of research and discussion

The aim of this introductory section is twofold. On the one hand, a short history of how MEMS technology first, and MEMS for RF, that is, RF-MEMS, then, were investigated by the scientific community is sketched. Subsequently, the focus is moved on RF-MEMS and their market evolution, from the first disappointed forecasts to the consolidation taking place nowadays, is going to be analyzed in details.

### 7.1.1 A brief history of MEMS technologies evolution

MEMS technology started to be ventured in the second half of the 1970s, with the aim of investigating silicon as a structural material with respect to its mechanical properties,

*New Materials and Devices Enabling 5G Applications and Beyond*
**DOI: https://doi.org/10.1016/B978-0-12-822823-4.00007-8**

beside the fast evolution it was having at that time as a semiconductor material. Nonetheless, studies around the characterization of the mechanical properties of silicon as well as of metal and insulating thin films were already conducted commencing from the 1960s (Beams et al., 1955; Blakely, 1964; Neugebauer, 1960). Relying on such a scientific background, several microdevices, like inertial sensors and pressure sensors, based on MEMS technology were discussed in literature, starting from mid-1970s (Bassous, 1978; Chen et al., 1979; Petersen, 1977, 1979; Roylance & Angell, 1979; Wen Hsiung Ko et al., 1979). To this end, a comprehensive and well-conceived article authored by Petersen and published in 1982 can be regarded as the first review sketching the state of the art of MEMS sensors at that time (Petersen, 1982).

A completely different path is that followed by the development of MEMS for RF passive components, commonly known as RF-MEMS, both in terms of timing and technologies. The first exploitations of microsystem technologies for the realization of RF passives are dated around mid-1990s, that is, roughly 20 years after the prototyping of the MEMS-based inertial and pressure sensors mentioned above. At an early stage, miniaturization of microwave and millimeter-wave transmission lines and their implementation in micromachining technologies based on silicon emerged as a quite promising research field (McGrath et al., 1993) thanks to the outstanding performance figures in terms of low loss and compactness compared to traditional solutions (Katehi et al., 1993). The possibility of integrating fixed RF signal manipulation functions, for example, through the realization of stubs (Weller & Katehi, 1995), appeared as an additional strength of silicon-based waveguides. Among the various families of transmission line configurations available, and well known for decades (Mahmoud, 1991), microfabrication technologies are particularly suited for planar devices. Therefore, most of the attention and interest concerning their miniaturization was around the coplanar waveguide (CPW) and microstrip implementations of transmission lines.

Notably, it should be highlighted how the just mentioned first examples of MEMS-based RF passives do not embody any feature of reconfiguration. In fact, it is just a few years later, across the end of the 1990s and the first 2000s, that the switching capability, made possible and implemented in MEMS technology, started to be ventured (Rebeiz & Muldavin, 2001), opening up the floor to the modern concept of RF-MEMS reconfigurable/tunable passive components. Since then, several classes of passives implemented in RF-MEMS technology have been studied and widely reported in literature, both encompassing lumped components and complex networks. Regarding the former, ohmic switches (Anuroop et al., 2018; Iannacci, 2018; Kageyama et al., 2017; Lee et al., 2010), switched capacitors (Belkadi et al., 2018, 2020; Molinero et al., 2019; Riverola et al., 2017), variable capacitors (i.e., varactors) (Cazzorla et al., 2015a, 2015b; Emami & Bakri-Kassem, 2016; Khodapanahandeh et al., 2018; Pu et al., 2016), and inductors (Choi et al., 2009; Xu et al., 2019; Zine-El-Abidine et al., 2003, 2004) were analyzed. Even more interestingly, starting from the just mentioned building blocks, diverse high-performance, broadband and largely reconfigurable/tunable complex RF passive networks have been developed and discussed. Among them, are certainly worth to be mentioned phase shifters (Bayraktar et al., 2015; Dey & Koul, 2014; Iannacci et al., 2020; Lampen et al., 2010; Psychogiou et al., 2011; Wu & Liu, 2018), tunable filters (Agaty et al., 2018; Jones & Daneshmand, 2020; Lin et al., 2018; Yang et al., 2020), impedance matching tuners (Iannacci, 2014; Morris et al., 2012; Singh et al., 2019; Vaha-Heikkila et al., 2005; Wipf et al., 2019), step attenuators (Iannacci et al., 2009; Iannacci, 2017a; Iannacci, Huhn, et al.,

2016a, 2016b; Iannacci, Tschoban, et al., 2016; Khaira et al., 2018; Sun & Li, 2016; Zhong et al., 2016), and switching matrices (Chan et al., 2012; Daneshmand & Mansour, 2005, 2006; Diaferia et al., 2014; Figur et al., 2012). Diverse RF-MEMS devices are depicted in the microphotograph in Fig. 7.1.

### 7.1.2 The market of RF-MEMS: from initial forecasts to current perspectives

As soon as RF-MEMS technology started to be discussed in literature, visions on its employment in emerging (at that time) applications were ignited. One among the most relevant prospects is certainly the one developed by Clark T-C Nguyen in the early 2000s (Nguyen, 2001, 2002). According to Nguyen, and bearing in mind the typical architecture of transceivers (transmitters/receivers) in the RF domain (Laskar et al., 2004), the availability of high-performance passives in RF-MEMS technology would have acted at different levels. In a first phase, MEMS-based low-complexity components, like switches, resonators, and filters, would have replaced the standard counterparts, thus improving the performance of the radio terminals (Nguyen, 2002, 2006). Subsequently, the spread of RF-MEMS technology was expected to be more prominent, leading to a redesign of transceivers architecture, leveraging multi-functional and widely reconfigurable monolithic high-complexity networks, entirely based on microsystems (Nguyen, 1998, 2007, 2013).

Given the scientific interest triggered by the outstanding performance and characteristics of RF-MEMS technology, the expectations around market breakthroughs used to increase accordingly in those years. As a matter of fact, for about one decade, from the early 2000s until 2013, the enthusiastic RF-MEMS market volume forecasts were systematically revised down, one after another. The graphs reported in Fig. 7.2 and released from 2006 to 2013 show how initial expectations were always overestimated (Iannacci, 2017b). It is of particular interest observing, when available, the time range predicted by one forecast compared against the subsequent forecast released after that range, as the latter one reports consolidated volumes. Such overlapped ranges are highlighted in light yellow. Just

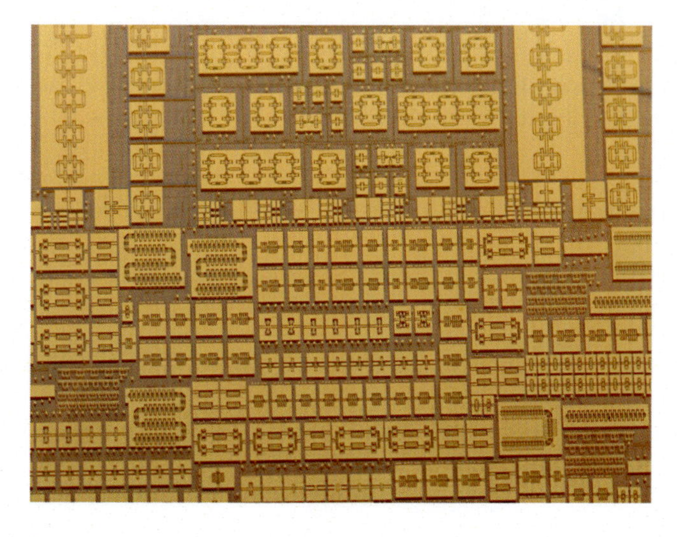

**FIGURE 7.1** Microphotograph of a variety of RF-MEMS passive components, both featuring lumped elements and complex networks.

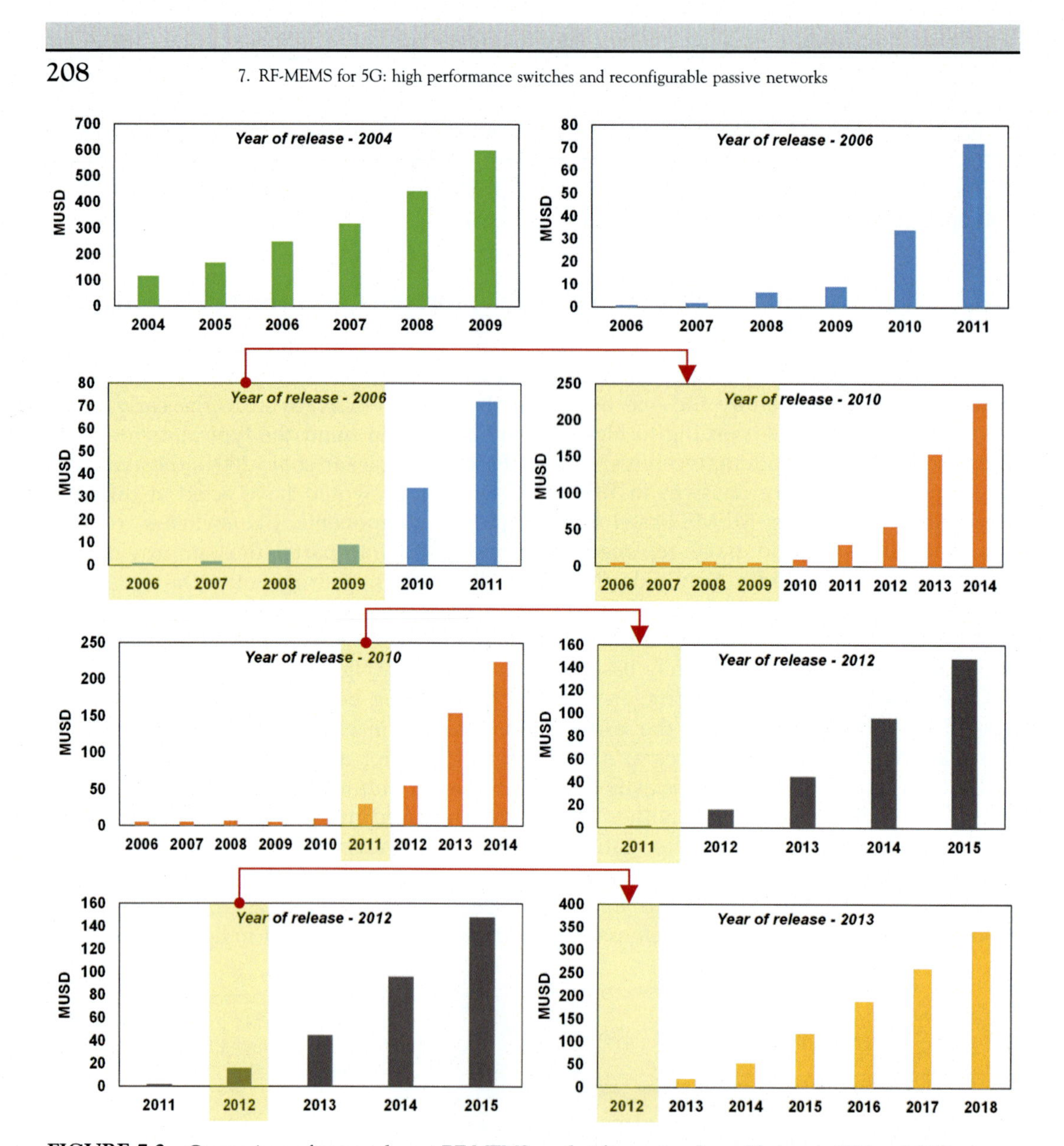


**FIGURE 7.2** Comparison of some relevant RF-MEMS market forecasts released between 2006 and 2013.

to mention one example, the forecast released in 2012 predicted for the same year a volume of 16 MUSD (Millions of US Dollars), while its actual size was less than 1 MUSD, as released in 2013.

Given the premises above, RF-MEMS technology, at a glance, started looking like a solution that failed expectation and that was not suitable for market exploitations, apart from niche applications, for example, in the space and defense sectors. In fact, a deeper analysis is necessary to fully embrace the motivations causing the missed success of RF-MEMS technology in the 2000s. Such reasons are to be sought both in intrinsic and extrinsic factors. The former are closely connected to technical aspects that used to make MEMS-based RF passives more difficult and expensive to integrate within subsystems and complex devices. More in details, intrinsic factors can be grouped around three main areas related to RF-MEMS, that is, the poor reliability they had in the early days (Chianrabutra et al., 2013; Hartzell et al., 2011; Iannacci, 2015b; Kawai et al., 2009; Khanna, 2010; Tavassolian et al., n.d.; van Gils et al., 2007), the need for ad-hoc packaging solutions (Cohn et al., 2002; Iannacci et al., 2008; Jourdain et al., 2003; Lau et al., 2010; Margomenos & Katehi, 2002, 2003; Yun-Kwon Park et al., 2002, 2003), and the difficulty of integration/hybridization with standard semiconductor technologies (e.g., CMOS) (De Silva & Hughes, 2003; Lu et al., 2005; Pacheco et al., 2004; Rijks et al., 2003; Zhang et al., 2006).

On a different plane of reference, extrinsic factors exerted an influence in leading RF-MEMS technology toward partial success that, most probably, was more crucial than that of the technical issues mentioned above. Many words could be spent in order to frame this matter, yet a very brief sentence can embody all the salient meaning: in the early 2000s, when RF-MEMS came on the landscape, mobile services (like 2G and 3G) and mass-market RF products did not have specific needs for the boosted performance enabled by such a technology (Iannacci, 2017b). This is the pillar factor deflecting industry from investing in addressing reliability, packaging, and integration aspects in the first phase of the RF-MEMS era (Iannacci, 2015a; Mulloni et al., 2015; Palego et al., 2010; Persano et al., 2016; Solazzi et al., 2011; Solazzi, Palego, et al., 2010; Solazzi, Tazzoli, et al., 2010; Tazzoli et al., 2009).

The whole context commenced to change after 2010 with the rising of smartphones era. With increasing integration of components within the handsets, as well as due to integration of antenna within the body of the smartphone itself, the quality of connection started heading over a decreasing trend, leading to more frequent call drops, nonoptimal data transfer rates, and increased power consumption ("RF MEMS Switches Are Primed For Mass-Market Applications," 2016). In practical terms, the fixed impedance match between the Tx/Rx antenna and the radio frequency front end (RFFE), which had been a standard technical solution in the first generations of cellular phones and communication protocols (up to 3G), was not anymore the optimal choice at the time of 4G-LTE (long term evolution) and of smartphones. The emerging feature needed was that of relying on adaptive (rather than fixed) impedance tuning, so that time variant fluctuations of antenna characteristic impedance could be followed by the impedance matching tuner, ensuring optimal real-time adaptation to the RFFE. The showing up of this need is what changed the approach to RF-MEMS from technology push to market pull. In other words, for the first time some of the main characteristics of RF-MEMS, that is, wide reconfigurability and ease in designing complex networks, met a specific market need. This combination of factors made in recent years MEMS-based analog impedance matching tuners the first successful breakthrough of RF-MEMS technology in the mass-market segment of smartphones and mobile communications (IHS iSuppli Teardown Analysis Service Identifies First Use of RF MEMS Part, Set to Be Next Big Thing in Cellphone Radios - Omdia, 2020; Staff, 2014).

Looking ahead and bearing in mind the lesson learned through what has been up to now, RF-MEMS carry relevant potential in terms of diverse market exploitations, especially when looking at emerging paradigms like the Internet of Things (IoT), including all its ramifications into industrial segments and smart things, and the 5G. The market report recently released by Yole Developpement (Status of the mems industry - Market update, 2020), starting from the consolidated 2019 revenues of the whole MEMS-related RF segments of 2210 MUSD, forecasts a volume of 5258 MUSD in 2025, with a CAGR (Compound Annual Growth Rate) of 15.5%. In particular, the report identifies RF-MEMS switches as an enabling building block that can be employed directly in more complex systems, or integrated within RF-MEMS networks (like the impedance matching tuners mentioned above), with reference to segments like telecom (base stations and 5G small cells), mobile phones, industrial ATE (Automated Test Equipment) instruments and appliances, space and defense. Beside the above-mentioned works, other relevant contributions are discussed in literature around the exploitation of RF-MEMS technology in the 5G scenario (hAnnaidh et al., 2018; Kourani et al., 2020; Ma et al., 2019; Peroulis, 2018; Zhang et al., 2018).

## 7.2 5G services characteristics distilled into passive components specifications

The discussion around what 5G is expected to be has been very broad and articulated since years. The common trait of most part of the roadmaps and technical reports produced up to here is the pronounced focus on system-level aspects and high-level specifications, without paying fully proper attention on how such drivers reflect on specifications and characteristics demanded to low-complexity level hardware components (Iannacci, 2017b). Therefore, prior to going into the details of RF-MEMS in the frame of 5G, it is wise developing a few considerations that bring system-level demands of 5G vertically down to characteristics and specification of hardware passives.

Starting from the highest level of abstraction, all the challenging performance expected from 5G in terms of services can be traced down and grouped according to three main drivers ("Evolving LTE to Fit the 5G Future," 2020; Flynn, 2020; Han et al., 2019; Henry et al., 2020), namely defined as EMBB (enhanced mobile broadband), URLLC (ultra-reliable low-latency communications), and MMTC (massive machine-type communications). In particular, the EMBB driver is related to the system capacity, aiming at 10 Gbps peak and a minimum of 10 Mbps per each user, as well as to the spectrum, that is, below 6 GHz and above 6 GHz 5G new radio (NR) communications. URLLC is demanded by applications relying on fast data transfer, like V2V (vehicle-to-vehicle) communications, aiming to the millisecond latency range. Finally, MMTC will lead to a significant increase of new connections to be supported by 5G services. Bearing in mind the three pillars of 5G mentioned above, a limited set of key performance indicators is derived, as reported in (Agyapong et al., 2014; Osseiran et al., 2014). It can be listed as follows:

1. Data volume increased up to 1000 times with a growth of indoor data traffic up to 70%.
2. Connected devices increased from 10 to 100 times.
3. Typical user data rate increased from 10 to 100 times.
4. Extended battery life up to 10 times for massive machine communication (MMC) devices.
5. End-to-end (E2E) latency reduced by 5 times.

Keeping in mind such performance indicators, their achievement implies acting at different levels, them including network infrastructure as well as hardware technologies and solutions. The discussion developed in Osseiran et al. (2014) suggests acting at the following four technical levels:

- **Radio links**. New transmission waveforms and new approaches for multiple access control and resource management.
- **Multi-node/multi-antenna transmissions**. Design of multi-antenna transmission and reception technologies, capitalizing on massive-MIMO (multiple-input-multiple-output) antennas.
- **Network dimension**. Novel approaches for efficient interference management in complex heterogeneous deployments.
- **Spectrum usage**. Operation in extended spectrum band and in new spectrum regimes, ranging from a few-GHz to millimeter-waves, that is, up to 60–100 GHz.

In light of the main drivers, application-level specifications, and technical macro-areas of action, it is now possible to forecast the classes of RF passive devices as well as their indicative performance that 5G applications will be demanding at low-complexity hardware component level (Iannacci, 2017b):

1. Broadband relays and switching units, like multiple pole multiple throws (MPMTs) characterized by very-low loss (when CLOSE), very-high isolation (when OPEN) and limited adjacent channels cross-talk, operable from 2–3 GHz up to 60–100 GHz.
2. Reconfigurable/tunable filters with good stopband rejection characteristics and low attenuation of the passed band.
3. Wideband multistate impedance matching tuners.
4. Digital and analog step attenuators with multiple configurations and very-flat characteristics over 60–70 GHz frequency spans, operable up to 100 GHz.
5. Wideband multistate digital and analog phase shifters/delay lines (from 2–3 GHz up to 100 GHz).
6. Hybrid devices with mixed phase shifting and programmable attenuation monolithically integrated in the same silicon chip (from 2–3 GHz up to 100 GHz).
7. Miniaturized antennas and arrays of antennas (e.g., MIMOs), possibly integrated with one or more of the devices described in the previous points (from 2–3 GHz up to 100 GHz).

## 7.3 Demand and supply: where RF-MEMS and 5G can meet

Capitalizing on the discussion developed up to now, a conclusive important step is going to be taken, closing the loop with the assessment of qualitative characteristics and quantitative reference specifications that RF passive components should comply in the 5G application scenario. To this end, Fig. 7.3 comes to aid, collecting the most relevant desirable features of passives, including their RF characteristics, reliability, size, power consumption, and integration.

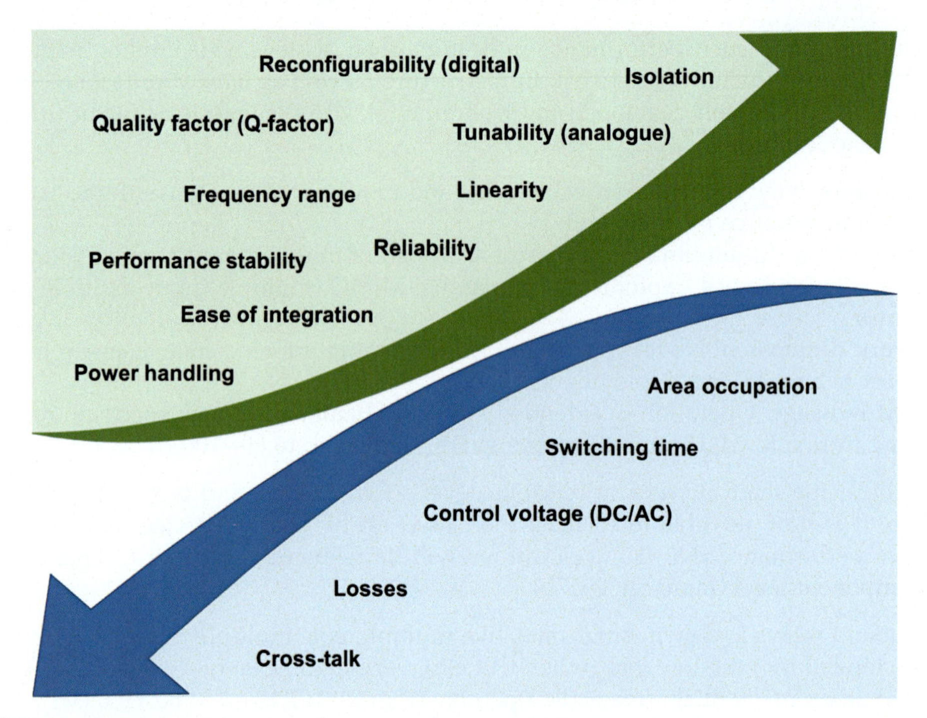


FIGURE 7.3 Summary of the most relevant specifications of 5G RF passive components, grouped depending on if they have to be maximized (upwards arrow) or minimized (downwards arrow).

Led by the aim of making the interpretation easy and immediate, the specifications are collected by two arrows oriented upwards and downwards, depending on if the goal of the items they group is that of maximizing of minimizing them, respectively. Given this frame of reference, an insight focused on the most relevant RF performance is going to be developed, bearing in mind the challenges posed by the 5G scenario. Per each characteristic, quantitative indications are also provided.

- **Frequency range**. Leveraging the discussion developed above, the 5G demands for pronounced diversification of the network infrastructure. To this end, evident divide is raising between the classical backhaul infrastructure (working in the sub-6 GHz range) and the so-called small cells, which will operate in the millimeter-wave (mm-Wave) range. Consequently, RF passives are demanded to show remarkable characteristics in a very wide range of frequencies, starting from 2–3 GHz and ranging up to 60–70 GHz (probably up to 100 GHz, as well). These considerations hold validity for the hardware (HW) dedicated both to the network infrastructure as well as to mobile handsets and devices, with particular importance for the latter ones, as mobile terminals will have to swap agilely between typical RF bands and mm-Waves.
- **Isolation**. Bearing in mind the prominent hardware reconfigurability that 5G demands to RFFEs, it is straightforward that switches are to be widely employed. When dealing with relays and microrelays, used both as lumped components as well as building blocks of more complex passive networks, a critical characteristic is their isolation,

when in the OPEN state, that is expected to be as higher as possible, that is, better than −30 dB up to the highest frequency possible. If designing switches with good isolation at few GHz is not a big deal, the target becomes very challenging when frequencies rise up to tens of GHz, due to the parasitic coupling effects (mainly capacitive) between the input and output terminations, allowing an increasing fraction of the RF signal to be transmitted, despite OPEN.

- **Losses**. Following up on the previous point, the other fundamental aspect is that of losses caused by the switching units, when in the CLOSE state, which reflect, on a more general reference plane, on the losses of RF passive components and networks, when configured in conducting state. Dually with respect to isolation, losses have to be as limited as possible, and a desirable mark is that of keeping them below −1 dB in the widest frequency range possible. As for isolation, keeping good performance in terms of losses is more and more challenging as the frequency rises, due to the increasing influence of parasitic effects and nonidealities. To this end, switches able to score good performance in the mm-Wave range should be radically redesigned.
- **Cross-talk**. While still keeping the reconfigurability demanded by the 5G at stake, the cross-talk happens to be another relevant performance indicator. It is closely related to isolation, and it quantifies the extent to which adjacent channels interfere with each other. As for isolation, cross-talk is desired to be as large as possible, and it is straightforward sketching its critical role in multi-path scenarios, like the one of massive MIMO technologies and advanced beamforming capability. It would be optimal keeping the cross-talks below −50/ − 60 dB over the widest frequency range possible. Similarly to the performance indicators discussed above, cross-talk degrades as frequency rises, and it is rather challenging achieving satisfactory numbers in the mm-Wave range.
- **Switching time**. Despite not directly related to RF characteristics, it is worth mentioning a couple of performance indicators that are in any case relevant to the usability and suitability of relays and RF passive components/networks featuring a certain number of switching units. The first characteristic at stake is the switching time, that is, the delay between when the application of the control signal is imposed to a relay, and the moment in which the latter reaches a stable condition transiting from CLOSE to OPEN state, or vice-versa. Still bearing in mind the reconfigurability 5G RFFEs and RF hardware are going to be demanded for, it results clear that idle time in commuting from one state to another must be kept as limited as possible. Picking a relevant example, one could reference the case of mobile devices hopping from sub-6 GHz up to mm-Wave small cell services, for example, during outdoor/indoor environment transition. Breaking down commutation time from system to component level, the critical device is the switch and its ON/OFF transition time. Thereafter, starting from the millisecond-range latency addressed by 5G services, the basic relay switching time has to be definitely lower, with a small fraction of millisecond (i.e., less than 100 μs) being a reasonable target.
- **Control signal**. Finally, the last non-RF characteristic here analyzed is that of control signals (typically DC/AC currents/voltages) necessary to drive the commutation of relays, which are closely linked to the power consumption demanded by their operation. To this regard, it has to be stressed that, if availability of power within the

frame of reference of ground infrastructure is an important yet not critical issue, it gains crucial centrality in mobile and battery-operated devices. In the latter case, the extent of DC/AC biasing signals (i.e., power consumption) in controlling RF passives should be kept as low as possible. It would also be desirable maintaining compliance with CMOS typical driving voltages, thus avoiding incorporation of additional ad-hoc voltage pumping circuitry. This means working with control voltages within 1–2 V.

## 7.4 An example of RF-MEMS technology platform

The focus of this section is that of discussing the opportunities offered by RF-MEMS-based passive components, after having sketched the 5G application scenario depicted in the previous pages. To this end, a specific RF-MEMS technology platform is taken as reference example and the typical performance/characteristics of a set of different RF passives (manufactured within it) are going to be shown and discussed.

### 7.4.1 RF-MEMS manufacturing process

The RF-MEMS technology that is going to be reported is the one available at Fondazione Bruno Kessler (FBK), in Italy. It leverages a surface micromachining process on 6 inch silicon wafers, with two passivated conducting layers, for DC biasing and RF signal underpasses, respectively, and two stacked gold electrodeposited films for the definition of the MEMS air gaps and of the CPWs (Giacomozzi & Iannacci, 2013; Giacomozzi et al., 2012). To this end, Fig. 7.4 shows the 3D schematic of a typical double-hinged (clamped-clamped) electrostatically actuated RF-MEMS series ohmic microrelay. In particular, the top image depicts all the layers, while in the bottom image the gold metallizations are hidden, in order to make the underneath electrodes visible.

Bearing in mind the RF-MEMS switch design in Fig. 7.4, its fabrication is going to be reviewed in details, step by step, in order to unroll the complete process flow. Starting from 6 inch high-resistivity silicon wafers, a 1 μm thick oxide layer is grown. Subsequently, a 630 nm thick poly-crystalline silicon (poly-Si) layer is grown and patterned with the first lithographic step. The poly-Si layer has a sheet resistance of about 1.5 kOhm/sq and it is used for providing DC biasing and control the actuation of the MEMS suspended membrane. The microphotograph in Fig. 7.5 shows the poly-Si bottom electrode and the feeding lines to apply DC bias and ground.

The next step consists in growing a 100 nm thick oxide to insulate electrically the poly-Si, and opening vias in the oxide where electrical continuity has to be established to the poly-Si. To this end, the microphotograph in Fig. 7.6 shows the vias opened in the oxide above the poly-Si.

Then, the second conductive layer is deployed. It consists of a 630 nm thick sputtered aluminum layer that is highly conductive, and thus suitable for RF signal underpasses. The microphotograph in Fig. 7.7 shows the aluminum layer defining the input/output RF terminations.

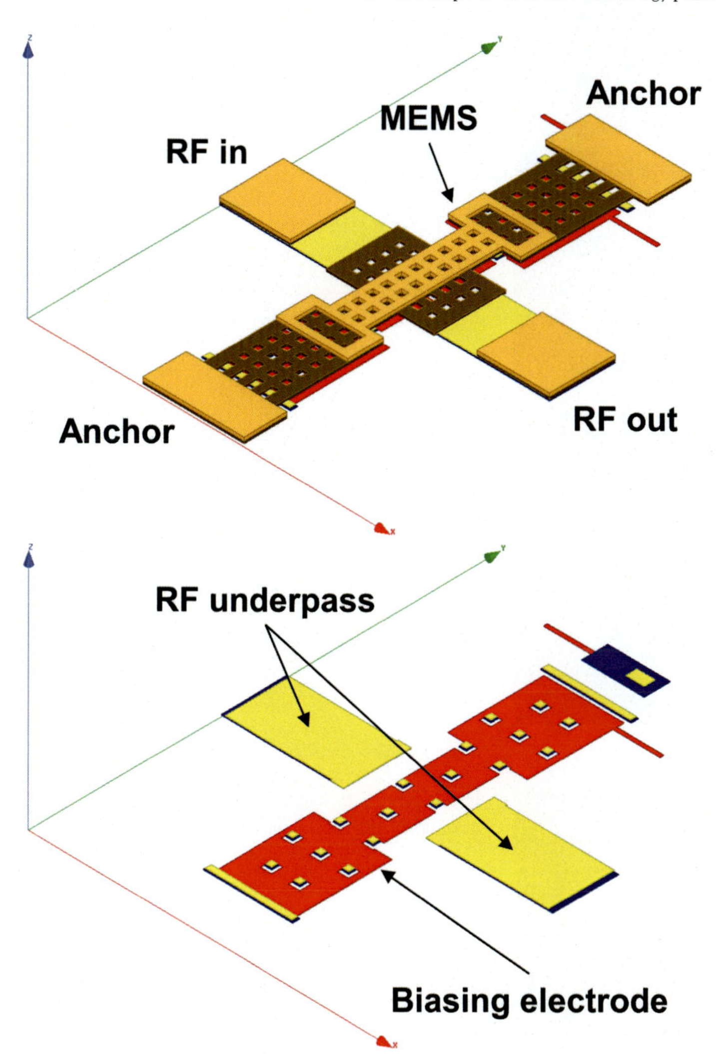


FIGURE 7.4 3D complete schematic of a clamped-clamped RF-MEMS series ohmic microrelay fabricated in the FBK technology. The top image reports the whole device, while in the bottom image the underneath DC electrodes and RF terminations are visible, as the MEMS gold layers are hidden.

A further 100 nm thick oxide layer passivates and insulates the aluminum layer. As before, where electrical continuity to the underneath aluminum layer is needed, vias are opened in the oxide, as visible in Fig. 7.8.

Thereafter, a 150 nm thick gold layer is evaporated above the vias openings to the underneath aluminum layer. Such a thin film is exploited to ensure better metal-to-metal contact resistance to the electrodeposited gold layers that are going to be structured later, as visible in Fig. 7.9.

From now on, the factual MEMS device structuring takes place. To this end, a 3 μm thick photoresist layer is spun over the wafer and patterned exclusively where the MEMS membrane is expected to be elevated (i.e., air gap). Therefore, such a photoresist acts as sacrificial layer and it is visible in the microphotograph reported in Fig. 7.10.

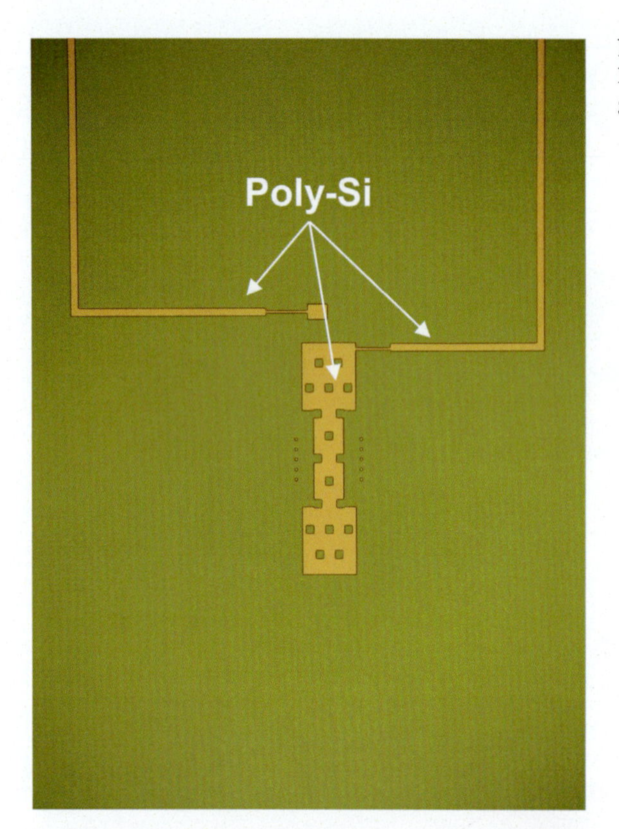


**FIGURE 7.5** Microphotograph of the poly-Si layer used for bringing and applying DC bias and ground to the MEMS suspended membrane.

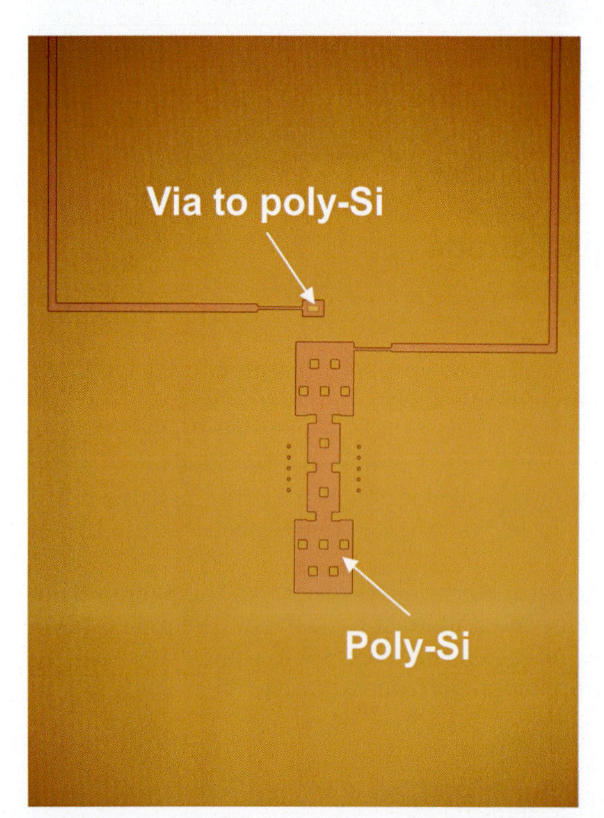


**FIGURE 7.6** Microphotograph of the vias opened in the oxide grown above the poly-Si.

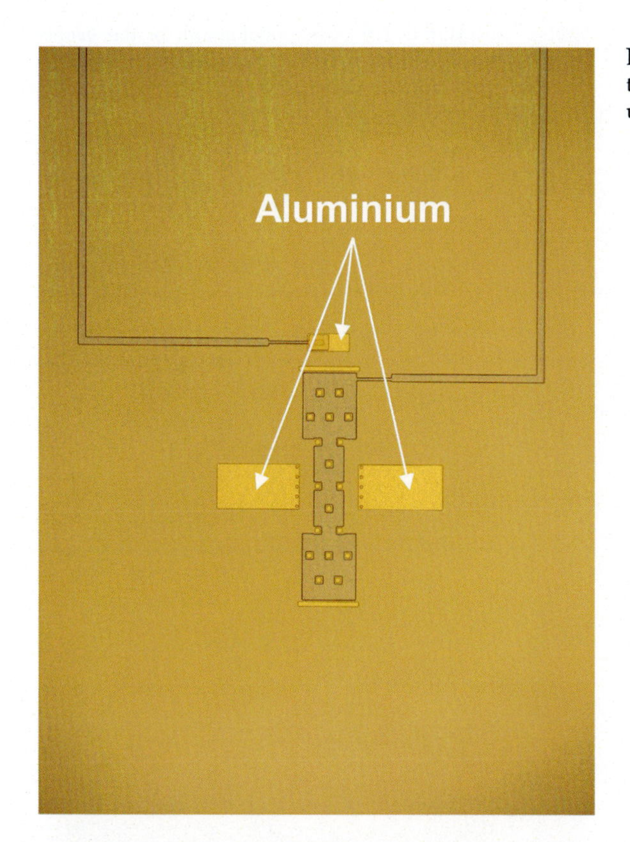


**FIGURE 7.7** Microphotograph of the sputtered aluminum layer used for RF signal underpasses.

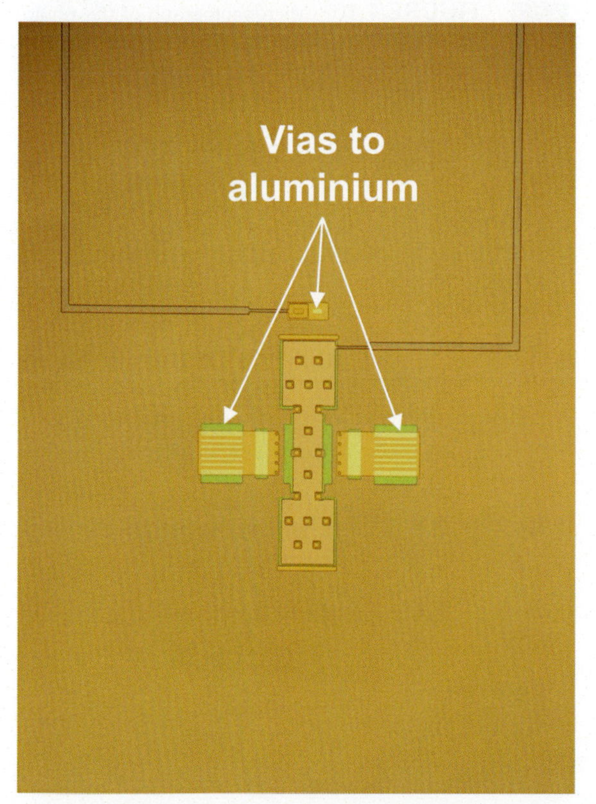


**FIGURE 7.8** Microphotograph showing the vias opened in the insulating oxide grown above the sputtered aluminum layer.

FIGURE 7.9 Microphotograph of the evaporated gold layer above contacts to the underneath aluminum RF underpass.

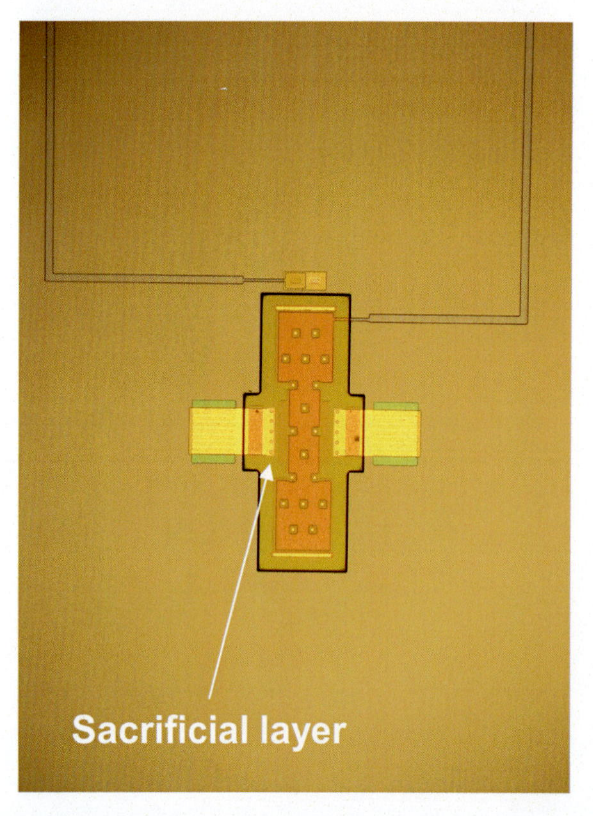


FIGURE 7.10 Microphotograph of the photoresist sacrificial layer patterned where the elevated MEMS membrane is expected to be structured.

The following fabrication step consists in the electrodeposition of a 2 μm thick gold layer. Wherever such a thin film lays over the substrate, it defines the RF lines (e.g., in CPW or microstrip configuration), the MEMS anchoring areas, and the pads for applying electrical signals. Differently, the gold deposited above the sacrificial layer will be the actual MEMS elevated membrane in the end. The microphotograph reported in Fig. 7.11 shows the CPW gold structure around the MEMS, as well as the elevated air gap.

Then, a second electrodeposition of gold structures with a 3 μm thick layer selectively defined above the first one. This metallization is needed to obtain more robust membranes where necessary, for example, on the CPW lines, in correspondence with the MEMS anchoring areas and in the portions of the air gaps that are expected to remain flat. The structuring of the second electrodeposited gold layer is visible in Fig. 7.12.

Now that the RF-MEMS series ohmic switch is complete, the very last step consists in removing the sacrificial layer underneath the air gap, in order to release the MEMS structure and allowing it to move when a DC bias is applied, thus reaching the pull-in and closing the microrelay (see Fig. 7.13).

Starting from the just reported example of an RF-MEMS series ohmic switch, the performance and characteristics of a few other RF passives realized in the same technology are going to be reviewed in the following subsections.

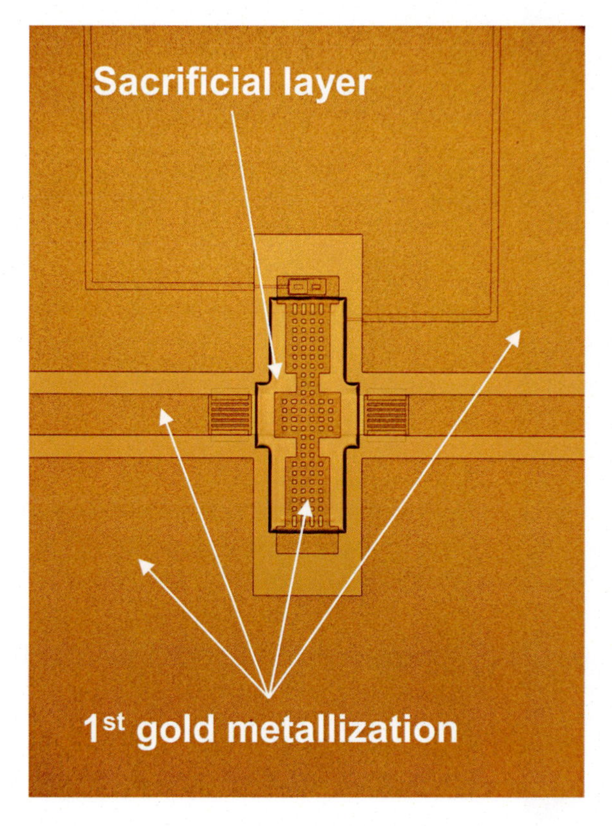


**FIGURE 7.11** Microphotograph of the first gold metallization defining the coplanar waveguide, the anchoring areas, and the MEMS elevated membrane.

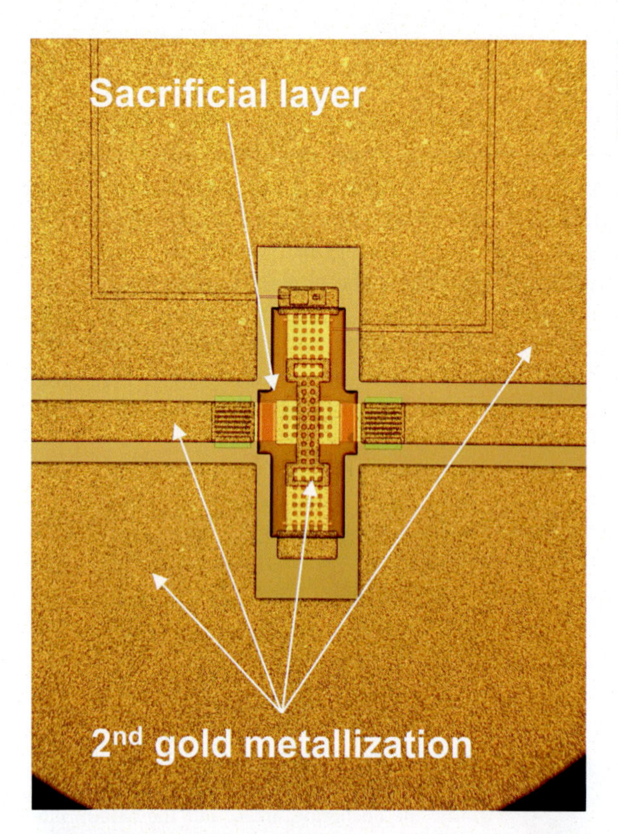


FIGURE 7.12 Microphotograph of the second gold electrodeposition over the coplanar waveguide and the areas of the MEMS structures that need to be rigid or to remain flat.

FIGURE 7.13 Microphotograph of the final released and operable RF-MEMS series ohmic switch.

## 7.4.2 Switching devices and solutions

Starting from the RF-MEMS technology platform discussed above, a few relevant examples of MEMS-based switching devices are now going to be reported in the following, along with measured and simulated data concerned to their RF characteristics. The first analyzed device is a series ohmic switch with a clamped-clamped MEMS membrane structure similar to that in Fig. 7.13. The measured and simulated (in Ansys HFSS) S-parameters, in the frequency range from nearly-DC up to 60 GHz, are plotted in Fig. 7.14.

Into more details, the top image in Fig. 7.14 shows the RF-MEMS microrelay behavior when not conducting, that is, OPEN switch and MEMS OFF (in its rest position). In this plot, the

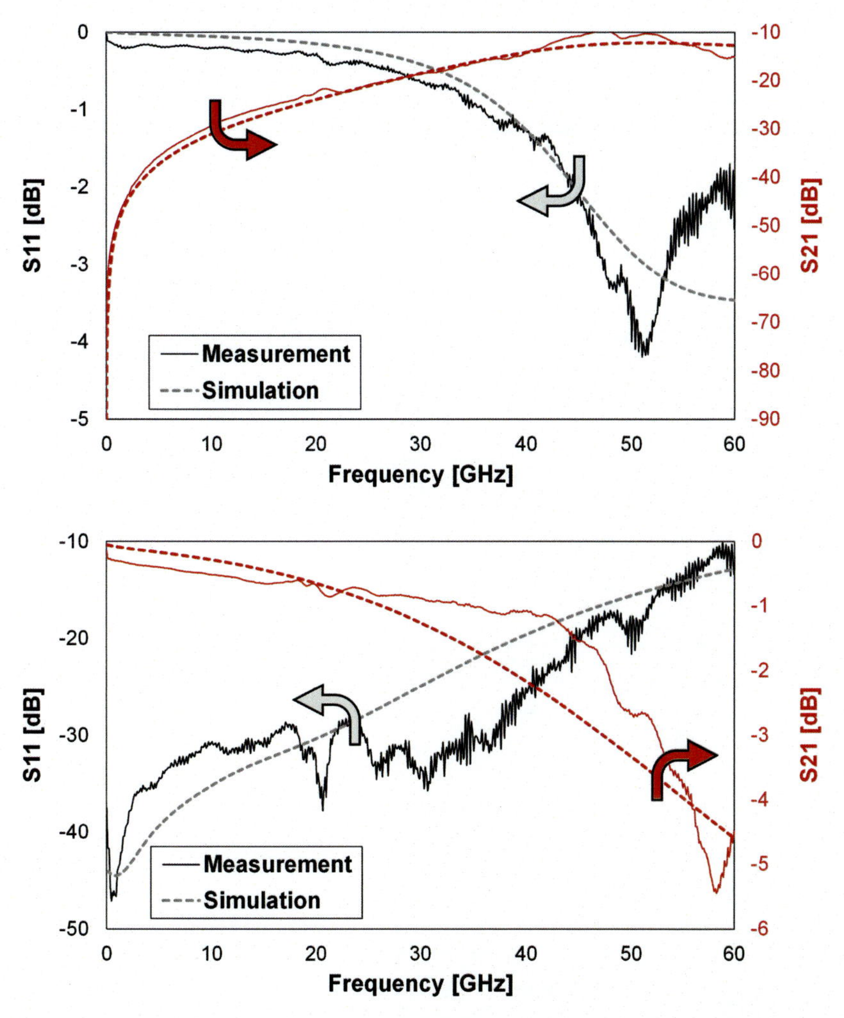


FIGURE 7.14 Measured and simulated S-parameters of the clamped-clamped RF-MEMS series ohmic switch. The top image shows the reflection (S11) and isolation (S21) when the switch is OPEN (i.e., the MEMS is OFF). The bottom image shows the reflection (S11) and transmission (S21) when the switch is CLOSE (i.e., the MEMS is ON).

relevant parameter to observe is the isolation (S21) that is better than −20 dB up to 30 GHz, and better than −10 dB up to 60 GHz. It has also to be noted that the simulated traces, obtained with the Ansys HFSS 3D finite element method (FEM) analysis tool, predict rather accurately the measured behavior of the physical device, both for what concerns the S21 (isolation) and S11 (reflection) parameters, over the whole frequency range. Looking now at the bottom image in Fig. 7.14, the S-parameters behavior of the switch is reported when it is conducting, that is, CLOSE switch and MEMS ON (actuated or pulled-in position). Starting from the reflection (S11) parameter, it exhibits values better than −30 dB up to around 40 GHz, better than −20 dB up to 50 GHz, and better than −10 dB up to 60 GHz. This means that the RF-MEMS conducting switch matches pretty well the input port impedance, allowing most part of the RF signal to travel through the device, towards the output port. This consideration is corroborated by looking at the transmission (S21) parameter. The transmitted power, in fact, is better than −1 dB up to 40 GHz, which is a rather low amount of losses for this type of switching device over a frequency span as large as 40 GHz. Also in this case, a rather accurate prediction of the curves produced by the simulated model is observable, both concerning the S11 and S21 parameters.

The next switching device to be analyzed is a shunt capacitive microrelay, also referred to as switched capacitor, as it realizes a two-state ON/OFF (i.e., HIGH/LOW) variable capacitance. The geometry is that shown in Fig. 7.15, where a clamped-clamped MEMS structure similar to the one of the previously reported ohmic switch is visible.

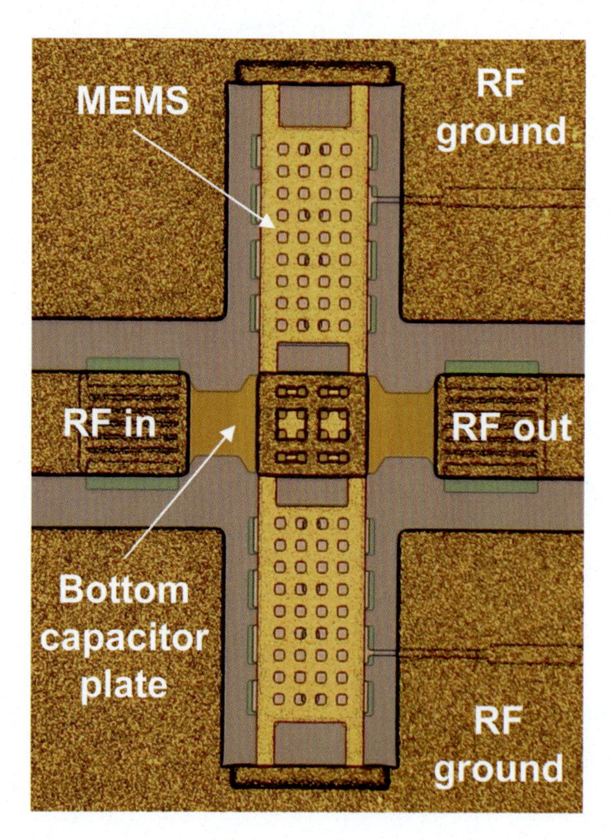


**FIGURE 7.15** Microphotograph of a clamped-clamped RF-MEMS switched shunt capacitor.

However, some important design modifications are embodied in this case. First, the MEMS suspended membrane is not isolated by the RF ground planes. Then, the underneath electrode is shaped in a radically different way, as it is continuous under the MEMS membrane, while in the ohmic device it is truncated (see Fig. 7.8). Moreover, vias are not opened above the underpass, thus leaving it passivated by oxide. This means that when the MEMS is OFF, that is, in its rest position, the RF signal travels across the device (i.e., the switch is CLOSE), encountering a very small capacitance to RF ground, due to the overlap of the MEMS membrane with the underneath electrode. Differently, when the MEMS is brought to actuation (i.e., it is ON), it collapses on the underneath oxide, thus realizing a high capacitance to RF ground. Such a high value of capacitance establishes a low impedance path to ground for the RF signal that is therefore shorted, not reaching the output (i.e., OPEN switch). In other words, the shunt capacitive switch in Fig. 7.15 is characterized by a behavior that is fully dual compared to the ohmic series switch in Fig. 7.13. First, it does not implement any metal-to-metal contact, but a variable capacitive coupling. Secondly, the switch is CLOSE when the MEMS is OFF and is OPEN when the MEMS is OFF, while in the ohmic case the vice versa takes place. The measured and simulated RF behavior of the shunt switched capacitor is reported in the plots in Fig. 7.16.

The top image in Fig. 7.16 shows the measured and simulated behavior of the S-parameters in the switch OPEN state (MEMS ON). It has to be highlighted that the measured traces in this case are available up to 40 GHz. However, given the accuracy of the 3D FEM model, the simulated curves are analyzed up to 60 GHz, therefore providing a better insight of the device behavior on a wider frequency range. Having said that, the most relevant parameter to study in the OPEN state is isolation (S21). Its qualitative behavior is radically different from that typical of an ohmic switch, as it must be kept in mind that in this case the element providing isolation is a capacitor which has a frequency-related characteristic. The typical MEMS ON state capacitance is in the range of several pF. Despite quite large, such a capacitance behaves as an OPEN when the frequency is low, therefore poorly shorting the RF signal to ground. This is the reason why isolation is not satisfactory up to around 8 GHz (see Fig. 7.16, top). Differently, for higher frequencies the S21 improves, being better than −20 dB from 8 to 23 GHz, and better than −13 dB up to 60 GHz. It has to be highlighted that the anti-resonance of the S21 parameter at around 15 GHz is due to the high-value capacitance to RF ground, realized by the pulled-in MEMS. Such a negative peak corresponds to the lowest level of impedance realized by the switched capacitor and it is where isolation scores better values (around −30 dB). This leads to an important consideration to be kept in mind. Differently from ohmic switches, capacitive switched capacitors are intrinsically less broadband because of their reactive frequency-dependent dominant behavior. Therefore, switched capacitors are more suitable for applications where the frequency range is not excessively wide. This allows significant margin for optimizing the design and scoring, in turn, characteristics in terms of isolation that can be comparable and even better than those typically achieved by series ohmic switches.

Stepping forth in the discussion, the bottom image in Fig. 7.16 reports the reflection (S11) and transmission (S21) parameters of the switched capacitor when the MEMS is OFF and the microrelay is conducting (CLOSE state). Despite measurements are quite noisy, the reflected power is visibly better than −20 dB up to 40 GHz. According to simulations,

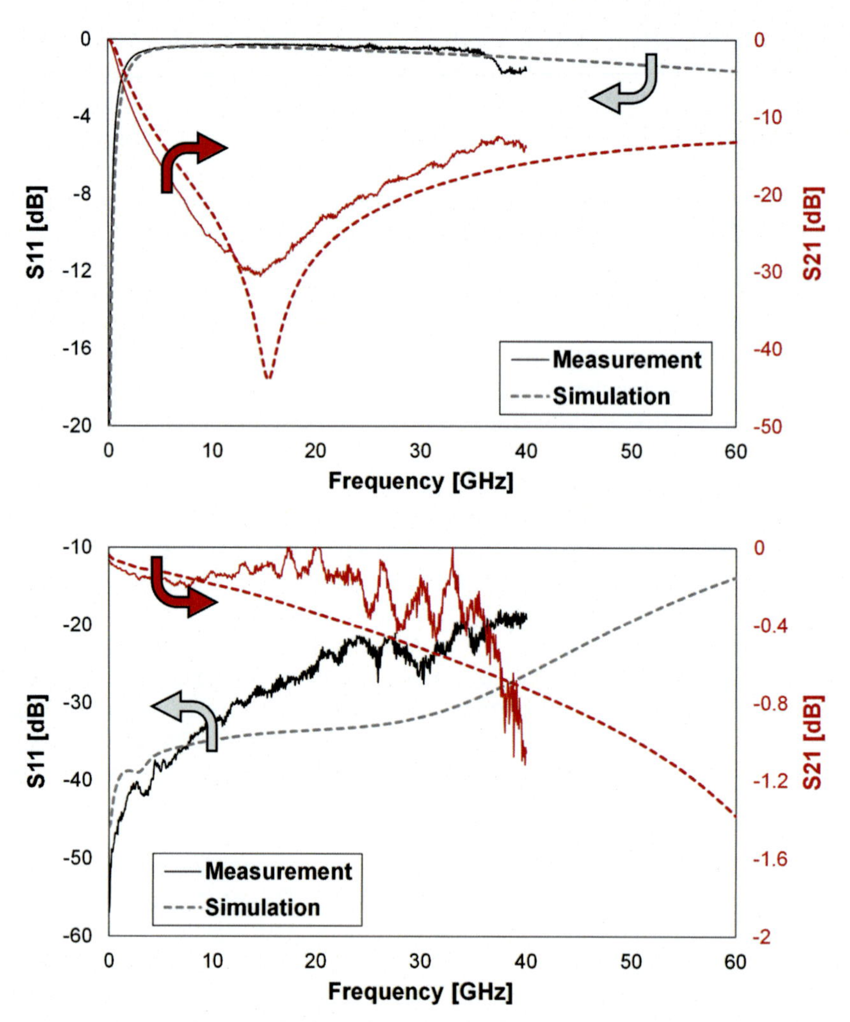


**FIGURE 7.16** Measured and simulated S-parameters of the clamped-clamped RF-MEMS shunt capacitive switch (i.e., switched capacitor). The top image shows the reflection (S11) and isolation (S21) when the switch is OPEN (i.e., the MEMS is ON). The bottom image shows the reflection (S11) and transmission (S21) when the switch is CLOSE (i.e., the MEMS is OFF).

the S11 is better than $-15$ dB up to 60 GHz. Concerning transmission (S21), the measured losses are better than $-0.2$ dB up to 25 GHz and better than $-0.8$ dB up to 40 GHz. FEM simulations predict losses better than $-1.4$ dB up to 60 GHz. From a general perspective, such characteristics are rather good and desirable for what concerns a switching unit. Once again, a few considerations have to be developed concerning the comparison with ohmic switches in CLOSE configuration. The shunt configuration microrelay design capitalizes on a continuous RF signal line that ensures very limited losses. On the other hand, series ohmic switches feature metal-to-metal contacts when conducting, therefore comprising contact resistances on the RF line that, depending on their extent, can jeopardize the transmission increasing the fraction of attenuated power. On the other hand, the shunt switch, when CLOSE, has the MEMS membrane in its rest position. This means that, despite small, there is always a capacitance to ground, implemented by the MEMS itself and the underneath fixed electrode. Such a capacitance, depending on the specific design, is typically in the order from 30–40 fF up to 100–150 fF. This means that with the frequency rising, the low-value capacitance implemented by the OFF MEMS membrane introduces losses that are less and less negligible.

The next switching solution that is going to be reported features the series ohmic solution once again, despite the structure of the MEMS membrane is different in comparison to the examples discussed above. It is known as cantilevered MEMS and the microphotograph of a sample is reported in Fig. 7.17.

Differently from the clamped-clamped switch, this cantilevered configuration is placed in line rather than in transverse position with respect to the RF path. Moreover, the membrane is anchored only on one end, while the opposite tip is free and suspended. Therefore, as the cantilever is placed along the RF line, its elevation in correspondence with the sacrificial layer defines the anchoring point of the air gap. In other words, the MEMS cantilever is simply an elevated portion of the RF line. The free tip, instead, is suspended above the contacts of the output RF line branch. This way, as it happens with clamped-clamped transverse switches, it realizes the OPEN state when the MEMS is OFF, while the switch is in CLOSE

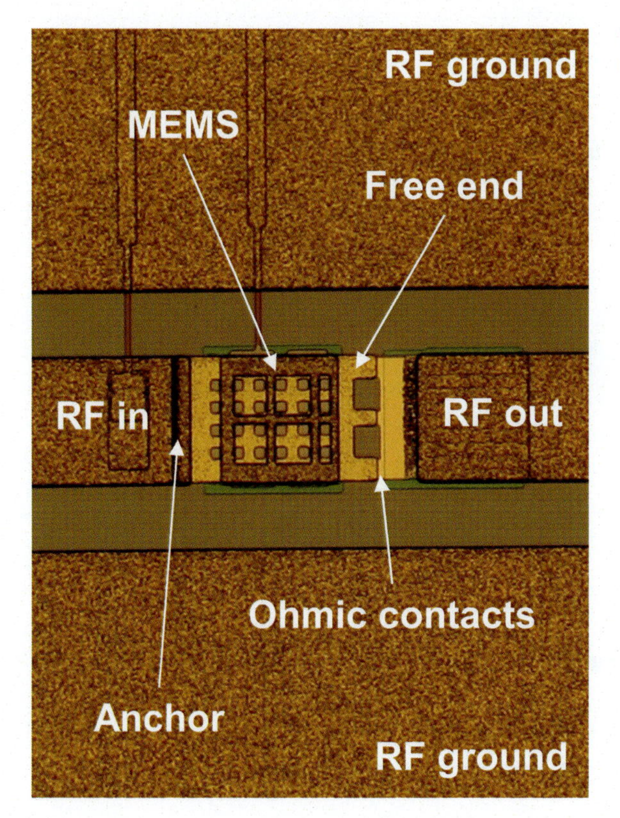


FIGURE 7.17 Microphotograph of a single-hinged cantilever RF-MEMS series ohmic switch.

configuration when the cantilevered MEMS is brought to actuation (ON state). The measured and simulated S-parameters characteristics of the OPEN/CLOSE cantilever-type RF-MEMS series ohmic switch are reported in Fig. 7.18. As before, the measured traces are available up to 40 GHz, while simulations are performed up to 60 GHz.

The top image in Fig. 7.18 plots the S-parameters in the OPEN switch configuration. Isolation (S21 parameter) is better than −20 dB up to 20 GHz, better than −10 dB up to 40 GHz, and slightly worse than −10 dB up to 60 GHz. Differently, when looking at the S-parameters in the CLOSE configuration (bottom image in Fig. 7.18), reflection (S11) ranges

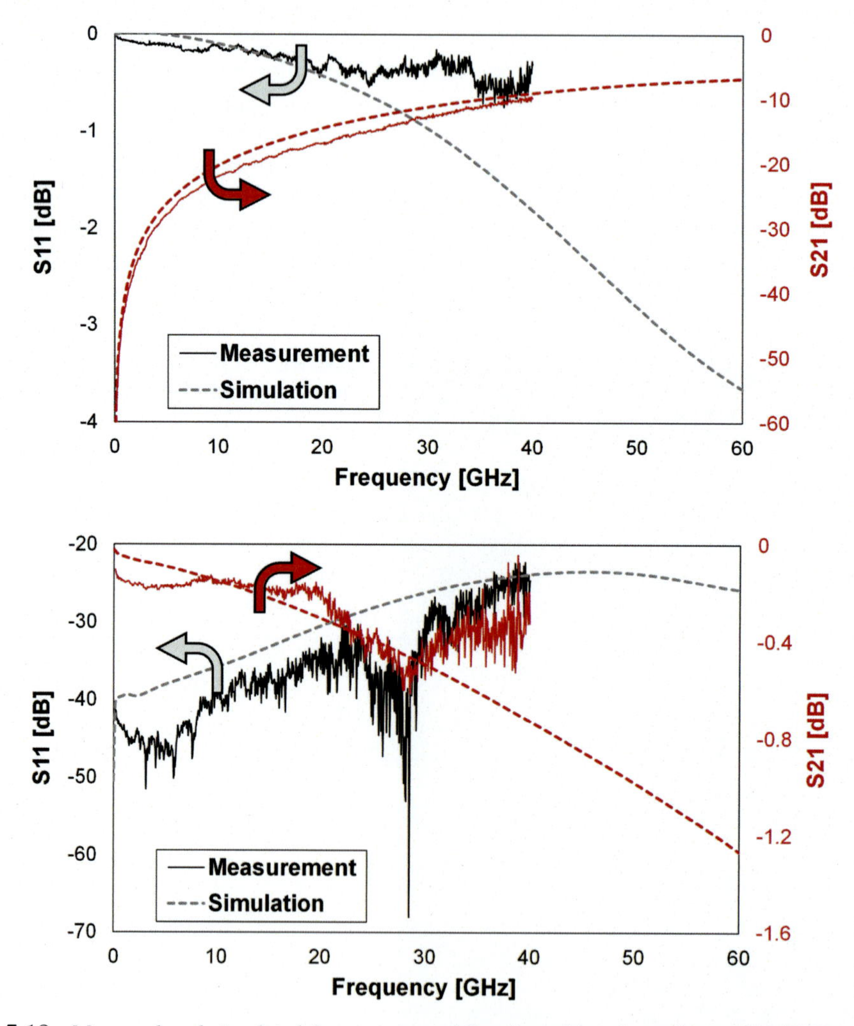


**FIGURE 7.18** Measured and simulated S-parameters of the single-hinged cantilever RF-MEMS series ohmic switch. The top image shows the reflection (S11) and isolation (S21) when the switch is OPEN (i.e., the MEMS is OFF). The bottom image shows the reflection (S11) and transmission (S21) when the switch is CLOSE (i.e., the MEMS is ON).

between −45 dB and −25 dB, from nearly DC up to 60 GHz. Also importantly, the transmission parameter (S21) ranges between −0.2 and −1.2 dB over the analyzed frequency span.

Now that also the behavior of the cantilever-type switch was covered, it is worth drawing some considerations arising from the comparison with the clamped-clamped series ohmic switch, previously reported in Fig. 7.13 and in Fig. 7.14. Concerning isolation (S21) in the OPEN switch configuration, the clamped-clamped microrelay scores numbers that are better of about 5−7 dB over the whole analyzed frequency range, as from looking at the top images in Fig. 7.14 and in Fig. 7.18, respectively. Differently, when referring to the CLOSE switch configuration, the single-hinged switch outperforms the clamped-clamped transverse one. In details, when comparing the bottom images in Fig. 7.14 and in Fig. 7.18, the S21 parameter in the former is better than −1 dB up to 40 GHz, while in the former it is better than −0.7 dB up to the same frequency. Moreover, the transmission of the clamped-clamped device worsens between −1 dB down to around −5 dB from 40 GHz up to 60 GHz. On the other hand, the S21 parameter of the cantilevered switch ranges between −0.7 and −1.3 dB, in the 40−60 GHz range.

Such differences in the performance of the two microrelays architectures can be explained in a straightforward fashion by taking into consideration their shape and placement. On the one hand, the clamped-clamped switch features a MEMS membrane suspended above the RF line truncated on both sides. This means that, when OFF, the suspended membrane realizes two equal parasitic capacitors connected in series, due to the overlap of the MEMS itself on the underneath contact areas. Differently, the cantilever is anchored directly on the RF line, and it features a unique contact area in correspondence to its free end. Therefore, if we assume an equal overlap area for each contact in the clamped-clamped device and for the sole one in the cantilever configuration, the latter one exhibits a series parasitic capacitance that is two times that of the clamped-clamped switch in the OFF state. This is the reason why the isolation (S21) of the cantilever is worse than that of the double-hinged microrelay.

From a different perspective, the same characteristic just discussed plays an opposite role when the switches are CLOSE (MEMS in the ON state). In fact, the double-hinged microrelay features two contact resistances connected in series on the RF line, while the cantilevered devices has just one contact resistance. Therefore, still assuming the same area for all the contacts, the cantilever has an unwanted series contact resistance that is half than that of the clamped-clamped one. This is the reason why the losses introduced by the single-hinged device are of minor entity over the whole frequency range. Eventually, still referring to what happens in the CLOSE conducting configuration, it is worth discussing an additional consideration. The cantilever-type switch, as already stated before, can be addressed as an elevated portion of the RF signal line. This means that the MEMS switch membrane does not introduce significant geometrical discontinuities encountered by the RF signal traveling across the device (when ON and conducting). Differently, the clamped-clamped microrelay is based on a transverse MEMS membrane, placed across the RF line (see Fig. 7.13). This means that the MEMS device, when actuated, introduces two metal patches orthogonal to the RF central metallization. Those extensions, if on one hand introduce discontinuities that can generate spurious reflections of the RF signal, on the other hand tend to behave as stubs, as much as the frequency increases, that is, because their physical length gets closer to fractions of wavelength. The consequence is that parasitic reactive elements start kicking in at high frequencies, worsening the characteristic impedance match of the microrelay and introducing additional losses. This is the reason why the transmission (S21) parameters divert so evidently referring to the bottom images in Fig. 7.14 and in Fig. 7.18 in the range from 40 to 60 GHz, with the cantilever performing significantly better than the clamped-clamped microrelay.

The overview of RF-MEMS based switches and switching principles reported up to now is already reasonably wide to provide an insight of the technology flexibility. Pros and cons existing among the switching principles (ohmic vs capacitive) and configurations (series vs shunt), as well as those related to the geometry of the MEMS membrane (clamped-clamped vs single-hinged) were also stressed and discussed. The examples reported above rely on switching solutions based on a unique MEMS reconfigurable membrane.

The wide flexibility of RF-MEMS technology in terms of degrees of freedom (DoFs) available to the designer comprises the strategy of capitalizing on redundancy as well. Given the substantial inexpensiveness of the technology, an additional path available in scoring the desired performance and characteristics is that of increasing the complexity of the design, duplicating, if necessary, the MEMS devices. To this end, a valuable example of this strategy is going to be reported before concluding this section.

Still having in mind basic switching devices, also referred to as SPSTs (Single Pole Single Throws), like those reported up to now, a possible driver could be that of improving isolation (S21) in the OPEN configuration, beyond the typical numbers achievable with the solutions discussed in the previous pages. Bearing this need in mind, it is possible demanding the approach to such a specification to two rather than one MEMS reconfigurable membrane. This is the case of the RF-MEMS switching unit shown in the microphotograph in Fig. 7.19. It must be highlighted that the device is an SPDT (Single Pole Double Throw), that is, a switching device that can selectively divert the input RF signal toward two output terminations (Casini et al., 2010; DiNardo et al., 2006a, 2006b; Farinelli et al., 2012). Nonetheless, for the purposes of this section, the device will be discussed referring to just one of the two output ports, as the behavior of the single switching unit is here at stake.

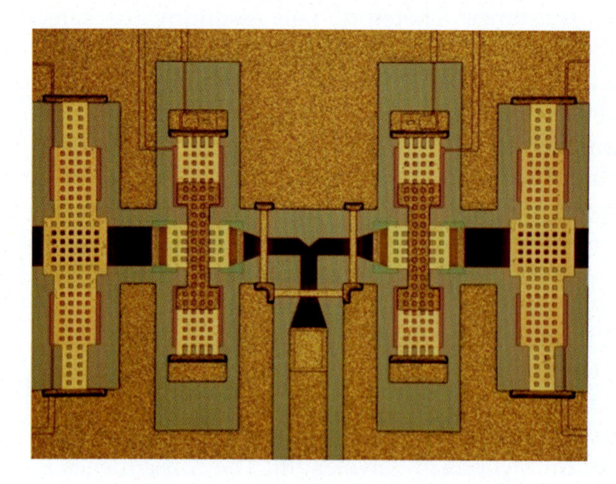

**FIGURE 7.19** Microphotograph of an SPDT switching unit relying on RF-MEMS series ohmic switches for opening/closing one of the two output channels and redundant shunt ohmic switches for boosting isolation when the corresponding branch is OPEN.

As visible, the switching devices placed on each output branch are two. The inner MEMS switches are series ohmic devices, and their design is the same previously shown in Fig. 7.13. The outer MEMS microrelays, instead, are shunt ohmic switches that have a similar behavior as compared to the devices previously reported in Fig. 7.15, despite they establish an ohmic contact rather than a capacitive coupling to RF ground when the MEMS is actuated (ON state). The latter devices behave as isolation boosters and they are to be activated when the corresponding branch is OPEN, that is, when the series ohmic switch in their vicinity is OFF. This way, the negative effects on isolation due to the parasitic series capacitive coupling of the RF-MEMS series ohmic switch in the OFF state are further mitigated by diverting the small fraction of RF passed signal to ground, thanks to the isolation booster. The comparison of the measured and simulated S-parameters in the OPEN and CLOSE state of a single output termination of the SPDT is reported in Fig. 7.20. In this case, when the output termination is OPEN, the corresponding isolation booster (i.e., shunt ohmic switch) is activated.

Looking at the top image in Fig. 7.20, a remarkable isolation better than −37 dB up to 40 GHz and better than −24 dB up to 60 GHz is visible. This performance is definitely better if compared to all the examples of single MEMS switches reported above in this section. On the other hand, the bottom image in Fig. 7.20 reports the switch CLOSE configuration. In this case, reflection (S11) is better than −20 dB up to 40 GHz and better than −10 dB up to 60 GHz. In addition, transmission (S21) is better than −1 dB up to 30 GHz, better than −2 dB up to 40 GHz and better than −4 dB up to 60 GHz.

Eventually, in order to provide a clearer evidence of the isolation booster effectiveness, the S-parameters of one output branch of the SPDT in the OPEN configuration are simulated with and without activating the corresponding shunt ohmic switch. Such a comparative plot is reported in Fig. 7.21.

By looking at the S21, the beneficial contribution of the isolation booster is evident, as it improves isolation of at least 10 dB from nearly DC up to 30 GHz. Above this frequency, the S21 improvement starts decreasing and the two curves overlap at around 50 GHz.

This is just an example of how the principles of HW redundancy and architectural complication can be leveraged in RF-MEMS technology when pursuing the achievement of challenging performance driven by demanding specifications. Along this direction, just to mention another circumstance in which redundancy is a valuable option, duplication of RF-MEMS microrelays can be exploited in order to improve the power handling capability of the whole switching device.

For completeness, a few additional examples of other switching devices developed and implemented in the same RF-MEMS technology platform are discussed and available in literature (Bartolucci et al., 2008; Cazzorla et al., 2015; Farinelli et al., 2008a, 2008b; Giacomozzi et al., 2007; Solazzi et al., 2012).

### 7.4.3 Reconfigurable passive networks enabled by RF-MEMS switching devices

As reported in the starting pages of this chapter, different types of reconfigurable RF-MEMS-based passive networks were demonstrated and discussed in literature. Beside those already mentioned before, like attenuators and phase shifters (Farinelli et al., 2013;

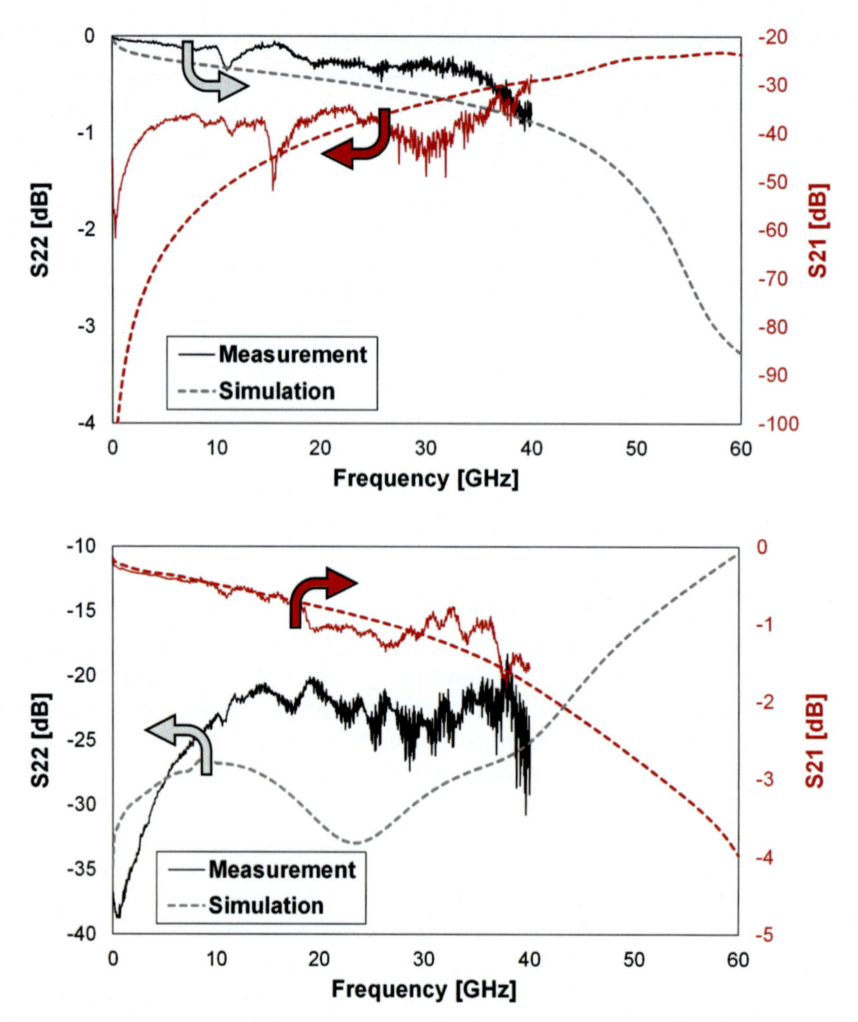


FIGURE 7.20 Measured and simulated S-parameters of one output termination of the SPDT. The top image shows the reflection (S11) and isolation (S21) when the series ohmic switch is OPEN (i.e., the MEMS is OFF) and the isolation booster (shunt ohmic switch) is activated (i.e., the MEMS is ON). The bottom image shows the reflection (S11) and transmission (S21) when the series switch is CLOSE (i.e., the MEMS is ON) and the isolation booster is OFF.

Marcaccioli et al., 2009), it is worth reporting LC tanks (Cazzorla et al., 2015a, 2015b), couplers (Marcaccioli et al., 2008), and filters (Farinelli et al., 2016; Gentili et al., 2012; Pelliccia et al., 2011, 2012; Pelliccia, Cacciamani, et al., 2015; Pelliccia, Farinelli, & Sorrentino, 2013; Pelliccia, Farinelli, et al., 2015; Pelliccia, Farinelli, Nocella, et al., 2013), as well.

The focus of this conclusive section is that of discussing into more details an example of RF-MEMS-based complex reconfigurable network realized in the same technology platform of the microrelays and other devices mentioned above. To this end, a multistate impedance matching network (IMN) is chosen here (Iannacci et al., 2011). The microphotograph of the device is shown in Fig. 7.22.

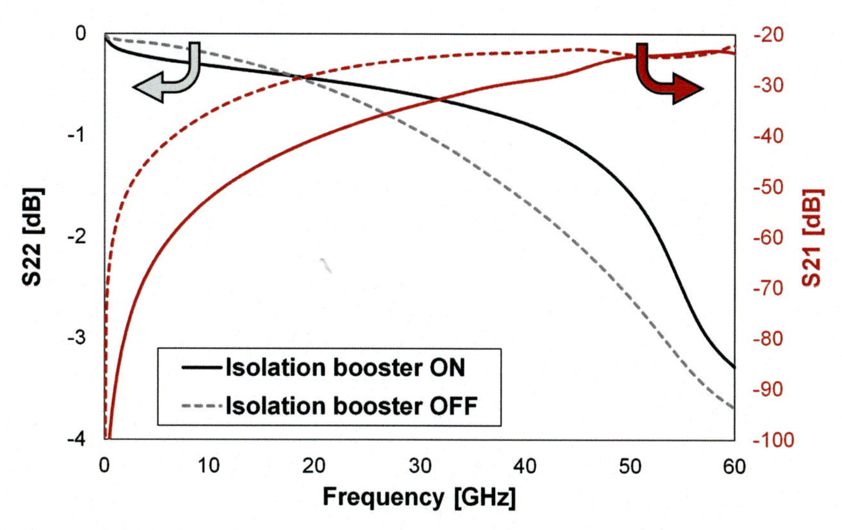


**FIGURE 7.21** Simulated S-parameters of one output termination of the SPDT in OPEN configuration when the corresponding isolation booster (i.e., the redundant shunt ohmic switch) is activated (ON) and deactivated (OFF).

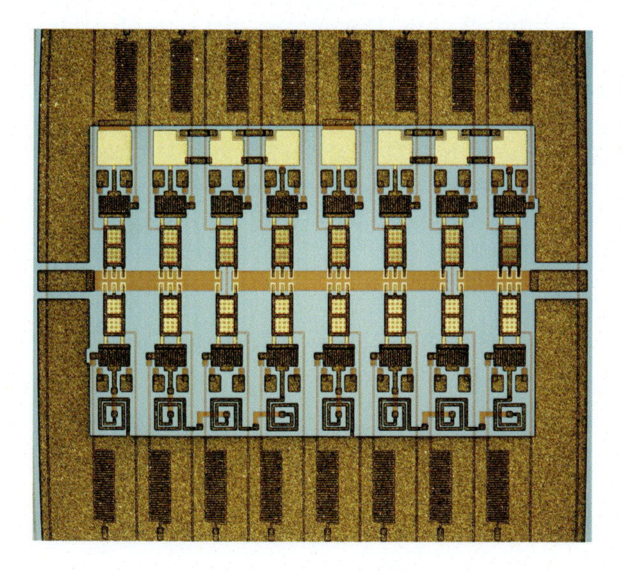

**FIGURE 7.22** Microphotograph of the RF-MEMS impedance matching network (IMN) featuring eight double channel switching stages, able to load the RF line with capacitive (top) and inductive (bottom) reactive elements, in series and/or shunt configuration, depending on the ON/OFF state of the corresponding cantilever-type MEMS microrelays.

The IMN features eight independent switching stages, implemented by means of cantilever-type ohmic switches. Such switching stages control two banks of reactive elements, that is, metal insulator metal (MIM) capacitors, visible in the top portion of the microphotograph, and suspended-in-air inductive coils, visible in the bottom portion of the microphotograph. The IMN realizes two redundant series/shunt LC sections, each of which implemented by four switching stages. Therefore, the detailed description just of the first four switching stages (left half of the IMN) is going to be developed, as the other four elements (right half of the IMN) behave exactly in the same fashion.

The first switching stage on the left controls a reactive shunt section. In particular, when the top switch is actuated, a shunt MIM capacitor loads the RF line, while when the bottom switch is actuated, a shunt inductor loads the RF line. Of course, the switches are independently controlled, in order to choose if one or both the reactive elements are to be included. In the latter case, the capacitor and the inductor are connected in parallel to RF ground.

The second switch is actuated together with the fourth one. This is true both for the capacitive and inductive elements. Considering the upper branch, when the second and fourth switches are actuated, a capacitance loads the RF line in series configuration. Such a capacitor is realized by the connection in parallel of the second, third, and fourth MIM capacitors visible in the top part of the IMN microphotograph. Differently, when the second and fourth switches are not actuated, the third series microrelay has to be pulled in, in order to let the RF signal flow across the two terminations.

Similar considerations are valid for the inductive bank, in the bottom part of the network. In this case, the only difference is that when the second and fourth switches are actuated, the corresponding three suspended inductors are connected in series rather than in parallel. However, like for capacitors, their total contribution in connected in series on the RF line. In summary, the whole IMN admits $6 + 6 = 12$ switching possibilities, which yield 256 different configurations.

The plot reported in Fig. 7.23 shows the simulated S-parameters of the RF-MEMS IMN up to 60 GHz, when no reactive elements load the RF line. This means that only the third and seventh switching stages are actuated, while all the other microrelays are OFF.

Looking at the transmission parameter (S21), it is evident that the loss becomes more and more important as the frequency raises. Because of this reason, in the following analysis the selected working frequency is 10 GHz, where the S21 is about −1.25 dB. Given this operating condition, the RF-MEMS IMN is simulated in its different configurations, and the results are plotted in Fig. 7.24.

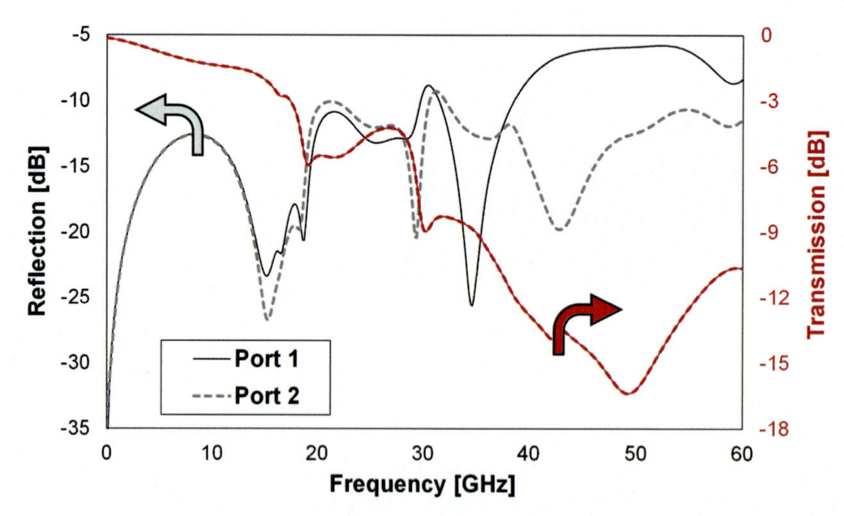


FIGURE 7.23 Simulated reflection (S11, S22) and transmission (S12, S21) parameters of the RF-MEMS IMN when no reactive element in inserted on the RF line (i.e., third and seventh switching stages actuated).

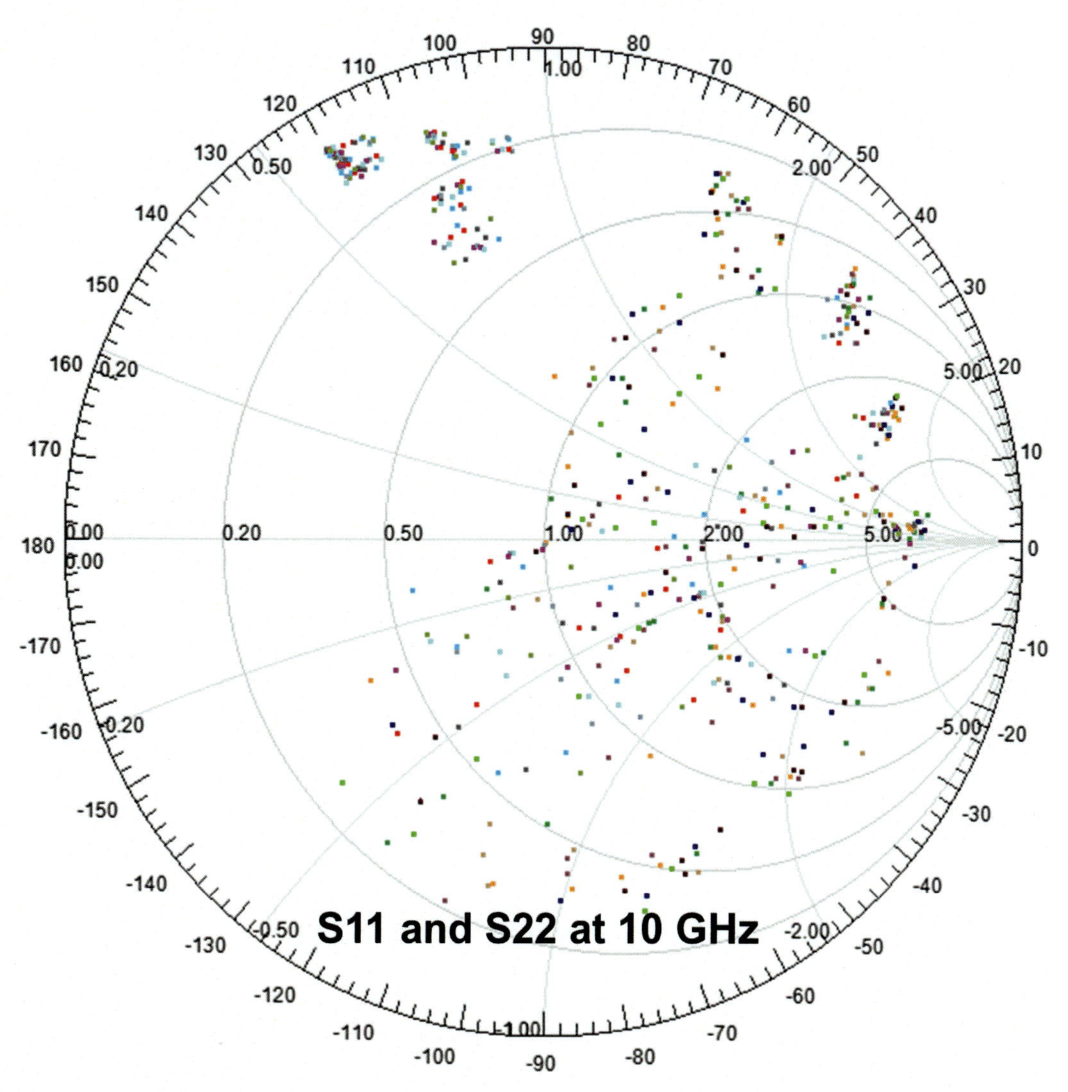


**FIGURE 7.24** Simulated Smith chart of the S11 (input impedance) and S22 (output impedance) parameters of the RF-MEMS IMN at 10 GHz for all the 256 network configurations.

The Smith chart reported in figure shows the S11 (input impedance) and S22 (output impedance) parameters simulated at 10 GHz in all the 256 different configurations allowed by the RF-MEMS IMN. First, a good coverage of the Smith chart is visible, meaning that the input and output impedance values and their transformations are quite diverse. This is particularly meaningful when observing the rather even distribution of dots in the higher (i.e., capacitive) and lower (i.e., inductive) half of the chart. On the other hand, it must be highlighted as well that most part of the realized characteristic impedance values lay in the right half of the Smith charts. This means that the IMN impedance states are characterized

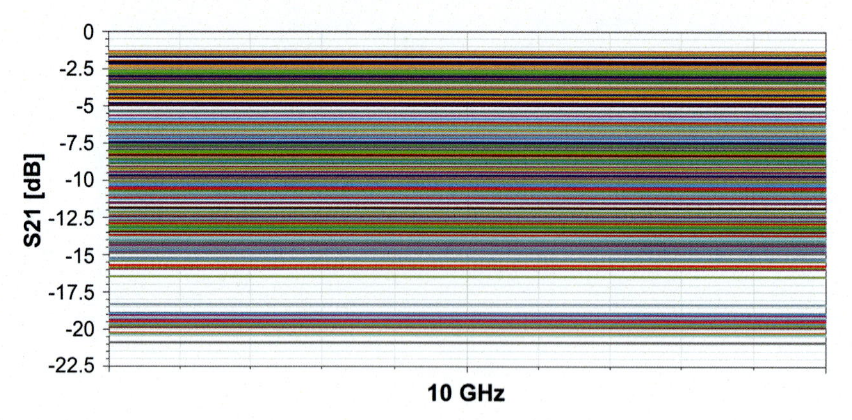


**FIGURE 7.25** Simulated S21 (transmission) parameter of the RF-MEMS IMN at 10 GHz for all the 256 network configurations.

by a resistance (i.e., real part of the impedance) always higher than the reference value of 50 Ohm. Differently, the left part of the Smith chart, where the resistance values are lower, is poorly covered by the IMN impedance configurations. This limitation is because in the RF-MEMS IMN preliminary design concept discussed here, the capacitors and inductors of the two LC series/shunt sections are shaped in order to be the same. Therefore, many impedance configurations are redundant, as can be either realized by the first and the second sections. Having said that, redesigns of the IMN leveraging diversification of the reactive elements incorporated would definitely improve the Smith chart coverage.

A few conclusive considerations have to be developed for what concerts the RF-MEMS IMN losses related to the just discussed impedance values and input/output transformations. To this end, the S21 parameter simulated at 10 GHz for all the 256 configurations is shown in Fig. 7.25.

Looking at this plot, important indications can be derived around which impedance transformations implemented by the RF-MEMS IMN can be effectively exploited and which, instead, should be neglected, with reference to the losses they introduce. The displayed S21 parameter at 10 GHz ranges between −1.3 and −21 dB. If we set −5 dB as threshold value for tolerable losses, it is straightforward that just a limited set of the implemented impedance transformations can be exploited. Therefore, this aspect should be carefully taken into consideration when further developing design concepts similar to the one discussed here as example, in order to avoid having several configurations that are, in fact, implemented by the RF-MEMS reconfigurable network but not exploitable due to the excessive loss they introduce.

## 7.5 Conclusions

The target of this chapter was that of framing the possibilities offered by RF-MEMS technology, that is, MEMS for RF applications, in realizing high-performance and wide-reconfigurable passive components suitable to meet the challenging requirements of 5G emerging applications.

First, a few considerations on the evolution of RF-MEMS technology, also in relation to how MEMS sensors and actuators started to be discussed in literature, were sketched. Then, the fluctuations characterizing for about an entire decade the market expectations and forecasts around RF-MEMS technology were reviewed and analyzed. Beside intrinsic factors linked to reliability, packaging, and integration that played against market absorption of RF-MEMS solutions, the key technology-push and market-pull factors were mentioned. In brief, in the emerging era of mobile communications, taking place with 2G and 3G, there were not actual needs for the pronounced performance and characteristics enabled by RF-MEMS. Differently, in more recent years, the widespread of smartphones and the rising of 4G-LTE started urging for RF passive networks with unprecedented capabilities in terms of reconfigurability/tunability. Such a transition from technology-push to market-driven philosophy started to mark the beginning of RF-MEMS breakthrough into mass-market applications.

Subsequently, the discussion started to embody the critical application field of 5G. To this end, a progressive top-down approach was adopted. In the very first place, a lack in deriving the key specifications that might be required to basic RF passive components by 5G applications was detected. Therefore, the main 5G high-level specifications, like data rate, data volume, and latency, were introduced and briefly discussed. Given such a starting point, an effort was taken in order to analyze how the just-mentioned characteristics reflect in infrastructure and HW characteristics, and, stepping further low in terms of complexity, which impact they exert on RF passive components. In this direction, a list of passives that are of interest for 5G was composed, for example, including high-performance wideband switches and multi-state digital attenuators, and a set of reference desired characteristics was reported, as well. Eventually, a common field between 5G applications and RF passive components was sketched, by reviewing, spec by spec, what 5G will demand and what RF-MEMS technology can offer, for example, in terms of isolation, loss, reconfigurability/tunability, and so on.

Then, a focused technical discussion on RF-MEMS technology and solutions started to be developed. In the first place, the surface micromachining technology platform available at FBK in Italy was taken as example. In order to explain the fabrication flow in a straightforward fashion, the case of an electrostatically actuated RF-MEMS series ohmic switch was chosen and its manufacturing was discussed, step by step, starting from the silicon substrate, until the release of the operable microrelay.

Once the technology background was consolidated, practical examples of RF-MEMS-based devices were reported, leveraging both experimental data and simulations of their electromagnetic characteristics (S-parameters) in a frequency range spanning from nearly DC up to 60 GHz. In details, different types of switches, that is, ohmic and capacitive, in series and shunt configuration, were reviewed, highlighting the pros and cons of each solution in terms of isolation, loss, wideband characteristics, and so on.

Eventually, the set of RF-MEMS reference examples was completed by introducing and discussing two more complex design solutions. The first one is related to a SPDT switch unit featuring redundant MEMS shunt ohmic switches needed for improving the isolation performance when the microrelay is not conducting (OPEN configuration). The second and concluding example is that of a multistate RF-MEMS-based IMN with several switching stages that was analyzed for what concerns its capability of covering the Smith chart plot, demonstrating several of the various impedance transformations it is able to implement, depending on the switches configuration.

# References

Agaty, M., Crunteanu, A., Dalmay, C., & Blondy, P. (2018). Ku band high-Q tunable cavity filters using MEMS and vanadium dioxide (VO2) tuners. In *2018 IEEE MTT-S international microwave workshop series on advanced materials and processes for RF and THz applications (IMWS-AMP)* (pp. 1–3). Available from https://doi.org/10.1109/IMWS-AMP.2018.8457138.

Agyapong, P. K., Iwamura, M., Staehle, D., Kiess, W., & Benjebbour, A. (2014). Design considerations for a 5G network architecture. *IEEE Communications Magazine*, *52*(11), 65–75. Available from https://doi.org/10.1109/MCOM.2014.6957145.

Anuroop, D. Bansal, Khushbu, Kumar, P., Kumar, A., & Rangra, K. (2018). Contact area design of ohmic RF MEMS switch for enhanced power handling. In *2018 12th international conference on sensing technology (ICST)* (pp. 91–95), 2156–8073. Available from https://doi.org/10.1109/ICSensT.2018.8603623.

Bartolucci, G., Marcelli, R., Catoni, S., Margesin, B., Giacomozzi, F., Lucibello, A., Mulloni, V., & Farinelli, P. (2008). Circuital modelling of shunt capacitive RF MEMS switches. In *2008 European microwave integrated circuit conference* (pp. 362–365). Available from https://doi.org/10.1109/EMICC.2008.4772304.

Bassous, E. (1978). Fabrication of novel three-dimensional microstructures by the anisotropic etching of. *IEEE Transactions on Electron Devices*, *25*(10), 1178–1185. Available from https://doi.org/10.1109/T-ED.1978.19249.

Bayraktar, O., Kobal, E., Sevinc, Y., Cetintepe, C., Comart, I., Demirel, K., Topalli, E. S., Akin, T., Demir, S., & Civi, O. A. (2015). RF MEMS based millimeter wave phased array for short range communication. 2015 9th European Conference on Antennas and Propagation (EuCAP), 1–5, 2164–3342.

Beams, J. W., Breazeale, J. B., & Bart, W. L. (1955). Mechanical strength of thin films of metals. *Physical Review*, *100* (6), 1657–1661. Available from https://doi.org/10.1103/PhysRev.100.1657.

Belkadi, N., Nadaud, K., Hallépée, C., Passerieux, D., & Blondy, P. (2018). Zero-level packaged 5 W CW RF-MEMS switched capacitors. In *2018 48th European microwave conference (EuMC)* (pp. 559–562). Available from https://doi.org/10.23919/EuMC.2018.8541766.

Belkadi, N., Nadaud, K., Hallepee, C., Passerieux, D., & Blondy, P. (2020). Zero-level packaged RF-MEMS switched capacitors on glass substrates. *Journal of Microelectromechanical Systems*, *29*(1), 109–116. Available from https://doi.org/10.1109/JMEMS.2019.2949949.

Blakely, J. M. (1964). Mechanical properties of vacuum-deposited gold films. *Journal of Applied Physics*, *35*(6), 1756–1759. Available from https://doi.org/10.1063/1.1713735.

Casini, F., Farinelli, P., Mannocchi, G., DiNardo, S., Margesin, B., De Angelis, G., Marcelli, R., Vendier, O., & Vietzorreck, L. (2010). High performance RF-MEMS SP4T switches in CPW technology for space applications. In *The 40th European microwave conference* (pp. 89–92). Available from https://doi.org/10.23919/EUMC.2010.5616406.

Cazzorla, A., Kaynak, M., Farinelli, P., & Sorrentino, R. (2015). A novel dual gap MEMS varactor manufactured in a fully integrated BiCMOS-MEMS process. In *2015 Asia-Pacific microwave conference (APMC)* (Vol. 3, pp. 1–3). Available from https://doi.org/10.1109/APMC.2015.7413373.

Cazzorla, A., Sorrentino, R., & Farinelli, P. (2015a). Double-actuation extended tuning range RF MEMS Varactor. In *2015 European microwave conference (EuMC)* (pp. 937–940). Available from https://doi.org/10.1109/EuMC.2015.7345918.

Cazzorla, A., Sorrentino, R., & Farinelli, P. (2015b). MEMS based LC tank with extended tuning range for Multi-band applications. In *2015 IEEE 15th mediterranean microwave symposium (MMS)* (pp. 1–4). Available from https://doi.org/10.1109/MMS.2015.7375433.

Chan, K. Y., Ramer, R., & Mansour, R. R. (2012). Novel miniaturized RF MEMS staircase switch matrix. *IEEE Microwave and Wireless Components Letters*, *22*(3), 117–119. Available from https://doi.org/10.1109/LMWC.2011.2170964.

Chen, P., Muller, R. S., Shiosaki, T., & White, R. M. (1979). WP-B6 silicon cantilever beam accelerometer utilizing a PI-FET capacitive transducer. *IEEE Transactions on Electron Devices*, *26*(11), 1857. Available from https://doi.org/10.1109/T-ED.1979.19782.

Chianrabutra, C., Jiang, L., Lewis, A. P., & McBride, J. W. (2013). Evaluating the influence of current on the wear processes of Au/Cr-Au/MWCNT switching surfaces. In *2013 IEEE 59th holm conference on electrical contacts (Holm 2013)* (pp. 1–6), 2158–9992. Available from https://doi.org/10.1109/HOLM.2013.6651411.

Choi, D., Lee, H. S., & Yoon, J. (2009). Linearly variable inductor with RF MEMS switches to enlarge a continuous tuning range. In *TRANSDUCERS 2009 - 2009 international solid-state sensors, actuators and microsystems conference* (pp. 573–576), 2164-1641 Available from https://doi.org/10.1109/SENSOR.2009.5285389.

Cohn, M. B., Roehnelt, R., Xu, J.-H., Shteinberg, A., & Cheung, S. (2002). MEMS packaging on a budget (fiscal and thermal). In *9th International conference on electronics, circuits and systems* 1 (Vol. 1, pp. 287–290). Available from https://doi.org/10.1109/ICECS.2002.1045390.

Daneshmand, M., & Mansour, R. R. (2005). Multiport MEMS-based waveguide and coaxial switches. *IEEE Transactions on Microwave Theory and Techniques, 53*(11), 3531–3537. Available from https://doi.org/10.1109/TMTT.2005.855738.

Daneshmand, M., & Mansour, R. R. (2006). Monolithic RF MEMS switch matrix integration. In *2006 IEEE MTT-S international microwave symposium digest* (pp. 140–143). Available from https://doi.org/10.1109/MWSYM.2006.249414 0149–645X.

De Silva, A. P., & Hughes, H. G. (2003). The package integration of RF-MEMS switch and control IC for wireless applications. *IEEE Transactions on Advanced Packaging, 26*(3), 255–260. Available from https://doi.org/10.1109/TADVP.2003.818056.

Dey, S., & Koul, S. K. (2014). 10–35-GHz frequency reconfigurable RF MEMS 5-bit DMTL phase shifter uses push-pull actuation based toggle mechanism. In *2014 IEEE international microwave and RF conference (IMaRC)* (pp. 21–24), 2377–9152. Available from https://doi.org/10.1109/IMaRC.2014.7038961.

Diaferia, F., Deborgies, F., Di Nardo, S., Espana, B., Farinelli, P., Lucibello, A., Marcelli, R., Margesin, B., Giacomozzi, F., Vietzorreck, L., & Vitulli, F. (2014). Compact 12 × 12 switch matrix integrating RF MEMS switches in LTCC hermetic packages. In *2014 44th European microwave conference* (pp. 199–202). Available from https://doi.org/10.1109/EuMC.2014.6986404.

DiNardo, S., Farinelli, P., Giacomozzi, F., Mannocchi, G., Marcelli, R., Margesin, B., Mezzanotte, P., Mulloni, V., Russer, P., Sorrentino, R., Vitulli, F., & Vietzorreck, L. (2006a). Broadband RF-MEMS based SPDT. In *2006 European microwave conference* (pp. 1727–1730). Available from https://doi.org/10.1109/EUMC.2006.281475.

DiNardo, S., Farinelli, P., Giacomozzi, F., Mannocchi, G., Marcelli, R., Margesin, B., Mezzanotte, P., Mulloni, V., Russer, P., Sorrentino, R., Vitulli, F., & Vietzorreck, L. (2006b). Broadband RF-MEMS based SPDT. In *2006 European microwave integrated circuits conference* (pp. 501–504). Available from https://doi.org/10.1109/EMICC.2006.282693.

Emami, N., & Bakri-Kassem, M. (2016). Slotted multi-step RF MEMS-CMOS parallel plate variable capacitor. In *2016 IEEE 59th international midwest symposium on circuits and systems (MWSCAS)* (pp. 1–4), 1558–3899. Available from https://doi.org/10.1109/MWSCAS.2016.7869990.

Evolving LTE to fit the 5G future. (2020). Ericsson ericssontechnologyreview With 5G research progressing at a rapid pace, the standardization process has started in 3GPP. As the most prevalent mobile broadband communication technology worldwide, LTE constitutes an essential piece of the 5G puzzle. Evolving LTE to fit the 5G future. https://www.ericsson.com/en/reports-and-papers/ericsson-technology-review/articles/evolving-lte-to-fit-the-5g-future.

Farinelli, P., El Ghannudi, H., Resta, G., Margesin, B., Erspan, M., & Sorrentino, R. (2012). High power SP4T MEMS switch for space applications. In *2012 42nd European microwave conference* (pp. 186–189). Available from https://doi.org/10.23919/EuMC.2012.6459296.

Farinelli, P., Montori, S., Fritzsch, C., Chiuppesi, E., Marcaccioli, L., Giacomozzi, F., Sorrentino, R., & Jakoby, R. (2013). MEMS and combined MEMS/LC technology for mm-wave electronic scanning. In *2013 6th UK, Europe, China millimeter waves and THz technology workshop (UCMMT)* (pp. 1–2). Available from https://doi.org/10.1109/UCMMT.2013.6641512.

Farinelli, P., Pelliccia, L., Margesin, B., & Sorrentino R. (2016). Ka-band surface-mountable pseudo-elliptic filter in multilayer micromachined technology for on-board communication systems. In *2016 IEEE MTT-S international microwave symposium (IMS)* (pp. 1–4). Available from https://doi.org/10.1109/MWSYM.2016.7540055.

Farinelli, P., Solazzi, F., Calaza, C., Margesin, B., & Sorrentino, R. (2008a). A wide tuning range MEMS varactor based on a toggle push-pull mechanism. In *2008 European microwave integrated circuit conference* (pp. 474–477). Available from https://doi.org/10.1109/EMICC.2008.4772332.

Farinelli, P., Solazzi, F., Calaza, C., Margesin, B., & Sorrentino, R. (2008b). A wide tuning range MEMS varactor based on a toggle push-pull mechanism. In *2008 38th European microwave conference* (pp. 1501–1504). Available from https://doi.org/10.1109/EUMC.2008.4751752.

Figur, S. A., Schoenlinner, B., Prechtel, U., Vietzorreck, L., & Ziegler, V. (2012). Simplified 16 × 8 RF MEMS switch matrix for a GEO-stationary data relay. In *2012 The 7th German microwave conference* (Vol. 1–4, pp. 2167–8022).

Flynn, K. (2020). The path to 5G: As much evolution as revolution https://www.3gpp.org/news-events/3gpp-news/1774-5g_wiseharbour.

Gentili, F., Pelliccia, L., Cacciamani, F., Farinelli, P., & Sorrentino, R. (2012). RF MEMS bandwidth-reconfigurable hairpin filters. In *2012 Asia Pacific microwave conference proceedings* (pp. 735–737), 2165–4743. Available from https://doi.org/10.1109/APMC.2012.6421720.

Giacomozzi, F., Calaza, C., Colpo, S., Mulloni, V., Collini, A., Margesin, B., Farinelli, P., Casini, F., Marcelli, R., Mannocchi, G., & Vietzorreck, L. (2007). Tecnological and design improvements for RF MEMS shunt switches. In *2007 International semiconductor conference* (Vol. 1, pp. 263–266), 2377-0678. Available from https://doi.org/10.1109/SMICND.2007.4519697.

Giacomozzi, F., & Iannacci, J. (2013). RF MEMS technology for next-generation wireless communications (pp. 225–257). Buckingham, England, UK: Woodhead Publishing. Available from https://doi.org/10.1533/9780857098610.1.225, http://www.sciencedirect.com/science/article/pii/B9780857092717500086.

Giacomozzi F., Mulloni V., Colpo S., Iannacci J., & Margesin B. (2012). A flexible fabrication process for RF MEMS devices. https://www.semanticscholar.org/paper/A-Flexible-Fabrication-Process-for-RF-MEMS-Devices-Giacomozzi-Mulloni/eca85e1cfe72a3554b96a0161191481de8360f4f?p2df.

Han M. S., Lee J. W., Kang C. G., & Rim M.J. (2019). System-level performance evaluation with 5G K-SimSys for 5G URLLC system. In *2019 16th IEEE annual consumer communications & networking conference (CCNC)* (pp. 1–5), 2331–9860. Available from https://doi.org/10.1109/CCNC.2019.8651885.

hAnnaidh B. O., Fitzgerald P., Berney H., Lakshmanan R., Coburn N., Geary S., & Mulvey B. (2018). Devices and sensors applicable to 5G system implementations. In *2018 IEEE MTT-S international microwave workshop series on 5G hardware and system technologies (IMWS-5G)*. (pp. 1–3). Available from https://doi.org/10.1109/IMWS-5G.2018.8484316.

Hartzell A. L., da Silva M. G., & Shea H. (2011). MEMS reliability | Allyson L. Hartzell | Springer. Springer US. Available from https://doi.org/10.1007/978-1-4419-6018-4, https://www.springer.com/gp/book/9781441960177.

Henry, S., Alsohaily, A., & Sousa, E. S. (2020). 5G is real: Evaluating the compliance of the 3GPP 5G new radio system with the ITU IMT-2020 requirements. *IEEE Access*, *8*, 42828–42840. Available from https://doi.org/10.1109/ACCESS.2020.2977406.

Iannacci, J., Bartek, M., Tian, J., Gaddi, R., & Gnudi, A. (2008). Electromagnetic optimization of an RF-MEMS wafer-level package. Special Issue: Eurosensors XX The 20th European Conference on Solid-State Transducers, 142(1), 434–441 Available from https://doi.org/10.1016/j.sna.2007.08.018, http://www.sciencedirect.com/science/article/pii/S0924424707006322.

Iannacci, J., Giacomozzi, F., Colpo, S., Margesin, B., & Bartek, M. (2009). A general purpose reconfigurable MEMS-based attenuator for Radio Frequency and microwave applications. *IEEE EUROCON, 2009*, 1197–1205. Available from https://doi.org/10.1109/EURCON.2009.5167788.

Iannacci, J., Huhn, M., Tschoban, C., & Pötter, H. (2016a). RF-MEMS technology for future mobile and high-frequency applications: Reconfigurable 8-bit power attenuator tested up to 110 GHz. *IEEE Electron Device Letters*, *37*(12), 1646–1649. Available from https://doi.org/10.1109/LED.2016.2623328.

Iannacci, J., Huhn, M., Tschoban, C., & Pötter, H. (2016b). RF-MEMS technology for 5G: Series and shunt attenuator modules demonstrated up to 110 GHz. *IEEE Electron Device Letters*, *37*(10), 1336–1339. Available from https://doi.org/10.1109/LED.2016.2604426.

Iannacci, J., Masotti, D., Kuenzig, T., & Niessner, M. (2011). A reconfigurable impedance matching network entirely manufactured in RF-MEMS technology. Smart sensors, actuators, and MEMS V. International society for optics and photonics (Vol. 8066, p. 80660X). Available from https://doi.org/10.1117/12.886186, https://www.spiedigitallibrary.org/conference-proceedings-of-spie/8066/80660X/A-reconfigurable-impedance-matching-network-entirely-manufactured-in-RF-MEMS/10.1117/12.886186.short?SSO = 1.

Iannacci, J., Resta, G., Bagolini, A., Giacomozzi, F., Bochkova, E., Savin, E., Kirtaev, R., Tsarkov, A., & Donelli, M. (2020). RF-MEMS monolithic K and Ka band multi-state phase shifters as building blocks for 5G and Internet of Things (IoT) applications. *Sensors.*, *20*(9), 1424–8220. Available from https://doi.org/10.3390/s20092612.

Iannacci, J., Tschoban, C., Reyes, J., Maaß, U., Huhn, M., Ndip, I., & Pötter, H. (2016). RF-MEMS for 5G mobile communications: A basic attenuator module demonstrated up to 50 GHz. *2016 IEEE Sensors*, 1–3. Available from https://doi.org/10.1109/ICSENS.2016.7808547.

Iannacci, J. (2014). RF-MEMS: A development flow driving innovative device concepts to high performance components and networks for wireless applications. In *2014 Microelectronic systems symposium (MESS)* (pp. 1–6). Available from https://doi.org/10.1109/MESS.2014.7010258.

Iannacci, J. (2015a). RF-MEMS: An enabling technology for modern wireless systems bearing a market potential still not fully displayed. *Microsystem Technologies*, *21*(10), 2039–2052. Available from https://doi.org/10.1007/s00542-015-2665-6.

Iannacci, J. (2015b). Reliability of MEMS: A perspective on failure mechanisms, improvement solutions and best practices at development level. *Advanced MEMS Technologies and Displays, 37*, 62–71. Available from https://doi.org/10.1016/j.displa.2014.08.003, http://www.sciencedirect.com/science/article/pii/S0141938214000602.

Iannacci, J. (2017a). *RF-MEMS technology for high-performance passives. The challenge of 5G mobile applications*. IOP Publishing. Available from http://doi.org/10.1088/978-0-7503-1545-6.

Iannacci, J. (2017b). RF-MEMS for high-performance and widely reconfigurable passive components – A review with focus on future telecommunications, Internet of Things (IoT) and 5G applications. *Journal of King Saud University - Science, 29*(4), 436–443. Available from https://doi.org/10.1016/j.jksus.2017.06.011, http://www.sciencedirect.com/science/article/pii/S1018364717304007.

Iannacci, J. (2018). RF-MEMS technology as an enabler of 5G: Low-loss ohmic switch tested up to 110 GHz. *Sensors and Actuators A: Physical*, 624–629. Available from https://doi.org/10.1016/j.sna.2018.07.005, http://www.sciencedirect.com/science/article/pii/S0924424718305004.

IHS iSuppli Teardown Analysis Service Identifies First Use of RF MEMS Part, Set to be Next Big Thing in Cellphone Radios – Omdia. (2020). https://technology.informa.com/389456/ihs-isuppli-teardown-analysis-service-identifies-first-use-of-rf-mems-part-set-to-be-next-big-thing-in-cellphone-radios.

Jones, T. R., & Daneshmand, M. (2020). Miniaturized folded ridged quarter-mode substrate integrated waveguide RF MEMS tunable bandpass filter. *IEEE Access, 8*, 115837–115847. Available from https://doi.org/10.1109/ACCESS.2020.3004116.

Jourdain, A., Ziad, H., De Moor, P., & Tilmans, H. A. C. (2003). Wafer-scale 0-level packaging of (RF-)MEMS devices using BCB. In *Symposium on design, test, integration and packaging of MEMS/MOEMS 2003* (pp. 239–244). Available from https://doi.org/10.1109/DTIP.2003.1287044.

Kageyama, T., Shinozaki, K., Zhang, L., Lu, J., Takaki, H., & Lee, S. (2017). An ohmic contact type RF-MEMS switch having Au-Au/CNTs contacts. In *2017 IEEE 12th international conference on nano/micro engineered and molecular systems (NEMS)* (pp. 287–290), 2474–3755. Available from https://doi.org/10.1109/NEMS.2017.8017026.

Katehi, L. P. B., Rebeiz, G. M., Weller, T. M., Drayton, R. F., Cheng, H. J., & Whitaker, J. F. (1993). Micromachined circuits for millimeter- and sub-millimeter-wave applications. *IEEE Antennas and Propagation Magazine, 35*(5), 9–17. Available from https://doi.org/10.1109/74.242171.

Kawai, T., Gaspar, J., Paul, O., & Kamiya, S. (2009). Prediction of strength and fatigue lifetime of MEMS structures with arbitrary shapes. In *TRANSDUCERS 2009 - 2009 international solid-state sensors, actuators and microsystems conference* (pp. 1067–1070), 2164-1641. Available from https://doi.org/10.1109/SENSOR.2009.5285943.

Khaira, N. K., Singh, T., & Mansour, R. R. (2018). RF MEMS based 60 GHz variable attenuator. In *2018 IEEE MTT-S international microwave workshop series on advanced materials and processes for RF and THz applications (IMWS-AMP)* (pp. 1–3). Available from https://doi.org/10.1109/IMWS-AMP.2018.8457154.

Khanna, V. K. (2010). Adhesion–delamination phenomena at the surfaces and interfaces in microelectronics and MEMS structures and packaged devices. *Journal of Physics D: Applied Physics, 44*(3), 034004. Available from https://doi.org/10.1088/0022-3727/44/3/034004.

Khodapanahandeh, M., Mirzajani, H., & Ghavifekr, H. B. (2018). A novel electrostatically actuated high Q RF MEMS tunable capacitor for UHF applications. In *Electrical engineering (ICEE), Iranian conference on* (pp. 11–16). Available from https://doi.org/10.1109/ICEE.2018.8472482.

Kourani, A., Yang, Y., Gong, S., & Ku-Band, A. (2020). Oscillator utilizing overtone lithium niobate RF-MEMS resonator for 5G. *IEEE Microwave and Wireless Components Letters, 30*(7), 681–684. Available from https://doi.org/10.1109/LMWC.2020.2996961.

Lampen, J., Majumder, S., Ji, C., & Maciel, J. (2010). Low-loss, MEMS based, broadband phase shifters. In *2010 IEEE international symposium on phased array systems and technology* (pp. 219–224). Available from https://doi.org/10.1109/ARRAY.2010.5613368.

Laskar, J., Matinpour, B., & Chakraborty, S. (2004). Modern receiver front-ends: Systems, circuits, and integration. Wiley. https://www.wiley.com/en-us/Modern + Receiver + Front + Ends%3A + Systems%2C + Circuits%2C + and + Integration-p-9780471474869.

Lau, J. H., Lee, C., Premachandran, C. S., & Aibin, Y. (2010). *Advanced MEMS packaging*. New York: McGraw-Hill Education. Available from https://www.accessengineeringlibrary.com/content/book/9780071626231.

Lee, Y., Jang, Y., Kim, J., & Kim, Y. (2010). A 50–110 GHz ohmic contact RF MEMS silicon switch with high isolation. In *2010 IEEE 23rd international conference on micro electro mechanical systems (MEMS)* (pp. 759–762). 1084–6999. Available from https://doi.org/10.1109/MEMSYS.2010.5442295.

Lin, T., Gao, L., Gaddi, R., & Rebciz, G. M. (2018). 400–560 MHz tunable 2-Pole RF MEMS bandpass filter with improved stopband rejection. In 2018 IEEE/MTT-S international microwave symposium – IMS (pp. 510–513), 2576–7216 Available from https://doi.org/10.1109/MWSYM.2018.8439541.

Lu, A. C. W., Chua, K. M., & Guo, L. H. (2005). Emerging manufacturing technologies for RFIC, antenna and RF-MEMS integration. In *2005 IEEE international wkshp on radio-frequency integration technology: Integrated circuits for wideband comm & wireless sensor networks* (pp. 142–146). Available from https://doi.org/10.1109/RFIT.2005.1598895.

Ma, L., Soin, N., Mohd Daut, M. H., & Wan Muhamad Hatta, S. F. (2019). Comprehensive study on RF-MEMS switches used for 5G scenario. *IEEE Access*, *7*, 107506–107522. Available from https://doi.org/10.1109/ACCESS.2019.2932800.

Mahmoud, S. F. (1991). Electromagnetic waveguides: Theory and applications. P. Peregrinus Ltd. on behalf of the Institution of Electrical Engineers, London.

Marcaccioli, L., Farinelli, P., Tentzeris, M. M., Papapolymerou, J., & Sorrentino, R. (2008). Design of a broadband MEMS-based reconfigurable coupler in Ku-band. In *2008 38th European microwave conference* (pp. 595–598). Available from https://doi.org/10.1109/EUMC.2008.4751522.

Marcaccioli, L., Montori, S., Gatti, R. V., Chiuppesi, E., Farinelli, P., & Sorrentino, R. (2009). RF MEMS-reconfigurable architectures for very large reflectarray antennas. In *2009 Asia Pacific microwave conference* (pp. 766–769). Available from https://doi.org/10.1109/APMC.2009.5384259, 2165–4743.

Margomenos, A., & Katehi, L. P. B. (2002). DC to 40 GHz on-wafer package for RF MEMS switches. In *2002 IEEE 11th topical meeting on electrical performance of electronic packaging* (pp. 91–94). Available from https://doi.org/10.1109/EPEP.2002.1057890.

Margomenos, A., & Katehi, L. P. B. (2003). High frequency parasitic effects for on-wafer packaging of RF MEMS switches. IEEE MTT-S International Microwave Symposium Digest, 2003, 3 3, 1931–1934. Available from https://doi.org/10.1109/MWSYM.2003.1210536, 0149–645X.

McGrath, W. R., Walker, C., Yap, M., & Tai, Y. (1993). Silicon micromachined waveguides for millimeter-wave and submillimeter-wave frequencies. *IEEE Microwave and Guided Wave Letters*, *3*(3), 61–63. Available from https://doi.org/10.1109/75.205665.

Molinero, D., Aghaei, S., Morris, A., & Cunningham, S. (2019). Linearity and RF power handling of capacitive RF MEMS switches. In *2019 IEEE MTT-S international microwave symposium (IMS)* (pp. 793–796). Available from https://doi.org/10.1109/MWSYM.2019.8700984, 2576–7216.

Morris, A. S., Natarajan, S. P., Gu, Q., & Steel, V. (2012). Impedance tuners for handsets utilizing high-volume RF-MEMS. In *2012 42nd European microwave conference* (pp. 193–196). Available from https://doi.org/10.23919/EuMC.2012.6459116.

Mulloni, V., Margesin, B., Farinelli, P., Marcelli, R., & De Angelis, G. (2015). Cycling reliability of RF-MEMS switches with gold-platinum multilayers as contact material. In *2015 Symposium on design, test, integration and packaging of MEMS/MOEMS (DTIP)* (pp. 1–5). Available from https://doi.org/10.1109/DTIP.2015.7161039.

Neugebauer, C. A. (1960). Tensile properties of thin, evaporated gold films. *Journal of Applied Physics*, *31*(6), 1096–1101. Available from https://doi.org/10.1063/1.1735751.

Nguyen, C. T. (1998). Microelectromechanical devices for wireless communications. In *Proceedings MEMS 98. IEEE. Eleventh annual international workshop on micro electro mechanical systems. An investigation of micro structures, sensors, actuators, machines and systems (Cat. No.98CH36176)* (pp. 1–7), 1084–6999. Available from https://doi.org/10.1109/MEMSYS.1998.659719.

Nguyen, C. T. (2001). Transceiver front-end architectures using vibrating micromechanical signal processors. In *2001 Topical meeting on silicon monolithic integrated circuits in RF systems*. Digest of Papers (IEEE Cat. No.01EX496) (pp. 23–32). Available from https://doi.org/10.1109/SMIC.2001.942335.

Nguyen, C. T. (2002). RF MEMS for wireless applications. 60th DRC. In *Conference digest device research conference* (pp. 9–12). Available from https://doi.org/10.1109/DRC.2002.1029485.

Nguyen, C. T. (2006). Integrated micromechanical circuits for RF front ends. In *2006 European solid-state device research conference* (pp. 7–16). Available from https://doi.org/10.1109/ESSDER.2006.307630, 2378–6558.

Nguyen, C. T. (2007). MEMS technology for timing and frequency control. *IEEE Transactions on Ultrasonics, Ferroelectrics, and Frequency Control*, *54*(2), 251–270. Available from https://doi.org/10.1109/TUFFC.2007.240.

Nguyen, C. T. (2013). MEMS-based RF channel selection for true software-defined cognitive radio and low-power sensor communications. IEEE Communications Magazine, 51(4), 110–119. Available from https://doi.org/10.1109/MCOM.2013.6495769.

Osseiran, A., Boccardi, F., Braun, V., Kusume, K., Marsch, P., Maternia, M., Queseth, O., Schellmann, M., Schotten, H., Taoka, H., Tullberg, H., Uusitalo, M. A., Timus, B., & Fallgren, M. (2014). Scenarios for 5G mobile and wireless communications: The vision of the METIS project. *IEEE Communications Magazine, 52*(5), 26–35. Available from https://doi.org/10.1109/MCOM.2014.6815890.

Pacheco, S., Zurcher, P., Young, S., Weston, D., & Dauksher, W. (2004). RF MEMS resonator for CMOS back-end-of-line integration. Digest of Papers. 2004 Topical meeting on silicon monolithic integrated circuits in RF systems 2004 (pp. 203–206). Available from https://doi.org/10.1109/SMIC.2004.1398203.

Palego, C., Solazzi, F., Halder, S., Hwang, J. C. M., Farinelli, P., Sorrentino, R., Faes, A., Mulloni, V., & Margesin, B. (2010). Effect of substrate on temperature range and power capacity of RF MEMS capacitive switches. In *The 40th European microwave conference* (pp. 505–508). Available from https://doi.org/10.23919/EUMC.2010.5617140.

Pelliccia, L., Cacciamani, F., Farinelli, P., & Sorrentino, R. (2015). High-$Q$ tunable waveguide filters using ohmic RF MEMS switches. *IEEE Transactions on Microwave Theory and Techniques, 63*(10), 3381–3390. Available from https://doi.org/10.1109/TMTT.2015.2459689.

Pelliccia, L., Cacciamani, F., Gentili, F., Farinelli, P., & Sorrentino, R. (2012). Compact ultra-wideband planar filter with RF-MEMS-based tunable notched band. In *2012 Asia Pacific microwave conference proceedings* (pp. 529–531). Available from https://doi.org/10.1109/APMC.2012.6421653, 2165–4743.

Pelliccia, L., Farinelli, P., Nocella, V., Cacciamani, F., Gentili, F., & Sorrentino, R. (2013). Discrete-tunable high-Q E-plane filters. In *2013 European microwave conference*. Available from https://doi.org/10.23919/EuMC.2013.6686882.

Pelliccia, L., Farinelli, P., & Sorrentino, R. (2011). MEMS-based high-Q reconfigurable E-plane filters. In *2011 41st European microwave conference* (pp. 369–372). https://doi.org/10.23919/EuMC.2011.6101855.

Pelliccia, L., Farinelli, P., & Sorrentino, R. (2013). Micromachined filters in multilayer technology for on-board communication systems in Ka-band. In *2013 IEEE MTT-S international microwave symposium digest (MTT)* (pp. 1–3). Available from https://doi.org/10.1109/MWSYM.2013.6697473, 0149–645X.

Pelliccia, L., Farinelli, P., Sorrentino, R., Cannone, G., Favre, G., & Coassini, P. (2015). K-band MEMS-based frequency adjustable waveguide filter for mobile back-hauling. In *2015 European microwave conference (EuMC)* (pp. 670–673). Available from https://doi.org/10.1109/EuMC.2015.7345852.

Peroulis, D. (2018). Tunable filter technologies for 5G communications. In *2018 IEEE international electron devices meeting (IEDM)* (pp. 14.6.1–14.6.4). Available from https://doi.org/10.1109/IEDM.2018.8614622, 2156-017X.

Persano, A., Quaranta, F., Capoccia, G., Proietti, E., Lucibello, A., Marcelli, R., Bagolini, A., Iannacci, J., Taurino, A., & Siciliano, P. (2016). Influence of design and fabrication on RF performance of capacitive RF MEMS switches. *Microsystem Technologies, 22*(7), 1741–1746. Available from https://doi.org/10.1007/s00542-016-2829-z.

Petersen, K. E. (1977). Micromechanical light modulator array fabricated on silicon. *Applied Physics Letters, 31*(8), 521–523. Available from https://doi.org/10.1063/1.89761.

Petersen, K. E. (1979). Micromechanical membrane switches on silicon. *IBM Journal of Research and Development, 23* (4), 376–385. Available from https://doi.org/10.1147/rd.234.0376.

Petersen, K. E. (1982). Silicon as a mechanical material. *Proceedings of the IEEE, 70*(5), 420–457. Available from https://doi.org/10.1109/PROC.1982.12331.

Psychogiou, D., Hesselbarth, J., Li, Y., Kuehne, S., & Hierold, C. (2011). W-band tunable reflective type phase shifter based on waveguide-mounted RF MEMS. In *2011 IEEE MTT-S international microwave workshop series on millimeter wave integration technologies* (pp. 85–88). Available from https://doi.org/10.1109/IMWS3.2011.6061894.

Pu, S. H., Darbyshire, D. A., Wright, R. V., Kirby, P. B., Rotaru, M. D., Holmes, A. S., Yeatman, E. M., & MEMS Zipping, R. F. (2016). Varactor with high quality factor and very large tuning range. *IEEE Electron Device Letters, 37*(10), 1340–1343. Available from https://doi.org/10.1109/LED.2016.2600264.

Rebeiz, G. M., & Muldavin, J. B. (2001). RF MEMS switches and switch circuits. *IEEE Microwave Magazine, 2*(4), 59–71. Available from https://doi.org/10.1109/6668.969936.

RF MEMS switches are primed for mass-market applications. (2016). Microwaves & RF while MEMS switches offer obvious countermeasures for performance issues in smartphones, the defense and test segments also are recognizing their benefits in an array of applications. RF MEMS switches are primed for mass-market applications. https://www.mwrf.com/technologies/active-components/article/21844989/rf-mems-switches-are-primed-for-massmarket-applications.

Rijks, T. G. S. M., van Beek, J. T. M., Ulenaers, M. J. E., De Coster, J., Puers, R., den Dekker, A., & van Teeffelen, L. (2003). Passive integration and RF MEMS: A toolkit for adaptive LC circuits. In *ESSCIRC 2004 - 29th European solid-state circuits conference (IEEE Cat. No.03EX705)* (pp. 269–272). Available from https://doi.org/10.1109/ESSCIRC.2003.1257124.

Riverola M., Uranga A., Torres F., Barniol N., Marigó E., & Soundara-Pandian M. (2017). A reliable fast miniaturized RF MEMS-on-CMOS switched capacitor with zero-level vacuum package. In *2017 IEEE MTT-S international microwave workshop series on advanced materials and processes for RF and THz applications (IMWS-AMP)* (pp. 1–3). Available from https://doi.org/10.1109/IMWS-AMP.2017.8247406.

Roylance, L. M., & Angell, J. B. (1979). A batch-fabricated silicon accelerometer. *IEEE Transactions on Electron Devices, 26*(12), 1911–1917. Available from https://doi.org/10.1109/T-ED.1979.19795.

Singh, T., Khaira, N. K., & Mansour, R. R. (2019). Monolithically integrated reconfigurable RF MEMS based impedance tuner on SOI substrate. In *2019 IEEE MTT-S international microwave symposium (IMS)* (pp. 790–792), 2576–7216. Available from https://doi.org/10.1109/MWSYM.2019.8701106.

Solazzi, F., Palego, C., Molinero, D., Farinelli, P., Colpo, S., Hwang, J. C. M., Margesin, B., & Sorrentino, R. (2012). High-power high-contrast RF MEMS capacitive switch. In *2012 7th European microwave integrated circuit conference* (pp. 32–35).

Solazzi, F., Palego, C., Halder, S., Hwang, J. C. M., Faes, A., Mulloni, V., Margesin, B., Farinelli, P., & Sorrentino, R. (2010). Electro-thermal analysis of RF MEM capacitive switches for high-power applications. In *2010 Proceedings of the European solid state device research conference* (pp. 468–471), 2378–6558. Available from https://doi.org/10.1109/ESSDERC.2010.5618174.

Solazzi, F., Resta, G., Mulloni, V., Margesin, B., & Farinelli, P. (2011). Influence of beam geometry on the dielectric charging of RF MEMS switches. In *2011 6th European microwave integrated circuit conference* (pp. 398–401).

Solazzi, F., Tazzoli, A., Farinelli, P., Faes, A., Mulloni, V., Meneghesso, G., & Margesin, B. (2010). Active recovering mechanism for high performance RF MEMS redundancy switches. In *The 40th European microwave conference* (pp. 93–96). Available from https://doi.org/10.23919/EUMC.2010.5616405.

Staff, S. X. (2014). RF MEMS: New possibilities for smartphones. *Phys.org*. Available from https://phys.org/news/2014-01-rf-mems-possibilities-smartphones.html.

Status of the mems industry - Market update. (2020). http://www.yole.fr/MEMSIndustry_MarketUpdate.aspx.

Sun, J., & Li, Z. (2016). A broadband 3-bit MEMS digital attenuator. In *2016 3rd International conference on electronic design (ICED)* (pp. 442–445). Available from https://doi.org/10.1109/ICED.2016.7804685.

Tavassolian, N., Koutsoureli, M., Papaioannou, G., Lacroix, B., & Papapolymerou, J. Dielectric charging in capacitive RF MEMS switches: The effect of electric stress. In *2010 Asia-Pacific microwave conference* (pp. 2165–4743). 1833–1836.

Tazzoli, A., Autizi, E., Barbato, M., Meneghesso, G., Solazzi, F., Farinelli, P., Giacomozzi, F., Iannacci, J., Margesin, B., & Sorrentino, R. (2009). Evolution of electrical parameters of dielectric-less ohmic RF-MEMS switches during continuous actuation stress. In *2009 Proceedings of the European solid state device research conference* (pp. 343–346), 2378–6558. Available from https://doi.org/10.1109/ESSDERC.2009.5331307.

Vaha-Heikkila, T., Varis, J., Tuovinen, J., & Rebeiz, G. M. (2005). A 20-50 GHz RF MEMS single-stub impedance tuner. *IEEE Microwave and Wireless Components Letters, 15*(4), 205–207. Available from https://doi.org/10.1109/LMWC.2005.845690.

van Gils, M., Bielen, J., & McDonald, G. (2007). Evaluation of creep in RF MEMS devices. In *2007 International conference on thermal, mechanical and multi-physics simulation experiments in microelectronics and micro-systems.* EuroSime 2007 (pp. 1–6). Available from https://doi.org/10.1109/ESIME.2007.360033.

Weller, T. M., & Katehi, L. P. B. (1995). Compact stubs for micromachined coplanar waveguide. In *1995 25th European microwave conference* (Vol. 2, pp. 589–593). Available from https://doi.org/10.1109/EUMA.1995.337029.

Wen Hsiung Ko., Hynecek, J., & Boettcher, S. F. (1979). Development of a miniature pressure transducer for biomedical applications. *IEEE Transactions on Electron Devices, 26*(12), 1896–1905. Available from https://doi.org/10.1109/T-ED.1979.19793.

Wipf, C., Sorge, R., Wipf, S. T., Göritz, A., Scheit, A., Kissinger, D., & Kaynak, M. (2019). RF-MEMS based V-band impedance tuner driven by integrated high-voltage LDMOS switch matrix and charge pump. In *2019 IEEE 19th topical meeting on silicon monolithic integrated circuits in RF systems* (SiRF) (pp. 1–3), 2474–9761. Available from https://doi.org/10.1109/SIRF.2019.8709116.

Wu, Z., & Liu, J. (2018). A new design of MEMS coplanar waveguide phase shifter. In *2018 International applied computational electromagnetics society symposium - China (ACES)* (pp. 1–2). Available from https://doi.org/10.23919/ACESS.2018.8669307.

Xu, T., Sun, J., Wu, H., Li, H., Li, H., & Tao, Z. (2019). 3D MEMS in-chip solenoid inductor with high inductance density for power MEMS device. *IEEE Electron Device Letters, 40*(11), 1816–1819. Available from https://doi.org/10.1109/LED.2019.2941003.

Yang, Z., Zhang, R., & Peroulis, D. (2020). Design and optimization of bidirectional tunable MEMS all-silicon evanescent-mode cavity filter. *IEEE Transactions on Microwave Theory and Techniques, 68*(6), 2398–2408. Available from https://doi.org/10.1109/TMTT.2020.2976011.

Yun-Kwon Park, Heung-Woo Park, Duck-Jung Lee, Jung-Ho Park, In-Sang Song, Chung-Woo Kim, Ci-Moo Song, Yun-Hi Lee, Chul-Ju Kim, Byeong Kwon Ju. (2002). A novel low-loss wafer-level packaging of the RF-MEMS devices. In *Technical digest. MEMS 2002 IEEE international conference. Fifteenth IEEE international conference on micro electro mechanical systems (Cat. No.02CH37266)* (pp. 681–684). Available from https://doi.org/10.1109/MEMSYS.2002.984362, 1084–6999.

Yun-Kwon Park, Yong-kook Kim, Hoon Kim, Duck-Jung Lee, Chul-Ju Kim, Byeong-Kwon Ju, Jong-Oh Park. (2003). A novel thin chip scale packaging of the RF-MEMS devices using ultra thin silicon. In *The sixteenth annual international conference on micro electro mechanical systems, 2003.* MEMS-03 Kyoto IEEE. (pp. 618–621), 1084-6999. Available from https://doi.org/10.1109/MEMSYS.2003.1189825.

Zhang, Q. X., Yu, A. B., Yang, R., Li, H. Y., Guo, L. H., Liao, E. B., Tang, M., Kumar, R., Liu, A. Q., Lo, G. Q., Balasubramanian, N., & Kwong, D. L. (2006). Novel monolithic integration of RF-MEMS switch with CMOS-IC on organic substrate for compact RF system. In *2006 International electron devices meeting* (pp. 1–4), 2156-017X. Available from https://doi.org/10.1109/IEDM.2006.346890.

Zhang, Y., Gong, Z., Guo, X., & Liu, Z. (2018). A High-Linearity SP12T RF MEMS Switch Using Parallel Dual-Cantilevers for 5G and Beyond Applications. *IEEE Electron Device Letters, 39*(10), 1608–1611. Available from https://doi.org/10.1109/LED.2018.2867841.

Zhong, Q., Guo, X., & Liu, Z. (2016). A DC-20GHz attenuator design with RF MEMS technologies and distributed attenuation networks. In *2016 8th IEEE international conference on communication software and networks (ICCSN)* (pp. 352–355). Available from https://doi.org/10.1109/ICCSN.2016.7586681.

Zine-El-Abidine, I., Okoniewski, M., & McRory, J. G. (2003). A new class of tunable RF MEMS inductors. In *Proceedings international conference on MEMS, NANO and smart systems* (pp. 114–115). Available from https://doi.org/10.1109/ICMENS.2003.1221976.

Zine-El-Abidine, I., Okoniewski, M., & McRory, J. G. (2004). A tunable RF MEMS inductor. In *2004 International conference on MEMS, NANO and smart systems (ICMENS'04)* (pp. 636–638). Available from https://doi.org/10.1109/ICMENS.2004.1509028.

CHAPTER

# 8

# Antenna-in-package design considerations for millimeter-wave 5G

*Bodhisatwa Sadhu, Duixian Liu and Xiaoxiong Gu*
**IBM Thomas J. Watson Research Center, Yorktown Heights, NY, United States**

## 8.1 Introduction

Driven by consumer demand, cellular data traffic has been skyrocketing over the last few decades. The spectrum crunch at lower frequencies has catalyzed the adoption of millimeter-wave (mm-wave) frequency bands for cellular applications. However, the widespread realization of mm-wave 5G mobile networks presents unprecedented technical challenges. From an antenna perspective, since antenna size shrinks quadratically with increasing frequency (antenna dimensions are proportional to the square of the signal wavelength), so does the fraction of incident energy that the antenna can capture. Moreover, signals at mm-wave frequencies are attenuated by atmospheric absorption to a greater degree than are signals at microwave frequencies. To meet the link budget requirements for reliable communications, it is necessary to identify technologies that maximize the effective isotropic radiated power (EIRP) of the mm-wave 5G transmitter (TX) and maximize the RF sensitivity of the mm-wave 5G receiver (RX) within the severe power consumption constraints demanded by mobile applications. In this chapter, we will discuss two key hardware components of the 5G mm-wave wireless network, namely, the handset antenna and the base-station antenna, in each case in conjunction with their respective associated circuits. In the context of base stations, the key mm-wave 5G antenna module research works that have been reported have focused primarily on planar phased-array modules (Dunworth, Homayoun, et al., 2018; Dunworth, Ku, et al., 2018; Gu et al., 2017; Kibaroglu et al., 2018; Park et al., 2020; Sadhu, 2022; Sadhu et al., 2017a,b; Sowlati et al., 2018). Versions of the circuits described in these published works have been deployed in the first wave of mm-wave 5G radio base-station products. Similarly, several papers related to mm-wave handset antenna designs have been published (Chen & Zhang, 2013; Futter & Soler, 2018; Hong et al., 2017; Koul & Karthikeya, 2020; Meena et al., 2019; Sadhu et al., 2020; Yang et al., 2016) focusing on design

*New Materials and Devices Enabling 5G Applications and Beyond*
DOI: https://doi.org/10.1016/B978-0-12-822823-4.00008-X

parameters such as gain, coverage, cellular handset effects, hardware integration, different antenna implementations, and cost. In the following sections, we will discuss key design considerations of mm-wave 5G antennas for handset applications along with an example design (Section 8.2), followed by key design considerations of mm-wave 5G antennas for base-station applications (Section 8.3).

## 8.2 Antenna design for mm-wave 5G handset applications

Antenna design for handset applications is a tightrope walk involving attempting to improve antenna gain and omnidirectionality within the constraints of the available physical space and environment. In the prior generations of commercial handsets, microwave frequency antennas, operating at sub-6 GHz frequencies, are designed to radiate with a fairly omnidirectional beam pattern. Unfortunately, the same antenna design methodology fails at mm-wave frequencies. To minimize the high interconnect loss between the integrated circuits (ICs) and the antenna at mm-wave frequencies, antennas and ICs are often co-integrated in the same package (Liu & Zhang, 2020). As a result, planar antenna topologies are the most popular. Unfortunately, planar antennas inherently exhibit sub-hemispherical beam coverage (Balanis, 2016). resulting in wireless coverage blind spots. To avoid such coverage blind spots, end-fire radiators can be added to increase the coverage for mm-wave handset applications (Ali et al., 2020; Huang et al., 2016; Liu et al., 2015; Ojaroudiparchin et al., 2016; Parchin et al., 2016; Samadi Taheri et al., 2019). Furthermore, the unpredictable mobility-driven changing position and orientation of the cellular handset also causes polarization mismatch, and needs to be considered in the wireless link budget calculations (Pan, 2011; Song et al., 2015).

Due to the above challenges in the design of antennas for handset applications, handsets often employ at least two antenna locations for reliable coverage (Chang et al., 2016; Hong et al., 2017). Fig. 8.1 shows one possible antenna layout employing a strategy with multiple antenna locations to improve antenna diversity and link robustness.

### 8.2.1 Handset antenna design considerations and trade-offs

This section discusses some of the considerations and challenges associated with antenna design for millimeter-wave 5G handset applications.

#### *8.2.1.1 Cellular handset environment*

Cellular handsets contain hundreds of components made from a variety of dielectric and metallic materials. During the antenna design process, these components can be collectively treated as stratified anisotropic media (Kim, 2016). The dielectric permittivity of major components in three typical handsets is discussed in Hong et al. (2014). Moreover, the placement of an antenna on the handset printed circuit board (PCB) affects antenna performance parameters such as the impedance bandwidth, radiation patterns, and gain values. To avoid radiation blockage and absorption loss, antennas must be placed close to the exterior of the cellular handset. The handset users also impact antenna performance

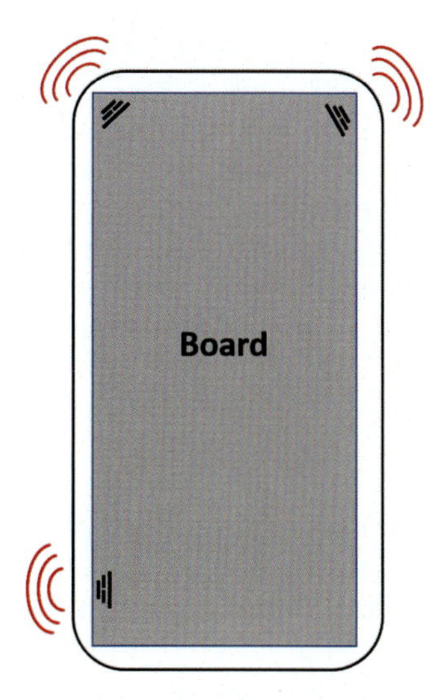


**FIGURE 8.1** Multiple antenna locations can help to ensure spatial coverage.

due to user body blockage (Liu et al., 2021) and mutual coupling between the user and the mm-wave 5G antenna. The impact of the user's hand on 5G cellular antennas in the 10-GHz frequency range is investigated in Buey et al. (2016). Experiments at 15 GHz suggest that the human body is a clear obstacle for signal propagation (Zhao et al., 2017). The effect of the user's hand at 39 GHz has been studied through electromagnetic (EM) simulations in (Fernandes, n.d.).

#### *8.2.1.2 Specific absorption rate*

For any handset device, ensuring the safety of the user is paramount. Prior to market release, a product needs to meet certification standards for human exposure to radiated fields. At sub-6-GHz frequencies, the existing specific absorption rate (SAR) standards apply. Sub-6-GHz frequencies have significant penetration into the human tissue (Hong, 2017; Koul & Karthikeya, 2020; Rutschlin, n.d.); as a result, SAR is a very important issue at these frequencies. At mm-wave frequencies, field penetration into the human body is very limited. Most of the energy is reflected, and the field penetrating the body dissipates within a few mm under the skin. As a result, SAR is not an appropriate measure of human exposure. Instead, the power flow on body surfaces at a certain distance from the device might provide a more suitable measure (Gustrau & Bahr, 2002; Hong, 2017; Lin, 2003; Wu et al., 2015).

#### *8.2.1.3 Handset antenna form factors*

Real estate inside current mobile handsets is extremely precious (Hong, 2017). As a result, there is a research push towards compact antenna solutions to optimize the space usage within the handset. In one example study (Hong et al., 2017), two-phased array antennas with dual polarized antennas are implemented in a phone with switchable polarization, thereby reusing the same antenna area for both polarizations. Antennas implemented on the handset cover (plastic or metal) have also been studied (Lee, Bang, et al., 2018; Enjiu, n.d.). However, sending signals to/from antennas on the phone cover will be challenging at mm-wave frequencies. Antennas can also be integrated behind plastic or glass covers by engineering the cover geometry to act locally as a lens, and even behind metal covers by including EM windows, based on the design principles of frequency selective surfaces (Rutschlin, n.d.; Zhao et al., 2018). Such an approach could improve the radiation pattern and scanning behavior. Moreover, it is important to remember that structures such as the phone cover are not electrically thin at mm-wave frequencies (unlike at lower frequencies), and can have a significant impact on the antenna performance. In this context, techniques for radome design can be employed in the design of mm-wave 5G handsets.

#### 8.2.1.4 Choice of substrate and antenna type

There are two practical categories for mm-wave 5G handset antenna designs: antenna in board (AiB) and antenna in package (AiP). Fig. 8.2 shows the implementation of these two types of antenna concepts. For the AiB designs, the antennas are implemented on the handset PCB directly (Martinez-Vazquez, 2020; Zhou, 2015). This design scheme provides the lowest cost among various antenna solutions and typically has the smallest footprint on the PCB. However, the PCB is usually less than 1 mm thick, and typically consists of 6 to 12 layers. As a result, there are constraints on the antenna design choice as well as on the antenna location choice. Antenna performance is further challenged by substrate losses in the PCB at mm-wave frequencies. In contrast, an AiP design allows more flexibility and can enable modular designs with better performance since the AiP design is independent of the PCB and can have relatively low loss substrates. Therefore, this chapter will concentrate on the AiP designs. Within the constraints of an AiP or AiB, a number of antenna structures can be used. Fig. 8.3 shows some example antenna types that can be used in AiP/AiB designs (Gu, Paidimarri, et al., 2021). For mm-wave phased-array applications, patch and magnetoelectric antennas can be used for broadside radiation while dipole and Yagi antennas can provide end-fire radiation. Grid antennas are useful for high gain fixed-beam applications. Substrate integrated waveguide antennas can be used in fixed beam or 1D phased array applications. For patch antennas, aperture-coupled (Pozar, 1992) or probe-fed (Pozar, 1992) patch antennas are a preferred choice since the antenna radiators and feed lines can be separated from each other by the antenna ground plane, allowing

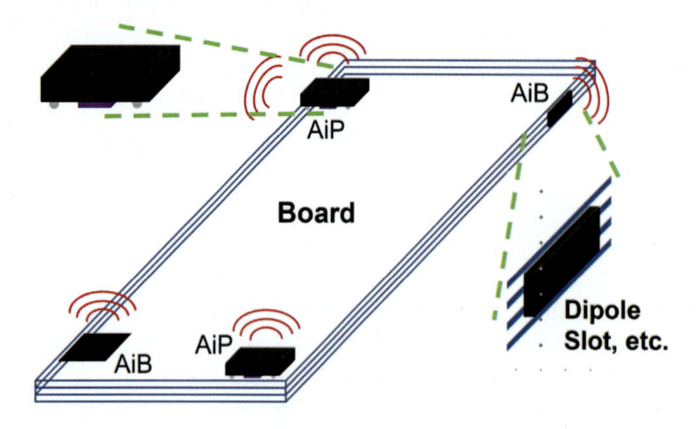


FIGURE 8.2 Example of antenna in package and antenna in board implementations on a mobile handset board.

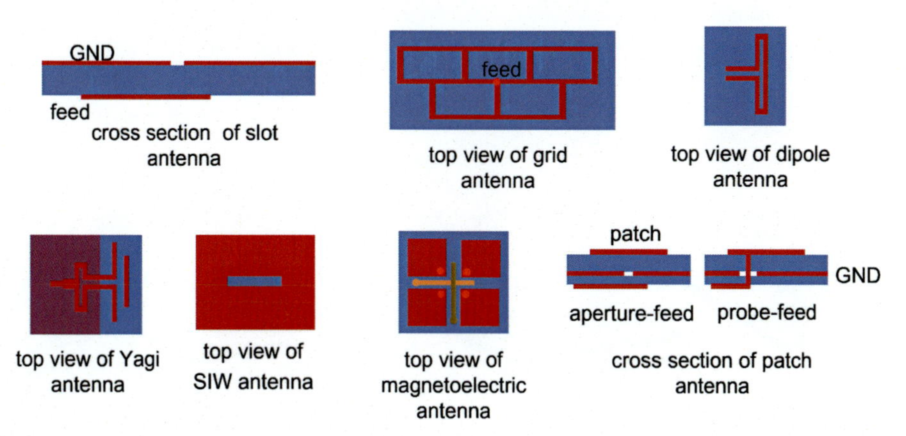


**FIGURE 8.3** Example of antenna designs suitable for antenna in package/antenna in board implementations. *Source: Modified from Gu, X., Liu, D., & Sadhu, B. (2021). Packaging and antenna integration for silicon-based millimeter-wave phased arrays: 5G and beyond. IEEE Journal of Microwaves, 1(1), 123–134. https://doi.org/10.1109/jmw.2020.3032891.*

the antenna radiators and feed lines to be optimized independently. From the studies in Liu et al. (2017), the probe-fed patch antenna has a narrow bandwidth in the package environment. As a result, it might be less suitable for wideband applications.

#### *8.2.1.5 Omni-directional antenna design for handset applications*

As discussed earlier, at frequencies <6 GHz, discrete antennas are utilized to create high-efficiency, low directivity beams (Dai et al., 2013). The discrete antennas are interfaced with packaged radio frequency integrated circuits (RFICs) using coaxial cables with tolerable insertion loss performance. However, at mm-wave frequencies, the insertion loss of such connecting coaxial cables makes the use of discrete antennas unattractive. As a result, AiP approaches co-integrating the RFIC and the mm-wave antennas are often desirable. Unfortunately, planar antennas integrated in packages typically feature a higher directivity, as shown in Fig. 8.4A. Phased arrays are able to steer beams within the beam range of a single antenna, and are therefore also handicapped by the directivity limitations of the individual antennas, as shown in Fig. 8.4B. One result of these directivity properties is the creation of blind spots in the antenna module, in turn resulting in a significant challenge in mobile applications where the user is expected to be able to use their device while pointing it in any direction. A combination of end-fire and broadside direction radiators, as shown in Fig. 8.4C, can overcome the blind spot issues.

### 8.2.2 An antenna design example for handset applications

This section discusses an example mm-wave AiP integrated with a low-power, high-performance 60-GHz CMOS transceiver IC for mobile handset applications (Sadhu et al., 2020). The package supports differential transmit and single-ended receive paths in both broadside and end-fire directions. Four antennas are integrated in the package including (1) single-ended patch, (2) differential patch, (3) single-ended Yagi, and (4) differential Yagi (Gu, Liu, et al., 2015). The multi-antenna configuration provides flexible link coverage

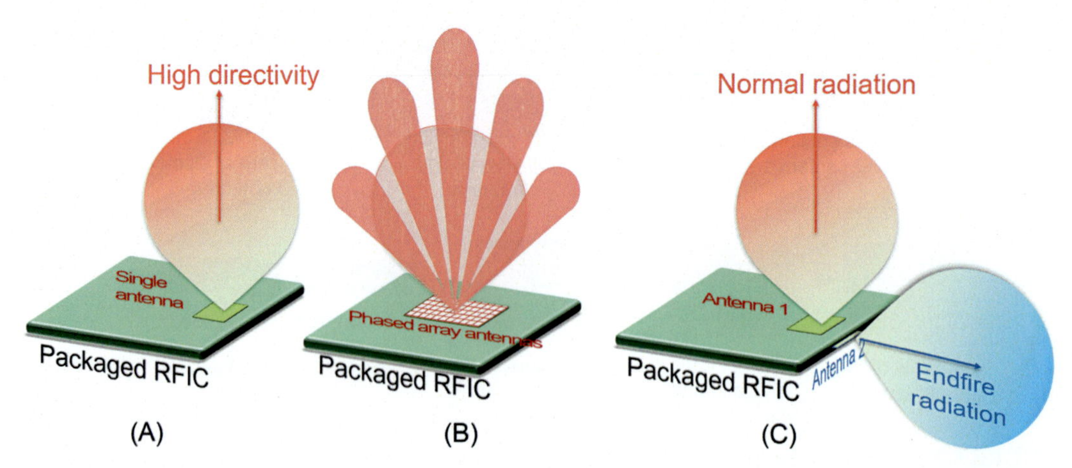


**FIGURE 8.4** Examples of antennas in antenna in package/antenna in board transceivers. (A) Single planar antenna with typical beam coverage, (B) phased array using multiple planar antennas provide beam steering within the beam coverage of a single antenna, (C) orthogonally pointed packaged antennas providing wide angular coverage. *Source: From Sadhu, B., Valdes-Garcia, A., Plouchart, J. O., Ainspan, H., Gupta, A. K., Ferriss, M., Yeck, M., Sanduleanu, M., Gu, X., Baks, C. W., Liu, D., & Friedman, D. (2020). A 250-mW 60-GHz CMOS transceiver SoC integrated with a four-element AiP providing broad angular link coverage. IEEE Journal of Solid-State Circuits, 55(6), 1516–1529. https://doi.org/10.1109/JSSC.2019.2943918.*

with a small form factor. A ball-grid array (BGA) provides the entire interface for digital I/O signals, transmit and receive differential baseband I/Q signals, phase locked loop (PLL) clock reference, power (6 power domains) and ground. The assembled packages are soldered down via a BGA to a compact application board for antenna pattern measurement. A cross-sectional view of the assembled package is shown in Fig. 8.5 (Gu, Liu, et al., 2015). The IC was flip-chip attached to the package using standard solder reflow and underfill. The patch antennas and Yagi antennas are on the top and bottom layers of the package, respectively. As shown in Fig. 8.5, ground planes from the package and board above and below the Yagi antennas are partially removed to improve their radiation efficiency. The M3 layer functions as the main ground plane. TX and RX baseband signal lines and control lines are implemented on the M4 layer. Power planes for different supply domains are implemented on the M2 layer.

Fig. 8.6 shows the layout and photos of the top and bottom of the package. Two different patch antennas are visible: a single-ended feed patch for RX and a differential feed patch for TX. In addition to the main patch, floating half-wavelength long parasitic patches are added to improve antenna radiation performance. The parasitic patches capture surface waves and radiate them so as to enhance the radiation from the main patch antennas (Li et al., 2005). While Li et al. (2005) use a grounded $\lambda/4$ parasitic patch, the packaging process used did not allow using buried vias to connect the floating patches to ground. To circumvent this problem, the design uses $\lambda/2$ floating parasitic patches. The $\lambda/2$ patches create a virtual ground at their center and work exactly as grounded $\lambda/4$ parasitic patches. For the normal direction, the antenna system uses two probe feeds for the differential TX patch antenna and a single feed for the single-ended RX patch antenna. For the

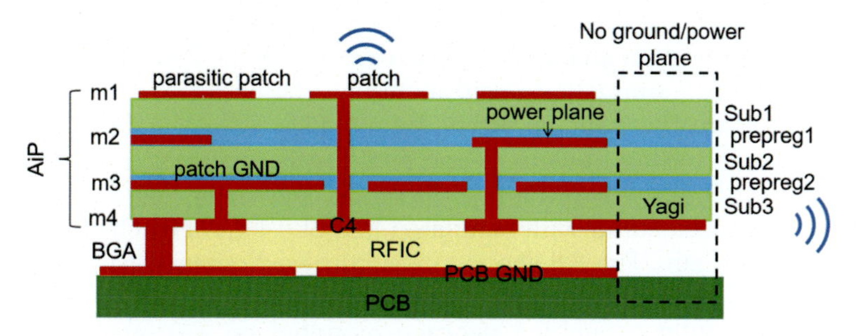


**FIGURE 8.5** Package cross-section showing the integration of the integrated circuit with the in-package antennas and printed circuit board. *Source: From Sadhu, B., Valdes-Garcia, A., Plouchart, J. O., Ainspan, H., Gupta, A. K., Ferriss, M., Yeck, M., Sanduleanu, M., Gu, X., Baks, C. W., Liu, D., & Friedman, D. (2020). A 250-mW 60-GHz CMOS transceiver SoC integrated with a four-element AiP providing broad angular link coverage. IEEE Journal of Solid-State Circuits, 55(6), 1516–1529. https://doi.org/10.1109/JSSC.2019.2943918.*

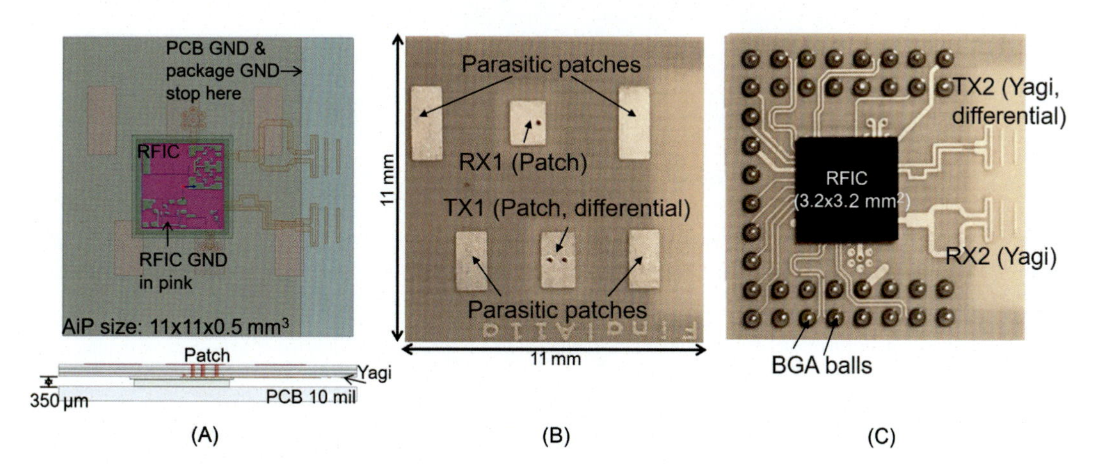


**FIGURE 8.6** (A) Package layout top view and cross-section showing integrated circuit (IC), package, antenna integration; the IC ground plane is highlighted in pink and used in package EM simulations, (B) annotated package photo top view showing 5G transmitter (TX) and 5G receiver (RX) patch antennas, (C) annotated package photo bottom showing the IC and TX and RX Yagi antennas on the bottom edge. *Source: From Sadhu, B., Valdes-Garcia, A., Plouchart, J. O., Ainspan, H., Gupta, A. K., Ferriss, M., Yeck, M., Sanduleanu, M., Gu, X., Baks, C. W., Liu, D., & Friedman, D. (2020). A 250-mW 60-GHz CMOS transceiver SoC integrated with a four-element AiP providing broad angular link coverage. IEEE Journal of Solid-State Circuits, 55(6), 1516–1529. https://doi.org/10.1109/JSSC.2019.2943918.*

end-fire direction, two Yagi antennas are used for TX and RX, respectively. A delay line balun converts a differential-fed Yagi to a single-ended feed for the RX port on chip. The package was modeled, including antenna structures, signal traces, and power/ground planes, as well as the adjacent PCB and RFIC ground planes in a full-wave EM simulator. The antennas were simulated in the package along with the ground plane of the IC, as highlighted in Fig. 8.6A, to capture the effect of the IC ground plane on antenna performance. Moreover, the end-fire Yagi antennas are simulated with the PCB ground plane to capture its impact on antenna performance (Liu et al., 2015).

Fig. 8.7 shows the AiP simulation setup including the package, and partially including IC and PCB to capture the antenna environment. Fig. 8.7 also tabulates the simulated antenna peak gains, including transitions and feed losses. Due to significant overlap of the differential Yagi feed line with the lossy silicon, as shown in Fig. 8.7, the single-ended Yagi (including the added balun with simulated loss of 0.4 dB at 60 GHz) and the differential Yagi, have similar gain. Fig. 8.8 shows the superposed antenna gains for the TX patch and Yagi antennas.

Fig. 8.9A shows a fully assembled package mounted on a compact test board using BGAs. Subminiature push-on connectors are populated on the opposite side of the board to provide reference clock and baseband signals to the RFIC. In addition, a 2 × 8 pin

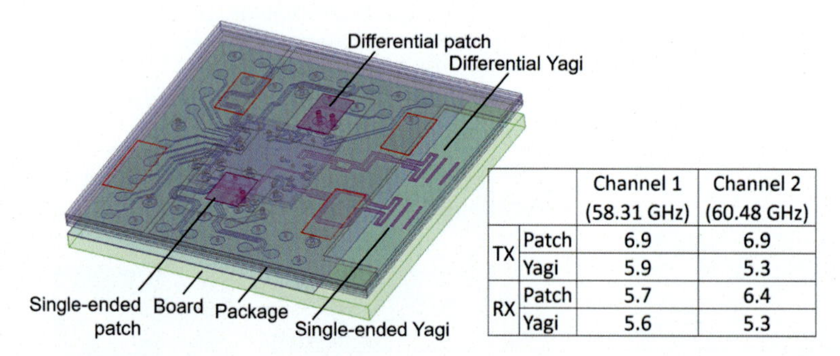


| | | Channel 1 (58.31 GHz) | Channel 2 (60.48 GHz) |
|---|---|---|---|
| TX | Patch | 6.9 | 6.9 |
| | Yagi | 5.9 | 5.3 |
| RX | Patch | 5.7 | 6.4 |
| | Yagi | 5.6 | 5.3 |

**FIGURE 8.7** Antenna in package simulation setup including the complete package, and partially including integrated circuit and printed circuit board to capture the antenna environment. Simulated antenna gains are tabulated in the inset. *Source: From Sadhu, B., Valdes-Garcia, A., Plouchart, J. O., Ainspan, H., Gupta, A. K., Ferriss, M., Yeck, M., Sanduleanu, M., Gu, X., Baks, C. W., Liu, D., & Friedman, D. (2020). A 250-mW 60-GHz CMOS transceiver SoC integrated with a four-element AiP providing broad angular link coverage. IEEE Journal of Solid-State Circuits, 55(6), 1516–1529. https://doi.org/10.1109/JSSC.2019.2943918.*

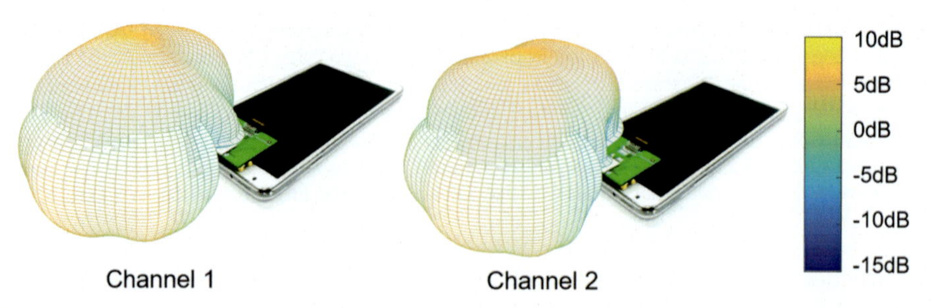


**FIGURE 8.8** Simulated superposed antenna gains for patch and Yagi antennas across angles. *Source: From Sadhu, B., Valdes-Garcia, A., Plouchart, J. O., Ainspan, H., Gupta, A. K., Ferriss, M., Yeck, M., Sanduleanu, M., Gu, X., Baks, C. W., Liu, D., & Friedman, D. (2020). A 250-mW 60-GHz CMOS transceiver SoC integrated with a four-element AiP providing broad angular link coverage. IEEE Journal of Solid-State Circuits, 55(6), 1516–1529. https://doi.org/10.1109/JSSC.2019.2943918.*

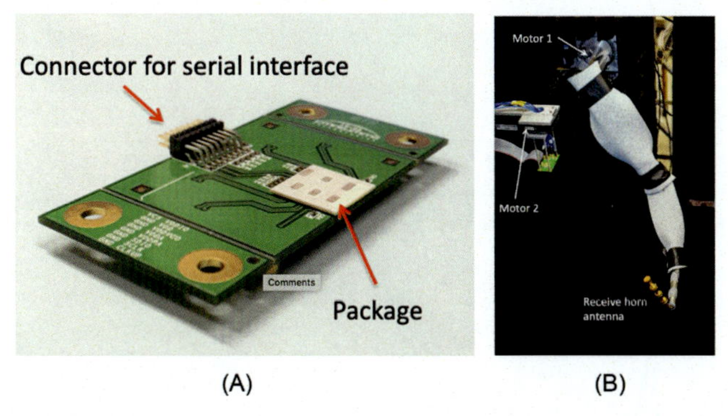


**FIGURE 8.9** (A) Side view of the assembly module on a printed circuit board. (B) Antenna radiation pattern measurement with active RFIC. *Source: From Gu, X., Liu, D., Baks, C., Sadhu, B., & Valdes-Garcia, A. (2015). A multi-layer organic package with four integrated 60 GHz antennas enabling broadside and end-fire radiation for portable communication devices. In* Proceedings - Electronic components and technology conference *(Vol. 2015-, pp. 1005–1009). Institute of Electrical and Electronics Engineers Inc. https://doi.org/10.1109/ECTC.2015.7159718.*

connector is used for power supply, ground, and a serial interface for digital control and programming of the RFIC. Fig. 8.9B illustrates the test setup of placing a fully assembled system in an antenna chamber for radiation pattern measurement. An field programmable gate arrays provides a serial interface to the test board through a ribbon cable along with the power supplies. PLL reference clock signal and baseband I/Q signals are supplied by two separate signal generators. Two motors are used to control the rotation of the package and the receiving horn (i.e., azimuth and elevation angles), respectively. The measurement is performed in TX mode at 60.25 GHz. The normalized radiation patterns are plotted in Fig. 8.10 based on active RFIC measurement in TX mode. Fig. 8.10A and B shows the patterns when only turning on end-fire direction and normal direction, respectively. In both cases, the received power varies less than 6 dB within $\pm 30$ degrees of the intended direction.

#### *8.2.2.1 Integrated circuit design for switched beam operation*

The IC is architected to switch between the normal and end-fire low-directivity beams described above without suffering significant loss of performance from the introduction of the switch (Sadhu et al., 2020). The proposed IC architecture is shown in Fig. 8.11. The IC includes TX and RX analog baseband circuits, PLL and local oscillator (LO) distribution, RF up/down conversion, power amplifier (PA) and low noise amplifier (LNA) circuits to interface with the antenna, and on-chip built-in self-test. Beam switching is enabled in both TX and RX modes by utilizing the two differential TX antennas and two single-ended RX antennas. Different antennas were used for TX and RX to avoid the performance degradation and additional power consumption that would be incurred due to a switch at the antenna. Antenna switching is achieved between the patch antenna (for broadside radiation) and the Yagi antenna (for end-fire radiation) by selecting one of two LNAs in the RX, or one of two PAs in the TX. In this architecture, the switching occurs after the LNA and before the PA; consequently, the performance impact of the switches is diminished. As an

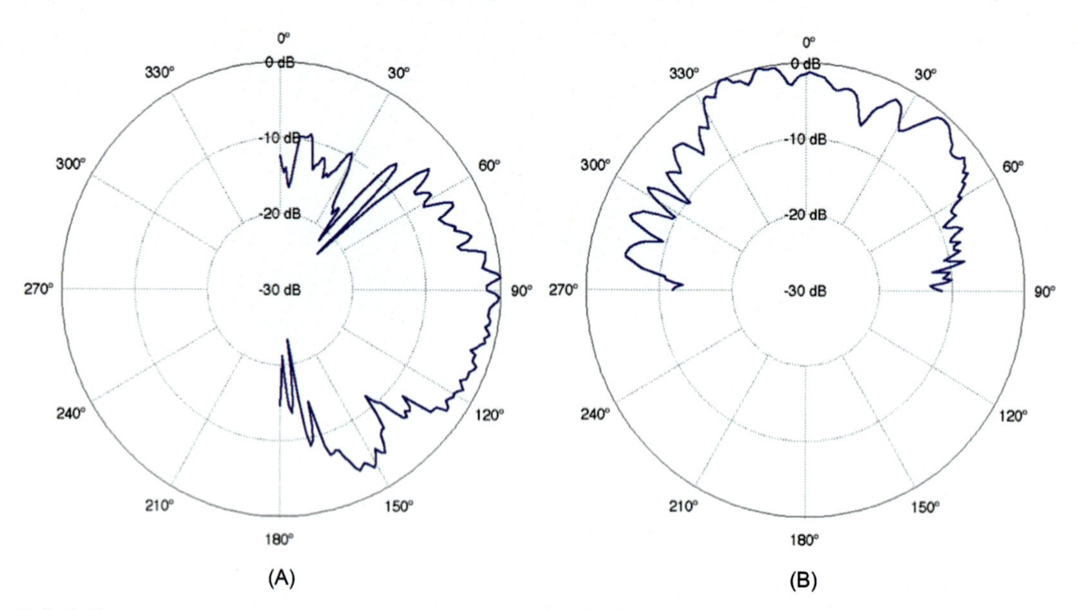


**FIGURE 8.10** Normalized radiation patterns in 5G transmitter mode measurement: (A) end-fire direction only; (B) normal direction only. *Source: From Gu, X., Liu, D., Baks, C., Sadhu, B., & Valdes-Garcia, A. (2015). A multi-layer organic package with four integrated 60 GHz antennas enabling broadside and end-fire radiation for portable communication devices. In* Proceedings - Electronic components and technology conference *(Vol. 2015, pp. 1005–1009). Institute of Electrical and Electronics Engineers Inc. https://doi.org/10.1109/ECTC.2015.7159718.*

example, for an on-chip switch with a typical loss of 2 dB at 60 GHz, the noise figure (NF) of the LNA in this work (intrinsic measured NF = 3.3 dB, gain = 21 dB) is degraded only by 0.01 dB, as compared to a potential 2 dB impact had the switch been at the antenna. Correspondingly, for the TX, the impact on output power is negligible with the proposed switching strategy at the input of the PA compared to a potential 2 dB impact had the switch been placed at the antenna (output of the PA). These savings come at the cost of an area penalty (due to the additional LNA and PA) of 13%.

The details of the TX, RX, and LO circuits are shown in Figs. 8.12, 8.13, and 8.14. Interested readers are referred to the following publications for further details of the circuits: overall transceiver (Sadhu et al., 2016, 2020); PA base circuit block (Ogunnika, 2016), LNA base circuit block (Plouchart et al., 2015), PLL (Ferriss et al., 2012), and voltage controlled oscillator (VCO) (Sadhu, 2012; Sadhu et al., 2013). Although a mixer based doubler (Sarmah et al., 2014) was used in this work, a harmonic extraction based doubler such as the one described in Sadhu et al. (2014); Sadhu et al. (2015) was also successfully used in a variant of the IC and can enable an even lower power and area solution. The 60-GHz IC also includes on-chip mm-wave test and self-healing infrastructure using a method of indirect sensing (Sun et al., 2014) where easy-to-measure circuit operating points (such as DC voltages) are used to estimate more difficult-to-measure circuit performance metrics (such as NF and phase noise) using an on-chip micro-controller unit (MCU) (Plouchart et al., 2014). An integrated 8051 MCU that can use data from >20 on-chip sensors digitized

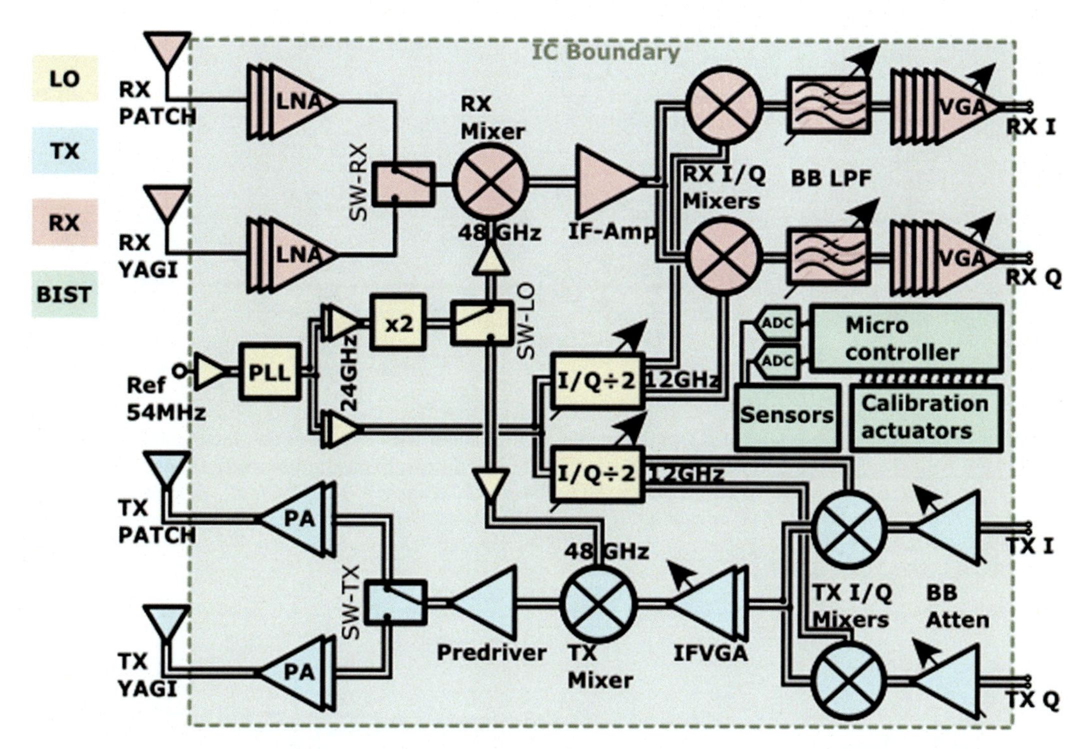


FIGURE 8.11 Integrated circuit block diagram of the switched-beam TRX architecture. *Source: From Sadhu, B., Valdes-Garcia, A., Plouchart, J. O., Ainspan, H., Gupta, A. K., Ferriss, M., Yeck, M., Sanduleanu, M., Gu, X., Baks, C. W., Liu, D., & Friedman, D. (2020). A 250-mW 60-GHz CMOS transceiver SoC integrated with a four-element AiP providing broad angular link coverage. IEEE Journal of Solid-State Circuits, 55(6), 1516–1529. https://doi.org/10.1109/JSSC.2019.2943918.*

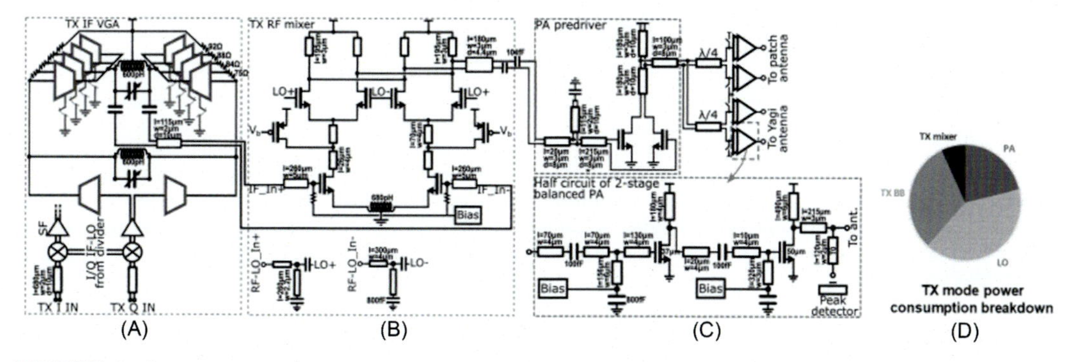


FIGURE 8.12 Schematic details of 5G transmitter (TX) path circuits: (A) RF up-conversion mixer, (B) RF chain with details of the power amplifier (PA) pre-driver, (C) two-stage PA schematic, (D) simulated power consumption breakdown in TX mode. *Source: From Sadhu, B., Valdes-Garcia, A., Plouchart, J. O., Ainspan, H., Gupta, A. K., Ferriss, M., Yeck, M., Sanduleanu, M., Gu, X., Baks, C. W., Liu, D., & Friedman, D. (2020). A 250-mW 60-GHz CMOS transceiver SoC integrated with a four-element AiP providing broad angular link coverage. IEEE Journal of Solid-State Circuits, 55(6), 1516–1529. https://doi.org/10.1109/JSSC.2019.2943918.*

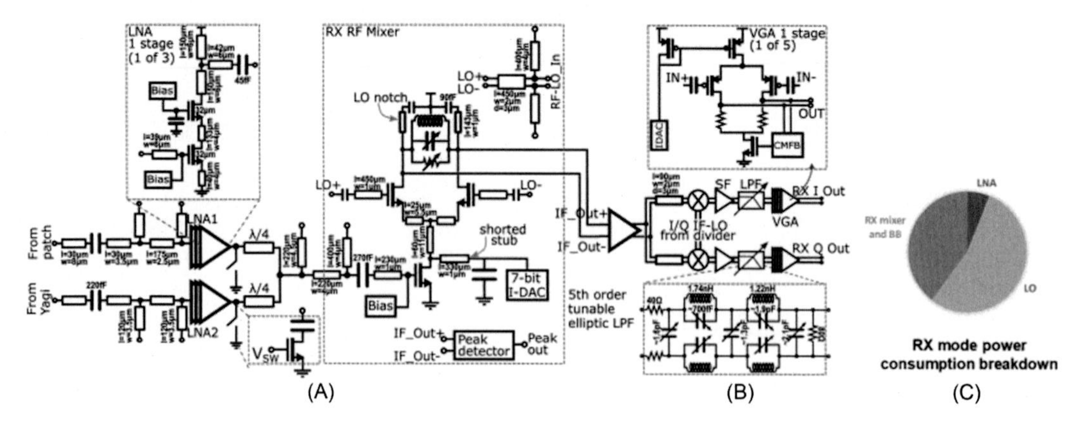


FIGURE 8.13 Schematic details of RX path circuits: (A) RF chain with details of RF down-conversion mixer, (B) IF chain with details of fifth order elliptic filter, (C) simulated power consumption breakdown in RX mode. *Source: From Sadhu, B., Valdes-Garcia, A., Plouchart, J. O., Ainspan, H., Gupta, A. K., Ferriss, M., Yeck, M., Sanduleanu, M., Gu, X., Baks, C. W., Liu, D., & Friedman, D. (2020). A 250-mW 60-GHz CMOS transceiver SoC integrated with a four-element AiP providing broad angular link coverage. IEEE Journal of Solid-State Circuits, 55(6), 1516–1529. https://doi.org/10.1109/JSSC.2019.2943918.*

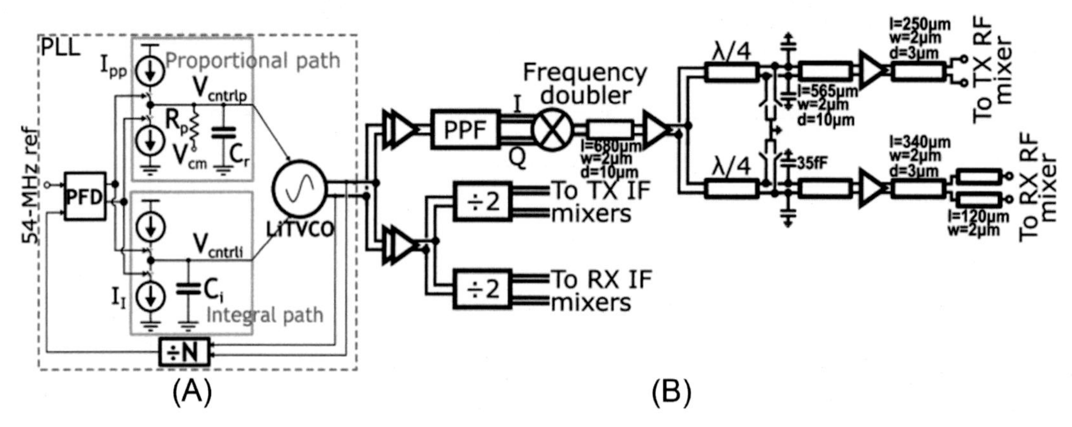


FIGURE 8.14 Schematic details of LO path circuits: (A) PLL architecture, (B) LO distribution chain. *Source: From Sadhu, B., Valdes-Garcia, A., Plouchart, J. O., Ainspan, H., Gupta, A. K., Ferriss, M., Yeck, M., Sanduleanu, M., Gu, X., Baks, C. W., Liu, D., & Friedman, D. (2020). A 250-mW 60-GHz CMOS transceiver SoC integrated with a four-element AiP providing broad angular link coverage. IEEE Journal of Solid-State Circuits, 55(6), 1516–1529. https://doi.org/10.1109/JSSC.2019.2943918.*

using on-chip SAR analog to digital converters (Sadhu et al., 2020) to estimate mm-wave performance metrics is utilized. Metrics that can be measured and improved include LNA NF (Plouchart et al., 2014), PLL bandwidth control (Ferriss et al., 2013), VCO phase noise (Sadhu et al., 2013), and local oscillator in-phase/quadrature-phase accuracy (Sadhu et al., 2020).

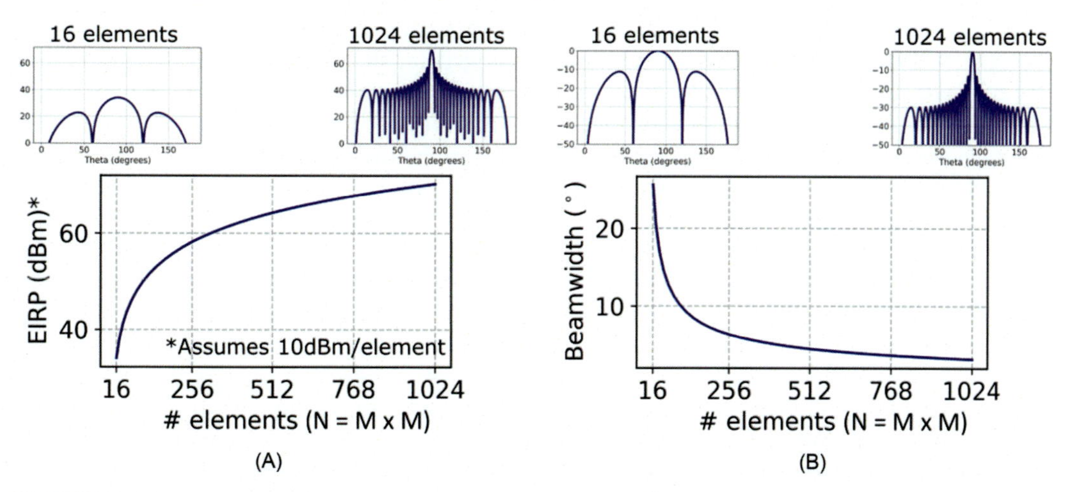


**FIGURE 8.15** (A) Effective isotropic radiated power grows as $N^2$ while (B) beamwidth reduces as $N$ versus number of phased array elements in a square array, $N = M \times M$, with an antenna spacing of = /2. *Source: From Sadhu, B., Gu, X., & Valdes-Garcia, A. (2019). The more (Antennas), the merrier: A survey of silicon-based mm-wave phased arrays using multi-IC scaling. IEEE Microwave Magazine, 20(12), 32–50. https://doi.org/10.1109/MMM.2019.2941632.*

## 8.3 Scalable phased arrays for base-station applications

Now that we have discussed the considerations for handset applications, we move on to design considerations for mm-wave 5G base-station applications. While some considerations such as co-packaging and power efficiency are common to both applications, many other challenges are unique to base-station designs.

To increase the communication distance, it is desirable to use a large number of antennas. Using many antennas reduces free-space path loss and increases communication range. For example, as shown in Fig. 8.15A, there is a large ($N^2$) improvement in the EIRP in a phased array TX containing N-antennas—N times higher power is focused into an interference pattern with an N times smaller beamwidth. Moreover, while the EIRP grows by $N^2$, the phased array has N elements and consumes only N times the power. Effectively, a phased array provides an efficient way to combine power in space. In a phased array RX, the signal to noise ratio (SNR) also improves as N by capturing signal from an aperture that is N times larger. The improved EIRP and SNR of large phased arrays effectively translates to an improvement in link range.

For example, while a typical 64-element 94-GHz silicon phased array can cover only tens of meters, a 1024-element silicon phased array can form a link at >10 km (Gu, Valdes-Garcia, et al., 2015).[1] While the narrow beamwidth of phased arrays can be a

[1] The calculation is based on the following assumptions: EIRP per element = 2 dBm, QPSK with Eb = N0 = 6 dB, link margin = 5 dB, data rate = 1 Gb/s, bandwidth = 800 MHz, LDPC code (1369, 1260) with code rate = 0.92, and BER = 1e-7 with 5 dB implementation loss and 0.4 dB/km atmospheric loss.

handicap when trying to point two beams at each other, it can provide spatial filtering. As seen in Fig. 8.15B, as the number of antennas grows, the array provides a proportionally focused beam characterized by a narrow main lobe. Moreover, while the first sidelobe has similar power, it moves closer to the desired pointing direction, and the sidelobe power in undesired directions is now much lower (Sadhu, 2019). The narrow beamwidth for large arrays requires the base stations to support a larger number of beams to support users in different directions. For example, a base station needs to support ~2000 beams for a 256 element phased array, and ~8000 beams for a 1024 element phased array to support all users over a $\pm 60$ span within a 1 degrees beamwidth. Moreover, 5G base-station arrays need to support fast beam switching among these beams to not waste precious network access time (Sadhu, 2022). Given the space and power consumption limitations in the typical handset, network designers usually put the onus on the base station to enable longer communication range, resulting in having many antennas included in base-station designs. Unfortunately, individual IC chips are limited in size due to foundry yield considerations, and can only contain a limited number of phased array signal chains. The limitation on the number of signal chains in silicon-based phased array ICs is overcome using multi-IC and multi-package scaling, first demonstrated at mm-wave in (Gu et al., 2014; Gu, Valdes-Garcia, et al., 2015; Valdes-Garcia et al., 2013). A single package can be designed to contain multiple identical silicon-based phased array ICs to create a phased array containing many more elements than could be fit in a single IC, as shown in Fig. 8.16. Moreover, the single package can be designed such that multiple such packages can be tiled to create an even larger phased array, as shown in Fig. 8.17 (Sadhu et al., 2019; Valdes-Garcia, Sadhu, Gu, Plouchart, et al., 2018; Valdes-Garcia, Sadhu, Gu, Tousi, et al., 2018). The general strategy for creating large phased arrays by combining repeatable circuit and/or antenna units is called phased array scaling.

The implementation of scaling shown in Fig. 8.17 is by far the most popular among multi-IC scaled silicon-based phased arrays (Dunworth, Homayoun, et al., 2018; Gu et al., 2014; Sadhu, 2022; Sadhu et al., 2017a,b; Shahramian et al., 2019; Yin et al., 2020). The design of large phased array antennas comes with its own challenges. The key challenges and popular solutions to these challenges are discussed in the next section.

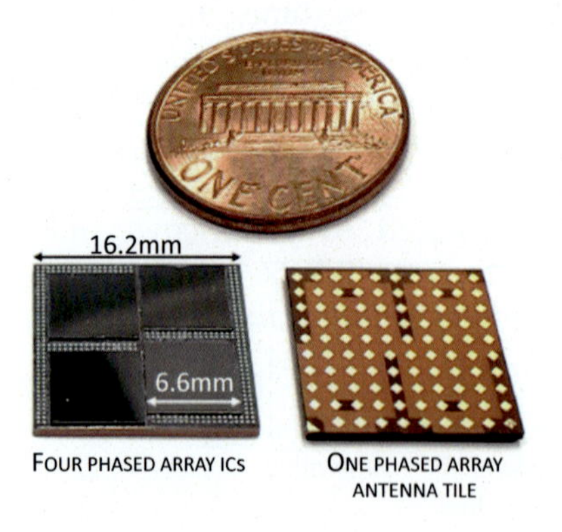


**FIGURE 8.16** The first silicon-based multi-integrated circuit phased array tile (left: bottom view; right: top view). *Source: From Sadhu, B., Gu, X., & Valdes-Garcia, A. (2019). The more (Antennas), the merrier: A survey of silicon-based mm-wave phased arrays using multi-IC scaling. IEEE Microwave Magazine, 20(12), 32–50. https://doi.org/10.1109/MMM.2019.2941632.*

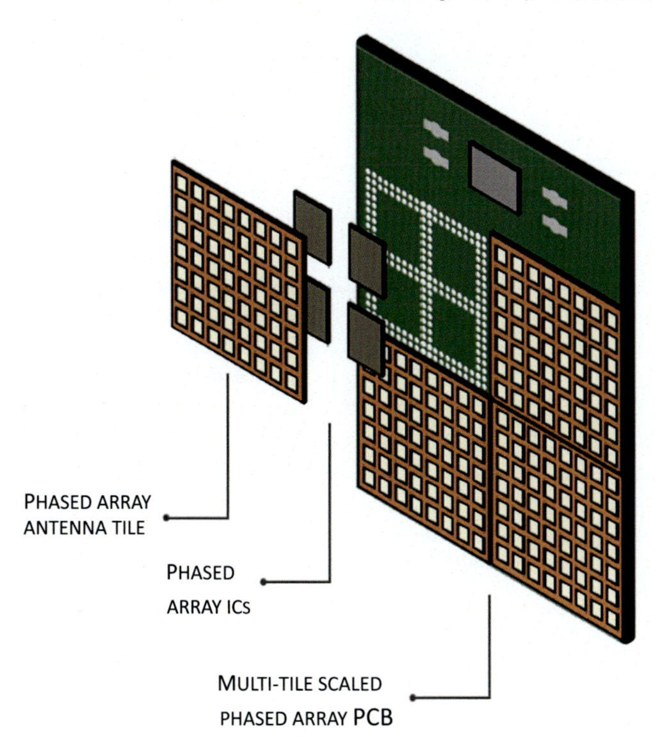


**FIGURE 8.17** Concept of phased array scaling using multi-integrated circuit phased array tiles, and multi-tile phased arrays. *Source: From Sadhu, B., Gu, X., & Valdes-Garcia, A. (2019). The more (Antennas), the merrier: A survey of silicon-based mm-wave phased arrays using multi-IC scaling. IEEE Microwave Magazine, 20(12), 32–50. https://doi.org/10.1109/MMM.2019.2941632.*

## 8.3.1 Scaled phased array design considerations and trade-offs

This section discusses some of the considerations, challenges, and trade-offs associated with phased array antenna design for 5G base-station applications.

### *8.3.1.1 Functional partitioning among integrated circuits*

Beamforming and beamsteering functions can be performed at different nodes in the radio transceiver chain: at RF, LO, intermediate frequency (IF), or even in digital (Hajimiri et al., 2005). Large phased arrays used in 5G base stations typically contain a large number of elements; as a result, the RF beamforming architecture, which minimizes the total amount of circuit hardware, is the most common choice. The circuit functions in an RF beamforming phased array are (1) RF phase shifting and combining/splitting, (2) frequency conversion, (3) analog to digital conversion, and (4) digital signal processing. These functions can be partitioned among multiple ICs. Such partitioning allows a modular architecture where each function can use the most appropriate process technology. Two partitions commonly adopted for 5G base stations are shown in Figs. 8.18 and 8.19. The first approach, shown in Fig. 8.18, includes a frequency conversion function in the front-end ICs that interface with the antenna. This partition results in the use of lower frequency signal traces on the package, improving signal integrity, and thus allowing the use of lower cost packaging substrates. Moreover, because the signal traces are carrying lower frequency signals, any potential coupling between the IC-to-IC signal connections and the

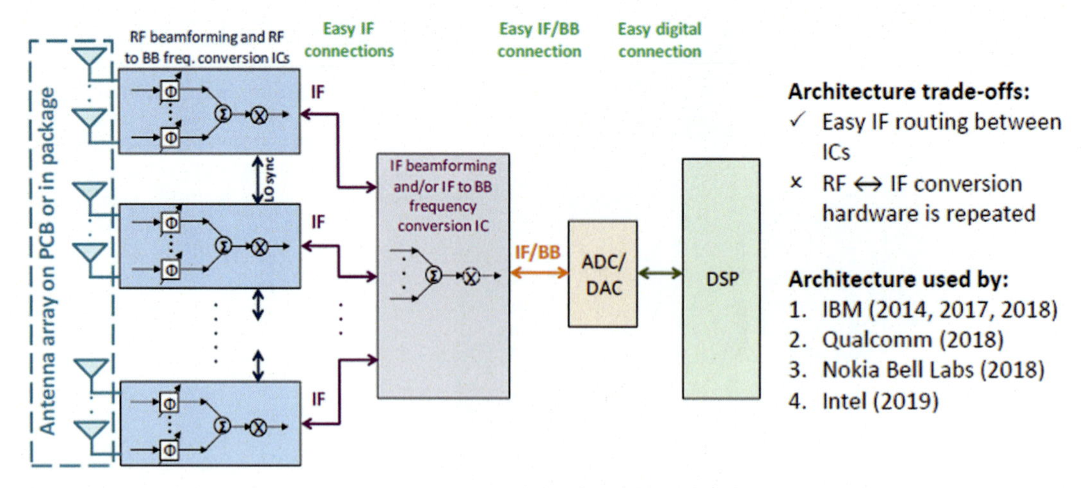


**FIGURE 8.18** An approach for functional partitioning among integrated circuits (IC) that includes frequency conversion in the front-end ICs. *Source: From Sadhu, B., Gu, X., & Valdes-Garcia, A. (2019). The more (Antennas), the merrier: A survey of silicon-based mm-wave phased arrays using multi-IC scaling. IEEE Microwave Magazine, 20(12), 32–50. https://doi.org/10.1109/MMM.2019.2941632.*

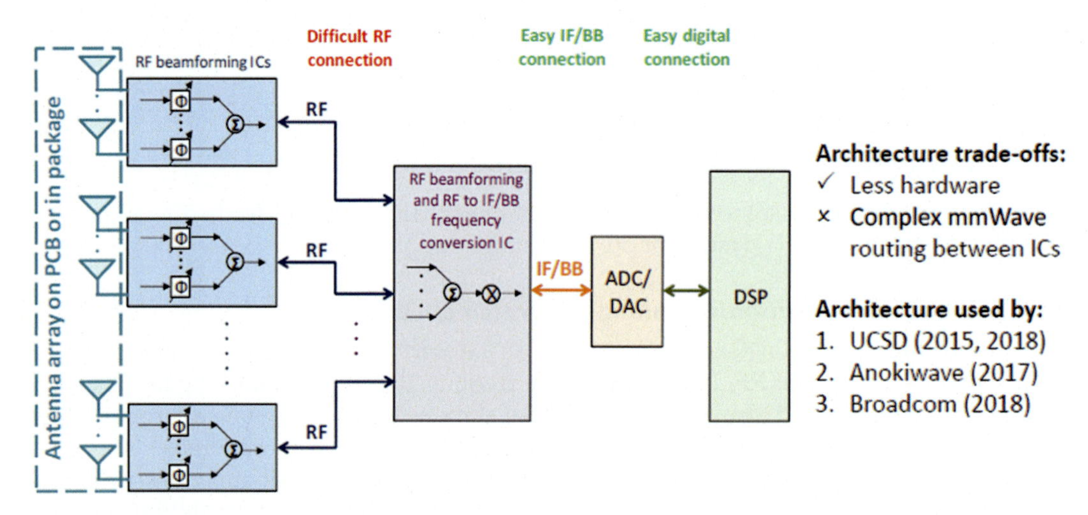


**FIGURE 8.19** An approach for functional partitioning among integrated circuits (IC) using a common frequency conversion IC for many RF-only front-end ICs. *Source: From Sadhu, B., Gu, X., & Valdes-Garcia, A. (2019). The more (Antennas), the merrier: A survey of silicon-based mm-wave phased arrays using multi-IC scaling. IEEE Microwave Magazine, 20(12), 32–50. https://doi.org/10.1109/MMM.2019.2941632.*

IC-to-antenna feed-line connections is less problematic than if higher frequency signals were required. However, such an approach has the drawback of increasing the amount of hardware by demanding that the frequency conversion circuitry be instantiated multiple times. The second approach, shown in Fig. 8.19, performs only the beamforming function in the front-end ICs. A separate frequency conversion IC is used to perform the frequency translation function for all the front-end ICs. This approach eliminates frequency conversion repetition, but demands the use of high-frequency connections among the ICs, which might increase signal losses and packaging substrate costs.

#### *8.3.1.2 Antenna array implementation*

On-wafer, In-package, on-PCB: Antenna arrays contain antennas on a fixed pitch, usually $\lambda/2$. At 5G mm-wave frequencies, this pitch corresponds to a few millimeters; as a result, it is usually necessary to integrate all the antennas into a single substrate. Moreover, to reduce the signal line length (and corresponding loss) between the IC and the antennas, at least the front-end ICs are often assembled on the same substrate. Different substrate materials can be used for this purpose; example materials appropriate for this application are shown in Fig. 8.20. The first, and usually lowest cost option, as shown in Fig. 8.20A, utilizes a PCB type material (Kibaroglu et al., 2018).

The low price tag does come with compromises, though the manufacturing tolerances and minimum feature sizes for PCB materials are large, helping to make it difficult to design high-efficiency wideband antennas in this process technology. To alleviate some of these challenges (Watanabe et al., 2021), a more expensive packaging substrate can be used, as shown in Fig. 8.20B. The packaging substrate can be selected depending on the requirements, and can range from low-temperature co-fired ceramic (LTCC) to glass, high density interconnect (HDI), and organic buildup substrate. The ICs are attached to the AiP (Zhang & Liu, 2009) using flip-chip technology; the AiP is then assembled onto a second-level PCB through BGAs (Dunworth, Ku, et al., 2018; Gu et al., 2019). Two examples are discussed below to describe the importance of co-design among package, RFIC and antenna elements. The first example is with respect to RFIC-package bump joint interface optimization. Fig. 8.21 (Kam et al., 2010) illustrates a three-dimensional (3D) EM model of a flip-chip bump interface on a 60-GHz phased array module. The model parameterizes the geometries of the bump joint between the RFIC chip and multilayered organic laminate package. The effects of transmission line width, pad size, pitch between adjacent signal and ground bumps, bump height, and bump diameter are studied within their

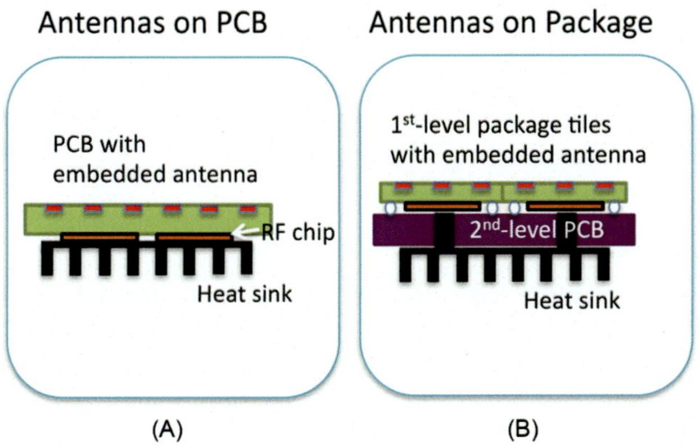

Antennas on PCB
PCB with embedded antenna
RF chip
Heat sink
(A)

Antennas on Package
1st-level package tiles with embedded antenna
2nd-level PCB
Heat sink
(B)

FIGURE 8.20 Antenna and package options for silicon-based scaled phased arrays. *Source: From Sadhu, B., Gu, X., & Valdes-Garcia, A. (2019). The more (Antennas), the merrier: A survey of silicon-based mm-wave phased arrays using multi-IC scaling. IEEE Microwave Magazine, 20(12), 32–50. https://doi.org/10.1109/MMM.2019.2941632.*

| Parameter | Best result | Worst result |
|---|---|---|
| Insertion Loss (dB) | 0.34 | 2.43 |
| T-line width (um) | 125 | 75 |
| Pad size (um) | 70 | 100 |
| Pitch (um) | 250 | 150 |
| Bump height (um) | 90 | 90 |
| Bump diameter (um) | 60 | 120 |

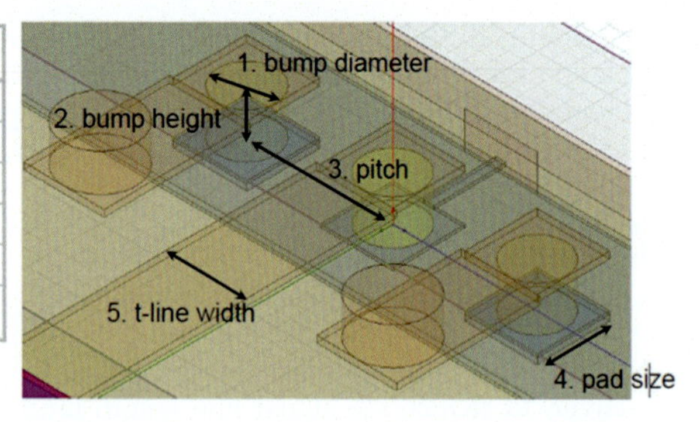


**FIGURE 8.21** Modeling and optimization of a C4 flip-chip bump interface for 60 GHz front ends (FEs). *Source: From Gu, X., Liu, D., & Sadhu, B. (2021). Packaging and antenna integration for silicon-based millimeter-wave phased arrays: 5G and beyond. IEEE Journal of Microwaves, 1(1), 123–134. https://doi.org/10.1109/jmw.2020.3032891.*

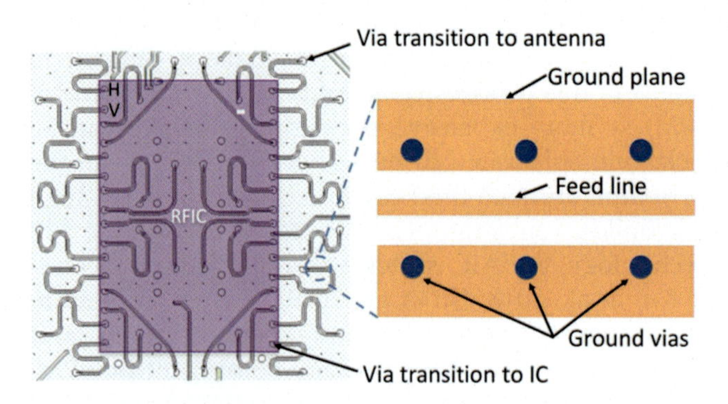


**FIGURE 8.22** Example of a symmetrical 28-GHz antenna feed-line fanout pattern inside the package. *Source: From Gu, X., Liu, D., Baks, C., Tageman, O., Sadhu, B., Hallin, J., Rexberg, L., Parida, P., Kwark, Y., & Valdes-Garcia, A. (2019). IEEE Transactions on Microwave Theory and Techniques, 67(7), 2975–2984. https://doi.org/10.1109/TMTT.2019.2912819.*

manufacturing tolerance range. The modeling results show that the insertion loss can be improved by over 2 dB by optimizing the transition.

Fig. 8.22 illustrates a second example of an antenna feed-line fan-out pattern from the front end (FE) positions of a given RFIC inside a 28-GHz phased array antenna module (Gu et al., 2019). It is important to length match all the antenna feed lines so that they have equal loss and phase responses across frequencies, thus enabling simplification or even elimination of front-end array calibration. In this particular case, the feed lines are implemented based on a stripline configuration. There is an additional ground plane on the same signal layer that is connected to the top and bottom grounds to further reduce the coupling between the RF signals.

#### *8.3.1.3 Element-to-element matching and calibration*

Phased arrays operate (ideally) on the principle that all antenna elements are identical and are configured using the application of phase and amplitude controls. In-field calibration of phased arrays remains an expensive endeavor; as a result, it is very helpful to design phased arrays that require minimal to no calibration. Calibration-free operation can be enabled if (1) all antenna elements are matched—antenna elements have similar directivity, radiation pattern, and feed-line loss and phase characteristics; (2) phase and gain control circuits provide accurate phase and gain; and (3) implemented phase and gain controls are orthogonal. For large phased arrays used in base stations, achieving good matching among elements and high accuracy in phase and gain control can be challenging. Both antenna routing and IC to IC routing need to be matched. Additionally, it is important to reduce antenna to antenna coupling. Coupling among antennas is minimized using various techniques including the use of dummy antennas to provide a uniform environment for all antennas (Dunworth, Ku, et al., 2018; Sadhu et al., 2017a,b), using ground rings and ground via fences (Liu et al., 2017), and using an air cavity (Gu et al., 2019), etc. Accurate and reliable phase control (for beam steering) and gain control (for beam tapering) are also key to controlling phased array beams (Balanis, 2016). Moreover, orthogonality between phase and gain control functions is also key. In other words, as the phase per element is being controlled, the gain for that element should remain constant so that the beam shape is maintained during beam steering. Similarly, when one needs to shape the beam using gain control per element, the phase per element should remain constant. As a result, it is necessary to implement gain invariant phase control and phase invariant gain control per element (Sadhu et al., 2017a,b). Combining antenna to antenna matching, accurate phase and gain control, and maintaining orthogonality between phase and gain functions, work such as (Sadhu et al., 2017a,b) has been able to demonstrate completely calibration-free beamforming and beam-steering functions. When calibration becomes necessary due to imperfections in any of the above-mentioned features, built-in self-calibration is usually desirable to keep calibration costs low (Sadhu et al., 2020). A generous use of on-chip sensors, and the use of algorithms to establish sensor to performance correlation, can be helpful to estimate difficult-to-measure parameters at mm-wave frequencies (Sadhu et al., 2020). Over-the-air loopback calibration from TX to RX is used as well, as shown in Shahramian et al. (2019).

#### 8.3.1.4 Thermal management

A typical silicon-based AiP phased array module incorporates one or multiple RFICs where the chips are mounted on a package substrate on the surface opposite the antenna aperture side. The packaging provides not only electrical connections to one or multiple RFICs and their associated antenna array, but also a thermal structure to remove the heat generated by the RFICs. It is often required to attach a heat spreader and a heat sink which are thermally coupled to the back side of the chip with thermal interface materials. Fig. 8.23 shows a thermal simulation and thermal measurement example for a 28-GHz scaled phased array. In this example, heat removal from the ICs is performed using heat sinks to conduct most of the heat out through holes in the PCB. Fig. 8.24 plots typical chip power density numbers in recent silicon-based phased array RFIC demonstrations. Such power density numbers, for example, in a range between 1 and 22 $W/cm_2$, are relatively low compared to densities associated with high-performance computing chips (Wei, 2008). In such configurations, the efficient heat transfer to the ambient air can be achieved through either natural convection or forced air cooling. Natural convection cooling occurs where the cooling air motion is caused by the density differences in the air occurring naturally due to temperature gradients in proximity to the heat sink. Forced air cooling works

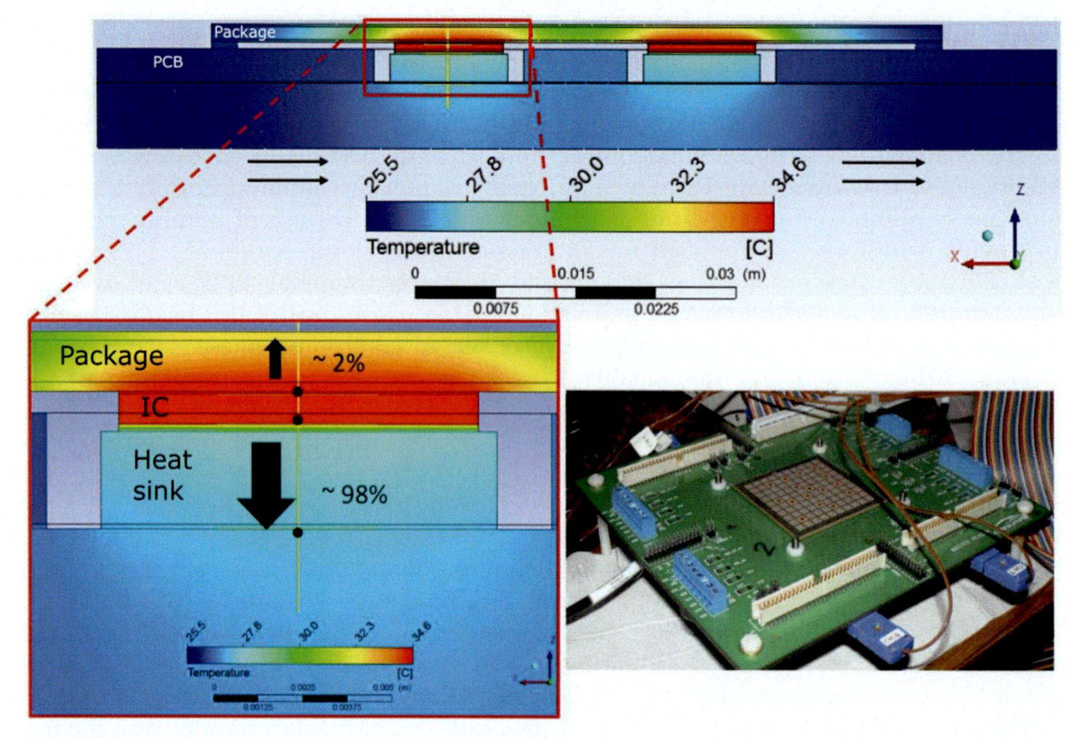


**FIGURE 8.23** An example thermal simulation and thermal measurement setup for a scaled phased array showing heat removal from the integrated circuits using heat sinks to conduct heat out through holes in the printed circuit board. *Source: From Sadhu, B., Gu, X., & Valdes-Garcia, A. (2019). The more (Antennas), the merrier: A survey of silicon-based mm-wave phased arrays using multi-IC scaling. IEEE Microwave Magazine, 20(12), 32–50. https://doi.org/10.1109/MMM.2019.2941632.*

in a similar fashion except that fans are required to force the air at desired velocities across the heat sink. The location and structure of a cooling fan can significantly influence thermal behavior.

#### *8.3.1.5 Electromagnetic simulation of antenna arrays*

Designing and verifying scaled mm-wave phased array antennas requires numerical EM simulation in every step of the process. To cope with a wide range of geometrical complexities, simulation and modeling can be performed at different levels, including unit cell level (i.e., a single antenna element), antenna package (tile) level, and array level (i.e., multiple packages). The simulations of individual antenna elements are used to explore and optimize the antenna design as a unit cell based on the package/PCB/wafer substrate material properties and manufacturing process rules. The model is also used to determine a layer stack-up to support the necessary routing (e.g., RF, IF, LO, and digital signals) required by the overall system architecture. Furthermore, interconnect effects on antenna feed signals such as trace loss and vertical via transitions are included in the model. Key antenna parameters obtained from element-level simulation

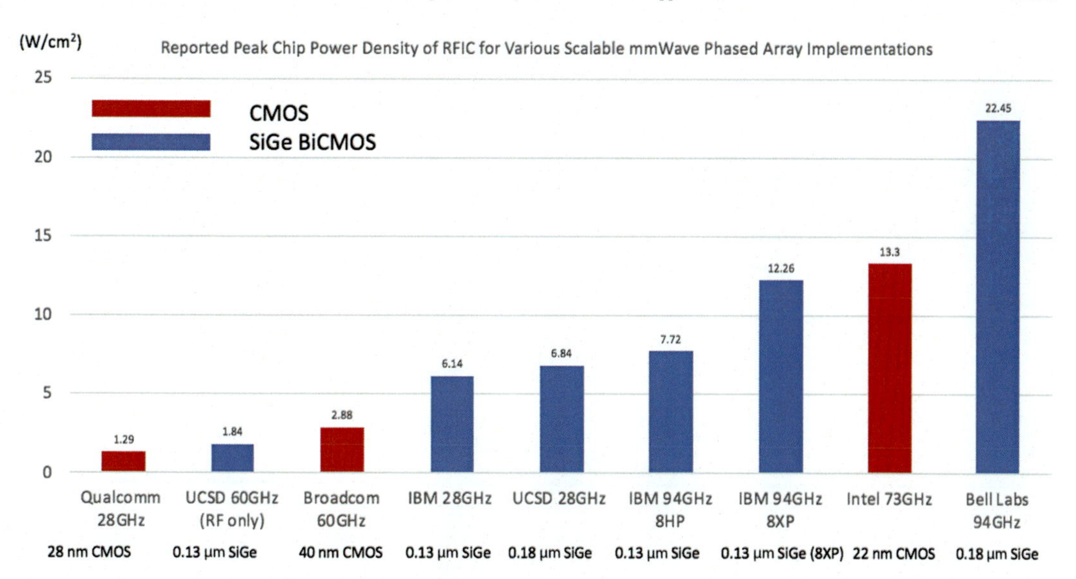


**FIGURE 8.24** A comparison of integrated circuit power density in silicon-based scaled phased arrays. *Source: From Sadhu, B., Gu, X., & Valdes-Garcia, A. (2019). The more (Antennas), the merrier: A survey of silicon-based mm-wave phased arrays using multi-IC scaling. IEEE Microwave Magazine, 20(12), 32–50. https://doi.org/10.1109/MMM.2019.2941632.*

include frequency matching bandwidth, gain across frequencies, antenna efficiency, and radiation patterns. Fig. 8.25A and B illustrates the breakdown of an antenna element model [V-band (Sowlati et al., 2018) and Ka-band (Liu et al., 2017), respectively]. The models include the entire feed signal path on the package from the bottom solder bump pads to the top radiator. Additional structures such as the auxiliary patches, grounded via fences, and grounded metal rings used to improve the radiation pattern by suppressing antenna-to-antenna coupling are included in the models as well. Other similar examples of such unit cell models have been found for 60 GHz LTCC antenna array (Emami et al., 2011), 28-GHz antenna prototyping in organic buildup packages (Gu et al., 2019), and W-band antennas in PCB (Shahramian et al., 2018). EM simulation of the entire antenna array tile package is often used for post-layout verification and cross-check with measurement data (e.g., far-field radiation patterns). This kind of simulation is significantly more computationally intensive than the unit-cell model. Fig. 8.25C–E illustrates several antenna array EM model examples for V-band (Kam et al., 2011; Natarajan et al., 2011), Ka-band (Gu et al., 2017), and W-band (Shahramian et al., 2018) applications, respectively.

For larger phase array configurations consisting multiple tile packages, it would be extremely challenging and even computationally prohibitive to achieve a direct EM solution if all design features are taken into account. The domain decomposition method provides an effective means to simulate large finite array apertures with high accuracy (Lee, 2017; Vouvakis et al., 2007). In other cases, the analysis of the antenna array often demands resorting to analytic formulas where antennas are treated like isotropic radiators (Gu, Valdes-Garcia, et al., 2015). Another numerical approximation that can be made is to replicate the

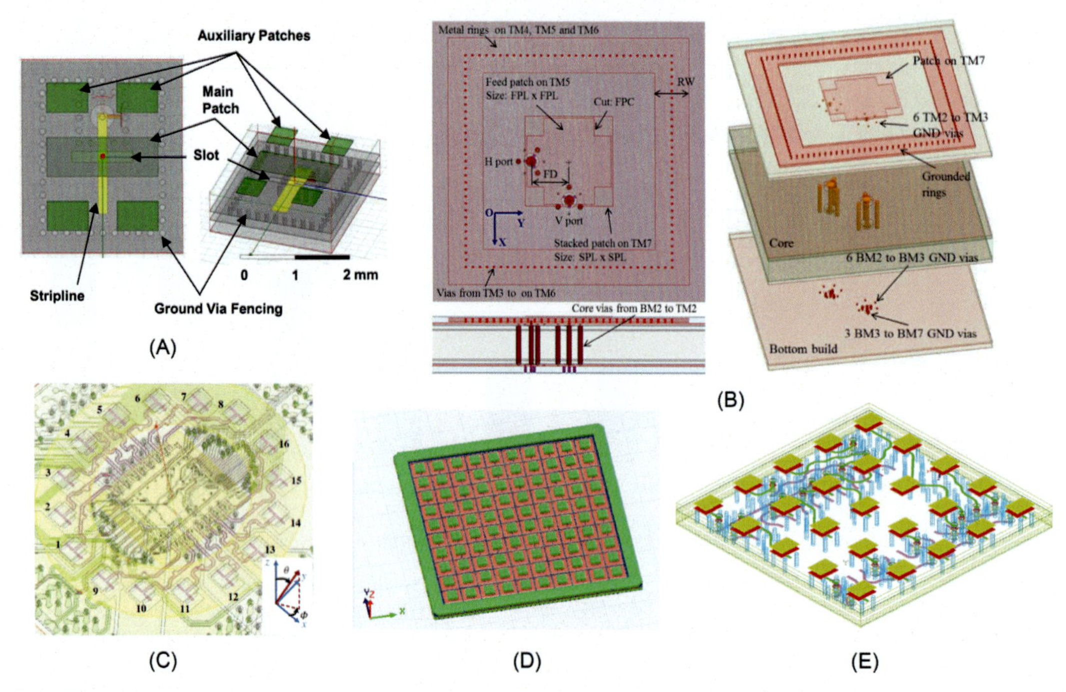


**FIGURE 8.25** Antenna element and array design and simulation examples. *Source: From Sadhu, B., Gu, X., & Valdes-Garcia, A. (2019). The more (Antennas), the merrier: A survey of silicon-based mm-wave phased arrays using multi-IC scaling. IEEE Microwave Magazine, 20(12), 32–50. https://doi.org/10.1109/MMM.2019.2941632.*

tile package EM solution as a unit macrocell in order to form the larger array. Such a simulation approach is used in Sowlati et al. (2018) to analyze the performance of a 60-GHz 288 antenna array.

## 8.4 Conclusions

This chapter discussed some of the key considerations for mm-wave antenna design as well as associated package and IC challenges in the context of 5G mm-wave. For handsets, robust operation at mm-wave often requires multi-antenna designs that provide larger angular coverage through antenna diversity. On the other hand, base stations need to support long distances at mm-wave and often utilize antenna arrays with a large number of antennas. As a result, both handset and base-station designs bring their own set of challenges. Creative engineering solutions have allowed us to overcome many of these challenges and made 5G mm-wave a reality, providing ultra-high data speeds on our mobile networks today. Moreover, mm-wave designs are expected to evolve further and support even higher data rates for 6G communications using a combination of higher frequencies (Lee, Plouchart, et al., 2018; Naviasky et al., 2021; Shahramian et al., 2019) and algorithm/software enhancements to aid antennas (Domae et al., 2021; Gu, Liu, et al., 2021; Guan et al., 2021; Klautau et al., 2018; Paidimarri & Sadhu, 2020; Sadhu et al., 2018; Valdes-Garcia et al., 2020).

## References

Ali, M., Watanabe, A. O., Lin, T. H., Okamoto, D., Pulugurtha, M. R., Tentzeris, M. M., & Tummala, R. R. (2020). Package-integrated, wideband power dividing networks and antenna arrays for 28-GHz 5G new radio bands. *IEEE Transactions on Components, Packaging and Manufacturing Technology, 10*(9), 1515–1523. Available from https://doi.org/10.1109/TCPMT.2020.3013725, http://cpmt.ieee.org/transactions-on-cpmt.html.

Balanis, C. A. (2016). *Antenna theory: analysis and design*. John Wiley & Sons, Inc.

Buey, C., Ferrero, F., Lizzi, L., Ratajczak, P., Benoit, Y., & Brochier, L. (2016). Investigation of hand effect on a handheld terminal at 11GHz. In *10th European conference on antennas and propagation, EuCAP 2016*. France: Institute of Electrical and Electronics Engineers Inc. Available from https://doi.org/10.1109/EuCAP.2016.7481536.

Chang, W. S., Yang, C. F., Chang, C. K., Liao, W. J., Cho, L., & Chen, W. S. (2016). Pattern reconfigurable millimeter-wave antenna design for 5G handset applications. In *10th European conference on antennas and propagation, EuCAP 2016*. Taiwan: Institute of Electrical and Electronics Engineers Inc. Available from https://doi.org/10.1109/EuCAP.2016.7481321.

Chen, Z., & Zhang, Y. P. (2013). FR4 PCB grid array antenna for millimeter-wave 5G mobile communications. *IEEE MTT-S international microwave workshop series on RF and wireless technologies for biomedical and healthcare applications, IMWS-BIO 2013 – Proceedings*. Singapore: IEEE Computer Society. Available from https://doi.org/10.1109/IMWS-BIO.2013.6756214.

Dai, X. W., Wang, Z. Y., Liang, C. H., Chen, X., & Wang, L. T. (2013). Multiband and dual-polarized omnidirectional antenna for 2G/3G/LTE application. *IEEE Antennas and Wireless Propagation Letters, 12*, 1492–1495. Available from https://doi.org/10.1109/LAWP.2013.2289743.

Domae, B. W., Li, R., & Cabric, D. (2021). Machine learning assisted phase-less millimeter-wave beam alignment in multipath channels. *arXiv*. Available from https://arxiv.org.

Dunworth, J., Ku, B. H., Ou, Y. C., Lu, D., Mouat, P., Homayoun, A., Chakraborty, K., Arnett, A., Liu, G., Segoria, T., Lerdworatawee, J., Park, J. W., Park, H. C., Hedayati, H., Tassoudji, A., Douglas, K., & Aparin, V. (2018). 28GHz phased array transceiver in 28nm bulk CMOS for 5G prototype user equipment and base stations. *IEEE MTT-S International Microwave Symposium Digest, 2018*, 1330–1333. Available from https://doi.org/10.1109/MWSYM.2018.8439517, Institute of Electrical and Electronics Engineers Inc. United States.

Dunworth, J. D., Homayoun, A., Ku, B. H., Ou, Y. C., Chakraborty, K., Liu, G., Segoria, T., Lerdworatawee, J., Park, J. W., Park, H. C., Hedayati, H., Lu, D., Monat, P., Douglas, K., & Aparin, V. (2018). A 28GHz Bulk-CMOS dual-polarization phased-array transceiver with 24 channels for 5G user and basestation equipment. In Digest of technical papers - IEEE international solid-state circuits conference (Vol. 61, pp. 70–72). United States: Institute of Electrical and Electronics Engineers Inc. Available from https://doi.org/10.1109/ISSCC.2018.8310188 .

Emami, S., Wiser, R. F., Ali, E., Forbes, M. G., Gordon, M. Q., Guan, X., Lo, S., McElwee, P. T., Parker, J., Tani, J. R., Gilbert, J. M., & Doan, C. H. (2011). A 60GHz CMOS phased-array transceiver pair for multi-Gb/s wireless communications. In *Digest of technical papers - IEEE international solid-state circuits conference* (pp. 164–165). United States: Institute of Electrical and Electronics Engineers Inc. Available from https://doi.org/10.1109/ISSCC.2011.5746265.

Enjiu, R. K. (n.d). Design of 5G mm-wave compatible covers for high end mobile phones NAFEMS World Congress.

Fernandes, F. (n.d). Millimeter-wave antenna implementation for 5G smartphones. Available from https://fenix.tecnico.ulisboa.pt/downloadFile/1689244997258800/RA.pdf.

Ferriss, M., Plouchart, J. O., Natarajan, A., Rylyakov, A., Parker, B., Babakhani, A., Yaldiz, S., Sadhu, B., Valdes-Garcia, A., Tierno, J., & Friedman, D. (2012). An integral path self-calibration scheme for a 20.1–26.7GHz dual-loop PLL in 32nm SOI CMOS. *IEEE Symposium on VLSI Circuits* (pp. 176–177). United States: Digest of Technical Papers. Available from https://doi.org/10.1109/VLSIC.2012.6243847.

Ferriss, M., Plouchart, J. O., Natarajan, A., Rylyakov, A., Parker, B., Tierno, J. A., Babakhani, A., Yaldiz, S., Valdes-Garcia, A., Sadhu, B., & Friedman, D. J. (2013). An integral path self-calibration scheme for a dual-loop PLL. *IEEE Journal of Solid-State Circuits, 48*(4), 996–1008. Available from https://doi.org/10.1109/JSSC.2013.2239114.

Futter, P. W., & Soler, J. (2018). Antenna design for 5g communications. In *IEEE 6th Asia-Pacific conference on antennas and propagation, APCAP 2017 – Proceeding* (pp. 1–3). United States: Institute of Electrical and Electronics Engineers Inc. Available from https://doi.org/10.1109/APCAP.2017.8420649, http://ieeexplore.ieee.org/xpl/mostRecentIssue.jsp?punumber = 8379743.

Gu, X., Liu, D., Baks, C., Valdes-Garcia, A., Parker, B., Islam, M. R., Natarajan, A., & Reynolds, S. K. (2014). A compact 4-chip package with 64 embedded dual-polarization antennas for W-band phased-array transceivers.

In *Proceedings - Electronic components and technology conference* (pp. 1272–1277). United States: Institute of Electrical and Electronics Engineers Inc. Available from https://doi.org/10.1109/ECTC.2014.6897455.

Gu, X., Liu, D., Baks, C., Sadhu, B., & Valdes-Garcia, A. (2015). A multilayer organic package with four integrated 60GHz antennas enabling broadside and end-fire radiation for portable communication devices. In *Proceedings - Electronic components and technology conference* (pp. 2015, pp. 1005–1009). United States: Institute of Electrical and Electronics Engineers Inc. Available from https://doi.org/10.1109/ECTC.2015.7159718.

Gu, X., Valdes-Garcia, A., Natarajan, A., Sadhu, B., Liu, D., & Reynolds, S. K. (2015). W-band scalable phased arrays for imaging and communications. *IEEE Communications Magazine, 53*(4), 196–204. Available from https://doi.org/10.1109/MCOM.2015.7081095.

Gu, X., Liu, D., Baks, C., Tageman, O., Sadhu, B., Hallin, J., Rexberg, L., & Valdes-Garcia, A. (2017). A multilayer organic package with 64 dual-polarized antennas for 28GHz 5G communication. *IEEE MTT-S international microwave symposium digest* (pp. 1899–1901). United States: Institute of Electrical and Electronics Engineers Inc. Available from https://doi.org/10.1109/MWSYM.2017.8059029.

Gu, X., Liu, D., Baks, C., Tageman, O., Sadhu, B., Hallin, J., Rexberg, L., Parida, P., Kwark, Y., & Valdes-Garcia, A. (2019). Development, implementation, and characterization of a 64-element dual-polarized phased-array antenna module for 28-GHz high-speed data communications. *IEEE Transactions on Microwave Theory and Techniques, 67*(7), 2975–2984. Available from https://doi.org/10.1109/TMTT.2019.2912819, http://ieeexplore.ieee.org/xpl/tocresult.jsp?isYear = 2009&isnumber = 4747395&Submit32 = View + Contents.

Gu, X., Liu, D., & Sadhu, B. (2021). Packaging and antenna integration for silicon-based millimeter-wave phased arrays: 5G and beyond. *IEEE Journal of Microwaves, 1*(1), 123–134. Available from https://doi.org/10.1109/jmw.2020.3032891.

Gu, X., Paidimarri, A., Sadhu, B., Baks, C., Lukashov, S., Yeck, M., Kwark, Y., Chen, T., Zussman, G., Seskar, I., & Valdes-Garcia, A. (2021). Development of a compact 28-GHz software-defined phased array for a city-scale wireless research testbed. *IEEE MTT-S International Microwave Symposium Digest, 2021*, 803–806. Available from https://doi.org/10.1109/IMS19712.2021.9574864, Institute of Electrical and Electronics Engineers Inc. United States.

Guan, J., Paidimarri, A., Valdes-Garcia, A., & Sadhu, B. (2021). 3-D imaging using millimeter-wave 5G signal reflections. *IEEE Transactions on Microwave Theory and Techniques, 69*(6), 2936–2948. Available from https://doi.org/10.1109/TMTT.2021.3077896, http://ieeexplore.ieee.org/xpl/tocresult.jsp?isYear = 2009&isnumber = 4747395&Submit32 = View + Contents.

Gustrau, F., & Bahr, A. (2002). W-band investigation of material parameters, SAR distribution, and thermal response in human tissue. *IEEE Transactions on Microwave Theory and Techniques, 50*(10), 2393–2400. Available from https://doi.org/10.1109/TMTT.2002.803445.

Hajimiri, A., Hashemi, H., Natarajan, A., Guan, X., & Komijani, A. (2005). Integrated phased array systems in silicon. *Proceedings of the IEEE, 93*(9), 1637–1655. Available from https://doi.org/10.1109/JPROC.2005.852231.

Hong, W. (2017). Solving the 5G mobile antenna puzzle: Assessing future directions for the 5G mobile antenna paradigm shift. *IEEE Microwave Magazine, 18*(7), 86–102. Available from https://doi.org/10.1109/MMM.2017.2740538.

Hong, W., Baek, K., Geon Kim, Y., Lee, Y., & Kim, B. (2014). MmWave phased-array with hemispheric coverage for 5th generation cellular handsets. *8th European Conference on Antennas and Propagation, EuCAP 2014* (pp. 714–716). South Korea: Institute of Electrical and Electronics Engineers Inc. Available from https://doi.org/10.1109/EuCAP.2014.6901859.

Hong, W., Baek, K. H., & Ko, S. (2017). Millimeter-wave 5G antennas for smartphones: Overview and experimental demonstration. *IEEE Transactions on Antennas and Propagation, 65*(12), 6250–6261. Available from https://doi.org/10.1109/TAP.2017.2740963.

Huang, T. C., Hsu, Y. W., & Lin, Y. C. (2016). End-fire Quasi-Yagi antennas with pattern diversity on LTCC technology for 5G mobile communications. *RFIT 2016 - 2016 IEEE international symposium on radio-frequency integration technology*. Taiwan: Institute of Electrical and Electronics Engineers Inc. Available from https://doi.org/10.1109/RFIT.2016.7578205.

Kam, D. G., Liu, D., Natarajan, A., Reynolds, S., & Floyd, B. A. (2010). Low-cost antenna-in-package solutions for 60-GHz phased-array systems. In *IEEE 19th conference on electrical performance of electronic packaging and systems, EPEPS 2010* (pp. 93–96). United States: IEEE Computer Society. Available from https://doi.org/10.1109/EPEPS.2010.5642554.

Kam, D. G., Liu, D., Natarajan, A., Reynolds, S. K., & Floyd, B. A. (2011). Organic packages with embedded phased-array antennas for 60-GHz wireless chipsets. IEEE Transactions on Components. *Packaging and Manufacturing Technology*, *1*(11), 1806–1814. Available from https://doi.org/10.1109/TCPMT.2011.2169064.

Kibaroglu, K., Sayginer, M., Phelps, T., & Rebeiz, G. M. (2018). A 64-Element 28-GHz phased-array transceiver with 52-dBm EIRP and 8–12-Gb/s 5G link at 300 meters without any calibration. *IEEE Transactions on Microwave Theory and Techniques*, *66*(12), 5796–5811. Available from https://doi.org/10.1109/TMTT.2018.2854174, http://ieeexplore.ieee.org/xpl/tocresult.jsp?isYear = 2009&isnumber = 4747395&Submit32 = View + Contents.

Kim, K. (2016). Invariant imbedding theory of wave propagation in stratified anisotropic media. *URSI International Symposium on Electromagnetic Theory, EMTS 2016* (pp. 269–271). South Korea: Institute of Electrical and Electronics Engineers Inc. Available from https://doi.org/10.1109/URSI-EMTS.2016.7571371.

Klautau, A., Batista, P., Gonzalez-Prelcic, N., Wang, Y., & Heath, R. W. (2018). 5G MIMO data for machine learning: Application to beam-selection using deep learning. *Information Theory and Applications Workshop, ITA 2018*. Brazil: Institute of Electrical and Electronics Engineers Inc. Available from https://doi.org/10.1109/ITA.2018.8503086, http://ieeexplore.ieee.org/xpl/mostRecentIssue.jsp?punumber = 8486613.

Koul, S. K., & Karthikeya, G. S. (2020). Compact antenna designs for future mmWave 5G smart phones. Available from https://www.microwavejournal.com/articles/34931-compact-antenna-designs-for-future-mmwave-5g-smart-phones.

Lee, J. (2017). *Computational electromagnetics: Domain decomposition methods and practical applications*. CRC Press.

Lee, H., Bang, J., Lee, S., & Choi, J. (2018). Phased array antenna with whole-metal-cover for MM-wave 5G mobile phone applications. *IEEE international workshop on electromagnetics: Applications and student innovation competition, iWEM 2018*. South Korea: Institute of Electrical and Electronics Engineers Inc. Available from https://doi.org/10.1109/iWEM.2018.8536690, http://ieeexplore.ieee.org/xpl/mostRecentIssue.jsp?punumber = 8517201.

Lee, W., Plouchart, J. O., Ozdag, C., Aydogan, Y., Yeck, M., Cabuk, A., Kepkep, A., Reynolds, S. K., Apaydin, E., & Valdes-Garcia, A. (2018). Fully integrated 94-GHz dual-polarized TX and RX phased array chipset in SiGe BiCMOS operating up to 105°C. *IEEE Journal of Solid-State Circuits*, *53*(9), 2512–2531. Available from https://doi.org/10.1109/JSSC.2018.2856254, http://ieeexplore.ieee.org/Xplore/home.jsp.

Li, R. L., DeJean, G., Tentzeris, M. M., Papapolymerou, J., & Laskar, J. (2005). Radiation-pattern improvement of patch antennas on a large-size substrate using a compact soft-surface structure and its realization on LTCC multilayer technology. *IEEE Transactions on Antennas and Propagation*, *53*(1 I), 200–208. Available from https://doi.org/10.1109/TAP.2004.840754.

Lin, J. C. (2003). Safety standards for human exposure to radio frequency radiation and their biological rationale. *IEEE Microwave Magazine*, *4*(4), 22–26. Available from https://doi.org/10.1109/MMW.2003.1266063.

Liu, D., & Zhang, Y. (2020). *Antenna-in-package technology and applications antenna-in-package technology and applications* (pp. 1–390). United States: wiley Available from. Available from https://onlinelibrary.wiley.com/doi/book/10.1002/9781119556671.

Liu, D., Gu, X., Baks, C., & Valdes-Garcia, A. (2015). 60GHz antennas in package for portable applications. *IEEE Antennas and Propagation Society, AP-S International Symposium (Digest)*, *2015*, 1536–1537. Available from https://doi.org/10.1109/APS.2015.7305157, Institute of Electrical and Electronics Engineers Inc. United States.

Liu, D., Gu, X., Baks, C. W., & Valdes-Garcia, A. (2017). Antenna-in-package design considerations for Ka-band 5G communication applications. *IEEE Transactions on Antennas and Propagation*, *65*(12), 6372–6379. Available from https://doi.org/10.1109/TAP.2017.2722873.

Liu, P., Syrytsin, I., Odum Nielsen, J., Frolund Pedersen, G., & Zhang, S. (2021). Characterization and modeling of the user blockage for 5G handset antennas. *IEEE Transactions on Instrumentation and Measurement*, *70*, 1–11. Available from https://doi.org/10.1109/tim.2020.3039644.

Martinez-Vazquez, M. (2020). PCB edge antennas for mm-wave communications user equipment. In *IEEE international symposium on antennas and propagation and North American radio science meeting, IEEECONF 2020 – Proceedings* (pp. 577–578). Germany: Institute of Electrical and Electronics Engineers Inc. Available from https://doi.org/10.1109/IEEECONF35879.2020.9329857, http://ieeexplore.ieee.org/xpl/mostRecentIssue.jsp?punumber = 9329358.

Meena, R. K., Avinash Dabhade, M. K., Srivastava, K., & Kanaujia, B. K. (2019). Antenna design for fifth generation (5G) applications. In *URSI Asia-Pacific radio science conference, AP-RASC 2019*. India: Institute of Electrical and Electronics Engineers Inc. Available from https://doi.org/10.23919/URSIAP-RASC.2019.8738212, http://ieeexplore.ieee.org/xpl/mostRecentIssue.jsp?punumber = 8732972.

Natarajan, A., Reynolds, S. K., Tsai, M. D., Nicolson, S. T., Zhan, J. H. C., Kam, D. G., Liu, D., Huang, Y. L. O., Valdes-Garcia, A., & Floyd, B. A. (2011). A fully-integrated 16-element phased-array receiver in SiGe BiCMOS for 60-GHz communications. *IEEE Journal of Solid-State Circuits, United States*, *46*(5), 1059–1075. Available from https://doi.org/10.1109/JSSC.2011.2118110.

Naviasky, E., Iotti, L., Lacaille, G., Nikolic, B., Alon, E., & Niknejad, A. (2021). 14.1A 71-to-86GHz packaged 16-element by 16-beam multi-user beamforming integrated receiver in 28nm CMOS 64. In *Digest of technical papers - IEEE international solid-state circuits conference* (pp. 218–220). United States: Institute of Electrical and Electronics Engineers Inc. Available from https://doi.org/10.1109/ISSCC42613.2021.9365999.

Ogunnika, O. T. (2016). *Millimeter-wave class-E PA design in CMOS*. Elsevier.

Ojaroudiparchin, N., Shen, M., & Pedersen, G. F. (2016). Design of Vivaldi antenna array with end-fire beam steering function for 5G mobile terminals. *23rd Telecommunications Forum, TELFOR 2015* (pp. 587–590). Denmark: Institute of Electrical and Electronics Engineers Inc. Available from https://doi.org/10.1109/TELFOR.2015.7377536.

Paidimarri, A., & Sadhu, B. (2020). Spatio-temporal filtering: Precise beam control using fast beam switching. *Digest of Papers - IEEE Radio Frequency Integrated Circuits Symposium, 2020*, 207–210. Available from https://doi.org/10.1109/RFIC49505.2020.9218275, Institute of Electrical and Electronics Engineers Inc. United States.

Pan, H. K. (2011). Dual-polarized Mm-wave phased array antenna for multi-Gb/s 60GHz communication. *IEEE Antennas and Propagation Society* (pp. 3279–3282). United States: AP-S International Symposium (Digest). Available from https://doi.org/10.1109/APS.2011.5997235.

Parchin, N. O., Shen, M., & Pedersen, G. F. (2016). End-fire phased array 5G antenna design using leaf-shaped bow-tie elements for 28/38GHz MIMO applications. *IEEE International Conference on Ubiquitous Wireless Broadband, ICUWB 2016*. Denmark: Institute of Electrical and Electronics Engineers Inc. Available from https://doi.org/10.1109/ICUWB.2016.7790538.

Park, H. C., Kang, D., Lee, S. M., Park, B., Kim, K., Lee, J., Aoki, Y., Yoon, Y., Lee, S., Lee, D., Kwon, D., Kim, S., Kim, J., Lee, W., Kim, C., Park, S., Park, J., Suh, B., Jang, J., ...Yang, S. G. (2020). A 39GHz-Band CMOS 16-channel phased-array transceiver IC with a companion dual-stream if transceiver IC for 5 G NR base-station applications. In *Digest of technical papers - IEEE international solid-state circuits conference* (Vol. 2020, pp. 76–78). South Korea: Institute of Electrical and Electronics Engineers Inc. Available from https://doi.org/10.1109/ISSCC19947.2020.9063006.

Plouchart, J. O., Parker, B., Sadhu, B., Valdes-Garcia, A., Friedman, D., Sanduleanu, M., Wang, F., Li, X., & Balteanu, A. (2014). Adaptive circuit design methodology and test applied to millimeter-wave circuits. *IEEE Design and Test*, *31*(6), 8–18. Available from https://doi.org/10.1109/MDAT.2014.2343192, http://ieeexplore.ieee.org/xpl/RecentIssue.jsp?punumber = 6221038.

Plouchart, J. O., Wang, F., Balteanu, A., Parker, B., Sanduleanu, M. A. T., Yeck, M., Chen, V. H. C., Woods, W., Sadhu, B., Valdes-Garcia, A., Li, X., & Friedman, D. (2015). A 18mW, 3.3dB NF, 60GHz LNA in 32nm SOI CMOS technology with autonomic NF calibration. *Digest of Papers - IEEE Radio Frequency Integrated Circuits Symposium, 2015*, 319–322. Available from https://doi.org/10.1109/RFIC.2015.7337769, Institute of Electrical and Electronics Engineers Inc. United States.

Pozar, D. M. (1992). Microstrip antennas. *Proceedings of the IEEE*, *80*(1), 79–91. Available from https://doi.org/10.1109/5.119568.

Rutschlin, M. (n.d). 5G antenna design for mobile phones. Available from https://blogs.3ds.com/simulia/5g-antenna-design-mobile-phones/.

Sadhu, B. (2012). A 21.8 27.5GHz PLL in 32nm SOI using Gm linearization to achieve 130dBc/Hz phase noise at 10MHz offset\nfrom a 22GHz carrier. *IEEE Radio Frequency Integrated Circuits Symposium*.

Sadhu, B. (2019). Chapter 9: Phased arrays for 5G millimeter-wave communications, Millimeter-Wave Circuits for 5G and Radar.

Sadhu, B. (2022). 27.3 a 24-to-30GHz 256-element dual-polarized 5G phased array with fast beam-switching support for > 30,000\nbeams. In *IEEE international solid- state circuits conference*.

Sadhu, B., Ferriss, M. A., Natarajan, A. S., Yaldiz, S., Plouchart, J. O., Rylyakov, A. V., Valdes-Garcia, A., Parker, B. D., Babakhani, A., Reynolds, S., Li, X., Pileggi, L., Harjani, R., Tierno, J. A., & Friedman, D. (2013). A linearized, low-phase-noise VCO-based 25-GHz PLL with autonomic biasing. *IEEE Journal of Solid-State Circuits*, *48* (5), 1138–1150. Available from https://doi.org/10.1109/JSSC.2013.2252513.

Sadhu, B., Ferriss, M., & Valdes-Garcia, A. (2014). A 46.4–58.1GHz frequency synthesizer featuring a 2nd harmonic extraction technique that preserves VCO performance. *Digest of Papers - IEEE Radio Frequency Integrated*

*Circuits Symposium* (pp. 173–176). United States: Institute of Electrical and Electronics Engineers Inc. Available from https://doi.org/10.1109/RFIC.2014.6851689.

Sadhu, B., Ferriss, M., & Valdes-Garcia, A. (2015). A 52GHz frequency synthesizer featuring a 2nd harmonic extraction technique that preserves VCO performance. *IEEE Journal of Solid-State Circuits, 50*(5), 1214–1223. Available from https://doi.org/10.1109/JSSC.2015.2414921, http://ieeexplore.ieee.org/Xplore/home.jsp.

Sadhu, B., Valdes-Garcia, A., Plouchart, J. O., Ainspan, H., Gupta, A. K., Ferriss, M., Yeck, M., Sanduleanu, M., Gu, X., Baks, C., Liu, D., & Friedman, D. (2016). A 60GHz packaged switched beam 32nm CMOS TRX with broad spatial coverage, 17.1dBm peak EIRP, 6.1dB NF at < 250mW. *Digest of Papers - IEEE Radio Frequency Integrated Circuits Symposium, 2016*, 342–343. Available from https://doi.org/10.1109/RFIC.2016.7508322, Institute of Electrical and Electronics Engineers Inc. United States.

Sadhu, B., Tousi, Y., Hallin, J., Sahl, S., Reynolds, S., Renstrom, O., Sjogren, K., Haapalahti, O., Mazor, N., Bokinge, B., Weibull, G., Bengtsson, H., Carlinger, A., Westesson, E., Thillberg, J.E., Rexberg, L., Yeck, M., Gu, X., Friedman, D., & Valdes-Garcia, A. (2017a). A 28GHz 32-element phased-array transceiver IC with concurrent dual polarized beams and 1.4 degree beam-steering resolution for 5G communication. In *Digest of technical papers - IEEE international solid-state circuits conference* (Vol. 60, pp. 128–129). United States: Institute of Electrical and Electronics Engineers Inc. Available from https://doi.org/10.1109/ISSCC.2017.7870294.

Sadhu, B., Tousi, Y., Hallin, J., Sahl, S., Reynolds, S. K., Renstrom, O., Sjogren, K., Haapalahti, O., Mazor, N., Bokinge, B., Weibull, G., Bengtsson, H., Carlinger, A., Westesson, E., Thillberg, J. E., Rexberg, L., Yeck, M., Gu, X., ... Ferriss, M. (2017b). A 28-GHz 32-element TRX phased-array IC with concurrent dual-polarized operation and orthogonal phase and gain control for 5G communications. *IEEE Journal of Solid-State Circuits, 52*(12), 3373–3391. Available from https://doi.org/10.1109/JSSC.2017.2766211, http://ieeexplore.ieee.org/Xplore/home.jsp.

Sadhu, B., Paidimarri, A., Ferriss, M., Yeck, M., Gu, X., & Valdes-Garcia, A. (2018). A software-defined phased array radio with mmWave to software vertical stack integration for 5G experimentation. *IEEE MTT-S International Microwave Symposium Digest, 2018*, 1323–1326. Available from https://doi.org/10.1109/MWSYM.2018.8439278, Institute of Electrical and Electronics Engineers Inc. United States.

Sadhu, B., Gu, X., & Valdes-Garcia, A. (2019). The more (Antennas), the merrier: A survey of silicon-based mm-wave phased arrays using multi-IC scaling. *IEEE Microwave Magazine, 20*(12), 32–50. Available from https://doi.org/10.1109/MMM.2019.2941632, https://ieeexplore.ieee.org/servlet/opac?punumber = 6668.

Sadhu, B., Valdes-Garcia, A., Plouchart, J. O., Ainspan, H., Gupta, A. K., Ferriss, M., Yeck, M., Sanduleanu, M., Gu, X., Baks, C. W., Liu, D., & Friedman, D. (2020). A 250-mW 60-GHz CMOS transceiver SoC integrated with a four-element AiP providing broad angular link coverage. *IEEE Journal of Solid-State Circuits, 55*(6), 1516–1529. Available from https://doi.org/10.1109/JSSC.2019.2943918, http://ieeexplore.ieee.org/Xplore/home.jsp.

Samadi Taheri, M. M., Abdipour, A., Zhang, S., & Pedersen, G. F. (2019). Integrated millimeter-wave wideband end-fire 5G beam steerable array and low-frequency 4G LTE antenna in mobile terminals. *IEEE Transactions on Vehicular Technology, 68*(4), 4042–4046. Available from https://doi.org/10.1109/TVT.2019.2899178, http://ieeexplore.ieee.org/xpl/tocresult.jsp?isnumber = 8039128&punumber = 25.

Sarmah, N., Heinemann, B., & Pfeiffer, U. R. (2014). 235–275 GHz (x16) frequency multiplier chains with up to 0dBm peak output power and low DC power consumption. *Digest of papers - IEEE radio frequency integrated circuits symposium* (pp. 181–184). Germany: Institute of Electrical and Electronics Engineers Inc. Available from https://doi.org/10.1109/RFIC.2014.6851691.

Shahramian, S., Holyoak, M. J., & Baeyens, Y. (2018). A 16-element W-band phased-array transceiver chipset with flip-chip PCB integrated antennas for multi-gigabit wireless data links. *IEEE Transactions on Microwave Theory and Techniques, 66*(7), 3389–3402. Available from https://doi.org/10.1109/TMTT.2018.2822304, http://ieeexplore.ieee.org/xpl/tocresult.jsp?isYear = 2009&isnumber = 4747395&Submit32 = View + Contents.

Shahramian, S., Holyoak, M. J., Singh, A., Baeyens, Y., & Fully, A. (2019). Integrated 384-element, 16-tile, W-band phased array with self-alignment and self-test. *IEEE Journal of Solid-State Circuits, 54*(9), 2419–2434. Available from https://doi.org/10.1109/JSSC.2019.2928694, http://ieeexplore.ieee.org/Xplore/home.jsp.

Song, J., Choi, J., Larew, S. G., Love, D. J., Thomas, T. A., & Ghosh, A. A. (2015). Adaptive millimeter wave beam alignment for dual-polarized MIMO systems. *IEEE Transactions on Wireless Communications, 14*(11), 6283–6296. Available from https://doi.org/10.1109/TWC.2015.2452263, http://ieeexplore.ieee.org/xpl/RecentIssue.jsp?puNumber = 7693.

Sowlati, T., Sarkar, S., Perumana, B. G., Chan, W. L., Papio Toda, A., Afshar, B., Boers, M., Shin, D., Mercer, T. R., Chen, W. H., Grau Besoli, A., Yoon, S., Kyriazidou, S., Yang, P., Aggarwal, V., Vakilian, N., Rozenblit, D., Kahrizi, M., ... Zhang, J. (2018). A 60-GHz 144-element phased-array transceiver for backhaul application. *IEEE Journal of Solid-State Circuits, 53*(12), 3640–3659. Available from https://doi.org/10.1109/JSSC.2018.2874048, http://ieeexplore.ieee.org/Xplore/home.jsp.

Sun, S., Wang, F., Yaldiz, S., Li, X., Pileggi, L., Natarajan, A., Ferriss, M., Plouchart, J. O., Sadhu, B., Parker, B., Valdes-Garcia, A., Sanduleanu, M. A. T., Tierno, J., & Friedman, D. (2014). Indirect performance sensing for on-chip self-healing of analog and RF circuits. *IEEE Transactions on Circuits and Systems I: Regular Papers, 61*(8), 2243–2252. Available from https://doi.org/10.1109/TCSI.2014.2333311, http://ieeexplore.ieee.org/xpl/RecentIssue.jsp?punumber = 8919.

Valdes-Garcia, A., Natarajan, A., Liu, D., Sanduleanu, M., Gu, X., Ferriss, M., Parker, B., Baks, C., Plouchart, J. O., Ainspan, H., Sadhu, B., Islam, M., & Reynolds, S. (2013). A fully-integrated dual-polarization 16-element W-band phased-array transceiver in SiGe BiCMOS. *Digest of Papers - IEEE Radio Frequency Integrated Circuits Symposium*, 375–378. Available from https://doi.org/10.1109/RFIC.2013.6569608, United States.

Valdes-Garcia, A., Sadhu, B., Gu, X., Tousi, Y., Liu, D., Reynolds, S. K., Haillin, J., Sahl, S., & Rexberg, L. (2018). Circuit and antenna-in-package innovations for scaled mmWave 5G phased array modules. In *IEEE custom integrated circuits conference, CICC 2018* (pp. 1–8). United States: Institute of Electrical and Electronics Engineers Inc. Available from https://doi.org/10.1109/CICC.2018.8357050, http://ieeexplore.ieee.org/xpl/mostRecentIssue.jsp?punumber = 8355238.

Valdes-Garcia, A., Sadhu, B., Gu, X., Plouchart, J. O., Yeck, M., & Friedman, D. (2018). Scaling millimeter-wave phased arrays: Challenges and solutions. *IEEE BiCMOS and compound semiconductor integrated circuits and technology symposium, BCICTS 2018* (pp. 80–84). United States: Institute of Electrical and Electronics Engineers Inc. Available from https://doi.org/10.1109/BCICTS.2018.8551062, http://ieeexplore.ieee.org/xpl/mostRecentIssue.jsp?punumber = 8536738.

Valdes-Garcia, A., Paidimarri, A., & Sadhu, B. (2020). Hardware-software co-integration for configurable 5GmmWave systems. *Digest of technical papers-Symposium on VLSI technology* (p. 2020) United States: Institute of Electrical and Electronics Engineers Inc. Available from https://doi.org/10.1109/VLSITechnology18217.2020.9265032.

Vouvakis, M., Zhao, K., Seo, S. M., & Lee, J. F. (2007). A domain decomposition approach for non-conformal couplings between finite and boundary elements for unbounded electromagnetic problems in R3. *Journal of Computational Physics, 225*(1), 975–994. Available from https://doi.org/10.1016/j.jcp.2007.01.014, http://www.journals.elsevier.com/journal-of-computational-physics/.

Watanabe, A. O., Ali, M., Sayeed, S. Y. B., Tummala, R. R., & Pulugurtha, M. R. (2021). A review of 5G front-end systems package integration. IEEE Transactions on Components. *Packaging and Manufacturing Technology, 11* (1), 118–133. Available from https://doi.org/10.1109/TCPMT.2020.3041412, http://cpmt.ieee.org/transactions-on-cpmt.html.

Wei, J. (2008). Challenges in cooling design of CPU packages for high-performance servers. *Heat Transfer Engineering, Japan, 29*(2), 178–187. Available from https://doi.org/10.1080/01457630701686727.

Wu, T., Rappaport, T. S., & Collins, C. M. (2015). Safe for generations to come: Considerations of safety for millimeter waves in wireless communications. *IEEE Microwave Magazine, 16*(2), 65–84. Available from https://doi.org/10.1109/MMM.2014.2377587.

Yang, Q. L., Ban, Y. L., Kang, K., Sim, C. Y. D., Wu, G., & Multibeam, S. I. W. (2016). Array for 5G mobile devices. *IEEE Access*, 2788–2796. Available from https://doi.org/10.1109/ACCESS.2016.2578458, http://ieeexplore.ieee.org/xpl/RecentIssue.jsp?punumber = 6287639.

Yin, Y., Zhang, Z., Kanar, T., Zihir, S., & Rebeiz, G. M. (2020). A 24–29.5GHz 256-element 5G phased-array with 65.5dBm peak EIRP and 256-QAM modulation. *IEEE MTT-S International Microwave Symposium Digest, 2020*, 687–690. Available from https://doi.org/10.1109/IMS30576.2020.9224031, Institute of Electrical and Electronics Engineers Inc. United States.

Zhang, Y. P., & Liu, D. (2009). Antenna-on-chip and antenna-in-package solutions to highly integrated millimeter-wave devices for wireless communications. *IEEE Transactions on Antennas and Propagation, 57*(10), 2830–2841. Available from https://doi.org/10.1109/TAP.2009.2029295.

Zhao, K., Helander, J., Sjöberg, D., He, S., Bolin, T., & Ying, Z. (2017). User body effect on phased array in user equipment for the 5G mmWave communication system. *IEEE Antennas and Wireless Propagation Letters, 16*, 864–867. Available from https://doi.org/10.1109/LAWP.2016.2611674, http://www.ieee.org.

Zhao, P., Zhang, Y., Sun, R., Zhao, W. S., Hu, Y., & Wang, G. (2018). Design of a novel miniaturized frequency selective surface based on 2.5-dimensional Jerusalem cross for 5G applications. *Wireless Communications and Mobile Computing, 2018*. Available from https://doi.org/10.1155/2018/3485208, https://www.hindawi.com/journals/wcmc/.

Zhou, H. (2015). Phased array for millimeter-wave mobile handsets and other devices US patent application.

CHAPTER

# 9

# Circuits for 5G applications implemented in FD-SOI and RF/PD-SOI technologies

*Lucas Nyssens, Martin Rack and Jean-Pierre Raskin*

**Electronics and Applied Mathematics (ICTEAM), Université Catholique de Louvain, Institute of Information and Communication Technologies, Louvain-la-Neuve, Belgium**

## 9.1 Introduction

Cellular communication has tremendously progressed in the last decades having a deep impact on today's connected society. Each successive cellular generation has brought dramatic increase in data rates. The deployment of 5G supporting the Internet-of-Things is further expected to bring a new transformation toward a society of connected people and connected objects. 5G's promises of very high data rates ($>$1 Gbps), low latency, increased spatial coverage, and massive machine-type communications will be met through several techniques, including multi-user multiple input, multiple output (MIMO) antennas with beamforming, use of millimeter (mm)-wave spectrum, small cells, low Earth orbit satellites, etc.

The suitability of RF-SOI (with partially depleted silicon-on-Insulator transistors, PD-SOI in short) for RF applications does not need to be demonstrated anymore. Its strong commercial success speaks for itself, mainly in RF switch banks equipping virtually a 100% of modern-day smartphones (Rack & Raskin, 2019). Many outstanding circuit demonstrators or fully integrated transceivers at mm-wave frequencies have recently been developed in RF-SOI technology and will be further discussed in this chapter. Fully depleted SOI (FD-SOI) technology offers comparable RF performance, with higher density for increased digital functionalities, at lower power. Furthermore, the presence of the back-gate (BG) terminal adds a tuning knob to the circuit for increased configurability. Therefore, FD-SOI is seen as a driving technology for low-power RF applications or smart mm-wave front-end modules (FEMs) using advanced signal processing (Wane et al., 2020).

*New Materials and Devices Enabling 5G Applications and Beyond*
**DOI: https://doi.org/10.1016/B978-0-12-822823-4.00009-1**

**TABLE 9.1** 5G NR operating frequencies in FR2 bands.

| 5G FR2 band | Frequency range (uplink and downlink) [GHz] | Bandwidth [MHz] | Duplex mode |
|---|---|---|---|
| n257 | 26.5–29.5 | 50–400 | TDD |
| n258 | 24.25–27.5 | 50–400 | TDD |
| n259 | 39.5–43.5 | 50–400 | TDD |
| n260 | 37–40 | 50–400 | TDD |
| n261 | 27.5–28.35 | 50–400 | TDD |

5G systems will make use of several frequency bands. Specifically, two frequency ranges, called FR1 and FR2, are defined. FR1 regroups the sub 6 GHz frequency bands and is envisaged to carry most of the traditional cellular mobile communications traffic (Electronics notes, 2022). FR2 lies in the mm-wave frequency region and focuses on short-range, high data rate capability. Table 9.1 describes the frequency bands of FR2 along with the duplex mode type and bandwidth (Dilli, 2020). Due to higher operating frequencies, more bandwidth is available in FR2 bands, thus increasing the achievable data rate. However, the path loss is also larger at such frequencies, thereby limiting the range. This issue is at least partially compensated by beamforming techniques, which alleviate the requirements on each FEM cell by combining together the power radiated by several antennas in a chosen direction. Complex modulation schemes such as 64 to 256QAM (or even higher) with orthogonal frequency division multiplexing (OFDM), requiring high linearity from the FEM elements to guarantee signal integrity, are also considered for increased data rate (3rd Generation Partnership Project, 2019; Shakib et al., 2016). In FR1 frequency bands, both frequency division duplexing and time division duplexing (TDD) are present, while only TDD is specified for FR2 bands. TDD has the advantages of easily supporting asymmetric traffic data (the asymmetry has increased recently up to a ratio of 10 in downlink with respect to uplink traffic (Chen & Zhao, 2014)), lower cost, lower complexity, and the possible benefit of better frequency availability (Chen & Zhao, 2014). The interested reader can find more information about FR2 frequency band definition, OFDM, and TDD in Chen and Zhao (2014) and Dilli (2020).

FR2 frequency bands in the mm-wave frequency range have different and stringent system requirements with respect to FR1 bands or 4G LTE. These requirements, which will be detailed hereunder, pose significant challenges in the design of RF FEM components, and thereby on device technology performance. RF-SOI technology has a lead on low-cost sub 6 GHz RF applications, but no silicon-based technology has gained a foothold in mm-wave applications. Therefore, this chapter focuses on circuits operating in the mm-wave frequency range.

In 5G, it is foreseen that each antenna is at least connected to a low noise amplifier (LNA) and a power amplifier (PA) via a switch. Then, come a phase shifter and variable gain amplifier (VGA) to enable beamforming as shown in Fig. 9.1 (Asbeck et al., 2019). Several of these RF front-end components are then connected to an RF transceiver responsible for signal generation, modulation, and demodulation. These multiple RF FEM cells are finally assembled together to create an antenna array, thereby enabling beamforming for increased radiated power and the use of MIMO.

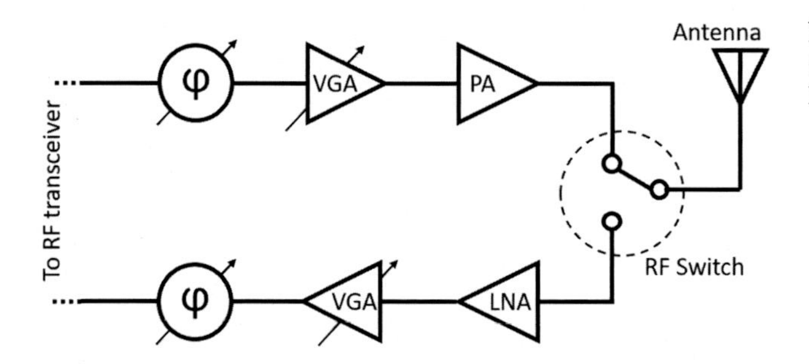


**FIGURE 9.1** Unit cell of radio frequency (RF) front-end module (FEM). Unit cell of RF FEM for 5G mm-wave frequency bands.

Such RF figure of merit (FoM) components perform time duplexing, whereby the receiving and transmitting periods of time do not overlap. The element enabling TDD is the switch, such that the receiver (RX) and transmitter (TX) share a common antenna. In the RX chain, the first element is the LNA, which amplifies the signal, without degrading too much the signal-to-noise ratio (SNR). In the TX chain, the last element is the PA, which amplifies the signal and delivers sufficient power to be radiated by the antenna (minimum power required such that the signal has a sufficient SNR after passing through the communication channel and being received at the intended receiver). The three aforementioned elements (switch, LNA, PA) are the most important and critical ones in an RF FEM.

PD-SOI and FD-SOI technologies are described at length in a previous chapter (Chapter 2), from device level to substrate and back-end-of-line (BEOL) aspects. In this chapter, we are going to focus on the realizations of three main RF functions (switch, LNA, PA) in PD-SOI and FD-SOI technologies. The first section provides a link budget analysis to derive the main FoMs for the three key circuit elements in the RX and TX paths: switch, LNA, and PA. Then, these switch, LNA, and PA elements will be discussed specifically in dedicated subsequent sections. In these three circuit sections, special focus is placed on two main points. First, the relation between device parameters and circuit performance is made, bringing insight of the intrinsic benefits and limitations related to SOI technologies. Second, a review of the state of the art for each component implemented in PD-SOI and FD-SOI technologies is presented.

## 9.2 Link budget analysis

The link budget is an accounting of a complete system's power gain and losses for both transmitter and receiver to quantify a communication's link performance. It is an application-dependent analysis, which is essential when designing a telecommunication system, as it provides some of the requirements on each component of the FEM. Our focus is on the most important elements of the transceiver, that is, switch, LNA, and PA. The main FoMs in which we are interested are insertion loss (IL) for switches, noise figure (NF) and gain for LNAs, output power and efficiency for PAs.

NF describes the amount of noise added to the signal coming from the device (here amplifier) itself, thus degrading the SNR. It is defined as the ratio of output SNR to input SNR. More insight on the NF is given in a subsequent paragraph.

In the link budget analysis described hereon, all powers are expressed in dBm and all gain/losses in dB. The link budget from an emitter to a receiver is written as

$$P_{RX} = \text{EIRP} - L_{\text{channel}} + G_{\text{ant,RX}} + 10\log_{10}(N_{\text{ant,RX}}), \tag{9.1}$$

where EIRP is the effective isotropic radiated power. $L_{\text{channel}}$ are the losses of the channel, including propagation losses, atmospheric losses, etc. $G_{\text{ant}}$ is the antenna gain. $N_{\text{ant}}$ is the number of antennas and accounts for the increased directionality with beamforming. $P_{RX}$ is the power available at the receiver. The EIRP can be expressed as (Asbeck et al., 2019)

$$\text{EIRP} = P_o + G_{\text{ant,TX}} + 10\log_{10}\left(N_{\text{ant,TX}}^2\right) - L_{TX}, \tag{9.2}$$

where $P_o$ is the output power of the individual PA. $L_{TX}$ are the losses at the transmitter side (including switch IL, interconnects losses, etc.). The squared term in $N_{\text{ant,TX}}$ is due to (1) $N_{\text{ant,TX}}$ PAs contributing together and (2) increased directionality with beamforming.

### 9.2.1 A realistic example for 5G telecommunications: campus scenario, the handset perspective

Let us focus on handsets application, a mass market application, for which the SOI technologies are most pertinent. Performing the link budget for this application will provide an estimation of the requirements for the main FoMs of the switch, LNA, and PA in a mobile device. The example detailed hereunder is summarized in Fig. 9.2. Two cases have to be analyzed: downlink and uplink to get representative FoMs for the LNA and PA. The $L_{\text{channel}}$ is the same for each case. The contribution from signal loss in free space depends on distance and frequency. It is given by Shakib et al. (2016)

$$L_{\text{space}} = 20\log_{10}(4\pi df/c) = 92.4 + 20\log_{10}f + 20\log_{10}d, \tag{9.3}$$

where $d$ is the distance between transmitter and receiver in km, $c$ is the speed of light, and $f$ is the operating frequency in GHz. The effect of a higher operating frequency on free space losses is direct: there are 29.9 dB of additional losses working at 28 GHz compared to 900 MHz. The total channel losses include the latter contribution, as well as atmospheric attenuation (which is highly frequency dependent). Scattering, refraction, and reflection on obstacles also increase the total channel attenuation. In general, the larger channel losses at the mm-wave range have to be compensated by some means, that is, beamforming with large antenna arrays, smaller cells, MIMO, etc.

In uplink, a representative value of average EIRP in handset application is ~30 dBm (Asbeck et al., 2019; Report ITU-R M.2376–0, 2015). The number of antennas is limited to 4–6 considering device compactness. With an antenna gain of 5 dBi (representative of a patch antenna) and 2 dB of $L_{TX}$ losses (accounting for the switch IL and interconnects; it could be much larger depending on package design), the average power to be delivered by each PA is 11 to 15 dBm. Another very important FoM for the PA is the average efficiency, which is represented by the power added efficiency (PAE) FoM in PAs. The PA is typically responsible for most of the power consumption of a wireless transceiver. Its average efficiency has a direct impact on battery life of mobile devices. Estimated target efficiency for 5G applications is around 20% (Asbeck et al., 2019).

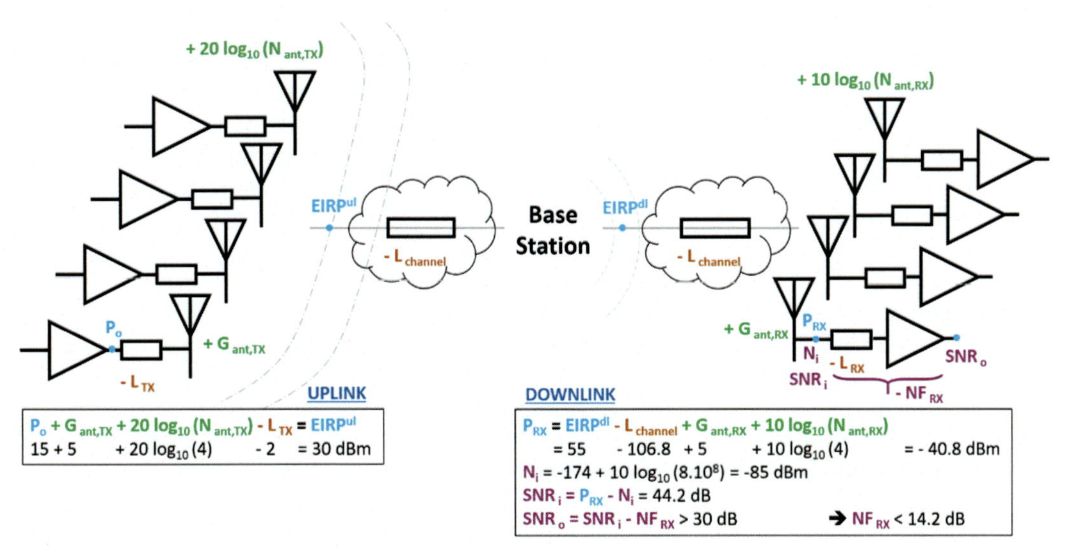


**FIGURE 9.2** Budget analysis example. Uplink EIRP ($EIRP^{ul}$) and downlink budget analysis example for user end to base-station communication.

In downlink, the base station is transmitting power. A detailed and more realistic link budget for 5G telecommunications involving MIMO can be found in Tuovinen et al. (2017). Here, some simplistic assumptions are made to have a rough estimation of the requirements at the handset receiver side. Due to the larger propagation losses at mm-wave bands and the lower penetration through materials and obstacles, small cells with a radius in the 100-m range are considered (Waveform, 2022). Considering the path loss model from Report ITU-R M.2376–0 (2015), which models the nonline-of-sight propagation (at 28 GHz) on a campus by the following relationship:

$$L_{\text{channel}} = 47.2 + 29.8 \times \log_{10} d, \tag{9.4}$$

where $d$ is the distance between transmitter and receiver in $m$. For a distance of 100 m, the channel losses are 106.8 dB. A representative value of EIRP from a base station is 55 dBm, including a 32–64 antenna array (Asbeck et al., 2019; Report ITU-R M.2376–0, 2015; Tuovinen et al., 2017). With an antenna gain of ~5 dBi, considering an array of four antennas ($N_{\text{ant,RX}}$) and a signal bandwidth (B) of 800 MHz, we can establish the link budget and compute the SNR at the input of the receiver:

$$P_{\text{RX}} = \text{EIRP} - L_{\text{channel}} + G_{\text{ant,RX}} + 10\log_{10}\left(N_{\text{ant,RX}}\right) = -40.8\text{dBm}. \tag{9.5}$$

Neglecting nonidealities at the TX chain (I-Q imbalance, signal distortion in the PA), thereby assuming the only noise contribution to the signal at the receiver input comes from thermal noise, the noise power ($N_i$) is

$$N_i = -174 \text{ dBm/Hz} + 10\log_{10} B = -85 \text{ dBm}, \tag{9.6}$$

and the SNR at the input of the receiver: $\text{SNR}_\text{i} = P_{\text{RX}} - N_i = 44.2$ dB. While the modulation order in uplink is limited to 64QAM due to challenging requirements at the handset transmitter chain (mainly at the PA level) (Asbeck et al., 2019; Shakib et al., 2016; Tuovinen et al., 2017), the modulation order can be as large as 256QAM or more (3rd Generation Partnership Project, 2019) in downlink to accommodate higher data rates. Assuming a 256QAM with OFDM signal, the SNR at the output of the receiver chain ($SNR_o$) has to be $\sim > 30$ dB (Tuovinen et al., 2017). Therefore, assuming the LNA has sufficient gain to neglect noise contribution from the demodulation part, the receiver NF ($NF_{RX}$) is limited to

$$NF_{RX,dB} = SNR_{i,dB} - SNR_{o,dB} < SNR_{i,dB} - 30dB = 14.2dB. \quad (9.7)$$

This receiver NF includes interconnect losses from the antenna to the LNA, switch IL, noise generated by the LNA, and the following of the RX chain, such that it is mainly determined by losses between the antenna and LNA, and the LNA itself, as long as sufficient gain is achieved.

The final $\sim 10$ dB value of receiver NF obtained above is the required NF for the particular case scenario (campus), using a specific modulation, with a fixed EIRP and TX-RX distance. If, for some reason, the SNR at the output of the RX chain cannot be achieved (e.g., too tight specifications on the NF, antenna gain, or not accounted additional losses), the requirements can be relaxed by considering a lower order modulation (still achieving Gbits/s-range data rates) (Tuovinen et al., 2017), smaller signal bandwidth, shorter TX-RX distances, or the stringent requirements in RX can be moved to the TX by increasing the EIRP. Similarly, if there is sufficient margin, we could consider working with a longer TX-RX range, with higher order modulation (and increased data rates), or simply reducing the EIRP and dissipating less power at the TX side. 10 dB of receiver NF is a realistic and typical specification for user-end (UE) 5G systems at 28 GHz, while $\sim 12$ dB of $NF_{RX}$ is required for the 39 GHz bands (International Wireless Industry Consortium, 2019; Mattisson, 2018).

Moreover, a design margin of $\sim 3$ dB is important to be considered (for process variability, environmental conditions such as temperature or supply voltage variation) (Mattisson, 2018), yielding a realistic target in NF of maximum $\sim 6$ dB for the RX chain. With $\sim 3$ dB of losses in interconnects and switch IL, the LNA exhibits a maximum NF of $\sim 3$ dB and a gain typically $>20$ dB to neglect the noise contribution from the demodulation part.

While a complete system specification is out of scope of this chapter, it is worth mentioning that other FoMs (in general nonlinear FoMs) are also necessary to specify. For the PA, signal distortion (usually characterized by AM–AM and AM–PM FoMs, i.e., output signal amplitude and phase variation due to input signal amplitude variation, respectively) can require to work with a larger back-off power to guarantee signal integrity, which leads to a lower average power and lower average efficiency. Adjacent channel leakage (described by the FoM adjacent channel power ratio, ACPR) at the PA, intermodulation product (described by the FoM input third order intercept point, IIP3) at the LNA (Didioui et al., 2012), large-signal distortion at the switch are other effects that can degrade signal integrity and increase the bit error rate. This can result in either tightening the requirements in the power link budget, reducing the coverage range of the small cell or preventing the use of higher order modulations.

The remaining of this chapter is devoted to a more detailed description of each of the aforementioned circuits (switch, LNA, PA). The main FoMs introduced above will be described in more detail and linked to the device-level performance. Some of the main circuit architectures used in SOI technologies will be briefly discussed showing how to push circuit performance to its best. Finally, a review of the state of the art for each circuit in SOI technologies will be presented, demonstrating the technologies' suitability for 5G applications.

## 9.3 Switch

Switch modules are utilized to toggle signal paths to and from different circuit blocks, such as antennas, filters, receivers, and emitters. A variety of switch modules exist for RF and mm-wave applications, routing signals from *pole* nodes to *throw* nodes. The most common switch for RF applications is the single-pole double-throw (SPDT) antenna switch, depicted in Fig. 9.3. It serves classically in handset devices to route the antenna to the TX path (through a PA) for emitting data, or to route the antenna to the RX path (through an LNA) for receiving data.

Switches are chosen for specific applications based on their technical FoMs, which are:

*Operating frequency and bandwidth*

Frequency range in which the switch has sufficient performance for a given application.

*In-band IL*

- Loss to a signal traveling between two ports of the switch through an on-state path. For wideband signals, the in-band flatness of the IL is also important to limit frequency-domain distortion.

*In-band isolation*

- Isolation between two ports of the switch through an off-state path.

*Return loss*

- Energy reflected back to a port when stimulated. It characterizes the matching in the on-state, and the reflective or absorptive nature of the switch in the off-state.

*Switching time*

- Time that it takes to toggle the switch from an on-state to an off-state, and vice versa.

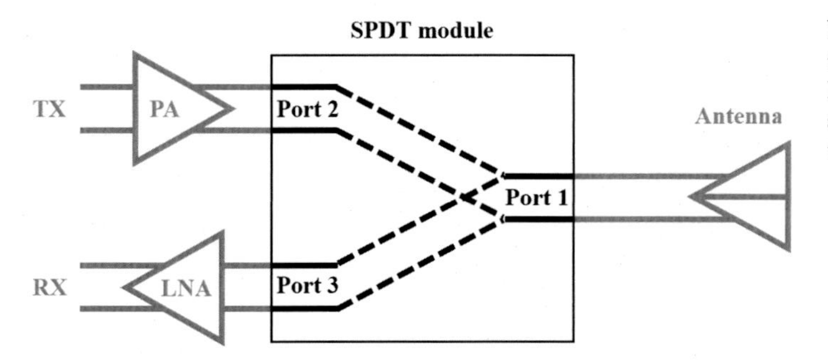


FIGURE 9.3 Single-pole double-throw antenna switch illustration. Illustration of a SPDT antenna switch inserted between the antenna to transmitter and receiver paths.

*Power handling*

- Maximum power that a signal presented to the switch may have without causing performance degradation or even damage.

*Linearity*

- Harmonic distortion due to the nonlinear voltage transfer function through real switches. Highly linear switches are crucial to preserving the integrity of the information contained in complex-modulated telecom signals.

*Durability*

- The stress that the switch can endure on average before damage ensues. For example, the number of cycles the device can be switched, or the number of years it can operate reliably.

*Number of pole and throw ports*

- The above metrics depend on the amount of overall switch ports, which can range from 2 up to several hundreds in RF applications.

Typically, RF switches are constructed based on transistors, pin-diodes, or electromechanical devices. For mass-market telecommunication applications, due to high manufacturing volume capabilities low unit cost, and co-integration with other functionalities, transistor-based integrated-circuit options are the solution of choice.

This section will then overview integrated switches, from the transistor device figures to the state-of-the-art RF and mm-wave switch performances, with a focus on SOI technology, that has been dominant in the RF switch market for the past 15 years. Indeed, starting from 2008, PD-SOI steadily displaced gallium-arsenide (GaAs) and silicon-on-sapphire technologies to become the mainstream technology for implementing RF switch banks in mobile applications. By 2014 PD-SOI accounted for more than 85% market share for mobile antenna switches (Raskin & Desbonnets, 2015).

Indeed, stacking advanced RF-SOI field-effect transistors (FETs) in series (to distribute the voltage drop on each FET to achieve higher power handling and better linearity) on a high-resistivity substrate provides high-quality RF switching modules that have steadily displaced rival technologies. Today, RF-SOI is the mainstream switch technology option, and can be found in virtually all modern-day smartphones in the switching circuits placed in the signal path to and from the antenna (Rack & Raskin, 2019).

Compared with planar bulk complementary metal oxide semiconductor (CMOS) technology, SOI achieves better electrostatic control of the transistor channel by isolating it from the silicon substrate by means of a buried oxide (BOX) layer (see Chapter 2.). The insulating BOX layer also greatly enhances lateral and vertical isolation, and reduces source and drain capacitances, minimizing the total dynamic operation power and/or increasing switching speed. Furthermore, these capacitances in SOI are more linear, due to the inexistence of PN junctions that constitute the substrate isolation scheme in bulk CMOS.

The well-known transistor FoM for RF switch applications is the $R_{on}C_{off}$ product. For a single device in switch configuration (RF-floating gate node), $R_{on}$ is the series device resistance between source and drain when biased in an on-state regime, and is inversely proportional to transistor total width $W_{tot}$. $C_{off}$ is the series device capacitance when biased in an off-state regime, and is proportional to transistor total width $W_{tot}$. Making a transistor wider then has the effect of reducing its $R_{on}$ metric and making it a better short circuit in the on-state, but this comes at the price of increasing its $C_{off}$ metric, making it a worse open circuit in the off-state. The $R_{on}C_{off}$ metric basically describes the contrast that a transistor can offer between its on- and off-states, with the lowest $R_{on}C_{off}$ metrics offering the best performance.

Nowadays, advanced transistor nodes (sub 100 nm gate lengths) are offering sub 100 fs $R_{on}C_{off}$ metrics. This is achieved by carefully engineering the source and drain accesses to minimize their contribution to the $R_{on}$ term. Furthermore, strained channels are also frequently employed to increase the intrinsic mobility towards $R_{on}$ reduction. In terms of $C_{off}$ capacitance, as nodes become shorter, the back-end-of-line metal accesses to the source and drain nodes become more tightly spaced, and contribute more strongly to the overall $C_{off}$ metric. Careful routing is employed in the BEOL metal accesses to achieve the best $R_{on}C_{off}$ metrics, such as stacking the metals in a staircase-like fashion (Li et al., 2017). A review of $R_{on}C_{off}$ metrics on different technologies is reported in Table 9.2. This benchmark demonstrates the competitiveness of SOI technology for switch applications, even competing with advanced III–V technologies despite the smaller channel conductivity.

## 9.3.1 Single-pole double-throw basic design: two FETs

The case of a symmetrical SPDT switch design based on a series-transistor topology employing only two FETs is depicted in Fig. 9.4A. The considered FET technology employed in this example has a characteristic $R_{on}C_{off}$ metric of 100 fs. To connect port 1 to port 2 and isolate port 1 from port 3, the FET M1 is biased to be a passing $R_{on}$ element and FET M2 is biased to be a blocking $C_{off}$ element (by applying the appropriate control voltage $V_{SW}$). Assuming we want to design a symmetrical switch (i.e., we want the performances between port 1 to 2 to be the same as between port 1 and 3), we define W as the width of both FETs M1 and M2.

At low frequency, the IL (-$S_{21}$ in dB) is determined by $R_{on}$ of M1. At high frequency, however, the $C_{off}$ of M2 provides a leakage path for RF power towards port 3 that effectively diverts RF current from the intended port and therefore increases the IL.

This behavior is depicted by the blue curve in Fig. 9.4B, showing the typical IL versus frequency trend of this circuit topology using a certain transistor width W1.

One could employ wider FETs to reduce $R_{on}$ of M1 and therefore the low-frequency IL, and this is illustrated by the red curve in Fig. 9.4B, where the FET widths are increased to $3W_1$. However, this comes at the price of increasing $C_{off}$ of M2, and hence reduces the cut-off frequency of the switch.

Conversely, one could utilize narrower FETs, and sacrifice low-frequency IL performance to push the cut-off frequency when the IL rolls off to a higher value. This is depicted by the green curve in Fig. 9.4B that employs FETs of widths $W_1/2$.

It is clear to see from Fig. 9.4B that $W_2$ ($=3W_1$) provides the best switch IL performance in the DC-5 GHz band, while W1 is the best choice for a DC-30 GHz switch, and $W_0$ ($=W_1/2$) should be chosen to work over the DC-65 GHz region. Note from Fig. 9.4B that these three switches all present better than 15 dB isolation over these respective bands.

Since the isolation metric (-$S_{31}$ in dB) is basically entirely determined by the $C_{off}$ of M2 (so long as the $R_{on}$ of M1 is significantly lower than the port impedances), the three

**TABLE 9.2** Benchmarking of $R_{on}C_{off}$ metrics on different integrated technologies.

| Ref | Yu et al. (2017) | GlobalFoundries (2017) | Balasubramaniyan and Bellaouar (2018) | Rack et al. (2020) | Foissey et al. (2021) | Ulusoy et al. (2014) | Thome et al. (2018) | Putnam et al. (1994) | Thome and Ambacher (2018) |
|---|---|---|---|---|---|---|---|---|---|
| Technology | 130 nm PD-SOI | 45 nm PD-SOI | 22 nm FD-SOI | 22 nm FD-SOI | 55 nm SiGe Diode | 130 nm SiGe HBT | 100 nm GaN HEMT | GaAs PIN Diode | 50 nm InGaAs mHEMT |
| $R_{on}C_{off}$ [fs] | 80 | 90 | 98 | 103 | 85 | 84[a] | 164 | 44 | 110 |

[a] *Simulated at 140 GHz.*

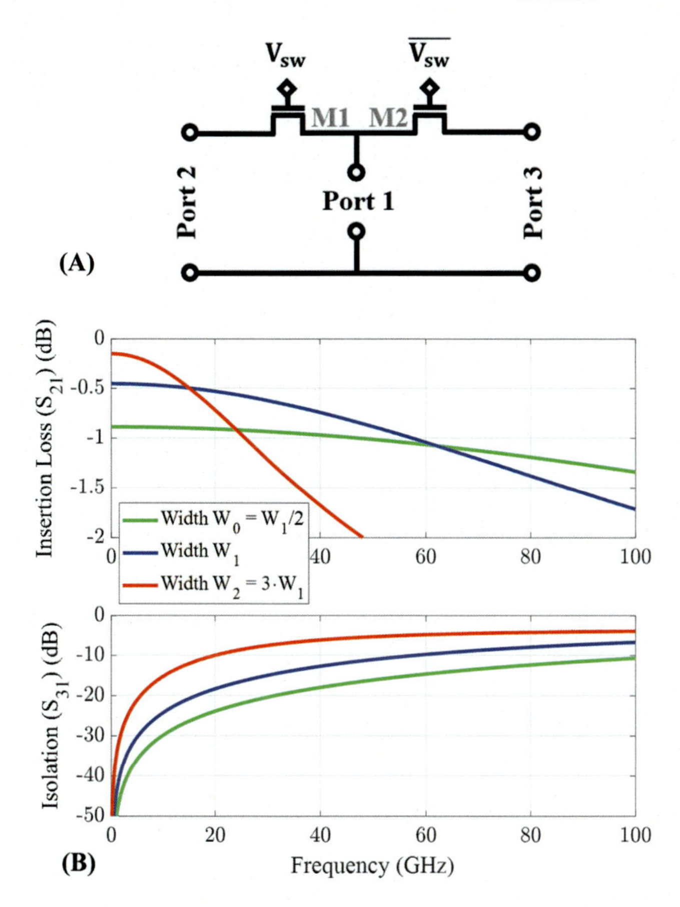


**FIGURE 9.4** Single-pole double-throw (SPDT) based on series FETs. Insertion loss and isolation (B) of a simple symmetrical SPDT based on two series FETs (A, $R_{on}C_{off}$ of 100 fs).

switches have different performances over frequency. The switches employing the widest FETs then have the poorest isolation. Isolation comes as somewhat of a secondary concern, since there is a simple technique to drastically improve it without deteriorating much the IL by adding shunt FETs to the topology, as will be discussed in the next section.

## 9.3.2 Adding shunt FETs for better isolation

Indeed, shunt FETs (M3 and M4) of widths $W_{sh}$ may be added to the topology as depicted in Fig. 9.5A. The idea is that when wanting to isolate port 3, FET M4 is biased as an $R_{on,sh}$ element which is effectively seen in parallel with port 3, attempting to short-circuit it, which strongly improves the isolation metric. In that scenario, FET M3 is biased as a $C_{off,sh}$ element. This element has the effect of diverting some RF current at port 2 (of reference impedance $Z_{ref}$), which increases mismatch and thereby degrades slightly the transferred power (IL) to port 1. $W_{sh}$ is then sized such to introduce little mismatch, that is, such that $j\omega C_{off,sh} << 1/Z_{ref}$.

Fig. 9.5B plots the IL and isolation for various sizings of $W_{sh}$, for a given sizing $W_1$ of the series transistors. The solid blue curves in Fig. 9.5 correspond to the blue data of

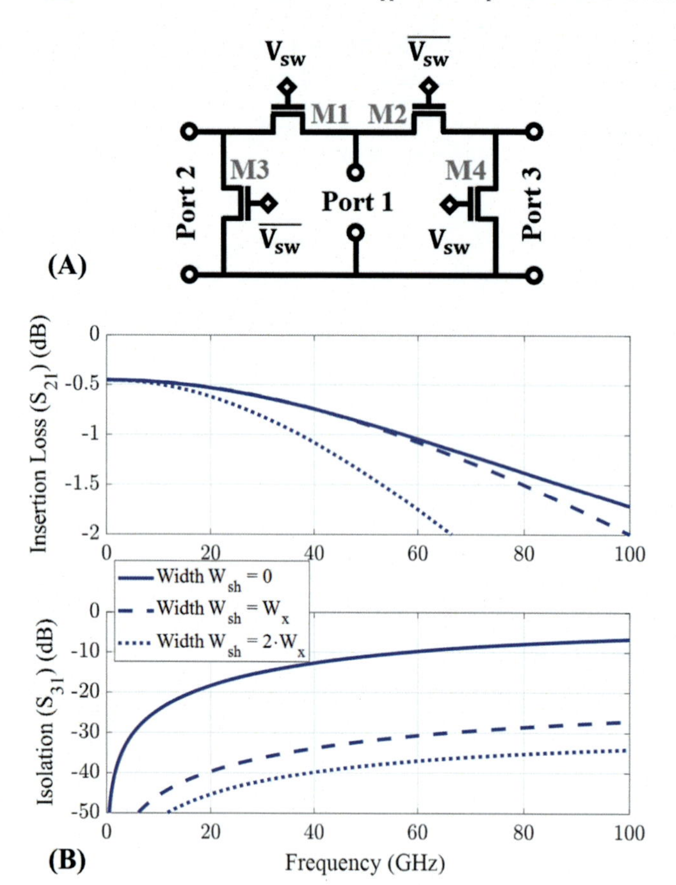


**FIGURE 9.5** Single-pole double-throw (SPDT) based on series-shunt FETs. Insertion loss and isolation (B) of a simple symmetrical SPDT based on a series-shunt topology (A, $R_{on}C_{off}$ of 100 fs), using a series FET width of W1.

Fig. 9.4, that is, without any shunt FETs included. It can be seen that adding such FETs has a strong impact on the isolation metric, improving it by over 20 dB thanks to the on-state M4 device in parallel to port 3. This comes at the price of a leakage capacitor, $\sim C_{off,sh}$ of M3 in parallel with port 2, which only degrades the IL at sufficiently high frequency (when $\omega C_{off,sh}$ approaches $1/Z_{ref}$). Fig. 9.5 clearly shows that adding shunt transistors of width $W_x$ barely impacts the IL below 60 GHz, while improving the isolation by over 20 dB.

## 9.3.3 Stacking FETs for higher power handling and better linearity

Integrated CMOS transistors mainly achieve high RF and mm-wave performance thanks to the downscaling of their physical dimensions (see Chapter 2). With the reduction in dimensions (gate length, field-oxide thickness, etc.) comes an inherent limitation in the voltage amplitudes that can excite the devices without causing wear or even damage. This is inconvenient for building switches that are required to control moderate-power RF

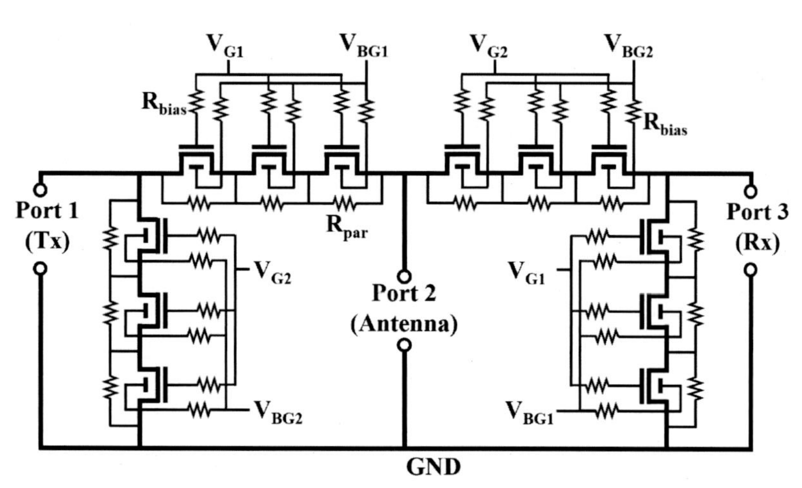


FIGURE 9.6 Series-shunt single-pole double-throw (SPDT) topology employing 3-stacked FET-branches. Series-shunt SPDT topology employing 3-stacked FET-branches for increased power handling.

signals, typically in the range of 15 to 45 dBm, that have sufficient power to damage a single advanced CMOS device.

To overcome this for switch applications, each FET branch depicted in Fig. 9.5A can be replaced by a *stacked-FET-branch* as depicted in Fig. 9.6 (stack of three in that figure). In this way, the RF signal's voltage drop over a single FET of a branch is reduced, allowing for the implementation of RF switches with higher power handling. Additionally, since each FET of a branch is excited by only a fraction of the overall RF signal's amplitude, they behave more linearly, allowing for highly stacked switches to achieve low harmonic distortion levels that are necessary to satisfy advanced high data-rate RF telecommunications specifications.

Fig. 9.6 also depicts various resistor elements associated to a full SPDT design. Gates and BGs (if applicable) are biased through large resistor elements (typically sized in the range of 10 to 100 kΩ) to define the transistor gate nodes as RF floating (since if they were RF grounded they would lead to RF leakage current and device mismatch through gate-source/drain capacitances). Additional high-valued resistors are also usually placed in parallel with each FET element, which serve to equalize any low-frequency and DC potential drop across each FET of a branch that is biased to an off-state, protecting them from overvoltaging.

## 9.3.4 Limits of the series $R_{on}C_{off}$ metric

At RF frequencies (i.e., below a few tens of GHz), the *series* terms $R_{on}$ and $C_{off}$ are quite sufficient to describe a device technology for switch applications (see the two-port equivalent circuit pi-model in Fig. 9.7). Shunt capacitance $C_{sh}$ terms can be neglected at RF, since the $j\omega C_{sh}$ admittance is quite low. Furthermore, any parasitic $C_{sh}$ can be compensated for using passive matching circuits. This is not the case for any shunt conductance terms $G_{sh}$ that will always be a source of additional degradation in switch IL, regardless of any passive matching scheme.

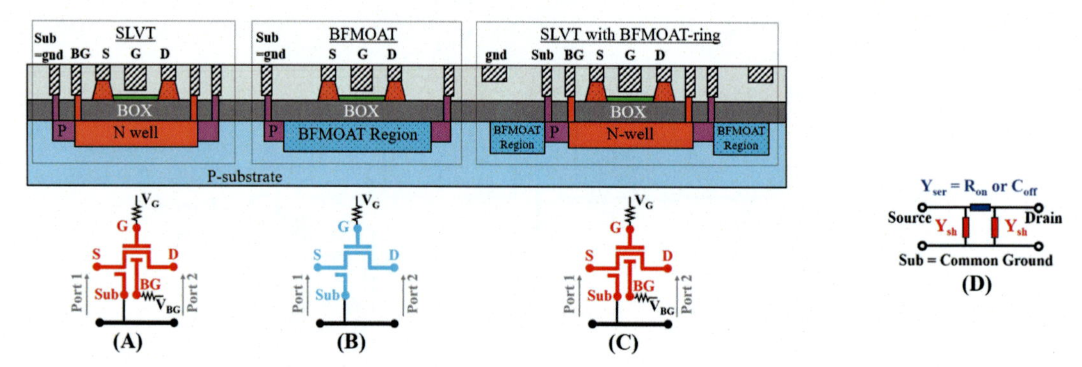


**FIGURE 9.7** Simple representation of different types of FD-SOI NFETs. Simple representation of different types of FD-SOI NFETs: SLVT (A), BFMOAT (B) and SLVT with BFMOAT-ring (C), and the two-port equivalent circuit pi-model (D), with $Y_{sh} = G_{sh} + j\omega C_{sh}$.

In the 22FDX node from GlobalFoundries, several RF devices are proposed for implementing switch devices, mainly the super-low threshold voltage (SLVT) and BFMOAT devices, that have their physical cross-sections depicted in Fig. 9.7. The SLVT device offers an $R_{on}C_{off}$ metric of the order of 100 fs (Balasubramaniyan & Bellaouar, 2018; Yadav et al., 2019), but is outperformed by the BFMOAT device for mm-wave switch applications (Rack et al., 2020; Yadav et al., 2019), even though the BFMOAT's $R_{on}C_{off}$ is substantially higher at around 160 fs.

In the following, the origin of BFMOAT's higher performance results, despite a significantly larger $R_{on}C_{off}$, will be exposed. Following from that, this FoM is then shown to be insufficient in general for benchmarking devices for switch applications in the high mm-wave spectrum and beyond.

### *9.3.4.1 Shunt loss analysis: SLVT versus BFMOAT*

As illustrated in Fig. 9.7, the SLVT incorporates a BG electrode below the thin (20 nm) BOX.

Applying a positive bias to the BG results in a more conductive channel in the on-state and to a lower value of $R_{on}$. The BFMOAT device sacrifices the BG functionality for reduced shunt losses $G_{sh}$ to the ground reference node. Both devices include a P-type substrate-tap ring ("sub" node in Fig. 9.7) to include shunt parasitics in the mm-wave models.

To highlight the impact of $R_{on}C_{off}$ and $G_{sh}$, single FETs are simulated in two-port configuration between source and drain.

The gate and BG are biased through large resistors, and these nodes can be considered as RF floating. Both FETs are 20 nm long, are referenced to the C3 (5th) metal layer, and have five fingers ($N_f = 5$), each with a width of $W_f = 5$ μm.

Fig. 9.8 plots the on- and off-state results pertaining to each FET. It plots the S-parameters and equivalent circuit elements of the two-port pi-model (see Fig. 9.7). The on-state is set by applying 0.9 V to the gate ($V_g$) and 3 V to the BG ($V_{bg}$), while in the off-state $V_g = -0.9$ V and $V_{bg} = 0$ V.

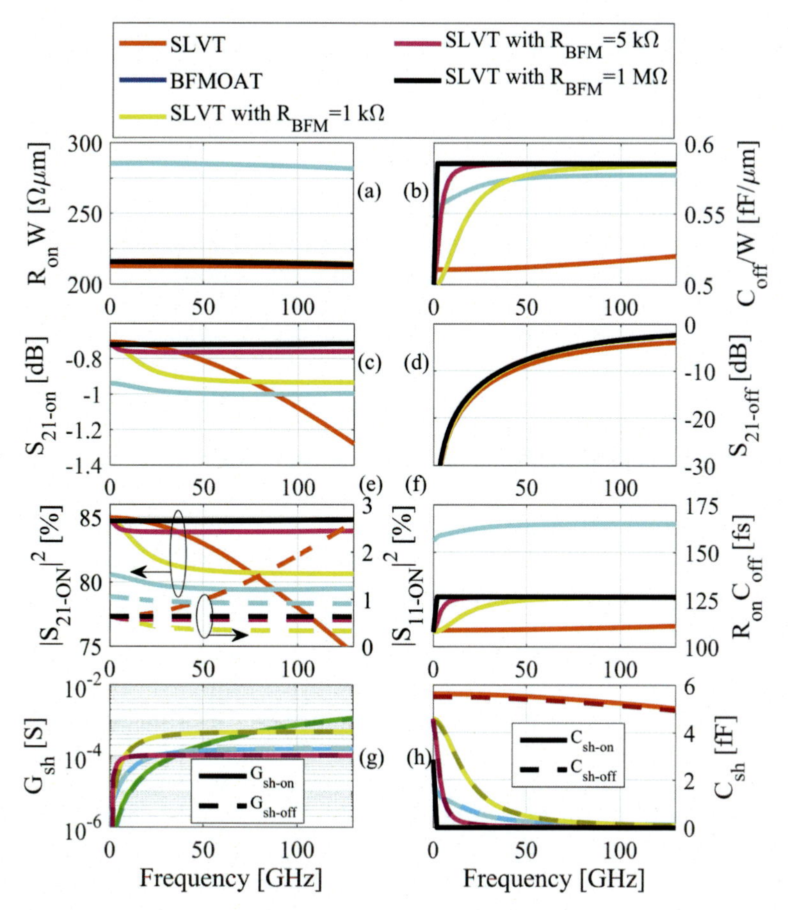


**FIGURE 9.8** Loss and pi-model analysis of SLVT and BFMOAT FETs from configuration in Fig. 10.7. Loss and pi-model analysis of SLVT and BFMOAT FETs simulated in two-port (switch) configuration (Fig. 10.7) in both ON and OFF states. All FETs are referenced to the C3 (5th) metal plane and have $W_f = 5\ \mu m$, $N_f = 5$ (total width $W = N_f W_f = 25\ \mu m$).

The $R_{on}$ element (extracted as $\Re\{-1/Y_{21}\}$) determines the low frequency loss; however, the real part of the shunt admittance $G_{sh} = \Re\{Y_{sh}\}$ increases with frequency and is responsible for substantial additional loss in the SLVT device, where $Y_{sh} = Y_{11} - Y_{21}$ ($= Y_{22} - Y_{21}$ in a symmetrical device). This is demonstrated in Fig. 9.8C which shows that $S_{21}$ appreciably decreases above ~30 GHz. Fig. 9.8E shows that the power transfer coefficient $|S_{21}|^2$ reduces by 11% over the considered frequency range, and that only 2% of this is due to reflection by increased mismatch (see $|S_{11}|^2$). This indicates that the $|S_{21}|^2$ reduction is mainly attributable to loss increase in the network.

Fig. 9.8G shows that the $G_{sh}$ term of the BFMOAT device saturates at a value of around 160 μS, which is reached at around 30 GHz. This explains the 1% degradation in $|S_{21}|^2$ at 30 GHz. The SLVT's $G_{sh}$ term reaches 1 mS at 110 GHz, and continues to increase with frequency. This term is responsible for the 9% increase in power loss in the SLVT at 110 GHz.

In Rack et al. (2021) a formulation for shunt-loss FoM to complement the series-$R_{on}C_{off}$-FoM was described. The factor $K$ accounts for the loss dissipated in the $G_{sh}$ term, and is expressed as follows, where $Y_0$ is the reference port impedance in the system:

$$K = \frac{Y_0}{Y_0 + G_{sh}}. \tag{9.8}$$

Accurate equivalent circuit extraction of the considered SLVT and BFMOAT devices can be obtained based on the simple lumped pi-model described in Table 1.

This simple model fits well to the simulated full process design kit models, and demonstrates the impact of the real part of the shunt admittance on the overall loss in the two FETs ($R_s = 5.5\ \text{k}\Omega$ for the BFMOAT and 82 Ω for the SLVT). High values of $R_s$ are desirable to avoid large shunt loss factors $K$.

### 9.3.4.2 *SLVT with BFMOAT-shunt-impedance-enhancing ring*

The effective shunt resistance to ground can be increased by adding some series impedances between the sub node of the devices and the common RF ground. This scheme is depicted in Fig. 9.7, where an SLVT device is proposed with a BFMOAT isolation ring defined around it, which will add some $R_{BFM}$ term toward the ground node.

Fig. 9.8 includes simulations of SLVT devices with such additional $R_{BFM}$ elements introduced with values of 1, 5 kΩ and 1 MΩ. The results (summarized in) show that the $S_{21\text{-}ON}$ data pertaining to a 1 MΩ $R_{BFM}$ achieve the lowest overall loss over the whole band, since: (1) the $R_{on}$ is low thanks to the inherent performance of a BG-biased SLVT core, and (2) since the overall $G_{sh}$ term is made extremely low for $R_{BFM} = 1\ \text{M}\Omega$. When $R_{BFM}$ is set to 5 kΩ, a small degradation in IL ($S_{21\text{-}ON}$ curves) is observed at around 6.7 GHz but the degradation is not severe ($K$ factor of around 0.08 dB). The frequency at which this transition occurs and the amplitude of the degradation are similar to those observed for the regular BFMOAT device ($K = 0.07$ dB). However, the low-frequency loss in the BFMOAT FET is higher due to a larger-valued $R_{on}$. The simulation employing an $R_{BFM}$ of 1 kΩ shows quite a substantial degradation, with a $K$ factor of 0.39 dB. A 1 kΩ value is then not considered high enough for comfortable operation. It is interesting to note that using an $R_{BFM}$ value of 1 kΩ results in an effective device that outperforms the regular SLVT above approximately 80 GHz, but actually has more loss below 80 GHz. This is due to the transition frequency being pushed higher for lower values of $R_s$. This is highlighted in Table 9.3, which lists the overall $R_s$ and $C_s$ for all equivalent FETs (with their equivalent circuit shown in Fig. 9.9), as well as giving the degradation $K$ value and the frequency at which the full degradation occurs (i.e., the high frequency at which the overall loss becomes frequency independent), when voltage begins to drop over $R_s$ as it starts to dominates over $\omega C_s$:

$$f_s = \frac{1}{2\pi R_s C_s}. \tag{9.9}$$

Though the degradation is more severe at high frequency when $R_{BFM}$ is zero, it only appears at frequencies that are relatively high. Therefore, when working below a few GHz, not adding any shunt resistive elements is the better choice.

TABLE 9.3 Analysis of on-state BFMOAT and SLVT FETs as switches.

| Model | | Equiv.-circuit | | | Performance | |
|---|---|---|---|---|---|---|
| FET | $R_{BFM}$ [Ω] | $R_{on}$ [Ω] | $C_s$ [fF] | $R_s$ [Ω] | $K$ [dB] (at $f \gg f_s$) | $f_s$ [GHz] |
| BFMOAT | 0 | 11.4 | 1.7 | 5530 | −0.07 | 16.9 |
| SLVT | 0 | 8.5 | 4.7 | 82 | −4.1 | 413 |
| SLVT | 1 k | 8.5 | 4.7 | 1082 | −0.39 | 31.3 |
| SLVT | 5 k | 8.5 | 4.7 | 5082 | −0.08 | 6.7 |
| SLVT | 1 M | 8.5 | 4.7 | 1 M | −0.0004 | 0.03 |

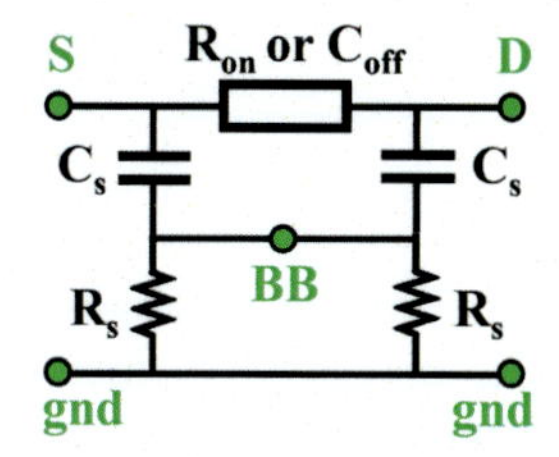


FIGURE 9.9 FET equivalent circuit for switch analysis. FET equivalent circuit for switch analysis.

Fig. 9.8B illustrates that the $C_{off}$ metric of the effective device is influenced by the shunt impedance value of $R_{BFM}$. This directly impacts the series-FoM of $R_{on}C_{off}$ (Fig. 9.8F), and the SLVT devices with increased shunt impedance elements demonstrate a 15% increase in $C_{off}$ and in $R_{on}C_{off}$. This phenomenon can be explained based on an analysis of the equivalent circuit of Table 9.3. As the elements $R_s$ tend to a short circuit (case for the usual SLVT device), the below-BOX node BB tends to an RF ground, and both $C_s$ elements contribute to pure shunt admittance. In that case ($R_s$ very low), the $Y_{21}$ parameter of the network becomes

$$Y_{21} = -j\omega C_{off}, \text{ for } R_s \approx 0. \tag{9.10}$$

The series capacitance between source and drain extracted from $Y_{21}$ is then simply the $C_{off}$ term related to the device network mainly above the BOX. However, as the $R_s$ elements tend to an open circuit (case for the usual BFMOAT device and for the SLVT devices with an added $R_{BFM}$ term), the BB node becomes floating from the reference ground node, and $C_s/2$ is effectively seen in parallel with $C_{off}$, as the $Y_{21}$ parameter of the network then becomes

$$Y_{21} = -j\omega\left(C_{off} + \frac{C_s}{2}\right), \text{ for } R_s \approx \infty. \tag{9.11}$$

The effective $C_{off}$ is then increased by $C_s/2$. Despite the resulting 15% increase in the $R_{on}C_{off}$ FoM when adding a large $R_{BFM}$ term to the SLVT device, the benefits in shunt loss for mm-wave applications are worthwhile. In that case, notice that the $C_{off}$ value is very close to that of the regular BFMOAT device, whose BB node is similarly isolated in the same way.

The analysis performed above suggests that substrate resistances of at least 5 kΩ to ground should be targeted for mm-wave switches in this technology. To evaluate the impedance of a BFMOAT ring placed around an SLVT device, electromagnetic simulations were run to extract the value of the equivalent $R_{BFM}$ element in Rack et al. (2022).

It was demonstrated in that study that a BFMOAT isolation ring of around 2 μm wide is sufficient to achieve an $R_{BFM}$ value of 5 kΩ. From these results, SLVT devices with 2 μm wide substrate isolation rings were designed with which to implement a DC-120 GHz SPDT switch, and those results are presented in the next section.

It is also worth noting that such considerations (importance of shunt losses on the overall switch performance) are pertinent for this FD-SOI technology in the considered (mm-wave) frequency range. It is not necessarily the case for other technologies or at other frequencies as mentioned above. Indeed, in PD-SOI technology with high-resistivity substrate for instance, the $R_s$ term from the above model (Fig. 9.8) is much larger (due to a larger substrate resistivity) and the resulting shunt losses become negligible.

### 9.3.5 State of the art

To demonstrate the device's performance for high-end mm-wave frequencies, a wideband SPDT switch was designed, fabricated, and measured (Rack et al., 2022) based on a three-stacked series-shunt topology (identical to that depicted in Fig. 9.6) targeting 20 dBm of power handling in the 22FDX technology. One port of the SPDT is loaded on-chip with a 50 Ω resistor.

The design is a re-used one that was based on BFMOAT-FETs (Rack et al., 2021), in which all BFMOAT FETs have been replaced by SLVT devices with a 2 μm wide substrate isolation ring to RF ground. The initial design was performed to achieve minimal on-state IL while maintaining at least 20 dB of isolation in the off-state at 80 GHz based on the BFMOAT FET.

The SPDT switches were fully characterized under small-signal conditions using an on-wafer set-up up to 130 GHz, and the S-parameter results are plotted in Fig. 9.10 (Rack et al., 2022). The BFMOAT-FET design achieved the desired performances, showing good operation over the DC-80 GHz band (Rack et al., 2021).

By substituting the BFMOAT FETs for the isolated SLVTs, a clear gain in performance is observed. Thanks to the lower $R_{on}$, there is less IL and better isolation over the entire band, and instead of the IL experiencing a strong roll-off starting from 70 GHz, it remains steady, below −2.4 dB, up to 120 GHz. The P1dB is also improved, as is expected when using the back-biased SLVT device with lower $V$th than the BFMOAT (Rack et al., 2020).

Good agreement is achieved between the post-layout simulations and the measured data for both SPDT switches.

Table 9.4 benchmarks these results against the published state-of-the-art SPDT switches covering frequencies close to 100 GHz, and demonstrates the competitiveness of the presented DC-120 GHz SVLT-based series-shunt switch.

Overall, Table 9.4 demonstrates the competitiveness of SOI designs for switch applications in the 100 GHz region. The FD-SOI designs from Rack et al. (2021, 2022) offer IL lower than 2.4 dB and isolation better than 22 dB up to 120 GHz, for up to 20 dBm of RF power handling. These results compete with designs in advanced III–V technologies, and usually for a lower form factor.

At lower mm-wave frequencies, Table 9.5 reviews the performances of SOI-based SPDT modules close to 28 and 39 GHz, along with the $R_{on}C_{off}$ metric and substrate options of the different considered SOI technologies.

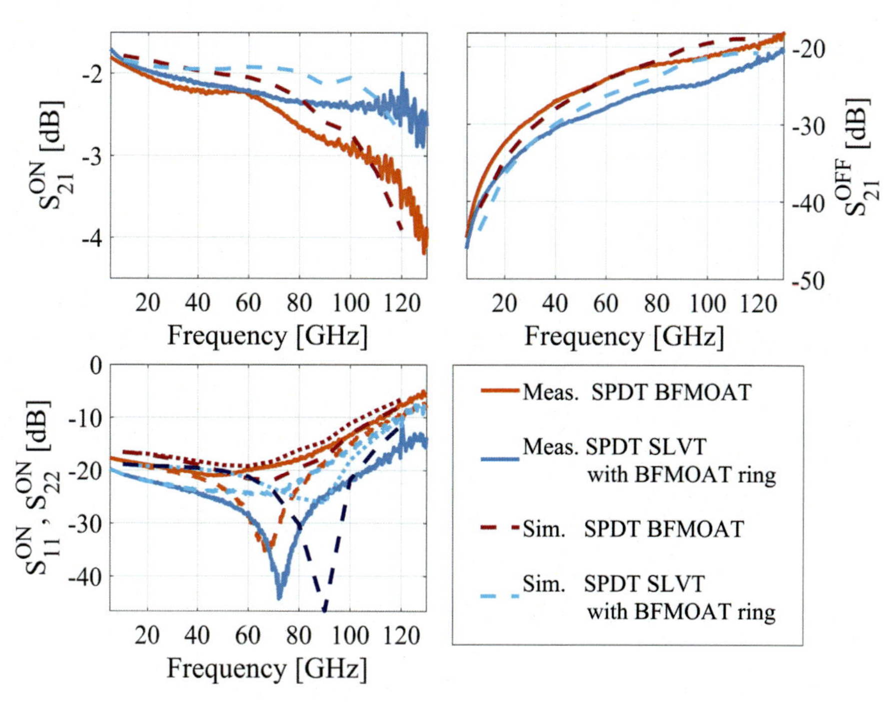


**FIGURE 9.10** Results of fabricated DC-80 and DC-120 GHz single-pole double-throw (SPDT) switches. Simulated and measured $S_{ij}$-on and $S_{ij}$-off of the fabricated SPDTs.

**TABLE 9.4** State-of-the-art single-pole double-throw switches operating close to 100 GHz.

| Work | Techno. | Topology | Freq. [GHz] | Insertion loss [dB] | Isolation [dB] | Area [$10^{-3}$ mm$^2$] |
|---|---|---|---|---|---|---|
| Tang et al. (2013) | 65 nm CMOS | Traveling wave | 17–100 | 2.8–4.5 | >15 | 420[a]<br>200[b] |
| Schmid et al. (2013) | 180 nm SiGe | Transform. | 90 | 2.7 | 14 | 43[c] |
| Shivan et al. (2019) | 800 nm InP | λ/4-Shunt | 90–170 | 3.0–5.0 | 42–55 | 650[b] |
| Thome et al. (2018) | 100 nm GaN | λ/4-Shunt | 68–134 | 1.1–2.1 | 17.6–21.5 | 308[c] |
| Thome et al. (2020) | 50 nm InGaAs | λ/4-Shunt | 50–75 | 1.0–1.6 | 31.6–32.8 | 348[c] |
| | | | 72–110 | 1.0–1.6 | 28.5–31.4 | 216[c] |
| Rack et al. (2021) | 22 nm FD-SOI | Series-Shunt | DC-80 | <2.6 | >22 | 14[c] |
| Rack et al. (2022) | | | DC-120 | <2.4 | >22 | 14[c] |

[a] *Including pads.*
[b] *Estimate excluding pads.*
[c] *Excluding pads.*

**TABLE 9.5** State-of-the-art mm-wave single-pole double-throw modules in silicon-on-insulator technology.

| | Parlak and Buckwalter (2011a) | Cardoso et al. (2014) | | Yu et al. (2017) | Chen et al. (2017) | Li, Ustundag, et al. (2018) | Elgaard et al. (2019) | Rack et al. (2020) | |
|---|---|---|---|---|---|---|---|---|---|
| Technology | 45-RFSOI | 180 nm SOI | | 130 nm SOI | 40 nm SOI | 45-RFSOI | 22FDX | 22FDX | |
| $R_{on}C_{off}$ [$f_s$] | / | / | | / | / | 90 | 98 | 103 | |
| Substrate | 13.5 Ωcm | >1 kΩcm | | >1 kΩcm | / | >1 kΩcm | >10 kΩcm | 10 Ωcm | |
| Switch type | SPDT | SPDT | | SPDT | SPDT | SPDT | SPDT | SPDT | |
| Topology | Series-Shunt with matching | Series-Shunt with matching | | Series-Shunt with matching | Series-Shunt | Series-Shunt | | Series-Shunt with matching | Series-Shunt |
| Band [GHz] | DC-60 | DC-40 | | DC-50 | DC-50 | DC-50 | 26.5–29.5 | DC-40 | |
| IL (28 GHz) [dB] | 1.4 | 1.1 | 2.1 | 1.0 | 1.4 | 0.7 | 1.6 | 1.66 | 2.29 |
| ISO (28 GHz) [dB] | 30 | 16.5 | 17.5 | 32 | 33 | 24 | 22 | 33.4 | |
| P1dB [dBm] | 7.1 | 11 | 15 | 14 | 21 | 29.5 | / | 20.5 | |
| IIP3 [dBm] | 18.2 | 25.8 | / | 27 | / | 46 | / | 38.5 | |
| Switching time [ns] | 0.35 | / | / | 10 | / | / | / | 0.25 | |
| Size [$10^{-3}$ $mm^2$] | 39 | 25 | / | 40 | / | 4 | 150 | 34 | 1.8 |

Table 9.5 clearly showcases the design trends and specifications that are achievable in SOI nodes for mm-wave SPDTs in those bands, with overall designs offering isolation levels in the range of 20 to 30 dB for IL values in the range of 1 to 2 dB for small overall area.

Such specifications given in Tables 9.4 and 9.5 are highly competitive, demonstrating SOI as a technology option of choice for mm-wave switch modules.

## 9.4 Low noise amplifier

As mentioned above, the two main FoMs in an LNA are NF and gain. As the first (amplifying) element after the switch, its NF, being the ratio of output and input SNR, directly contributes to the total receiver NF, which has to be minimized. In addition, the signal received at the input must be sufficiently amplified to avoid that subsequent circuitry (mixers) adds significant noise to the signal and further degrades the receiver SNR. This is a direct outcome from Frii's formula of cascaded devices total noise.

Gain is determined by the $f_t$ & $f_{max}$ of the transistor, along with matching network. As shown in Chapter 2, $f_t$ & $f_{max}$ in the 300–400 GHz range are attained in CMOS SOI thanks to the shrunk gate length and optimized device accesses. Although most of the $f_t$ & $f_{max}$ reported in the literature refers to a reference plane at the first metal layer, with proper care in interconnects, $f_t$ & $f_{max}$ from 200 to 300 GHz can still be achieved (Gao et al., 2020; Inac et al., 2014; Nyssens et al., 2022; Torres et al., 2020) including extrinsic parasitics of interconnects up to a higher-level metal reference plane, enabling sufficient gain for operation in the mm-wave range.

### 9.4.1 Transistor noise models

The minimum NF is usually written as follows (Fukui formula (Fukui, 1979))

$$NF_{min} = 1 + K_f \frac{f}{f_t} \sqrt{g_m (R_g + R_s)}, \tag{9.12}$$

where $K_f$ is a fitting parameter. From Eq. (10.12), it is plain to understand that the $NF_{min}$ increases with frequency.[1] To reduce $NF_{min}$, the cutoff frequency has to be maximized and the series resistances must be minimized, particularly $R_g$ (usually significantly larger than $R_s$).

There are many noise models for MOSFETs, namely the PRC, Van der Ziel, and Pospieszalski models (Niknejad et al., 2008). The above equation can be derived from the PRC model (Cappy, 1988; Pucel et al., 1975), a similar expression can be derived from the Pospieszalski model (Asgaran et al., 2004; Niknejad et al., 2008). The Pospieszalski model shown in Fig. 9.11A is a simple model with good prediction at mm-wave frequencies and is commonly used at such high frequencies (Kane et al., 2019; Niknejad et al., 2008; Waldhoff et al., 2008). It contains a drain current noise source ($i_{nd}$) and a gate current noise source that

[1] Actually, the noise factor (i.e., NF in natural units) is $F = 1 + Na/(Ni \times G)$, where Ni is the noise level at the input, Na is the additional noise introduced by the LNA and G is the gain. From here, it is plain to see how NFmin is proportional to $f/f_t$, which is actually proportional to the gain. Therefore, as frequency increases, the gain decreases and thereby the noise factor increases.

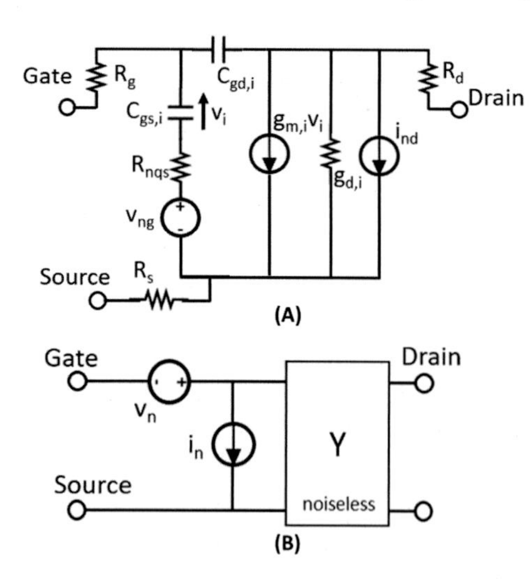


**FIGURE 9.11** Summary of high-frequency noise models. (A) Pospieszalski noise model, complemented with noise sources from extrinsic resistances. (B) Equivalent noiseless two-port with two noise sources: a current (in) and a voltage ($v_n$) noise source.

are assumed to be uncorrelated. The gate current noise source is modeled as thermal noise from the non-quasi-static (NQS) resistance ($R_{nqs}$). The main contribution to the drain current noise source is channel thermal noise at high frequencies. A secondary noise source contribution to the drain at such frequencies comes from majority carriers in the substrate under the BOX (Kushwaha et al., 2016). Moreover, thermal noise sources for the extrinsic resistances ($R_g$, $R_s$, and $R_d$) must also be included (Chan et al., 2019; Kushwaha et al., 2017) (additionally to Fig. 9.11A, not shown here). The thermal noise from the gate resistance has an important contribution to the minimum NF as expressed in Eq. (10.12).

## 9.4.2 Total noise figure and equivalent noise resistance

In practice, it is convenient to represent the complex noisy two-port by a noiseless two-port preceded by two (voltage, $v_n$ and current, $i_n$) correlated input noise sources (cfr Fig. 9.11B), which are characterized by an equivalent noise resistance ($R_n$), noise conductance ($G_n$), respectively, and a correlation term ($G_c$). This representation is often used for device characterization based on measurement data. The total NF is a function of the source impedance ($Z_s$ or admittance $Y_s$) presented before the device. It can be expressed as (Niknejad et al., 2008; Pozar, 1997)

$$NF = NF_{min} + \frac{R_n}{G_s}\left|Y_s - Y_{opt}\right|^2, \tag{9.13}$$

where $Y_{opt}$ is the optimal source impedance leading to the $NF_{min}$ and $Y_s = G_s + jB_s$ is the source admittance. By neglecting the gate current noise source compared with the drain current noise source (valid as long as $\omega R_g\ (C_{gs} + C_{gd}) << 1$), a simple expression of $R_n$ is (Asgaran et al., 2004)

$$R_n = R_g + \frac{S_{id}}{4k_B T g_m^2}, \tag{9.14}$$

where $S_{id}$ is the drain current noise power spectral density, $k_B$ is the Boltzmann constant, and $T$ is ambient temperature. From Eq. (10.13), $R_n$ indicates how much NF increases from $NF_{min}$ in case of noise mismatch, that is, when $Y_s \neq Y_{opt}$. Therefore, $R_n$ should be kept as low as possible to achieve the lowest NF when noise mismatch is present at the input of the device. $R_n$ can be conveniently expressed as a function of the noise sheet resistance ($R_{nsh}$), which is defined as (Chen et al., 2012)

$$R_{nsh} \equiv \frac{S_{id}}{4k_B T g_m^2} \cdot \left(\frac{W_{tot}}{L}\right), \tag{9.15}$$

such that $R_n = R_g + R_{nsh} \cdot (L/W_{tot})$. $R_{nsh}$ (with the normalization geometry factor) is a process-dependent FoM facilitating noise performance comparison among different technologies. $R_{nsh}$ depends on several physical parameters, such as threshold voltage, effective mobility, critical field along the channel above which carriers travel at their saturation velocity and gate oxide capacitance per unit area. Although theoretically constant with device scaling, $R_{nsh}$ increases as gate length scales down in sub 100-nm technologies due to short channel effects. As $R_{nsh}$ has a complex dependence on geometry parameters (the relationship varies according to technology), the overall $R_n$ relationship with device dimensions will not be discussed in this chapter. The interested reader can find more information about $R_{nsh}$ and $R_n$ and their values for different technologies in Chen et al. (2012) and Chen (2017).

## 9.4.3 $NF_{min}$ dependence with transistor's cutoff frequency and geometry

As explained in Chapter 2, there is a trade-off between $f_t$ and $R_g$ and thus between $f_t$ and $NF_{min}$ due to opposite dependence on finger width, as shown in Fig. 9.12. Usually in mm-wave technologies, the $W_f$ leading to a minimum in achievable device $NF_{min}$ does not lead to too strong reduction in $f_t$ for proper operation at mm-wave frequencies, and such technologies are therefore still able to offer close to optimal $NF_{min}$ designs above 100 GHz. Indeed, $NF_{min}$ has benefitted from devices scaling down. $NF_{min}$ improved thanks to gate length scaling with increasing $f_t$ and a controlled $R_g$ at the same time as can be understood

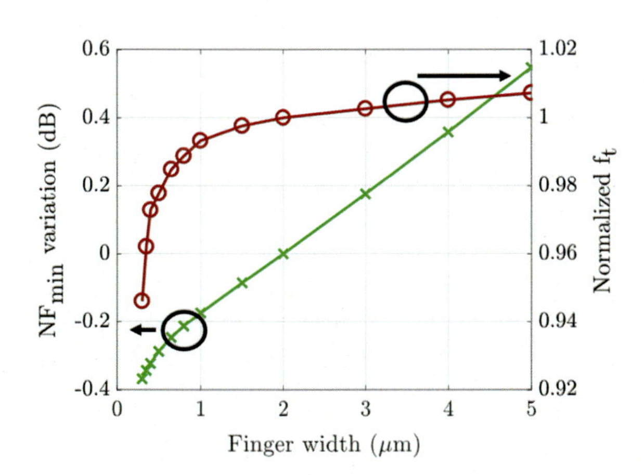


**FIGURE 9.12** $NF_{min}$ and $f_t$ variation with finger width variation of $NF_{min}$ at 28 GHz (left axis) and normalized $f_t$ (right axis) for various finger widths. Peak $f_t$ is typically in the 300–350 GHz range in advanced CMOS technologies and state-of-the-art values of $NF_{min}$ are reported in Table 10.6.

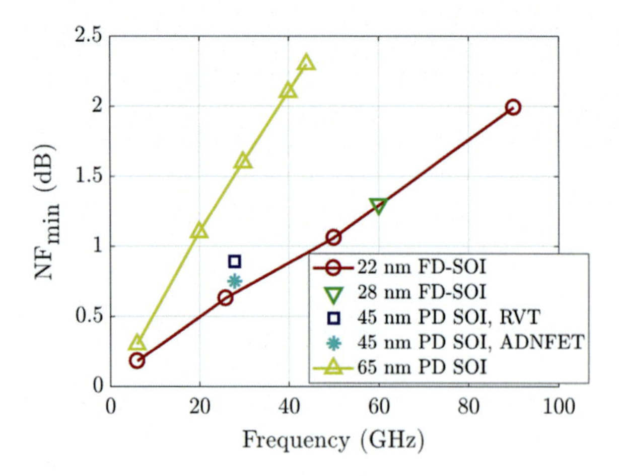


**FIGURE 9.13** NF $_{min}$ versus frequency for recent FD-SOI and PD-SOI technologies $NF_{min}$ versus frequency for recent FD-SOI and PD-SOI technologies.

from Eq. (10.12). Fig. 9.13 summarizes $NF_{min}$ values reported in the literature at mm-wave frequencies for recent FD-SOI and PD-SOI technologies (Cathelin, 2017; Chan et al., 2019; Jain et al., 2021; Tagro et al., 2012).

In order for the final LNA's overall NF to be equal to $NF_{min}$, a proper impedance has to be presented to the LNA's input (corresponding to $1/Y_{opt}$ in Eq. (10.13)). Such impedance does not necessarily correspond to the maximum power transfer impedance (conjugate impedance matching). This represents the second main trade-off between NF and gain ($S_{21}$), which can be relaxed with an $R_n$ as low as possible. Nevertheless, it is sometimes possible to bring the two optimum impedances (for noise and for gain) close to each other, with a careful choice of total transistor width.

### 9.4.4 State-of-the-art mm-wave low noise amplifiers: topology and performance review

mm-wave LNAs usually adopt a traditional architecture of common-source configuration for minimum noise; or a cascode or common gate topologies when noise can be traded-off with gain and/or bandwidth. Inductance degeneration is often implemented to improve stability and ease input matching at the cost of some gain reduction. Despite the gain reduction, inductance degeneration is still implemented in LNAs working in the high mm-wave spectrum (Gao et al., 2018, 2020). Single-ended as well as differential topologies can be found in recent literature. The differential topology usually includes neutralization capacitors to mitigate the feedback through $C_{gd}$ and thereby improve stability at the cost of doubled power consumption (Ding et al., 2019). Spiral inductors are typically implemented as part of the matching network in circuit design up to roughly 60 GHz while transmission-line based matching is typically preferred above around 100 GHz (Karaca et al., 2017). Lumped metal-on-metal capacitors are still used with high-quality factors up to 100 GHz (Gao et al., 2020; Hietanen et al., 2021). Transformers and baluns are commonly found in differential topologies, but not only

**TABLE 9.6** Best-in-class low noise amplifiers covering 5G FR2 bands in CMOS silicon-on-insulator technologies.

| Ref | Technology | Frequency [GHz] | Peak gain [dB] | Min NF [dB] | NF [dB] | $P_{dc}$ [mW] |
|---|---|---|---|---|---|---|
| Li, El-Aassar, et al. (2018) | 45 nm PD-SOI | 14–31 | 12.8 | 1.4 (24–28 GHz) | 1.4–1.7 (23–30 GHz) | 15 |
| Zhang et al. (2019) | 22 nm FD-SOI | 19–34 23–40 | 12 12.6 | 1.46 1.35 | 1.46–1.7 1.35–1.5 | 9.8 13 |
| Mohsen et al. (2019) | 28 nm FD-SOI | 24–27[a] | 13[a] | 2.8 | 2.8–3[a] | / |
| Konidas et al. (2019)[b] | 28 nm FD-SOI | 29.5–31.5[a] | 24.1 | 3.7 | 3.7–4[a] | 9.3 |
| Parlak and Buckwalter (2011b) | 45 nm PD-SOI | 40–50 | 18.5 | 2.9 | 2.9–4 | / |
| Chauhan and Floyd (2018) | 45 nm PD-SOI | 24–44 | 20 | 4.2 | 4.2–5.5 | 58 |
| El-Aassar and Rebeiz (2020) | 22 nm FD-SOI | 21.6–32.8 19.5–29 23–27 | 7.8 10.2 16.9 20.1 23.2 28.5 | 2.65[c] 2.2[c] 2.18[c] 2.08[c] 2.38[c] 2.25[c] | / / / / / / | 6 15 3.2 9.6 5.5 20 |
| Cui and Long (2020) | 22 nm FD-SOI | 22–32 22–32 | 21.5 17 | 1.7 2.1 | 1.7–2.2 2.1–2.9 | 17.3 5.6 |
| Gao and Rebeiz (2019) | 22 nm FD-SOI | 24–43 24–43 | 23 18.2 | 3.1 3.4 | 3.1–3.7 3.4–4.3 | 20.5 12.1 |
| Xu et al. (2021) | 22 nm FD-SOI | 26.6–31.6/ 36.4–39.3 dual band | 19.3/ 24 | 4.5[a] | 6–4.5[a]/ 4.9–5.5[a] | 11.4 |
| Xu et al. (2020) | 22 nm FD-SOI | 27.1–29.4/ 37.7–43.9 dual band | 24.8/ 22.4 | 3.6 | 3.6–4.9 | 13.6 |
| Nyssens et al. (2022) | 22 nm FD-SOI | 36.7–43.3 36.7–42.5 | 19.9 14.8 | 2.5 3.2 | 2.5–2.6 3.2–3.4 | 20.8 7.4 |
| Hu and Chi (2021) | 45 nm PD-SOI | 27–46 | 21.2 | 2.4 | 2.4–4.2 | 25.5 |
| Li et al. (2021) | 45 nm PD-SOI | 22–33 37–40 | 23 25 | 2.7 2.3 | / / | 28 18.6 |
| Ma et al. (2015) | 0.25 μm SiGe | 25–34 | 26.4 | 2.1 | 2.1–3.5 | 134 |
| Chen et al. (2018) | 0.25 μm SiGe | 29–37 | 28.5 | 3.1 | 3.1–4.1 | 80 |
| Micovic et al. (2016) | 20 nm GaN | 30–39.3 30–37 | 27 14.5[a] | 1 1.9 | 1–1.6 1.9–2.2 | 82 5 |

[a]*Estimated from figure.*
[b]*Simulated results.*
[c]*mean value of NF within the 3-dB bandwidth.*

(Cui & Long, 2020; Gao et al., 2020; Gao & Rebeiz, 2019). Transformers can compact layouts at very high frequencies compared with pieces of transmission line to implement inductive elements. For some specific designs, requiring ultrawide bandwidth or at very high frequencies, more sophisticated architectures can be preferred, such as transformer feedback for ultra-wideband behavior (Cui & Long, 2020) or gain-boosting techniques with a transformer at the cascoded gate and drain sides to boost gain over a large bandwidth (Gao et al., 2018, 2020).

CMOS SOI technologies are very attractive for LNA implementation at mm-wave frequencies, able to fulfill the constraining requirements for 5G telecommunications, at competitively low power consumption. LNAs at the state of the art designed in 45RFSOI, 22FDX, and 28 nm FD-SOI technologies demonstrate LNA performances superior to SiGe and close to III–V technologies over the 5G mm-wave bands as reported in Table 9.6. Moreover, recent publications have been focusing on developing LNAs with incredibly low-power consumptions (El-Aassar & Rebeiz, 2020), or wideband designs covering multiple frequency bands used or foreseen for 5G, targeting the required specifications in several mm-wave 5G frequency bands (see Table 9.6), such as from 24 to 42 GHz (Chauhan & Floyd, 2018; Cui & Long, 2020; Gao & Rebeiz, 2019; Li et al., 2021; Xu et al., 2020).

Thanks to the high oscillation frequencies (reported in Chapter 2.) and low-noise performance of advanced SOI nodes (shown in Fig. 9.13), achieved through front-end optimizations (reduced RF parasitics, notably at the gate access), and to the quality of available passives in the BEOL (thick metals, away from the substrate), the high-performance mm-wave LNAs in Table 9.6 have been achieved, well establishing these nodes as well suited for LNA FEM blocks.

## 9.5 Power amplifier

The last element in the TX chain (before the switch) is the PA. This circuit block faces tremendous design challenges due to the stringent requirement of being able to deliver large amount of power at a high efficiency and with high linearity. Indeed, the transceiver's overall power consumption is usually dominated by the PA, which therefore has a major impact on battery life in portable applications (such as mobile handsets).

The high-order digital modulations envisaged for 5G have a large peak-to-average power ratio (PAPR), of the order of 8–9 dB. This can be visually understood as a wide probability distribution function (PDF) such as illustrated in Fig. 9.14C, therefore requiring a high degree of linearity for a low bit error rate and error vector magnitude (EVM)[2] (Acar, 2022; Padaki & Reed, 2014). Due to the large signal bandwidth, digital predistortion (DPD) is power hungry and is not a solution of immediate interest, that is, the

[2] EVM is a system-level metric to quantify the performance of a transmitter-receiver system. It is frequently defined for systems with digital modulation signals. EVM is closely related to the BER of a system. EVM is mainly influenced by channel noise, noise phase from the oscillator as well as system nonlinearities, more details can be found in the reference immediately after this footnote. Indeed, system nonlinearities can degrade EVM because of intermodulation products at the receiver, which can fall within the signal bandwidth.

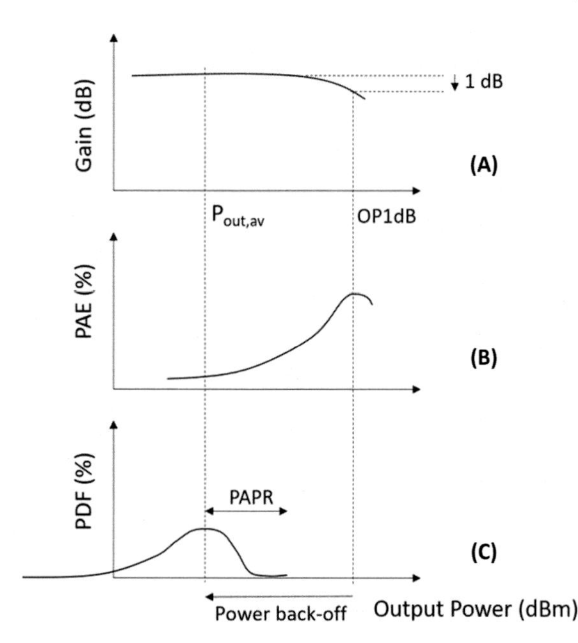


**FIGURE 9.14** Sketch of typical gain, power added efficiency (PAE), and probability density function versus P $_{out}$ curves of high-order modulation sketch of typical gain versus $P_{out}$ (A), PAE versus $P_{out}$ (B) curves, and probability density function of high-order modulation (C).

requirements for 5G systems should be met without the use of digital predistortion (Asbeck et al., 2019). With more investigation in this topic, DPD could be envisaged in the future to improve 5G systems.

To achieve such high linearity, the average signal power is therefore backed-off (from the 1-dB compression point), since the PA is always more linear the further away its regime is from compression: the average power of the PDF in Fig. 9.14C is shifted to lower values such that the PDF lies across a linear region in the gain versus $P_{out}$ curve (shown in Fig. 9.14A). Using OFDM instead of single carrier also intensifies the need for high linearity, which is translated in an additional 1.5 to 2 dB of power back-off. The main drawback of working in power back-off is the loss in average efficiency of the PA, which is maximum at compression and decreases at lower power as shown in Fig. 9.14B. Despite the obvious advantages of Si-based technologies for high integration and the high performance demonstrated above in LNAs and switches, the PA still remains the bottleneck for these technologies due to their intrinsic device limitations to being able to deliver large amount of power at high efficiency. Detailed explanations are given below, revealing why GaN technologies are more suited to these circuits.

## 9.5.1 Power amplifier basics—the loadline concept

First, let us introduce the loadline concept useful to understand the main limitations. Let us consider the simple example depicted in Fig. 9.15 showing a common-source transistor loaded by a resistor $R_L$ biased for class A operation.

The loadline concept is illustrated in Fig. 9.16 on top of the DC output characteristics of the transistor, that is, $V_{ds}$ versus $I_d$ for different $V_{gs}$. At a given input power, the drain

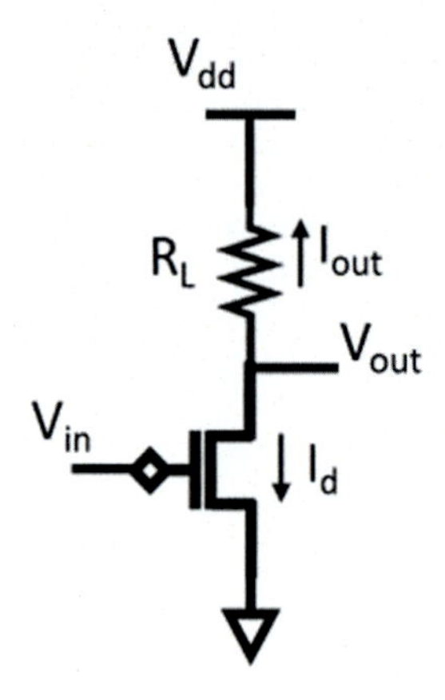


FIGURE 9.15 Common-source FET loaded by a resistor $R_L$ as a basic power amplifier (PA) configuration. Common-source FET loaded by a resistor $R_L$ as a basic PA configuration.

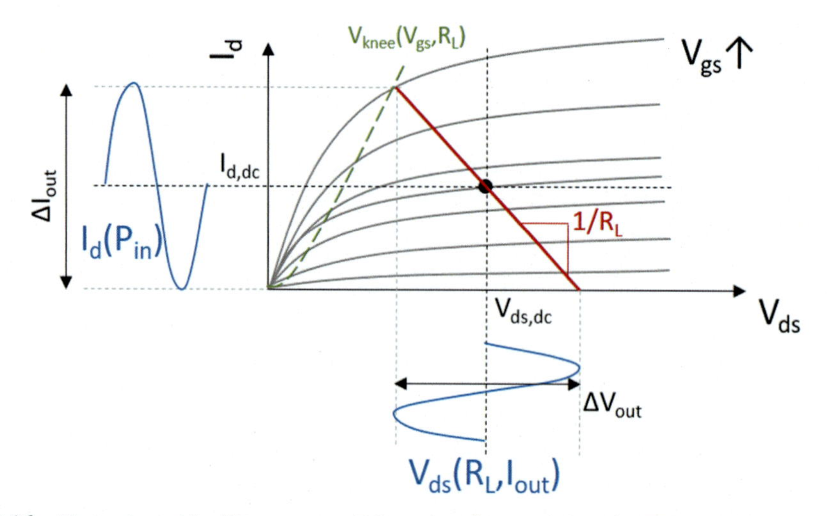


FIGURE 9.16 Illustration of loadline concept. DC output characteristics of a FET and illustration of loadline concept.

current follows the input signal and the $V_{ds}$ swings accordingly with an amplitude that depends on the load as expressed below.

$$V_{ds} = V_{dd} - I_d R_L. \tag{9.16}$$

This $V_{ds}(I_d)$ is a straight line due to the purely resistive load, and is highlighted in red in Fig. 9.16. For high input power, the output $V_{ds}$ will be clipped at a minimum value, where the transistor enters in the triode region, called knee volgate ($V_{knee}$, in green in Fig. 9.16) and the maximum value of $V_{ds}$ is determined by the bias point and $R_L$ when $I_d$ reaches 0 A.

The output root mean square (RMS) power is given by the product of the current and voltage RMS amplitudes ($P_{out} = \Delta V_{out} \Delta I_{out}/8$[3]). The $\Delta V_{out}$ and $\Delta I_{out}$ are a function of the bias and load, the output power can then be optimized by a careful choice of ($V_{ds,dc}$, $I_{d,dc}$) and $R_L$. A class A PA is biased at the point ($V_{ds,dc}$, $I_{d,dc}$) that corresponds to half of the maximum allowed $I_d$ ($I_{d,max}$ is fixed by electromigration rules or corresponding maximum $V_{gs}$ preventing oxide breakdown) and $V_{ds,dc}$ is half of the maximum allowed $V_{ds}$ ($V_{ds,max}$ is a portion of the breakdown voltage, guaranteeing reliable operation throughout the circuit lifetime). The corresponding optimum load maximizes the output voltage and current swings and is given by $R_L = V_{ds,max}/I_{d,max}$.

### 9.5.2 Power added efficiency: a detailed contributors analysis

The PAE FoM metric of a PA is given by

$$\text{PAE} = \frac{P_{out} - P_{in}}{P_{dc}} = \frac{P_{out}}{P_{dc}}\left(1 - \frac{1}{G}\right), \tag{9.17}$$

where G is the forward gain of the amplifier: $G = P_{out}/P_{in}$.

The PAE can be re-written as a product of several factors (Thomas et al., 2018):

$$\text{PAE} = F_{gain} F_{matching,Q} F_{waveform} F_{Vswing}. \tag{9.18}$$

$F_{waveform}$ describes the overlap function of drain voltage and current. Depending on the class of operation (A, B, switching classes), its value can be from 0.5 to 1 for ideal switching. It depends on the biasing (class of operation) as well as harmonic matching and therefore not on technology (except perhaps for losses in harmonic termination, but this is accounted for in the $F_{matching,Q}$ term). In the class A example provided above, by assuming the PA is biased at an optimum point and an optimum impedance is presented to the FET, and by assuming $V_{knee} = 0$ V (thus $V_{ds,dc} = \Delta V_{out}/2$, $I_{ds,dc} = \Delta I_{out}/2$, and $R_L = \Delta V_{out}/\Delta I_{out}$), then the RF power delivered at the load at maximum power is

$$\left|P_{out}(t)\right| = \left|\frac{\Delta V_{out} \Delta I_{out} \cos^2(\omega t)}{4}\right|, \tag{9.19}$$

since $I_{out} = - I_d$ and the load and the RF $V_{ds}$ and $V_{out}$ are equal. Under these assumptions, the dc power is $V_{ds,dc} I_{ds,dc} = \Delta V_{out} \Delta I_{out}/4$, and the RMS $P_{out}$ at the load is $\Delta V_{out} \Delta I_{out}/8$. Therefore, it is plain to deduce that a maximum value of 0.5 in PAE ($P_{out}/P_{dc}$) is attained by class A PAs. This value can be enhanced to 0.78 with class B operation and even theoretically to 1 with class F or switching-mode amplifiers (class D and E). In these cases, the

[3] In general, $P_{out} = \Delta V_{out} \Delta I_{out} \cos(\theta)/8$, where $\theta$ is the phase difference between $I_{out}$ and $V_{out}$, if the load has a reactive element. $P_{out}$ is obviously maximized when $\theta = 0$, that is, presenting a purely real load. As frequency increases, the parasitic capacitances (access and even extrinsic elements of the FET) induce a non-negligible reactive component to the load that must be compensated. Indeed, at mm-wave frequencies, the optimal load generally has a positive imaginary part to compensate for the capacitive parasitics. For simplicity, we neglected such considerations related to reactive elements in this example.

waveforms are such that the device does not dissipate power by having $I_d = 0$ on half of a period and $V_{ds} = 0$ on the other half. Switching-mode amplifiers are highly nonlinear and therefore not suited to 5G waveforms, while class AB, B, C, and F amplifiers require some harmonic matching. Due to the increased losses associated to additional passive elements required for harmonic matching, harmonic engineering beyond the second harmonic is usually not implemented at mm-wave frequencies.

$F_{matching,Q}$ is related to the losses in the output matching network, harmonic termination, and power combining, thus to the quality factors ($Q$) of the matching elements. Provided the last stage features enough gain, the losses in matching circuits from input matching or previous stages will not affect the PAE. From Eq. (10.17), if the gain of the last stage is large enough, PAE $\approx P_{out}/P_{dc}$ and does not depend on the input power delivered to the last stage. It implies that a reduction of power delivered at the input of the last stage (for instance due to low-Q matching circuits before the last stage) has little effect on the overall PAE of the PA. On the contrary, any losses at the output of the last stage will decrease the power delivered at the load ($P_{out}$) and thereby directly decrease the overall PAE. As explained in Chapter 2, the quality factor of passive devices has greatly been improved in CMOS SOI technologies in reducing losses by (1) using thick Cu and Al metal layers in technologies targeting RF and mm-wave applications and (2) replacing the underlying standard resistivity wafers (with a nominal resistivity of around 10 Ωcm) with high-resistivity [trap-rich (TR)] silicon wafer (resistivity of several kΩcm).

The term $F_{gain}$ in Eq. (10.18) corresponds to the $(1-1/G)$ term in Eq. (10.17). It accounts for the amount of RMS input power required to drive the PA, which might be non-negligible. The gain $G$ is determined by the $f_t$ & $f_{max}$ of the transistor, along with matching network. As shown in Chapter 2, $f_t$ & $f_{max}$ in the 300 GHz range can be found in CMOS SOI technologies thanks to the shrunk gate length, enabling sufficient gain for operation in the mm-wave frequency range.

The term $F_{Vswing}$ in Eq. (10.18) describes the loss in efficiency due to the nonzero value of the knee voltage. $F_{Vswing}$ is roughly equal to $1 - V_{knee}/V_{dd}$, where $V_{dd}$ is the supply voltage (Fig. 9.16). This term can be evaluated from the dc $I_d - V_{ds}$ curves of the transistor and the load resistance (cfr. green dashed curve in Fig. 9.16). $V_{knee}$ is related to the ON-resistance ($R_{on}$) of the transistor in linear region and the maximum current density. Decreasing $V_{knee}$ can be achieved with a sharp $I_d$-$V_{ds}$ transition in the linear region (i.e., low $R_{on}$) or with a low maximum current. The efficiency is maximized by using a $V_{dd}$ as high as possible, as long as reliability is guaranteed (breakdown, etc.). This $F_{Vswing}$ is governed by a FoM similar to Baliga's first FoM (Asbeck et al., 2019; Baliga, 1982):

$$V_{bk}^2 R_{on} = \alpha \mu \varepsilon E_{bk}^3 A_c. \tag{9.20}$$

$A_c$ is the effective area for current flow, $E_{bk}$ the breakdown electric field, which depends on the size of the bandgap. $\varepsilon$ is the permittivity, $\mu$ the mobility, and $\alpha$ is a numerical coefficient of order unity. The inferiority of CMOS in PA application is clearly evidenced here due to low breakdown electric field (and mobility), with respect to III–V technologies, in particular GaN (Lie et al., 2020).

## 9.5.3 Techniques for increased output power

The other important requirement or FoM of the PA is the maximum or saturated output power ($P_{sat}$). Increasing the output power can be realized by (1) increasing the voltage swing and (2) increasing the current swing (thus maximum current). The voltage swing is mainly limited by reliability issues linked to breakdown voltage of the transistor. The $I_{d,max}$ (and current swing) can be increased by increasing the transistor width. Despite the fact that increasing much the FET width reduces $f_t$ & $f_{max}$ (due to additional parasitics), the resulting optimal load resistance to be presented at the output decreases proportionally, to the point of making the impedance matching networks very difficult. Low values of load resistances below 10–20 Ω are difficult to implement, which has the effect of increasing losses originating in complex matching networks, thus reducing the power improvement and mostly the efficiency.

### 9.5.3.1 *Transistor stacking*

FET stacking is a technique that circumvents the low breakdown voltage of individual FETs and enables an increase in the output voltage swing as illustrated in Fig. 9.17. With a proper choice of the gate capacitances ($C_{gk}$, $k = 2, 3, \ldots, N$), the drain-source voltage swing across each FET ($V_{ds,k}$, $k = 1, 2, 3, \ldots, N$) is equal such that the total output voltage swing is multiplied by the number of stacked-FETs ($N$). Besides an increased power due to larger voltage swing, for a constant load impedance ($R_{opt}$), the FET widths can also be multiplied, such that the gain in power is proportional to $N^2$.

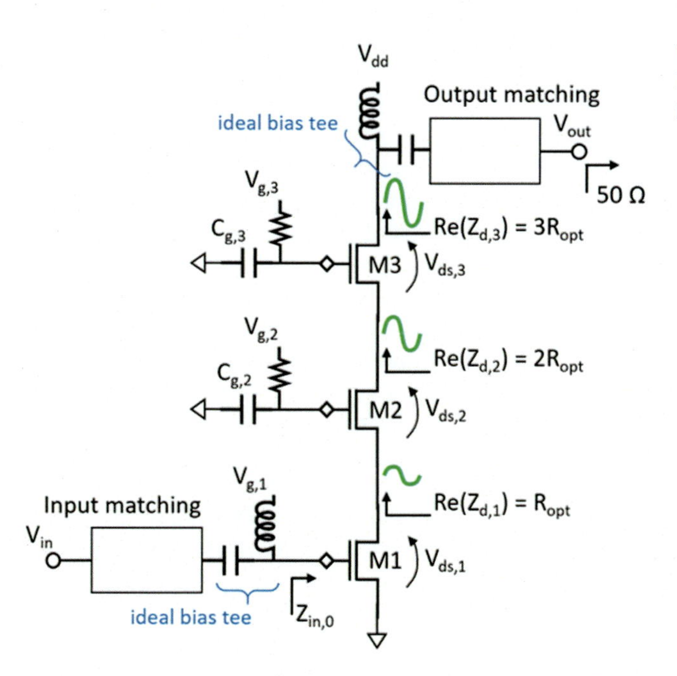


**FIGURE 9.17** FET stacking concept. FET stacking concept to increase output power (with ideal bias tee).

Unfortunately, the maximum number of FETs that can be stacked is limited to 2 or 3 in practice. First, by increasing the transistor width, additional parasitics lower the $f_t$ & $f_{max}$, thus gain, and decreases the input impedance ($Z_{in,0}$), therefore complexifying the input matching. Second, due to resistive losses in the transistor, and to phase misalignment of the drain voltages, the effective power gain is limited in practice beyond $N=3$ (Dabag et al., 2013). Phase misalignment arises from the fact that $C_{gk}$ tunes out the real part of the impedance seen at each drain ($Re(Z_{dk})$, to achieve equal voltage drop across each FET), but not the imaginary part (an inductive load should be presented to compensate the total drain capacitance at each FET, whereas the above stacked transistor presents a capacitive load). Additional passive elements can be added to compensate the phase misalignment, but at the cost of additional losses through them. The interested reader can find more information in Dabag et al. (2013), which provides a detailed theoretical analysis of FET stacking for PA. SOI technologies are particularly suited to FET stacking compared to bulk thanks to reduced and more linear junction capacitances (reduced phase misalignment).

#### 9.5.3.2 *Power combining*

FET stacking is a common technique used, either alone or combined with other techniques, in practically all SOI-based PAs in recent literature. Many demonstrator circuits employing FET stacking have been published (Aikio et al., 2019; Dabag et al., 2013; Jayamon et al., 2016; Rostomyan, Ozen, et al., 2018; Thomas et al., 2018). In PD-SOI (notably in 45RFSOI node at mm-waves), large $P_{sat}$ values are achieved with good or outstanding peak PAE, for instance (1) $P_{sat}$ going up to 24.8 dBm for 29% peak PAE at 29 GHz of a 4-stack PA (Jayamon et al., 2016) or (2) $P_{sat}$ of 19.5 dBm with 46.7% peak PAE at 27 GHz of a 2-stack pMOSFET PA (Thomas et al., 2018) and (3) $P_{sat}$ of 18.9 dBm with 40.5% peak PAE at 26.75 GHz of a 2-stack nMOSFET PA (Rostomyan, Ozen, et al., 2018). In Aikio et al. (2019), the 3-stack PA implemented in 22FDX (FD-SOI technology) achieves 18.8 dBm $P_{sat}$ for a peak PAE of 23.4% at 28.5 GHz. Due to their shorter gate lengths FD-SOI nodes are limited to lower breakdown voltages compared with PD-SOI technologies. In FD-SOI conventional FET stacking is employed for single PA cells, then, beyond that, the outputs of multiple PA cells are combined together using passive combiners to further increase the overall output power.

Power combining can be implemented in many different ways. A (pseudo-) differential configuration is a very common and efficient technique, since it has the advantage of improving stability thanks to capacitance neutralization and can be combined with baluns (often implemented with transformers (Cheung et al., 2004; Zhao & Long, 2012) in a very compact way (Çelik & Reynaert, 2019)). Additional (or distinct) power combining can be achieved with current (Zong et al., 2020) or voltage combining (Gu et al., 2012). A balanced PA topology also increases output power, with additional benefits in terms of third-order intermodulation product and voltage standing wave ratio robustness (Torres et al., 2020). In the case of an on-chip antenna, power combining can also be realized by feeding the antenna through multiple paths (Chi et al., 2017). In practice, however, power combining is limited by (1) losses in the power combiner elements and (2) area.

## 9.5.4 State-of-the-art power amplifiers with 5G modulations, the challenge of high efficiency at high power back-off

In complex modulations such as the ones used or envisaged in 5G, the high PAPR of the signal (additionally to the OFDM) requires the PA to work with a large amount of back-off, thus operating less efficiently on average to meet signal integrity requirements. The 4-stack PA in 45RFSOI presented in Jayamon et al. (2016), for instance, featuring a $P_{sat}$ of 24.8 dBm and 29% peak PAE, has a mean $P_{out}$ and PAE of 15 dBm and 14%, respectively, with a 64-QAM OFDM signal. Several techniques are able to increase average PAE in back-off operation. The goal is not to present them in details here, but only to give an overview of their implementations in PD-SOI and FD-SOI technologies. For more details, the interested reader is invited to consult (Cripps, 2006) for instance. The Doherty architecture is composed of a main PA, always active, and an auxiliary PA that turns on to give extra power close to compression. It has a much better efficiency at back-off, with a theoretical peak PAE at 6 dB of power back-off. Several PAs with this architecture have already been published in both PD-SOI (Rostomyan, Özen, et al., 2018; Rostomyan et al., 2019; Wang et al., 2019) and FD-SOI technologies (Zong et al., 2021). A Chireix outphasing architecture was implemented in Li, Chi, et al. (2018) using two 45RFSOI chips and performing the power combination and outphasing at the antenna feed level on package. Their performance with 5G modulation schemes is reported in Table 9.7.

Supply modulation is also a technique used to improve the average efficiency when the PA is fed by a complex modulated signal. Two variations of supply modulation exist, envelope elimination and restoration (EER) and envelope tracking (ET). EER is more difficult to implement than ET due to several factors described in Modi et al. (2012). ET consists of continuously changing the $V_{dd}$ supply bias, following the signal envelope, to continuously work close to the PA's compression region—that is modulated through time according to the envelope—where the PAE is high. Its implementation in 4G LTE telecommunication is already nonstraightforward due to the large signal bandwidth, which reduces the supply modulation efficiency and overall system efficiency. Although several variations exist and have been applied to LTE signals (Asbeck & Popovic, 2016; Modi et al., 2012; Popovic, 2017), there has not yet been any ET PA implementation supporting $\geq 200$ MHz modulation bandwidth for 5G signals, to the best of the authors' knowledge.

### *9.5.4.1 Back-gate biasing—an opportunity for radio frequency designers*

One possible direction to enhance the current state of the art of FD-SOI is through the BG biasing. As detailed in the previous section, the BG terminal is a tuning knob that could enable novel techniques. It has attracted the attention of the RF community only recently, with a few publications that will be detailed in the following. Several studies of the PA static response to BG bias variation have been reported. In Torres et al. (2020), the BG tunes the gain of the PA, yielding a variable gain PA with roughly constant large-signal power performances. In Mayeda et al. (2019) and Rusanen et al. (2019), linearity is studied and more precisely AM-PM (output signal phase variation due to input signal amplitude variation) distortion. It is shown that by varying the BG bias, the PA changes its class of operation from deep class AB to class A and the AM-PM curves change

**TABLE 9.7** Summary of state-of-the-art CMOS-silicon-on-insulator-based power amplifier performance in continuous-wave (CW) mode and with 64QAM on a single-carrier (SC) or on orthogonal frequency division multiplexing.

| References | Technology | Frequency [GHz] | $P_{sat}$ [dBm] | P1dB [dBm] | Peak PAE [%] | QAM modulation | Data rate [Gb/s] | $EVM_{rms}$ [dB] | $P_{ave}$ @ $EVM_{rms}$ [dBm] | Efficiency @ $EVM_{rms}$ [%] |
|---|---|---|---|---|---|---|---|---|---|---|
| Zong et al. (2021) | 22 nm FD-SOI | 28 | 22.5 | 21.1 | 28.5 | 64 (SC-/) | 2.4 | −25.1 | 10.9 | 9.2 |
| Wang et al. (2019) | 45 nm PD-SOI | 27 | 23.3 | 22.4 | 40.1 | 64 (SC-/) | 6 15 | −25.3 to 24 | 15.9 15 | 29.1 26.4 |
| Li, Chi, et al., 2018 | 45 nm PD-SOI | 28 | 17.1 | / | 56 | 64 (SC-/) | 6 15 | −29.7 to 27.5 | 10.7 10.4 | 36 34 |
| Zong et al. (2020) | 22 nm FD-SOI | 28 | 21.7 | 19.1 | 27.1 | 64 (SC-/) | 3 | −25 | 12.9 | 9 |
| Rostomyan et al. (2019) | 45 nm PD-SOI | 26 | 25 | 23.2/ 24.2 | 31 | 64 (SC-50 MHz) | / | −25.2 | 17.2 | 23 |
| Rostomyan, Özen, et al. (2018) | 45 nm PD-SOI | 28 | 22.4 | 21.5 | 40 | 64 (OFDM-800 MHz) | / | −25.2 | 13 | 16.8 |
| Jayamon et al. (2016) | 45 nm PD-SOI | 29 | 23.5 | / | 30 | 64 (OFDM-800 MHz) | / | −25.2 | 15 | 14 |
| Thomas et al. (2018) | 45 nm PD-SOI | 27 | 19.5 | / | 46.7 | 64 (OFDM-800 MHz) | / | −25.2 | 9.2 | 17 |

accordingly. In Zong et al. (2021), the BG bias is used in the adaptive gate biasing of the auxiliary PA (in a Doherty configuration) to control its turn-on point and power gain flatness. Application of the dynamic variation of BG bias is more scarcely found. In Kim et al. (2021), adaptive BG biasing is implemented to improve linearity thanks to a reduction of the third harmonic, hence third-order intermodulation, and is used to reduce the amount of back-off power for high-order modulations, thereby increasing the average efficiency.

One can see that the state of the art in PD-SOI technologies have demonstrated PAs with sufficient average power ($\sim$15 dBm) and efficiency ($\sim$20%) for a successful integration in FEMs for 5G applications. PAs in FD-SOI technologies are currently still lacking a bit of power and have lower efficiency, but a strong effort is currently in progress in this research area to fill this gap, which might yet be overcome in a near future.

## 9.6 Conclusion

Even though SOI technologies have been under development for more than 20 years to improve CMOS scaling in the digital domain, its specific characteristics are particularly suited to analog and RF applications. PD-SOI and FD-SOI technologies feature nowadays the best RF FoMs among CMOS technologies, comparable to SiGe, GaAs and getting closer to other III–V technologies. They further benefit from low-cost manufacturability for mass market applications, being full deployed in 300 mm diameter process flows. The high $f_t$ & $f_{\max}$ in the 350 GHz range reported for the advanced PD-SOI and FD-SOI CMOS technologies as well as the availability of low-loss passive circuit elements makes them great contenders for mm-wave applications. Indeed, the electrical isolation provided by the BOX provides a unique feature compared to other CMOS technologies, which allows for the separate engineering of the device characteristics and the substrate performance. In particular, the TR high-resistivity Si substrate offers great performances in terms of small-signal losses, crosstalk isolation, and nonlinearity generation. Even though the TR substrate is not necessarily compatible with utlra-thin body and BOX (UTBB) FD-SOI devices, other solutions (mentioned in Chapter 2) exist (Rack et al., 2019a, 2019b; Scheen et al., 2020), which also exhibit great improvements compared with the standard lossy Si substrate.

The low-cost production for mass market applications, along with their excellent RF FoMs, makes SOI technologies particularly interesting for the on-going and upcoming deployment of 5G handsets. The great $R_{\text{on}}C_{\text{off}}$ FoM of both PD-SOI and FD-SOI technologies yields low-loss wideband switches based on well-established standard topologies. mm-wave LNAs in SOI technologies demonstrate excellent noise performances ($\sim$1.4–2.3 dB of NF), meeting the $<$2–3 dB noise requirements in a receiver for FR2 frequency bands (24, 28, 39 GHz) 5G handset applications set by the link budget. Furthermore, state-of-the-art mm-wave PA in PD-SOI technologies have demonstrated sufficient average power ($\sim$15 dBm) with decent efficiency ($\sim$20%) using representative 5G modulated signals (64QAM OFDM), enabling a successful integration in FEMs for handset applications. Although PAs in FD-SOI technologies are currently lacking a bit of power and efficiency due to shorter channels withstanding lower voltages, strong effort in this research field might yet overcome that weakness in a near future. Integrating FD-SOI technologies on high-resistivity, low-loss substrate is another path to be investigated to narrow the performance gap with PD-SOI-based PAs. Indeed, reducing the passives losses can directly improve the overall PA efficiency or opens the way to additional power combining through low-loss passives to reach a higher output power. Furthermore, the feature of BG biasing—an RF tuning knob, unique to UTBB FD-SOI technologies—might be the track for SOI to leapfrog ahead of other (Bi)CMOS technologies. Indeed, the great integration of the advanced UTBB nodes coupled with their outstanding mm-wave performances opens the door to fully integrated, robust, programmable, and low-cost FEMs for 5G handset applications in the mm-wave spectrum.

## References

3rd Generation Partnership Project. (2019). Unpublished content 3GPP TR 21.915. Release 15. Available from https://www.3gpp.org/lte-2.

Acar, E. (2022). How error vector magnitude (EVM) measurement improves your system-level performance. Available from https://www.analog.com/en/technical-articles/how-evm-measurement-improves-system-level-performance.html.

Aikio, J. P., Hietanen, M., Tervo, N., Rahkonen,T., & Pärssinen, A. (2019). Ka-band 3-stack power amplifier with 18.8 dBm Psat and 23.4% PAE using 22nm CMOS FDSOI technology. In: *2019 IEEE topical conference on RF/microwave power amplifiers for radio and wireless applications (PAWR)* (pp. 1–3). Available from https://doi.org/10.1109/PAWR.2019.8708719.

Asbeck, P. M., Rostomyan, N., Ozen, M., Rabet, B., & Jayamon, J. A. (2019). Power amplifiers for mm-Wave 5G applications: Technology comparisons and CMOS-SOI demonstration circuits. *IEEE Transactions on Microwave Theory and Techniques*, *67*(7), 3099–3109. Available from https://doi.org/10.1109/TMTT.2019.2896047.

Asbeck, P., & Popovic, Z. (2016). ET comes of age: Envelope tracking for higher-efficiency power amplifiers. *IEEE Microwave Magazine*, *17*(3), 16–25. Available from https://doi.org/10.1109/MMM.2015.2505699.

Asgaran, S., Deen, M. J., & Chen, C.-H. (2004). Analytical modeling of MOSFETs channel noise and noise parameters. *IEEE Transactions on Electron Devices*, *51*(12), 2109–2114. Available from https://doi.org/10.1109/TED.2004.838450.

Balasubramaniyan, A., & Bellaouar, A. (2018). RF/mmWave front-end module switch in 22nm FDSOI process. In: *2018 IEEE SOI-3D-subthreshold microelectronics technology unified conference (S3S)* (pp. 1–2). Available from https://doi.org/10.1109/S3S.2018.8640171.

Baliga, B. J. (1982). Semiconductors for high-voltage, vertical channel FETs. *Journal of Applied Physics*, *53*(3), 1759–1764. Available from https://doi.org/10.1063/1.331646.

Cappy, A. (1988). Noise modeling and measurement techniques (HEMTs). *IEEE Transactions on Microwave Theory and Techniques*, *36*(1), 1–10. Available from https://doi.org/10.1109/22.3475.

Cardoso, A. S., Saha, P., Chakraborty, P. S., Fleischhauer, D. M., & Cressler, J. D. (2014). Low-loss, wideband SPDT switches and switched-line phase shifter in 180-nm RF CMOS on SOI technology. In: *2014 IEEE radio and wireless symposium (RWS)* (pp. 199–201). Available from https://doi.org/10.1109/RWS.2014.6830161.

Cathelin, A. (2017). Fully depleted silicon on insulator devices CMOS: The 28-nm node is the perfect technology for analog, RF, mmW, and Mixed-Signal System-on-Chip Integration. *IEEE Solid-State Circuits Magazine*, *9*(4), 18–26. Available from https://doi.org/10.1109/MSSC.2017.2745738.

Çelik, U., & Reynaert, P. (2019). An E-band compact power amplifier for future array-based backhaul networks in 22nm FD-SOI. In: *2019 IEEE radio frequency integrated circuits symposium (RFIC)* (pp. 187–190). Available from https://doi.org/10.1109/RFIC.2019.8701866.

Chan, L. H. K., Ong, S. N., Oo, W. L., Chew, K. W. J., Zhang, C., Bellaouar, A., Chow, W. H., Chen, T., Rassel, R., Wong, J. S., Lim, C. K., Wan, C. W. F., Kim, J., Seet, W. H., & Harame, D. (2019). 22nm Fully-depleted SOI high frequency noise modeling up to 90GHz enabling ultra low noise millimetre-wave LNA design. In: *2019 IEEE radio frequency integrated circuits symposium (RFIC)* (pp. 31–34). Available from https://doi.org/10.1109/RFIC.2019.8701875.

Chauhan, V., & Floyd, B. (2018). A 24–44 GHz UWB LNA for 5G cellular frequency bands. In: *2018 11th Global symposium on millimeter waves (GSMM)* (pp. 1–3). Available from https://doi.org/10.1109/GSMM.2018.8439672.

Chen, C., Lee, R., Tan, G., Chen, D. C., Lei, P., & Yeh, C. (2012). Equivalent sheet resistance of intrinsic noise in sub-100-nm MOSFETs. *IEEE Transactions on Electron Devices*, *59*(8), 2215–2220. Available from https://doi.org/10.1109/TED.2012.2198651.

Chen, C., Xu, X., & Yoshimasu, T. (2017). A DC-50 GHz, low insertion loss and high P1dB SPDT switch IC in 40-nm SOI CMOS. In: *2017 IEEE Asia Pacific microwave conference (APMC)* (pp. 5–8). Available from https://doi.org/10.1109/APMC.2017.8251363.

Chen, S., & Zhao, J. (2014). The requirements, challenges, and technologies for 5G of terrestrial mobile telecommunication. *IEEE Communications Magazine*, *52*(5), 36–43. Available from https://doi.org/10.1109/MCOM.2014.6815891.

Chen, X. (2017). *Noise characterization and modeling of nanoscale MOSFETs* (PhD dissertation). Hamilton, Ontario: McMaster University, Canada Unpublished content.

Chen, Z., Gao H., LeenaertsD., MilosevicD., & Baltus P. (2018). A 29–37 GHz BiCMOS low-noise amplifier with 28.5 dB Peak Gain and 3.1–4.1 dB NF. In: *2018 IEEE radio frequency integrated circuits symposium (RFIC)* (pp. 288–291). Available from https://doi.org/10.1109/RFIC.2018.8429020.

Cheung T. S. D., Long J. R., Tretiakov Y. V., & Harame D.L. (2004). A 21–27GHz self-shielded 4-way power-combining PA balun. In: *Proceedings of the IEEE 2004 custom integrated circuits conference (IEEE Cat. No.04CH37571)* (pp. 617–620). Available from https://doi.org/10.1109/CICC.2004.1358901.

Chi T., Wang F., Li S., Huang M.-Y., Park J. S., & Wang H. (2017). 17.3 A 60GHz on-chip linear radiator with single-element 27.9dBm Psat and 33.1dBm peak EIRP using multifeed antenna for direct on-antenna power combining. In: *2017 IEEE international solid-state circuits conference (ISSCC)* (pp. 296–297). Available from https://doi.org/10.1109/ISSCC.2017.7870378.

Cripps, S. C. (2006). *RF power amplifiers for wireless communications* (2nd). Boston: Artech House.

Cui, B., & Long, J. R. (2020). A 1.7-dB minimum NF, 22–32-GHz low-noise feedback amplifier with multistage noise matching in 22-nm FD-SOI CMOS. *IEEE Journal of Solid-State Circuits*, *55*(5), 1239–1248. Available from https://doi.org/10.1109/JSSC.2020.2967548.

Dabag, H.-T., Hanafi, B., Golcuk, F., Agah, A., Buckwalter, J. F., & Asbeck, P. M. (2013). Analysis and design of stacked-FET millimeter-wave power amplifiers. *IEEE Transactions on Microwave Theory and Techniques*, *61*(4), 1543–1556. Available from https://doi.org/10.1109/TMTT.2013.2247698.

Didioui A., Bernier C., Morche D., & Sentieys O. (2012). Impact of RF front-end nonlinearity on WSN communications. In: *2012 International symposium on wireless communication systems (ISWCS)* (pp. 875–879). Available from https://doi.org/10.1109/ISWCS.2012.6328493.

Dilli R. (2020). Analysis of 5G wireless systems in FR1 and FR2 frequency bands. In *2020 2nd International conference on innovative mechanisms for industry applications (ICIMIA)* (pp. 767–772). Available from https://doi.org/10.1109/ICIMIA48430.2020.9074973.

Ding Y., Vehring S., & Boeck G. (2019). Design of 24 GHz high-linear high-gain low-noise amplifiers using neutralization techniques. In: *2019 IEEE MTT-S international microwave symposium (IMS)* (pp. 944–947). Available from https://doi.org/10.1109/MWSYM.2019.8700859.

El-Aassar, O., & Rebeiz, G. M. (2020). Design of low-power sub-2.4 dB mean NF 5G LNAs using forward body bias in 22 nm FDSOI. *IEEE Transactions on Microwave Theory and Techniques*, *68*(10), 4445–4454. Available from https://doi.org/10.1109/TMTT.2020.3012538.

Electronics notes. (2022). 5G Frequency bands, channels for FR1 & FR2. Available from https://www.electronics-notes.com/articles/connectivity/5g-mobile-wireless-cellular/frequency-bands-channels-fr1-fr2.php.

Elgaard, C., Andersson, S., Caputa, P., Westesson, E., & Sjöland, H. (2019). A 27 GHz adaptive bias variable gain power amplifier and T/R switch in 22nm FD-SOI CMOS for 5G antenna arrays. In: *2019 IEEE radio frequency integrated circuits symposium (RFIC)* (pp. 303–306). Available from https://doi.org/10.1109/RFIC.2019.8701819.

Foissey, O., Gianesello, F., Gidel, V., Durand, C., Gauthier, A., Guitard, N., Chevalier, P., Hello, M., Azevedo Goncalves, J., Gloria, D., Velayudhan, V., & Lugo, J. (2021). 85 fs RON × COFF and CP1dB@28GHz >25dBm innovative PIN diode integrated in 55 nm BiCMOS technology targeting mmW 5G and 6G front end module. In: *2021 IEEE 20th topical meeting on silicon monolithic integrated circuits in RF systems (SiRF)* (pp. 40–43). Available from https://doi.org/10.1109/SiRF51851.2021.9383348.

Fukui, H. (1979). Optimal noise figure of microwave GaAs MESFET's. *IEEE Transactions on Electron Devices*, *26*(7), 1032–1037. Available from https://doi.org/10.1109/T-ED.1979.19541.

Gao, L., Ma, Q., & Rebeiz, G. M. (2018). A 4.7 mW W-Band LNA with 4.2 dB NF and 12 dB gain using drain to gate feedback in 45nm CMOS RFSOI technology. In: *2018 IEEE radio frequency integrated circuits symposium (RFIC)* (pp. 280–283). Available from https://doi.org/10.1109/RFIC.2018.8428986.

Gao, L., & Rebeiz, G. M. (2019). A 24–43 GHz LNA with 3.1–3.7 dB noise figure and embedded 3-pole elliptic high-pass response for 5G applications in 22 nm FDSOI. In: *2019 IEEE radio frequency integrated circuits symposium (RFIC)* (pp. 239–242). Available from https://doi.org/10.1109/RFIC.2019.8701782.

Gao, L., Wagner, E., & Rebeiz, G. M. (2020). Design of E- and W-band low-noise amplifiers in 22-nm CMOS FD-SOI. *IEEE Transactions on Microwave Theory and Techniques, 68*(1), 132–143. Available from https://doi.org/10.1109/TMTT.2019.2944820.

GlobalFoundries. (2017). Delivering on the promises of 5G. SOI RF semiconductor solutions for the next wave of data. Available from http://soiconsortium.eu/wp-content/uploads/2017/02/GLOBALFOUNDRIES-SOI-Consortium-Tokyo-June-2017-PRabbeni.pdf.

Gu, Q. J., Xu, Z., & Chang, M.-C. F. (2012). Two-way current-combining W-band power amplifier in 65-nm CMOS. *IEEE Transactions on Microwave Theory and Techniques, 60*(5), 1365–1374. Available from https://doi.org/10.1109/TMTT.2012.2187536.

Hietanen, M., Singh, S. P., Rahkonen, T., & Pärssinen, A. (2021). Noise consideration of radio receivers using silicon technologies towards 6G communication. In: *2021 Joint European conference on networks and communications & 6G summit (EuCNC/6G Summit)* (pp. 514–519). Available from https://doi.org/10.1109/EuCNC/6GSummit51104.2021.9482508.

Hu, Y., & Chi, T. (2021). A 27–46-GHz low-noise amplifier with dual-resonant input matching and a transformer-based broadband output network. *IEEE Microwave and Wireless Components Letters, 31*(6), 725–728. Available from https://doi.org/10.1109/LMWC.2021.3059592.

Inac, O., Uzunkol, M., & Rebeiz, G. M. (2014). 45-nm CMOS SOI technology characterization for millimeter-wave applications. *IEEE Transactions on Microwave Theory and Techniques, 62*(6), 1301–1311. Available from https://doi.org/10.1109/TMTT.2014.2317551.

International Wireless Industry Consortium. (2019). 5G Millimeter wave frequencies and mobile networks. A technology whitepaper on key features and challenges. Available from https://www.skyworksinc.com/-/media/SkyWorks/Documents/Articles/IWPC_062019.pdf.

Jain, S. H., Lederer, D., Kumar, A., Saroop, S., Prindle, C., Srinivasan, P., Liu, W., Achanta, R., Kaltalioglu, E., Moss, S., Freeman, G., & Colestock, P. (2021). Novel mmWave NMOS device for high pout mmWave power amplifiers in 45RFSOI. In: *ESSDERC 2021 – IEEE 51st European solid-state device research conference (ESSDERC)* (pp. 199–202). Available from https://doi.org/10.1109/ESSDERC53440.2021.9631775.

Jayamon, J. A., Buckwalter, J. F., & Asbeck, P. M. (2016). Multigate-cell stacked FET design for millimeter-Wave CMOS power amplifiers. *IEEE Journal of Solid-State Circuits, 51*(9), 2027–2039. Available from https://doi.org/10.1109/JSSC.2016.2592686.

Kane, O. M., Lucci, L., Scheiblin, P., Lepilliet, S., & Danneville, F. (2019). 22nm Ultra-thin body and buried oxide FDSOI RF noise performance. In: *2019 IEEE radio frequency integrated circuits symposium (RFIC)* (pp. 35–38). Available from https://doi.org/10.1109/RFIC.2019.8701740.

Karaca, D., Varonen, M., Parveg, D., Vahdati, A., & Halonen, K. A. I. (2017). A 53–117 GHz LNA in 28-nm FDSOI CMOS. *IEEE Microwave and Wireless Components Letters, 27*(2), 171–173. Available from https://doi.org/10.1109/LMWC.2016.2646912.

Kim, K., Lee, K., Cho, S., Shin, G., & Song, H.-J. (2021). A 28–34-GHz stacked-FET power amplifier in 28-nm FD-SOI with adaptive back-gate control for improving linearity. *IEEE Solid-State Circuits Letters, 4*, 52–55. Available from https://doi.org/10.1109/LSSC.2021.3054885.

Konidas, G., Gkoutis, P., Kolios, V., & Kalivas, G. (2019). A 30 GHz low power & high gain low noise amplifier with Gm-boosting in 28nm FD-SOI CMOS technology. In: *2019 8th International conference on modern circuits and systems technologies (MOCAST)* (pp. 1–4). Available from https://doi.org/10.1109/MOCAST.2019.8741860.

Kushwaha, P., Dasgupta, A., Sahu, Y., Khandelwal, S., Hu, C., & Chauhan, Y. S. (2016). Characterization of RF Noise in UTBB FD-SOI MOSFET. *IEEE Journal of the Electron Devices Society, 4*(6), 379–386. Available from https://doi.org/10.1109/JEDS.2016.2603181.

Kushwaha, P., Lin, Y.-K., Duarte, J. P., Agarwal, H., Chang, H.-L., Dabhi, C. K., Khandelwal, S., Jandhyala, S., Paydavosi, N., Chauhan, Y., Sriramkumar, V., Lin, C.-H., Dunga, M., Lu, D., Yao, S., Niknejad, A., Hu, C. (2017). BSIM-IMG 102.9.1. Independent multi-Gate MOSFET compact model. Technical Manual. Unpublished content. Available from http://bsim.berkeley.edu/models/bsimimg/.

Li, C., El-Aassar, O., Kumar, A., Boenke, M., & Rebeiz, G.M. (2018). LNA design with CMOS SOI process-l.4dB NF K/Ka band LNA. In: *2018 IEEE/MTT-S international microwave symposium - IMS* (pp. 1484–1486). Available from https://doi.org/10.1109/MWSYM.2018.8439132.

Li, C., Freeman, G., Boenke, M., Cahoon, N., Kodak, U., & Rebeiz, G. (2017). 1W < 0.9dB IL DC-20GHz T/R switch design with 45nm SOI process. In: *2017 IEEE 17th topical meeting on silicon monolithic integrated circuits in RF systems (SiRF)* (pp. 57–59). Available from https://doi.org/10.1109/SIRF.2017.7874370.

Li, C., Ustundag, B., Kumar, A., Boenke, M., Kodak, U., & Rebeiz, G. (2018). < 0.8dB IL 46dBm OIP3 Ka band SPDT for 5G communication. In: *2018 IEEE 18th topical meeting on silicon monolithic integrated circuits in RF systems (SiRF)* (pp. 1–3). Available from https://doi.org/10.1109/SIRF.2018.8304213.

Li, S., Chi, T., Nguyen, H.T., Huang, T.-Y., & Wang, H. (2018). A 28GHz packaged chireix transmitter with direct on-antenna outphasing load modulation achieving 56%/38% PA efficiency at Peak/6dB back-off output power. In: *2018 IEEE radio frequency integrated circuits symposium (RFIC)* (pp. 68–71). Available from https://doi.org/10.1109/RFIC.2018.8429015.

Li, S., Huang, T.-Y., Liu, Y., Yoo, H., Na, Y., Hur, Y., & Wang, H. (2021). A millimeter-wave LNA in 45nm CMOS SOI with over 23dB peak gain and Sub-3dB NF for different 5G operating bands and improved dynamic range. In: *2021 IEEE radio frequency integrated circuits symposium (RFIC)* (pp. 31–34). Available from https://doi.org/10.1109/RFIC51843.2021.9490455.

Lie, D. Y. C., Mayeda, J. C., & Lopez, J. (2020). Highly-efficient broadband millimeter-wave 5G power amplifiers in GaN, SiGe, and CMOS-SOI. In: *2020 IEEE international symposium on radio-frequency integration technology (RFIT)* (pp. 169–171). Available from https://doi.org/10.1109/RFIT49453.2020.9226215.

Ma, Q., Leenaerts, D. M. W., & Baltus, P. G. M. (2015). Silicon-based true-time-delay phased-array front-ends at Ka-Band. *IEEE Transactions on Microwave Theory and Techniques*, *63*(9), 2942–2952. Available from https://doi.org/10.1109/TMTT.2015.2458326.

Mattisson, S. (2018). An overview of 5G requirements and future wireless networks: Accommodating scaling technology. *IEEE Solid-State Circuits Magazine*, *10*(3), 54–60. Available from https://doi.org/10.1109/MSSC.2018.2844606.

Mayeda, J. C., Tsay, J., Lie, D. Y. C., & Lopez, J. (2019). Effective AM-PM cancellation with body bias for 5G CMOS power amplifier design in 22nm FD-SOI. In: *2019 IEEE international symposium on circuits and systems (ISCAS)* (pp. 1–4). Available from https://doi.org/10.1109/ISCAS.2019.8702159.

Micovic, M., Brown, D., Regan, D., Wong, J., Tai, J., Kurdoghlian, A., Herrault, F., Tang, Y., Burnham, S.D., Fung, H., Schmitz, A., Khalaf, I., Santos, D., Prophet, E., Bracamontes, H., McGuire, C., & Grabar, R. (2016). Ka-band LNA MMIC's realized in $F_{max}$ >580 GHz GaN HEMT technology. In: *2016 IEEE compound semiconductor integrated circuit symposium (CSICS)* (pp. 1–4). Available from https://doi.org/10.1109/CSICS.2016.7751051.

Modi, S. S., Balsara, P. T., & Eliezer, O. E. (2012). Reduced bandwidth class H supply modulation for wideband RF power amplifiers. In: *WAMICON 2012 IEEE wireless & microwave technology conference* (pp. 1–7). Available from https://doi.org/10.1109/WAMICON.2012.6208439.

Mohsen, A., Harb, A., Deltimple, N., & Kassem, A. (2019). Variable gain differential low noise power amplifier in 28-nm FD-SOI. In: *2019 31st international conference on microelectronics (ICM)* (pp. 206–209). Available from https://doi.org/10.1109/ICM48031.2019.9021538.

Niknejad, A. M., Emami, S., Doan, C., Heydari, B., & Bohsali, M. (2008). *mm-Wave silicon technology. 60 GHz and beyond design and modeling of active and passive devices integrated circuits and systems* (1st). New York: Springer. Available from https://doi.org/10.1007/978-0-387-76561-7.

Nyssens, L., Rack, M., Wane, S., Schwan, C., Lehmann, S., Zhao, Z., Lucci, L., Lugo-Alvarez, J., Gaillard, F., Raskin, J.-P., & Lederer, D. (2022). A 2.5–2.6 dB noise figure LNA for 39 GHz band in 22 nm FD-SOI with back-gate bias tunability. In: *2022 17th European microwave integrated circuits conference (EuMIC)* (pp. 60–63). Available from https://doi.org/10.23919/EuMIC54520.2022.9923552.

Padaki, A. V., & Reed, J. H. (2014). Impact of intermodulation distortion on spectrum preclusion for DSA: A new figure of merit. In: *2014 IEEE international symposium on dynamic spectrum access networks (DYSPAN)* (pp. 358–361). Available from https://doi.org/10.1109/DySPAN.2014.6817814.

Parlak, M., & Buckwalter, J. F. (2011a). A 2.5-dB insertion loss, DC-60 GHz CMOS SPDT switch in 45-nm SOI. In: *2011 IEEE compound semiconductor integrated circuit symposium (CSICS)* (pp. 1–4). Available from https://doi.org/10.1109/CSICS.2011.6062463.

Parlak, M., & Buckwalter, J. F. (2011b). A 2.9-dB noise figure, Q-band millimeter-wave CMOS SOI LNA. In: *2011 IEEE custom integrated circuits conference (CICC)* (pp. 1–4). Available from https://doi.org/10.1109/CICC.2011.6055321.

Popovic, Z. (2017). Amping Up the PA for 5G: Efficient GaN power amplifiers with dynamic supplies. *IEEE Microwave Magazine*, *18*(3), 137–149. Available from https://doi.org/10.1109/MMM.2017.2664018.

Pozar, D. M. (1997). *Microwave engineering* (2nd). Wiley.

Pucel, R. A., Haus, H. A., & Statz, H. (1975). Signal and noise properties of gallium arsenide microwave field-effect transistors. *Advances in Electronics and Electron Physics, 38*, 195–265. Available from https://doi.org/10.1016/S0065-2539(08)61205-6.

Putnam, J., Fukuda, M., Staecker, P., & Yun, Y.-H. (1994). A 94 GHz monolithic switch with a vertical PIN diode structure. In: *Proceedings of 1994 IEEE GaAs IC symposium* (pp. 333–336). Available from https://doi.org/10.1109/GAAS.1994.636996.

Rack, M., Nyssens, L., Courte, Q., Lederer, D., & Raskin, J.-P. (2021). Impact of device shunt loss on DC-80 GHz SPDT in 22 nm FD-SOI. In: *ESSDERC 2021 - IEEE 51st European solid-state device research conference (ESSDERC)* (pp. 195–198). Available from https://doi.org/10.1109/ESSDERC53440.2021.9631835.

Rack, M., Nyssens, L., Courte, Q., Lederer, D., & Raskin, J.-P. (2022). A DC-120 GHz SPDT switch based on 22 nm FD-SOI SLVT NFETs with substrate isolation rings towards increased shunt impedance. In: *2022 IEEE radio frequency integrated circuits symposium (RFIC)* (pp. 83–86). Available from https://doi.org/10.1109/RFIC54546.2022.9863217.

Rack, M., Nyssens, L., & Raskin, J.-P. (2019a). Low-loss Si-substrates enhanced using buried PN junctions for RF applications. *IEEE Electron Device Letters, 40*(5), 690–693. Available from https://doi.org/10.1109/LED.2019.2908259.

Rack, M., Nyssens, L., & Raskin, J.-P. (2019b). Silicon-substrate enhancement technique enabling high quality integrated RF passives. In: *2019 IEEE MTT-S international microwave symposium (IMS)* (pp. 1295–1298). Available from https://doi.org/10.1109/MWSYM.2019.8701095.

Rack, M., Nyssens, L., Wane, S., Bajon, D., & Raskin, J.-P. (2020). DC-40 GHz SPDTs in 22 nm FD-SOI and back-gate impact study. In: *2020 IEEE radio frequency integrated circuits symposium (RFIC)* (pp. 67–70). Available from https://doi.org/10.1109/RFIC49505.2020.9218317.

Rack, M., & Raskin, J.-P. (2019). (Invited) SOI technologies for RF and millimeter wave applications. *ECS Transactions, 92*(4), 79–94. Available from https://doi.org/10.1149/09204.0079ecst.

Raskin, J.-P., & Desbonnets, E. (2015). High resistivity SOI wafer for mainstream RF system-on-chip. In: *2015 IEEE 15th topical meeting on silicon monolithic integrated circuits in RF systems* (pp. 33–36). Available from https://doi.org/10.1109/SIRF.2015.7119866.

Report ITU-R M.2376-0. Technical feasibility of IMT in bands above 6 GHz. (2015). 2022 8 16 Report ITU-R M.2376-0. Technical feasibility of IMT in bands above 6 GHz. Available from https://www.itu.int/pub/R-REP-M.2376-2015.

Rostomyan, N., Özen, M., & Asbeck, P. (2018). 28 GHz Doherty power amplifier in CMOS SOI with 28% back-off PAE. *IEEE Microwave and Wireless Components Letters, 28*(5), 446–448. Available from https://doi.org/10.1109/LMWC.2018.2813882.

Rostomyan, N., Özen, M., & Asbeck, P. (2019). A Ka-band asymmetric dual input CMOS SOI Doherty power amplifier with 25 dBm output power and high back-off efficiency. In: *2019 IEEE topical conference on RF/microwave power amplifiers for radio and wireless applications (PAWR)* (pp. 1–4). Available from https://doi.org/10.1109/PAWR.2019.8708739.

Rostomyan, N., Ozen, M., & Asbeck, P. (2018). Comparison of pMOS and nMOS 28 GHz high efficiency linear power amplifiers in 45 nm CMOS SOI. In: *2018 IEEE topical conference on RF/microwave power amplifiers for radio and wireless applications (PAWR)* (pp. 26–28). Available from https://doi.org/10.1109/PAWR.2018.8310058.

Rusanen, J., Hietanen, M., Sethi, A., Rahkonen, T., Pärssinen, A., & Aikio, J.P. (2019). Ka-band stacked power amplifier on 22 nm CMOS FDSOI technology utilizing back-gate bias for linearity improvement. In: *2019 IEEE nordic circuits and systems conference (NORCAS): NORCHIP and international symposium of System-on-Chip (SoC)* (pp. 1–4). Available from https://doi.org/10.1109/NORCHIP.2019.8906915.

Scheen, G., Tuyaerts, R., Rack, M., Nyssens, L., Rasson, J., Nabet, M., & Raskin, J.-P. (2020). Post-process porous silicon for 5G applications. *Solid-State Electronics, 168*107719. Available from https://doi.org/10.1016/j.sse.2019.107719.

Schmid, R. L., Song, P., & Cressler, J. D. (2013). A compact, transformer-based 60 GHz SPDT RF switch utilizing diode-connected SiGe HBTs. In: *2013 IEEE bipolar/BiCMOS circuits and technology meeting (BCTM)* (pp. 111–114). Available from https://doi.org/10.1109/BCTM.2013.6798156.

Shakib, S., Park, H.-C., Dunworth, J., Dunworth, V., Aparin, V., & Entesari, K. (2016). A highly efficient and linear power amplifier for 28-GHz 5G phased array radios in 28-nm CMOS. *IEEE Journal of Solid-State Circuits, 51*(12), 3020–3036. Available from https://doi.org/10.1109/JSSC.2016.2606584.

Shivan, F., Hossain, M., Doerner, R., Schulz, S., Johansen, T., Boppel, S., Heinrich, W., & Krozer, V. (2019). Highly linear 90–170 GHz SPDT switch with high isolation for fully integrated InP transceivers. In: *2019 IEEE MTT-S international microwave symposium (IMS)* (pp. 1011–1014). Available from https://doi.org/10.1109/MWSYM.2019.8700974.

Tagro, Y., Lecavelier des Etangs-Levallois, A., Poulain, L., Lepilliet, S., Gloria, D., Raynaud, C., Dubois, E., & Danneville, F. (2012). High frequency noise potentialities of reported CMOS 65 nm SOI technology on flexible substrate. In: *2012 IEEE 12th topical meeting on silicon monolithic integrated circuits in RF systems* (pp. 89–92). Available from https://doi.org/10.1109/SiRF.2012.6160147.

Tang, X.-L., Pistono, E., Ferrari, P., & Fournier, J.-M. (2013). A traveling-wave CMOS SPDT using slow-wave transmission lines for millimeter-wave application. *IEEE Electron Device Letters, 34*(9), 1094–1096. Available from https://doi.org/10.1109/LED.2013.2274452.

Thomas D., Rostomyan, N., & Asbeck, P. (2018). A 45% PAE pMOS power amplifier for 28GHz applications in 45nm SOI. In: *2018 IEEE 61st international midwest symposium on circuits and systems (MWSCAS)* (pp. 680–683). Available from https://doi.org/10.1109/MWSCAS.2018.8624062.

Thome, F., & Ambacher, O. (2018). Highly isolating and broadband single-pole double-throw switches for millimeter-wave applications up to 330 GHz. *IEEE Transactions on Microwave Theory and Techniques, 66*(4), 1998–2009. Available from https://doi.org/10.1109/TMTT.2017.2777980.

Thome, F., Leuther, A., & Ambacher, O. (2020). Low-loss millimeter-wave SPDT switch MMICs in a metamorphic HEMT technology. *IEEE Microwave and Wireless Components Letters, 30*(2), 197–200. Available from https://doi.org/10.1109/LMWC.2019.2958209.

Thome, F., Ture, E., Brückner, P., Quay, R., & Ambacher, O. (2018). W-band SPDT switches in planar and tri-gate 100-nm gate-length GaN-HEMT technology. In: *2018 11th German microwave conference (GeMiC)* (pp. 331–334). Available from https://doi.org/10.23919/GEMIC.2018.8335097.

Torres, F., Cathelin, A., & Kerhervé, E. (2020). *The fourth terminal. benefits of body-biasing techniques for FDSOI circuits and systems millimeter-wave power amplifiers for 5G applications in 28 nm FD-SOI technology* (1st, pp. 169–222). Cham: Springer. Available from https://doi.org/10.1007/978-3-030-39496-7.

Tuovinen, T., Tervo, N., & Pärssinen, A. (2017). Analyzing 5G RF system performance and relation to link budget for directive MIMO. *IEEE Transactions on Antennas and Propagation, 65*(12), 6636–6645. Available from https://doi.org/10.1109/TAP.2017.2756848.

Ulusoy, A. Ç., Song, P., Schmid, R. L., Khan, W. T., Kaynak, M., Tillack, B., Papapolymerou, J., & Cressler, J. D. (2014). A low-loss and high isolation D-band SPDT switch utilizing deep-saturated SiGe HBTs. *IEEE Microwave and Wireless Components Letters, 24*(6), 400–402. Available from https://doi.org/10.1109/LMWC. 2014.2313529.

Waldhoff, N., Andrei, C., Gloria, D., Danneville, F., & Dambrine, G. (2008). Small signal and noise equivalent circuit for CMOS 65 nm up to 110 GHz. In: *2008 38th European microwave conference* (pp. 321–324). Available from https://doi.org/10.1109/EUMC.2008.4751453.

Wane, S., Huard, V., Rack, M., Nyssens, L., Kieniewicz, B., Bajon, D., & Raskin, J.-P. (2020). Broadband smart mmWave front-end-modules in advanced FD-SOI with adaptive-biasing and tuning of distributed antenna-arrays. In: *2020 IEEE Texas symposium on wireless and microwave circuits and systems (WMCS)* (pp. 1–5). Available from https://doi.org/10.1109/WMCS49442.2020.9172398.

Wang, F., Li, T.-W., Hu, S., & Wang, H. (2019). A super-resolution mixed-signal Doherty power amplifier for simultaneous linearity and efficiency enhancement. *IEEE Journal of Solid-State Circuits, 54*(12), 3421–3436. Available from https://doi.org/10.1109/JSSC.2019.2937435.

Waveform. (2022). *5G's faster data rates and Shannon's Law*. Available from https://www.waveform.com/a/b/guides/5g-and-shannons-law.

Xu, X., Li, S., Szilagyi, L., Testa, P.V., Carta, C., & Ellinger, F. (2020). A 28 GHz and 38 GHz dual-band LNA using gain peaking technique for 5G wireless systems in 22 nm FD-SOI CMOS. In: *2020 IEEE Asia-Pacific microwave conference (APMC)* (pp. 98–100). Available from https://doi.org/10.1109/APMC47863.2020.9331333.

Xu, X., Schumann, S., Ferschischi, A., Finger, W., Carta, C., & Ellinger, F. (2021). A 28 GHz and 38 GHz high-gain dual-band LNA for 5G wireless systems in 22 nm FD-SOI CMOS. In: *2020 15th European microwave integrated circuits conference (EuMIC)* (pp. 77–80). Available from https://ieeexplore.ieee.org/document/9337432.

Yadav, S., Bellaouar, A., Wong, J. S., Chen, T., Sekine, S., Schwan, C., Chin, M. S., Workman, G., Chew, K. W. J., & Chow, W. H. (2019). Demonstration and modelling of excellent RF switch performance of 22nm FD-SOI technology for millimeter-wave applications. In: *ESSDERC 2019 - 49th European solid-state device research conference (ESSDERC)* (pp. 170–173). Available from https://doi.org/10.1109/ESSDERC.2019.8901823.

Yu, B., Ma, K., Meng, F., Yeo, K. S., Shyam, P., Zhang, S., & Verma, P. R. (2017). Ultra-wideband low-loss switch design in high-resistivity trap-rich SOI with enhanced channel mobility. *IEEE Transactions on Microwave Theory and Techniques*, *65*(10), 3937–3949. Available from https://doi.org/10.1109/TMTT.2017.2696944.

Zhang, C., Zhang, F., Syed, S., Otto, M., & Bellaouar, A. (2019). A low noise figure 28GHz LNA in 22nm FDSOI technology. In: *2019 IEEE radio frequency integrated circuits symposium (RFIC)* (pp. 207–210). Available from https://doi.org/10.1109/RFIC.2019.8701831.

Zhao, Y., & Long, J. R. (2012). A wideband, dual-path, millimeter-wave power amplifier with 20 dBm output power and PAE above 15% in 130 nm SiGe-BiCMOS. *IEEE Journal of Solid-State Circuits*, *47*(9), 1981–1997. Available from https://doi.org/10.1109/JSSC.2012.2201275.

Zong, Z., Tang, X., Khalaf, K., Yan, D., Mangraviti, G., Nguyen, J., Liu, Y., & Wambacq, P. (2021). A 28-GHz SOI-CMOS Doherty power amplifier with a compact transformer-based output combiner. *IEEE Transactions on Microwave Theory and Techniques*, *69*(6), 2795–2808. Available from https://doi.org/10.1109/TMTT.2021.3064022.

Zong, Z., Tang, X., Nguyen, J., Khalaf, K., Mangraviti, G., Liu, Y., & Wambacq, P. (2020). A 28GHz two-way current combining stacked-FET power amplifier in 22nm FD-SOI. In: *2020 IEEE custom integrated circuits conference (CICC)* (pp. 1–4). Available from https://doi.org/10.1109/CICC48029.2020.9075906.

CHAPTER

# 10

# Power amplifiers monolithic microwave integrated circuit design for 5G applications

*Xin Liu*[1,2], *Guansheng Lv*[1], *Dehan Wang*[3] *and Wenhua Chen*[1]

[1]Department of Electronic Engineering, Tsinghua University, Beijing, P.R. China [2]School of Microelectronics, Xidian University, Xi'an, Shaanxi, P.R. China [3]ZTE Corporation, Shenzhen, P.R. China

## 10.1 Introduction

To support a drastically increasing number of subscribers, multiple-input-multiple-output (MIMO) systems have been widely used in modern wireless communication systems. In fact, MIMO techniques can increase data rates, coverage of service areas, and communication reliability without additional radio frequencies. Massive MIMO (mMIMO) is a critical technology in 5G, which significantly increases network capacity and spectral efficiency, and reduces wireless network interference. In the recent proposals for the fifth-generation (5G) systems, the number of radio frequency (RF) chains in mMIMO RF front ends can reach up to 256 with bandwidth up to 800 MHz per chain (Gao et al., 2015; Harris et al., 2017; Larsson et al., 2014). Broader bandwidth is expected in the upcoming 5G system to accommodate for the growing user demand for much higher data rate. The spectrum allocation is quite crowded already, and operators are seeking a new radio spectrum resource. Owing to their lower path loss, low-frequency bands remain crucial for maintaining the extensive coverage of 5G networks. Therefore, the 3.3–5.0 GHz is receiving significant attention as a frequency range for 5G that can provide significant bandwidth. Initial studies and prototyping efforts conducted at millimeter wave (mmWave) frequency have demonstrated that mmWave bands would also be promising for fulfilling the requirements of 5G. Therefore, multiple mmWave bands, including spectra around 26, 28, 37, and 39 GHz, have been opened for 5G development.

For the sake of better spectrum efficiency, beamforming (BF) architectures have attracted intensive investigation in mMIMO systems. Full digital BF can yield optimal performance,

*New Materials and Devices Enabling 5G Applications and Beyond*
DOI: https://doi.org/10.1016/B978-0-12-822823-4.00010-8

widely adopted in the sub 6 GHz band. Multipath effects are quite abundant in the sub 6 GHz band, and full digital BF architecture can achieve higher capacity and flexibility. However, due to the severe path loss at mmWave, the advantage of full digital BF architecture is not straightforward; alternatively, the hybrid BF architecture stands out as a good tradeoff between performance and system complexity. Moreover, the antenna element dimension in mmWave mMIMO is smaller; thus, more RF chains can be integrated into a single chipset. This chapter is organized as follows. Firstly, transmitter architectures for mMIMO are analyzed, and the required power amplifier (PA) architectures and specifications are given. Then, recent research progress of sub 6 GHz and mmWave Doherty PA (DPA) is introduced in detail.

## 10.2 Transmitter architectures for massive multiple-input-multiple-output

As one of the key devices in RF transmitters, PAs are characterized as the most power-hungry components and their performances directly affect the efficiency of RF transmitters. The semiconductor technology and the architectures must be carefully selected when designing 5G PAs to meet the output power, efficiency, and linearity requirements, in particular for mMIMO transmitters. Therefore, in this section, we will start by discussing the mMIMO structures in sub 6 GHz and mmWave bands, and then discuss the corresponding requirements for PAs.

### 10.2.1 Massive multiple-input-multiple-output for sub 6 GHz

Measurements carried out in recent years regarding the sub 6 GHz mMIMO channels revealed some fundamental differences from what is currently being employed in small-scale MIMO systems. These measurements indicate that the user equipment (UE) signals received from the different antenna elements of the transmitter tend to become space orthogonal with the increase in the number of antenna elements, and hence this MIMO state channels offers a favorable propagation scenario (Marzetta, 2016). In addition, the multitudes of scattering clusters at sub 6 GHz lead to multipath propagation, which is beneficial in supporting multiple subscribers. Generally, one mMIMO base station at sub 6 GHz can cover more than 10 users simultaneously (Bai & Heath, 2014). Ideally, the number of simultaneously served subscribers increases with the expansion of the array scale. However, considering the transmitter volume and form factor in the base station, the array size has to be kept within reasonable dimensions. For example, a 64-element's antenna array distributed as an $8 \times 8$ uniform rectangle array with half-wavelength element spacing at 3.5 GHz occupies at least $35 \times 35\ \text{cm}^2$ of space. As the array dimensions are moderate in sub 6 GHz mMIMO, a full digital BF architecture is appropriate to make full use of the rich multipath channels (Bai & Heath, 2014; Hu et al., 2017). Since mmWave transmission is not suitable for long-range outdoor communications, sub 6 GHz frequencies are expected to serve both outdoor and indoor coverage, and support high-mobility UE. This will require high output power from the transmitters to serve long-range, large cells. Taking into account hardware cost and power consumption, 64 channels associated with digital BF as shown in Fig. 10.1A are the optimum array size for a sub 6 GHz mMIMO transmitting system, as is already utilized in mMIMO measurements and prototypes.

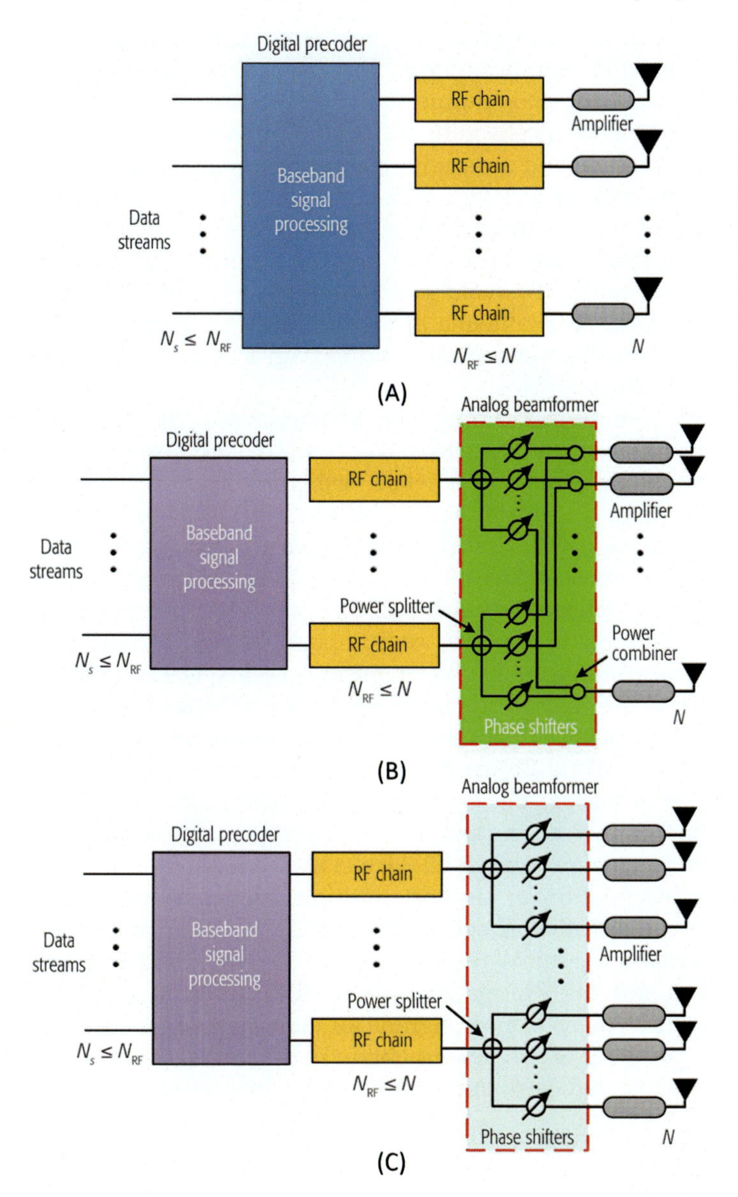


**FIGURE 10.1** Massive multiple-input-multiple-output (mMIMO). Massive MIMO transmitter architectures (A) digital beamforming, (B) fully connected hybrid beamforming, (C) sub-array based hybrid beamforming (reprinted from Gao et al., 2018). *Source: From Gao, X., Dai, L., & Sayeed, A. M. (2018). Low RF-complexity technologies to enable millimeter-wave MIMO with large antenna array for 5G wireless communications. IEEE Communications Magazine, 56(4), 211–217. https://doi.org/10.1109/MCOM.2018.1600727.*

## 10.2.2 Massive multiple-input-multiple-output for mmWave

The development of mmWave frequency band is remarkable to relieve the global spectrum scarcity and bandwidth shortage (Pi & Khan, 2011), which can provide GHz bandwidth and tens of Gigabit data transmission rate to expand significantly the transmission scale and enhance service quality. However, the propagation condition is unfavorable for mmWave communication due to its considerable propagation loss. The legitimate concerns regarding the propagation characteristics at mmWave frequencies, such as higher rain attenuation, atmosphere absorption, and significant loss through foliage, are the main obstacles for mmWave carriers in long-range communications (Rappaport & Deng, 2015). However, when one considers the small sized cells in the urban environment (on the order of 200 m), one realizes that rain attenuation and atmosphere absorption will not create significant path loss for mmWave frequency, especially for 28 and 38 GHz (Rappaport et al., 2013). Thus the employment of small cell and high-density base stations can mitigate the attenuation issue. However, this approach may cause other interference to neighboring UE due to the small coverage of the base station. To avoid such situations, a highly directional beam should be synthesized by the large-scale antenna array with high array gain to mitigate this interference, and, simultaneously, compensate for the path loss and extend the range of communications (Niu et al., 2016).

According to the aforementioned mmWave propagation characteristics, the mmWave mMIMO scale should be much larger than that in the sub 6 GHz band (i.e., 256 antennas compared to 64 antennas) in order to achieve high array gain, which will pose severe challenges for its realization in practice. One challenging problem is that in a sub 6 GHz MIMO system with digital BF structure, each antenna requires one dedicated RF chain, including digital-to-analog converters, mixers, PAs, and passive components (Heath et al., 2016). This will result in unaffordable hardware costs and power consumption in mmWave MIMO systems, due to the significant number of antennas as well as the high-power consumption of the RF chain (e.g., 250 mW at mmWave frequencies compared with 30 mW at microwave frequencies). On the other hand, analog BF only requires one RF chain, resulting in relatively low hardware cost and power consumption; however, it only supports single digital stream and therefore not applicable in multiuser scenarios. Therefore, the hybrid BF, as shown in Fig. 10.1B and C, which combine the merits of both digital and analog BF to support multiuser and multistream transmission while requiring moderate hardware cost and power consumption, is considered to be the most appropriate architecture for mmWave mMIMO (Han et al., 2015).

Note that although mmWave hybrid BF can support multiple users, the number of simultaneously served users is lower than that of sub 6 GHz digital BF (10 or more users in sub 6 GHz compared to 1–4 users in the mmWave band) as the result of hardware limitations and mmWave channel conditions (Rappaport et al., 2013; Roh et al., 2014). However, as long as the number of RF chains is more significant than the rank of the channel matrix, the small-scale digital precoder is still able to obtain near-optimal performance compared to fully digital BF, which also indicates the feasibility of hybrid BF for mmWave mMIMO (Gao et al., 2016).

### 10.2.3 The PA architecture and specification

A large-scale array involving hundreds of antenna elements leads to different requirements for PAs, namely highly integrated and lower output power capability. Fig. 10.2 presents the output power of a single PA in the mmWave and sub 6 GHz array, varying with array size and the resulting process selection. For simple analysis, the antennas in the array are anticipated to have roughly 6 dB gain for each single antenna and to be spaced half a wavelength for each adjacent element (Shakib et al., 2016). It is intuitive to see in

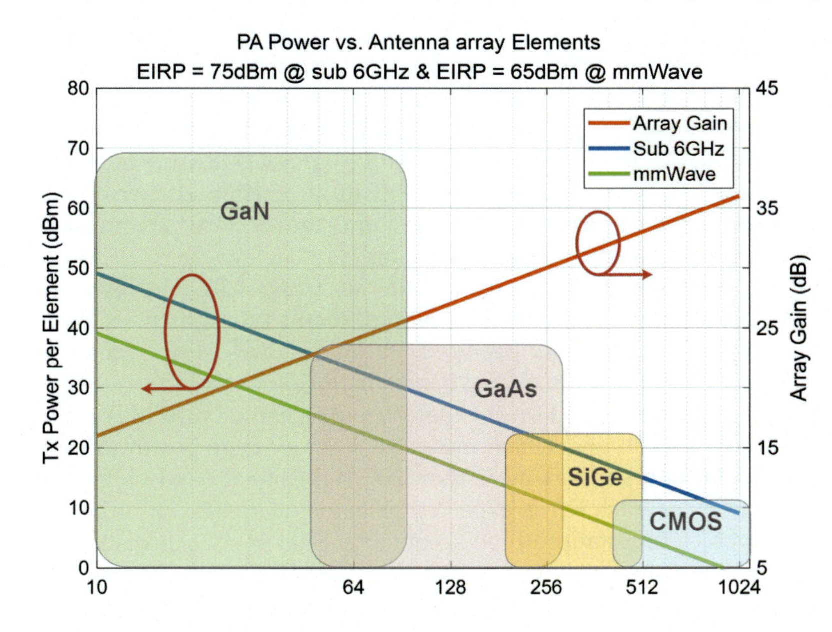


FIGURE 10.2 Power amplifier (PA) specification. PA specification analysis in sub 6 GHz and mmWave.

Fig. 10.2 that the array gain increases significantly with the expansion of the array size. When the array size reaches 500 antennas or more (this is also expected to be the most common mmWave large array size), the antenna array can provide a more than 30 dB array gain, which reduces the output power requirement for each PA in the array. Unlike the high output power (above 30 dBm) in the sub 6 GHz band, the required output power of single PA in mmWave mMIMO is only 10–20 dBm, where the high output technologies, such as gallium nitride (GaN) and gallium arsenide (GaAs), are not needed. Most importantly, the highly integrated low power requirement suggests that complementary metal oxide semiconductor (CMOS) or SiGe technologies would be more appropriate to fabricate the PAs.

Since PA is the most power-hungry component in the transmitter, its efficiency plays a key role in the overall system efficiency performance. Due to the complex modulation scheme, modern communication signal usually exhibits a high peak to average power ratio (PAPR). In order to achieve a high average efficiency under the modulation signal excitation, PA should have a high back-off efficiency. Many class-AB PAs with state-of-the-art performance have been reported, including GaN monolithic microwave integrated circuit (MMIC) PAs in the sub 6 GHz band (Giofré et al., 2015; Liu et al., 2018; Quaglia, Camarchia, Pirola et al., 2014), and Si-based or GaAs PAs in the mmWave band (Ali et al., 2017, 2018; Li et al., 2018; Lv et al., 2018; Nguyen & Pham, 2016; Sarkar et al., 2017). Despite their high saturated efficiency, the back-off efficiency is relatively low. It is lower than 30% for sub 6 GHz GaN MMIC PAs, and lower than 20% for mmWave PAs. Therefore, high-efficiency PA architecture must be employed to further improve the back-off efficiency. Among the efficiency enhancement techniques, Doherty and envelope tracking (ET) architectures are successfully adopted in 4G transmitting systems. The ET architecture is a good solution for multiband and multimode energy-efficient transmitters, a solution which has been widely employed in mobile terminals; however, it suffers from the bandwidth limitation of envelope amplifiers. For 5G systems, the signal bandwidth requirement exceeds 200 MHz in sub 6 GHz or mmWave; therefore, it is impractical to utilize ET in mMIMO transmitters. In contrast, DPA architecture is a promising technique for moderate-to-high efficiency enhancement with high PAPR signals. DPA architecture almost dominates the PA market in base stations because of its promising efficiency and relatively low circuit complexity. In the upcoming 5G system, the Doherty architecture offers a true, demonstrated and energy-efficient advantage for 5G PAs.

The array's gain increases along with the expansion of the array scale. However, in practice, the array scale is also limited by technological factors, such as equipment volume, power consumption, efficiency, etc. A possible PA specification analysis is illustrated in Fig. 10.2 and outlined in Table 10.1. In the sub 6 GHz band, the target equivalent isotropic radiated power (EIRP) is above 75 dBm. Based on the literature, 64-channel mMIMO architectures have been used in demo prototypes, which means that the average power of each PA should be around 33–38 dBm and support more than 200 MHz signal bandwidth. For 5G systems, the average efficiency of each PA is expected to better than 40%, in order for the overall efficiency of the whole base station to be acceptable. Shakib et al. investigated the optimum carrier frequency and output power of mmWave mMIMO transmitters and UEs, considering the array scale, path loss, and other factors (Shakib et al., 2016). Fig. 10.2 shows the total power consumption for the analog beamformer and the array's gain as a function of the number of antenna elements at 28 GHz, under the assumption of 200 m non-line-of-sight transmission for 65 dBm EIRP and 800 MHz bandwidth.

## 10.3 Design of a sub 6 GHz Doherty power amplifier

The sub 6 GHz bands can achieve wider coverage if the total transmit power is large enough since the path loss is relatively small at the low-frequency bands. As a result, the saturation output power of each PA may reach up to 20 W, and so GaN devices have been recognized as a good candidate technology due to their high power density and excellent efficiency. The DPAs, implemented using a printed circuit board (PCB) and which are available in the literature, show, in general, high performance. However, these designs

**TABLE 10.1** Power amplifier specifications for 5G.

| Spec. | Sub 6 GHz | mmWave |
|---|---|---|
| Frequency | 3.4 ~ 3.8 GHz/4.8–5.0 GHz | 24.75–27.5 GHz/37–42.5 GHz |
| Bandwidth | > 200 MHz | > 800 MHz |
| Pout@sat | > 43 dBm | > 25 dBm (III–V)/17 dBm (Si) |
| Average efficiency | > 40% | > 20% |
| ACPR | < −45 dBc | < −27.5 dBc |
| Error vector magnitude (EVM) | < 5% (64QAM)/ < 3% (256QAM) | < 5% (64QAM)/ < 3% (256QAM) |

are too complicated for deployment in a mMIMO system due to their large size (Chen et al., 2016; Huang et al., 2018; Probst et al., 2017). Fortunately, DPA GaN MMICs for microwave backhaul applications and small-cell base-stations are available (Ayad et al., 2017; Camarchia, Fang et al., 2013; Camarchia, Rubio et al., 2013; Giofre et al., 2015; Gustafsson et al., 2013, 2014, 2016; Jee et al., 2015; Kim et al., 2014; Lee, Lim, Bae et al., 2017; Lee, Lim, Lee et al., 2017; Park et al., 2015; Piazzon et al., 2014), and such DPAs are excellent references for the design of DPAs in 5G transmitters. DPA GaN MMICs for 5G application have also been reported in recent years (Ishikawa et al., 2018; Li et al., 2018; Lv, Chen, Chen, & Feng, 2019; Lv, Chen, Chen & Ghannouchi, & Feng, 2019; Lv, Chen, Liu & Ghannouchi, & Feng, 2019; Maroldt & Ercoli, 2017), but further research is still required.

To enable a compact communication system, fully integrated GaN MMICs using only one process are the best solutions, as no off-chip matching components are required and the chip can be easily packaged. However, the fully integrated DPAs end up in most cases to be costly due to their large footprint on the GaN wafer, the fabrication of which is still considered to be an expensive process. An alternative solution is hybrid integration, which means that only active devicesand power bars are fabricated in the GaN process, and all the matching and biasing circuits are realized in other less-expensive integrated passive device (IPD) processes. Then, different dies can be connected with bonding wires and integrated into one package. Partial integration is also investigated to reduce the die size, where transistors and partial matching circuits are implemented in the GaN process, while some passive devices are placed outside the chip. The performances of the typical works referenced in this section are summarized in Table 10.2.

### 10.3.1 A Doherty power amplifier monolithic microwave integrated circuit with full integration

A fully integrated C-band GaN DPA MMIC for 5G mMIMO application is presented in Lv, Chen, Liu & Ghannouchi, & Feng (2019) using a 0.25-um GaN-high electron mobility transistor (HEMT) process from WIN semiconductor. A low-Q output network is employed to broaden the bandwidth, and on-chip transmission lines (TLs) are used as drain bias inductors for achieving low insertion loss. Reversed uneven power splitting and back-off input matching are proposed for gain enhancement. Fig. 10.3 shows the chip photo and measurement results. The fabricated DPA occupies a size of $2.2 \times 2.1$ mm, requiring no off-chip components. A saturated output power of 40.4–41.2 dBm and a 6-dB back-off drain efficiency (DE) of 47%–50% are achieved across a wide bandwidth from 4.5 to 5.2 GHz. Using a 40-MHz long term evolution (LTE) signal, the measured adjacent channel power ratio (ACPR) is $-29$ dBc at the output power of 33 dBm, and is improved to $-46$ dBc after applying digital predistortion (DPD).

Since there is only one stage, the power gain of the DPA in Lv, Chen, Liu & Ghannouchi, and Feng (2019) is relatively low. To improve the power gain, two stages are adopted in Lv, Chen, Chen, and Feng (2019). Single driver and dual-driver topologies are compared, and it is demonstrated that the former can significantly degrade the back-off power added efficiency (PAE) when the gain of the power stage is not high enough. Thus, to minimize the impact of the driver stage on PAE, a dual-driver topology is employed, where the driver stage is used in both the main and auxiliary branches. Impedance

**TABLE 10.2** Sub 6 GHz Doherty power amplifier monolithic microwave integrated circuit performance.

| References | Freq (GHz) | BW (GHz) | Psat (dBm) | Back-off PAE | Gain (dB) | Integration | Process | GaN Die Size (mm$^2$) |
|---|---|---|---|---|---|---|---|---|
| Camarchia, Fang et al. (2013) | 6.8–8.5 | 1.7 | 35 | 24%–37% @ 9 dB | 9 | Full | GaN | 3.2 |
| Piazzon et al. (2014) | 7 | 0.35 | 37 | 47% @ 7 dB[a] | 10 | Full | GaN | 21.6 |
| Gustafsson et al. (2013) | 7 | 1 | 38 | 36% @ 7 dB | 10.5 | Full | GaN | 25 |
| Gustafsson et al. (2014) | 5.8–8.8 | 3 | 36 | 31%–39% @ 9 dB | 9 | Full | GaN | 8.4 |
| Giofre et al. (2017) | 7 | 0.6 | 38 | 41% @ 6 dB | 16 | Full | GaN | 9 |
| Gustafsson et al. (2016) | 6.5 | NA | 42 | 21% @ 9 dB | 18 | Hybrid | GaN + GaAs | 4.1 |
| Ayad et al. (2017) | 5.5–6.5 | 1 | 43.5 | 23% @ 9 dB | 11 | Hybrid | GaN + GaAs | 2.6 |
| | | | 44[b] | 35% @ 9 dB[b] | 12.5[b] | | | |
| Kim et al. (2014) | 2.14 | NA | 40.5 | 52.2% @ 7.3 dB[a] | 15.7 | Partial | GaN + PCB | 8.6 |
| Jee et al. (2015) | 2.1–2.7 | 0.6 | 41 | 55% @ 8 dB[a] | 14 | Partial | GaN + PCB | 5 |
| Lee, Lim, Bae et al. (2017) | 2.6 | NA | 44 | 54.2% @ 6 dB[a] | 13.7 | Full | GaN | 4.7 |
| Lee, Lim, Lee et al. (2017) | 2.6 | NA | 40.5 | 52.2% @ 6.5 dB[a] | 13.8 | Full | GaN | 3.1 |
| Maroldt and Ercoli (2017) | 3.3–3.6 | 0.3 | 44.3 | 44% @ 6 dB | 28 | Hybrid | GaN + GaAs | NA |
| Ishikawa et al. (2018) | 4.0–4.5 | 0.5 | 36 | 44% @ 6 dB | 17 | Full | GaN | 3.5 |
| Li et al. (2018) | 5.1–5.9 | 0.8 | 38.7 | 49.5% @ 6 dB | 14.4 | Full | GaN | 1.9 |
| Lv, Chen, Chen, and Feng (2019) | 4.4–5.1 | 0.7 | 42 | 47% @ 6 dB | 17 | Full | GaN | 7.5 |
| Lv, Chen, Liu, Ghannouchi, and Feng (2019) | 4.5–5.2 | 0.7 | 41.2 | 45% @ 6 dB | 11.6 | Full | GaN | 2.9 |
| Lv, Chen, Liu et al. (2019) | 3.3–3.8 | 0.5 | 42.6 | 51% @ 6 dB[a] | 12 | Full | GaN | 5.8 |

[a]*Drain efficiency.*
[b]*Dual input.*

transformers in the output network are realized by lumped high-pass π-type networks to reduce the occupied area. Fig. 10.4A shows the proposed output network. The shunt inductors at the same node can be merged into a much smaller inductor, and there are only three inductors in the final output network. Because of the limited DC current capacity and low Q-factor of the on-chip inductors, two drain bias inductors are realized by on-chip TLs, as shown in Fig. 10.4B. The fabricated DPA exhibits a power gain of 17 dB, a saturated output power of 41.3 dBm, a saturated PAE of 61%, and a 6-dB back-off PAE of 47% at 4.7 GHz, with a compact size of 3 × 2.5 mm. Moreover, the 6-dB back-off PAE is better than 40% across a large bandwidth from 4.4 to 5.1 GHz. Fig. 10.5 presents the chip photo and the measurement results.

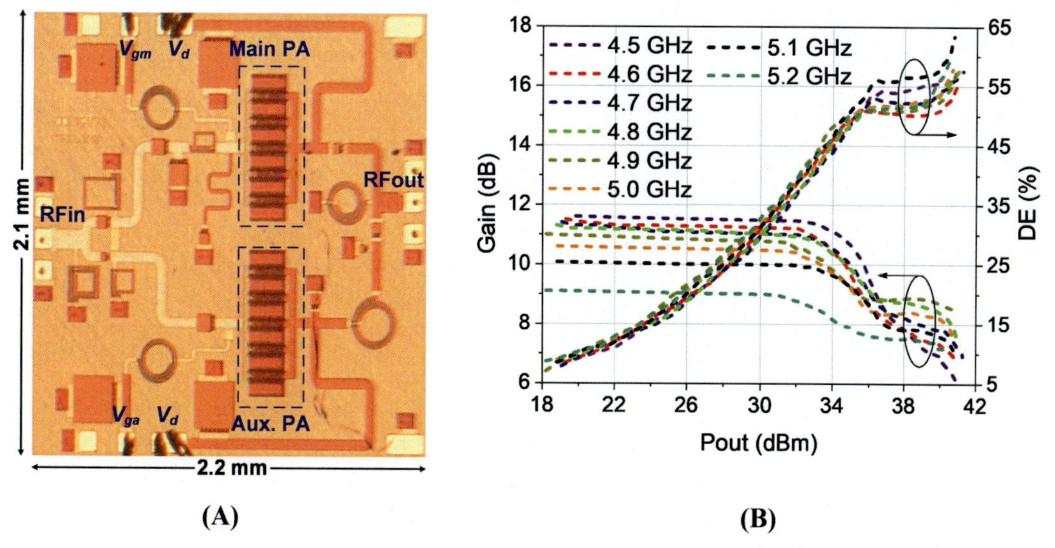


**FIGURE 10.3** Gallium nitride (GaN) Doherty power amplifier (DPA) monolithic microwave integrated circuit (MMIC). A fully integrated GaN DPA MMIC: (A) chip photo, (B) measurement results. *Source: From Lv, G., Chen, W., Liu, X., Ghannouchi, F. M., & Feng, Z. (2019). A fully integrated C-band GaN MMIC Doherty power amplifier with high efficiency and compact size for 5G application. IEEE Access, 7, 71665–71674. https://doi.org/10.1109/ACCESS.2019.2919603. (© 2019 IEEE).*

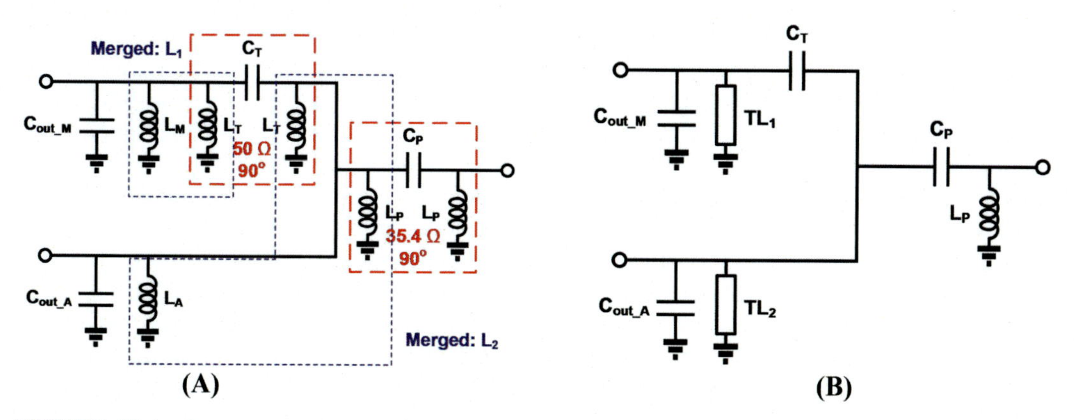


**FIGURE 10.4** Output network. Compact output network: (A) schematic before simplification, and (B) final schematic with high-Q transmission lines. *Source: From Lv, G., Chen, W., Chen, L., & Feng, Z. (2019). A fully integrated C-band GaN MMIC Doherty power amplifier with high gain and high efficiency for 5G application. In* IEEE MTT-S international microwave symposium digest *(Vols. 2019, pp. 560–563). Institute of Electrical and Electronics Engineers Inc. https://doi.org/10.1109/mwsym.2019.8701103.*

A dual-band DPA with hybrid operating modes for 5G mobile communication and vehicle network is demonstrated in Lv, Chen, Liu et al. (2019). A 5G communication network is a very big network composed of many different networks, such as a mobile communication network, internet of things, and internet of vehicles, rather than just for mobile communication.

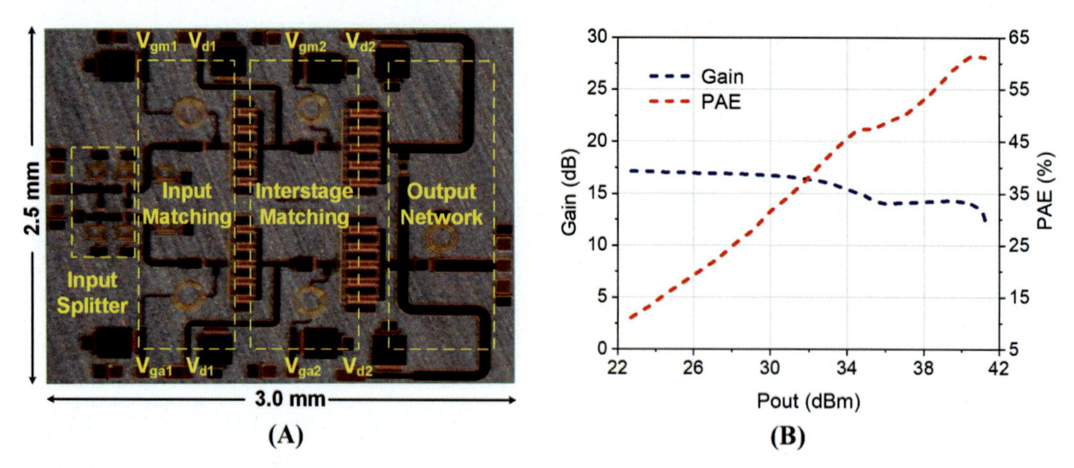


**FIGURE 10.5** Two-stage Doherty power amplifier (DPA). Two-stage DPA: (A) chip photo, and (B) measurement results at 4.7 GHz (reprinted from Lv, Chen, Chen, & Feng, 2019). *Source: From Lv, G., Chen, W., Chen, L., & Feng, Z. (2019). A fully integrated C-band GaN MMIC Doherty power amplifier with high gain and high efficiency for 5G application. In* IEEE MTT-S international microwave symposium digest *(Vols. 2019, pp. 560–563). Institute of Electrical and Electronics Engineers Inc. https://doi.org/10.1109/mwsym.2019.8701103.*

The performance requirements of PAs for the different networks are usually not consistent. For example, a 5.8-GHz dedicated short-range communication system for the vehicle network adopts a simple modulation scheme and requires high saturation efficiency, while mobile communication employs complex modulation and high back-off efficiency is required because of the large PAPR. The proposed dual-band PA in Lv, Chen, Liu et al. (2019) operates at Doherty mode in the 3.5-GHz band for a high back-off efficiency, and operates at harmonic tuned class-AB mode at 5.8 GHz for a high saturated efficiency. At Doherty mode, the main PA is biased at class-AB and the auxiliary PA is biased at class-C. A 6-dB back-off efficiency of 51% is realized at 3.5 GHz, as shown in Fig. 10.6A. At class-AB mode, both the main PA and the auxiliary PA are biased at class-AB, and second harmonic tuning is employed in the output matching network of the auxiliary PA. A saturated efficiency of 55% is achieved at 5.8 GHz, as shown in Fig. 10.6B.

It is worth highlighting that one of the main points of differentiation between PCB technology and MMIC technology is that, for the latter, we need to strictly comply with the design rules of the process that require the access to and use of process design kits. In addition, the MMIC process offers a limited DC current density through the metal layers. As a result of the high drain bias voltage, the optimal load impedance is often larger than 100 Ω, if the required output power is only several watts. The impedance inverting network of a DPA with such a high characteristic impedance requires impractically narrow TLs. To overcome this limitation, a compact T-network of TLs with feasible line widths is employed in Li et al. (2018). Shunt inductors are required to neutralize extra output capacitance of the main PA transistor and auxiliary PA transistor. Then, the π-type inductor network is converted to a T-type network. The implemented DPA, as shown in Fig. 10.7, demonstrates a state-of-the-art performance with a 6-dB back-off PAE up to 49.5%. In addition, the measured X-parameters are employed to investigate the DPA nonlinear characteristics and verify the accuracy of conventionally used PA characterization/measurement methods for system-level design and testing applications.

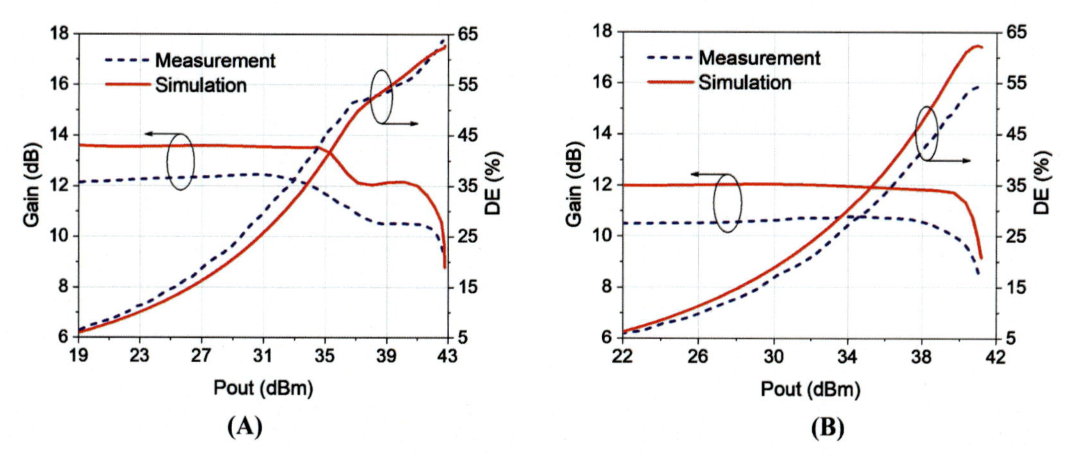


FIGURE 10.6 Dual-band Doherty power amplifier (DPA). Performance of the dual-band DPA with hybrid modes: (A) 3.5 GHz Doherty mode, and (B) 5.8 GHz class-AB mode. *Source: From Lv, G., Chen, W., Liu, X., & Feng, Z. (2019). A dual-band GaN MMIC power amplifier with hybrid operating modes for 5G application. IEEE Microwave and Wireless Components Letters, 29(3), 228–230. https://doi.org/10.1109/LMWC.2019.2892837.*

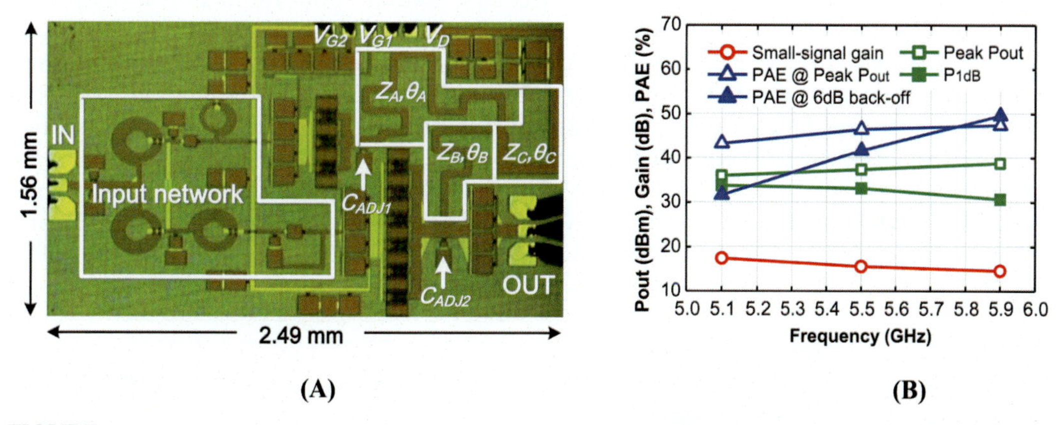


FIGURE 10.7 T-network. gallium nitride Doherty power amplifier monolithic microwave integrated circuit based on a T-network: (A) layout, (B) experimental results. *Source: From Li, S. H., Hsu, S. S. H., Zhang, J., & Huang, K. C. (2018). Design of a compact GaN MMIC doherty power amplifier and system level analysis with X-parameters for 5G communications. IEEE Transactions on Microwave Theory and Techniques, 66(12), 5676–5684. https://doi.org/10.1109/TMTT.2018.2876255.*

## 10.3.2 A Doherty power amplifier monolithic microwave integrated circuit with hybrid integration

To reduce the cost of commercial chips and maintain the benefits of GaN technology, a 6.5 GHz hybrid DPA for microwave links was designed in Gustafsson et al. (2013). The GaN active devices are fabricated in a 0.25 μm GaN HEMT process from United Monolithic Semiconductors, while all passive circuits are fabricated in integrated GaAs process with high dielectric and low loss. The main PA and auxiliary PA are both composed of two stages, and all dies are mounted in a single 9 × 9 mm quad-flat no-leads

(QFN) package. The output network is designed together with bonding wires and device's parasitics to realize an equivalent $\lambda/4$ TL behind the main PA and an equivalent $\lambda/2$ TL behind the auxiliary PA. However, the measurements exhibit a frequency offset and a reduction of bandwidth compared to the simulations, caused mainly by the model inaccuracy of the bonding wires. An output power of 42 dBm and a PAE of 21% at 9-dB power back-off (PBO) are achieved at 6. 5 GHz, and the GaN die area is only 4.1 $mm^2$

A quasi-MMIC DPA operating in 1.8–3.2 GHz is presented in Quaglia et al. (2019). The power cells are based on Qorvo's AlGaN/GaN HEMT 0.25-μm high voltage technology on SiC substrate, whereas the passive matching networks are realized on the Qorvo's IPC3 passive component process on GaAs substrate. To increase the bandwidth, a two-stage high-pass filter is used to connect the main and auxiliary outputs with an impedance inverter equivalent behavior, whereas an additional output matching is adopted to improve the matching of the auxiliary at saturation. The design is based on a dual-input Doherty architecture. The lumped components values are adjusted to absorb the devices' capacitance and the bond wires effects, whereas the splitting ratio between the main and auxiliary inputs is considered as an additional degree of freedom during the design. The measured efficiency and gain at some frequencies for dual-input configuration are shown in Fig. 10.8. The output power is higher than 42 dBm and the 6-dB back-off efficiency is higher than 38% over the 1.8- to 3.2-GHz frequency band.

Two 3.5-GHz MMIC DPAs for 5G mMIMO base-station applications are developed in Maroldt and Ercoli (2017). High-gain two-stage GaN PA MMICs are designed as cores of the DPAs. Area-consuming passive networks, such as input splitters and output combiners, are implemented on IPD in the same QFN package. Fig. 10.9 presents the block diagram. To maximize the efficiency, the output combiner uses a parallel circuit class-E like matching topology to properly terminate the output harmonics. A symmetrical and an asymmetrical version of a DPA are both designed. The symmetrical DPA is fully integrated and yields a saturation output power of 44.3 dBm with 52% maximum PAE, and 44% PAE at 6-dB PBO. The asymmetrical DPA maintains an excellent efficiency of 40%–44% at a high PBO of 8.5–9 dB from 3.4 to 3.6 GHz.

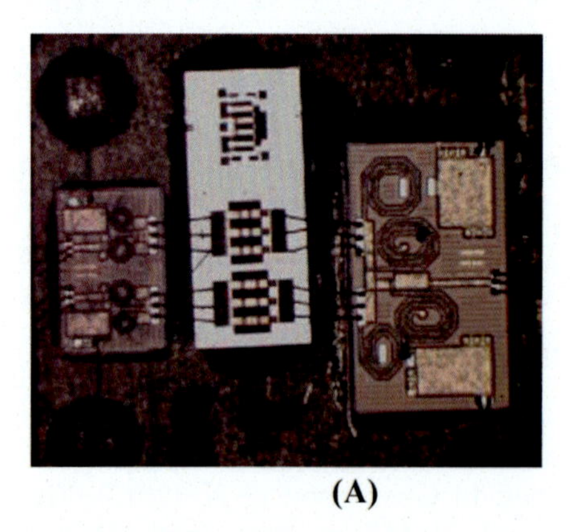

(A)

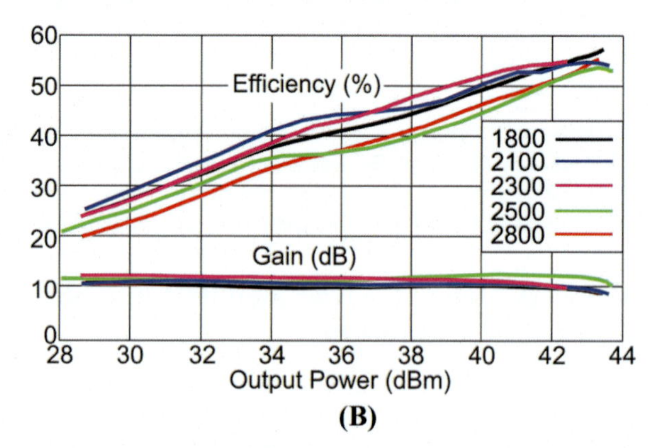


(B)

**FIGURE 10.8** Quasi- monolithic microwave integrated circuit (MMIC) Doherty power amplifier (DPA). 1.8–3.2 GHz Quasi-MMIC DPA: (A) chip photo, (B) measurement results. *Source: From Quaglia, R., Greene, M. D., Poulton, M. J., & Cripps, S. C. (2019). A 1.8–3.2-GHz Doherty power amplifier in quasi-MMIC technology. IEEE Microwave and Wireless Components Letters, 29(5), 345–347. https://doi.org/10.1109/LMWC.2019.2904883.*

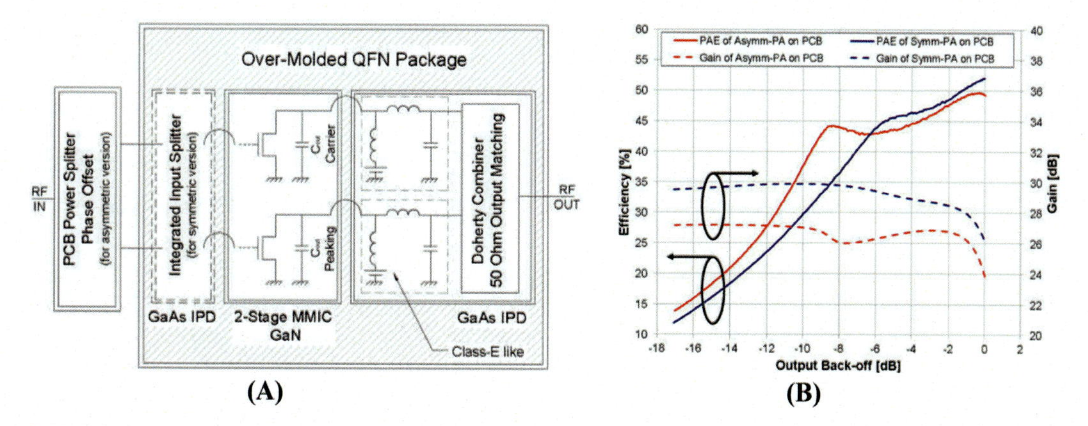


FIGURE 10.9 Hybrid integration. 3.5 GHz Doherty power amplifier with hybrid integration: (A) block diagram, (B) measurement results. *Source: From Maroldt, S., & Ercoli, M. (2017). 3.5-GHz ultra-compact GaN class-E integrated Doherty MMIC PA for 5G massive-MIMO base station applications. In* 2017 12th European microwave integrated circuits conference, EuMIC 2017 *(Vols. 2017-, pp. 196–199). Institute of Electrical and Electronics Engineers Inc. https://doi.org/10.23919/EuMIC.2017.8230693.*

## 10.3.3 A Doherty power amplifier monolithic microwave integrated circuit with partial integration

A compact 2.14 GHz DPA was developed for small-cell base stations in Kim et al. (2014). The circuit is designed using lumped passives for size reduction. It should be noted that lumped passives on MMICs such as spiral inductors usually degrade the efficiency due to their ohmic and substrate losses. Therefore, the major part of the PA is integrated on the MMIC die to achieve a small footprint while low-loss surface-mounted off-chip inductors are utilized to enhance the efficiency. Fig. 10.10 shows the circuit diagram. High-Q chip inductors from Coilcraft are mounted on a low-loss Taconic TLY PCB. The chip size is $3.3 \times 2.6\ mm^2$, and the total PA has a footprint of $1.1 \times 1.4\ cm^2$ including all the off-chip components. At 7.3-dB PBO, a DE of 52.2% and a power gain of 15.7 dB are achieved.

A 2.14 GHz two-stage DPA GaN MMIC was presented in Lee et al. (2014). The quadrature coupler based on a lumped element magnetic coupler with capacitors is matched directly to the driver PA's optimum impedance, rather than matched to 50 Ω first and then to the optimum impedance. Thus, the size of the interstage matching network is reduced. Input matching network, interstage matching network, and all the power cells are integrated in a GaN die, while the output combiner and biasing networks are implemented on the PCB. Fig. 10.11 presents the chip layout. Since the PCB-based output network exhibits a lower loss, a higher efficiency can be achieved. In addition, the size of the GaN die can be reduced, resulting in a lower cost. The fabricated DPA shows a saturated power of 41.2 dBm, a peak PAE of 56.2%, and a gain of 19.7 dB. For a down link wideband code division multiple access signal, the PAE of 39.6% is achieved at an output power of 35.3 dBm. The area of the GaN die is only $2.5 \times 2.7\ mm^2$.

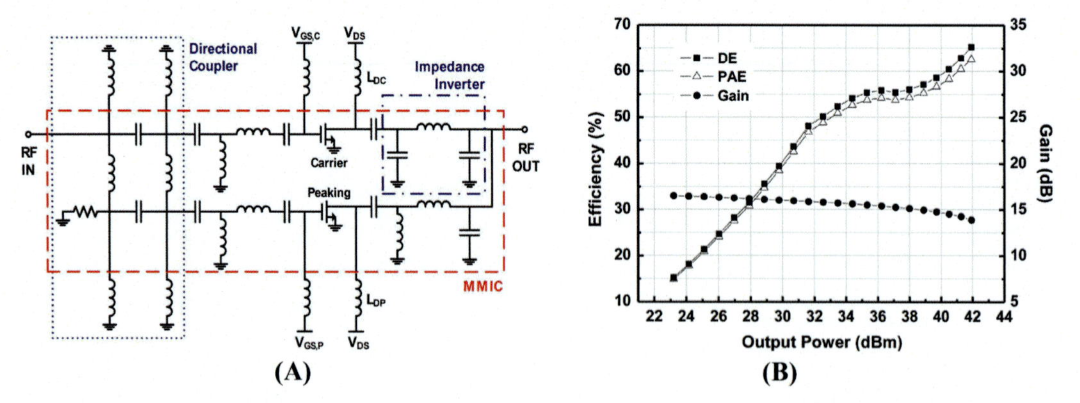


FIGURE 10.10 2.14 GHz gallium nitride (GaN) Doherty power amplifier (DPA) monolithic microwave integrated circuit (MMIC). 2.14 GHz GaN DPA MMIC with off-chip components: (A) circuit diagram, (B) measurement results. *Source: From Kim, C. H., Jee, S., Jo, G. D., Lee, K., & Kim, B. (2014). A 2.14-GHz GaN MMIC Doherty power amplifier for small-cell base stations. IEEE Microwave and Wireless Components Letters, 24(4), 263–265. https://doi.org/10.1109/LMWC.2014.2299536.*

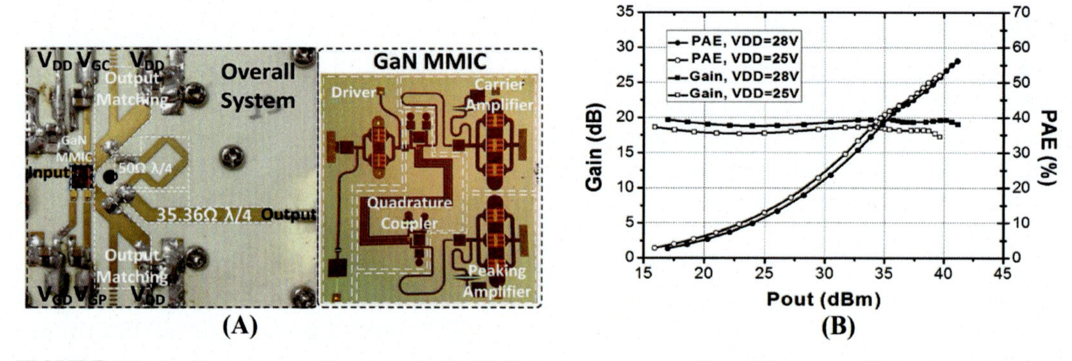


FIGURE 10.11 Two-stage gallium nitride (GaN) Doherty power amplifier (DPA) monolithic microwave integrated circuit (MMIC). Two-stage GaN DPA MMIC with printed circuit board based output network: (A) layout, (B) measurement results. *Source: From Lee, J., Lee, D. H., & Hong, S. (2014). A Doherty power amplifier with a GaN MMIC for femtocell base stations. IEEE Microwave and Wireless Components Letters, 24(3), 194–196. https://doi.org/10.1109/LMWC.2013.2292926.*

## 10.4 Design of a mmWave Doherty power amplifier

The power requirement of the unit PA in mmWave MIMO phased array transmitters is much lower than that in sub 6 GHz MIMO transmitters, and this is mainly due to the large number of antenna elements in mmWave MIMO transmitters. In fact, the reported output power values of a single PA in such MIMO mmWave systems range from 10 to 30 dBm.

Therefore, GaAs, SiGe, and CMOS technologies can comfortably deliver such power and hence they would be valuable technology candidates for the design and manufacturing of mmWave PA units. The design method of a PA in a compound semiconductor is very different from that in silicon-based technology, so the research progress of mmWave DPAs is described in the following two subsections. The state-of-the-art mmWave DPAs are summarized in Table 10.3.

## 10.4.1 A mmWave Doherty power amplifier based on III-V semiconductors

A Ka-band DPA implemented in a 0.15-um E-mode GaAs process is reported in Nguyen et al. (2017). Based on the process' low-K dielectric crossovers, the authors propose a meander broadside coupler with minimal size. As a result, an ultra-compact DPA with the size of 2.2 × 1.3 mm is realized. At 28 GHz, the measured saturated power is

**TABLE 10.3** mmWave Doherty PA performance.

| References | Freq (GHz) | BW (GHz) | Psat (dBm) | PeakPAE | Back-offPAE | Gain (dB) | Process | Area (mm$^2$) |
|---|---|---|---|---|---|---|---|---|
| Curtis et al. (2013) | 26.4 | NA | 25.1 | 38% | 27% @ 6 dB | 10.3 | D-mode 0.15 μm GaAs | 25 |
| Quaglia, Camarchia, Jiang et al. (2014) | 24 | 2.4 | 30.9 | 38% | 20% @ 6 dB | 12.5 | D-mode 0.15 μm GaAs | 4.29 |
| Nguyen et al. (2017) | 28 | NA | 26 | 40% | 29% @ 6 dB | 12 | E-mode 0.15 μm GaAs | 2.86 |
| Nguyen, Curtis et al. (2018) | 29.5 | 1.0 | 27 | 38% | 32% @ 6 dB | 10.5 | E-mode 0.15 μm GaAs | 4.59 |
| Nguyen, Pham et al. (2018) | 28 | 1.25 | 28.7 | 37% | 27% @ 6 dB | 14.4 | E-mode 0.15 μm GaAs | 4.93 |
| Lv et al. (2018) | 31.1 | 2.8 | 26.3 | 35% | 28% @ 7 dB | 14 | E-mode 0.15 μm GaAs | 3.57 |
| Lv, Chen, Chen, Ghannouchi, and Feng (2019) | 29 | 2 | 25.4 | 33% | 22% @ 6 dB | 16 | D-mode 0.1 μm GaAs | 3.08 |
| | 46 | 2 | 25.2 | 25% | 17% @ 6 dB | 10.5 | | |
| Guo et al. (2018) | 26 | NA | 32 | 21.7% | 20% @ 6 dB | 13.6 | 0.15 μm GaN on SiC | 5 |
| Nakatani et al. (2018) | 28.5 | 2 | 35.6 | 25.5% | 22.7% @ 6 dB | 15.8 | 0.15 μm GaN on SiC | 4.32 |
| Giofre et al. (2019) | 28 | 1 | 32 | 30% | 30% @ 6 dB | 13 | 0.1 μm GaN on Si | 6 |
| Agah et al. (2012) | 42 | NA | 18 | 23% | 17% @ 6 dB | 7 | 45 nm CMOS SOI | 0.64 |

(Continued)

TABLE 10.3 (Continued)

| References | Freq (GHz) | BW (GHz) | Psat (dBm) | PeakPAE | Back-offPAE | Gain (dB) | Process | Area ($mm^2$) |
|---|---|---|---|---|---|---|---|---|
| Kaymaksut et al. (2015) | 72 | 19 | 21 | 13.6% | 7% @ 6 dB | 18.5 | 40 nm CMOS | 0.96 |
| Hu et al. (2017) | 28 | NA | 16.8 | 20.3% | 13.9% @ 5.9 dB | 18.2 | 0.13 μm SiGe | 1.76 |
| | 37 | NA | 17.1 | 22.6% | 16.6% @ 6 dB | 17.1 | | |
| | 39 | NA | 17 | 21.4% | 12.6% @ 6.7 dB | 16.6 | | |
| Chen et al. (2017) | 60 | 7 | 13.2 | 22%[a] | 10.6% @ 6 dB[a] | NA | 65 nm CMOS | 0.43 |
| Özen et al. (2017) | 30 | 1 | 21 | 17% | 24.3% @ 6 dB | 6.7 | 0.13 μm SiGe | 1.87 |
| Indirayanti and Reynaert (2017) | 32 | 5 | 19.8 | 21% | 12.8% @ OP1 dB | 22 | 28 nm CMOS | 1.79 |
| Chen et al. (2018) | 60 | 8 | 14.9 | 16.8% | 8.7% @ 6 dB | 18 | 65 nm CMOS | 0.195 |
| Nguyen, Chi et al. (2018) | 65 | 6 | 19.4 | 27.5% | 20.1% @ 6 dB | 12.5 | 45 nm CMOS SOI | 3.23 |
| Rostomyan et al. (2018) | 28 | 6 | 22.4 | 40% | 28% @ 6 dB | 10 | 45 nm CMOS SOI | 0.63 |

[a]*Drain efficiency.*

26 dBm, and the 6-dB back-off PAE is up to 29%. To further improve the back-off efficiency, an asymmetrical DPA using a novel load–pull-based design technique is demonstrated in Nguyen, Curtis et al. (2018). A new load modulation scheme is proposed to optimize the efficiency at a desired back-off level. The asymmetrical ratio is determined based on empirical load–pull data, instead of the current profile in the conventional design approach. A record PAE of 32% at 6-dB PBO and 28.5% at 8-dB PBO are achieved at 29.5 GHz. A similar load-pull-based based design method is also employed in Lv et al. (2018), but the offset line behind the main PA is removed by choosing an appropriate output matching network, which results in a compact size of 2.1 × 1.7 mm and a large bandwidth of 2.8 GHz. The 7-dB back-off PAE is better than 21% in the frequency of 29–31.8 GHz, and a maximum value of 28% is observed at 31.1 GHz, as shown in Fig. 10.12. A Ka/Q dual-band DPA was implemented in a 0.1-um GaAs process in Lv, Chen, Chen, and Ghannouchi, and Feng (2019). Dual-band TLs employed in the sub 6-GHz dual-band DPA are not suitable to be used as the offset lines in mm-wave dual-band DPAs due to their large sizes and insert losses. An in-depth analysis reveals that the phase requirement of the offset lines can be relaxed, and the DPA exhibits a reasonable performance in a certain phase-shift range. A novel design method is proposed to realize offset lines using simple TLs, which can satisfy the phase-shift ranges in dual bands by

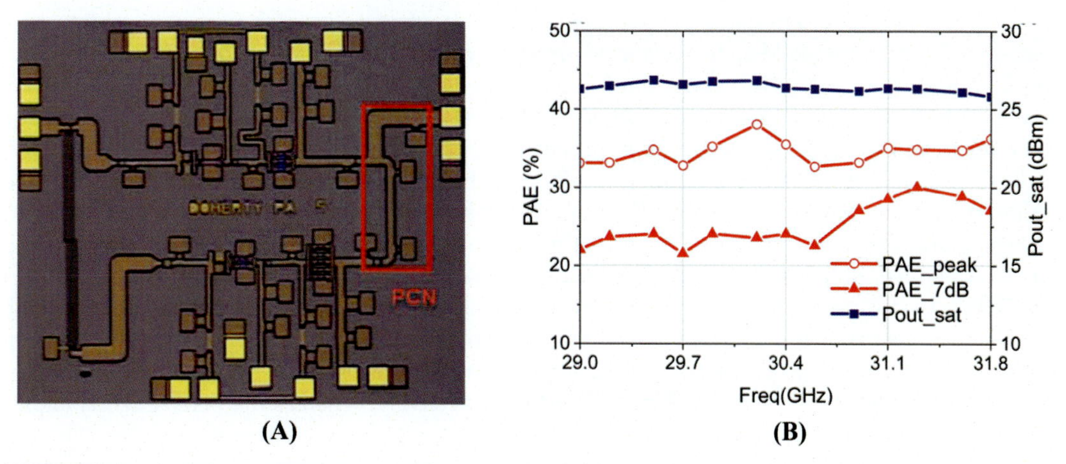


**FIGURE 10.12** Broadband Ka-band gallium arsenides (GaA) Doherty power amplifier (DPA) monolithic microwave integrated circuit (MMIC). Broadband Ka-band GaAs DPA MMIC: (A) chip photo, (B) measurement results. *Source: From Lv, G., Chen, W., & Feng, Z. (2018). A compact and broadband Ka-band asymmetrical GaAs Doherty power amplifier MMIC for 5G communications. In* IEEE MTT-S international microwave symposium digest *(Vols. 2018-, pp. 808–811). Institute of Electrical and Electronics Engineers Inc. https://doi.org/10.1109/MWSYM.2018.8439219.*

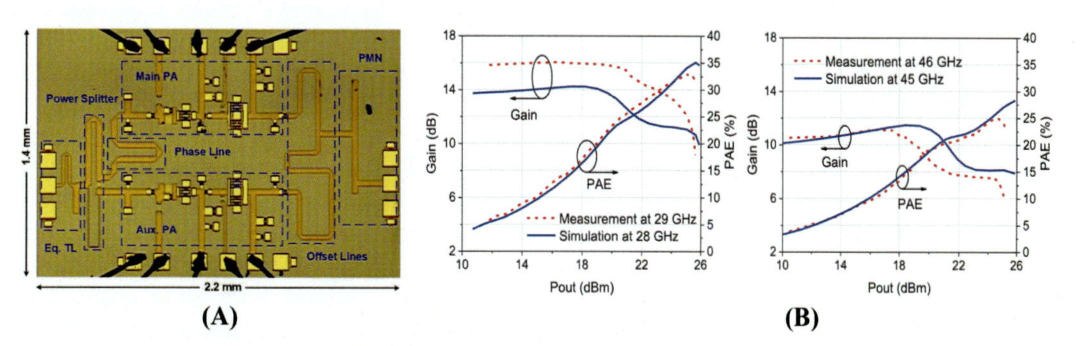


**FIGURE 10.13** Ka/Q dual-band gallium arsenides (GaA) Doherty power amplifier (DPA) monolithic microwave integrated circuit (MMIC). Ka/Q dual-band GaAs DPA MMIC: (A) chip photo, (B) measurement results. *Source: From Lv, G., Chen, W., Chen, X., Ghannouchi, F. M., & Feng, Z. (2019). A compact Ka/Q dual-band GaAs MMIC Doherty power amplifier with simplified offset lines for 5G applications. IEEE Transactions on Microwave Theory and Techniques, 67(7), 3110–3121. https://doi.org/10.1109/TMTT.2019.2908103.*

choosing a proper electrical length. To enhance the gain of the DPA, a reversed uneven power splitter is adopted to deliver more power to the main PA. The fabricated DPA achieves an output power of 25.4/25.2 dBm, a peak PAE of 33%/25%, and a 6-dB back-off PAE of 22%/17% at 29/46 GHz, respectively, with a compact size of $2.2 \times 1.4\ mm^2$. It should be noted that this work is the first demonstration of mm-wave dual-band DPAs that do not require any additional switching or reconfiguration. Chip photo and measurement results are shown in Fig. 10.13.

GaN technology is attractive for 5G applications in view of its high-power capability and high efficiency; nevertheless, the reported mm-wave GaN DPAs are scarce (Campbell et al., 2012; Giofre et al., 2019; Guo et al., 2018; Nakatani et al., 2018; Valenta et al., 2018). A K-band GaN DPA is presented in Campbell et al. (2012). At 23 GHz, the PAE at 8-dB input power back-off from the 1-dB gain compression point is 25%, while the saturated power is over 5 W. A 26-GHz GaN MMIC DPA in Guo et al. (2018) exhibits a 6-dB back-off PAE of 20%, but the saturated output power is only 32 dBm. The Ka-band DPA in Nakatani et al. (2018) demonstrates a saturation output power of 35.6 dBm and a 6-dB back-off PAE of 22.7% at 28.5 GHz. Those designs are all based on a 0.15-μm GaN on SiC process. In Giofre et al. (2019), a 28 GHz DPA based on a 0.1-μm GaN on Si process is demonstrated. GaN on Si allows to exploit the key features of the most common GaN-SiC technologies, while assuring significant benefits especially in terms of production costs, thanks to the larger wafers. At 28 GHz, measured gain and output power are close to 13 dB and 32 dBm, respectively, whereas the PAE remains higher than 30% in 6-dB output power back-off.

### 10.4.2 A mmWave Doherty power amplifier based on a Si technology

A 45-GHz DPA implemented in 45 nm CMOS silicon-on-insulator (SOI) is reported in Agah et al. (2012). Two-stack field effect transistor (FET) amplifiers are used as main and auxiliary amplifiers to enhance the output power. The use of slow-wave coplanar waveguides improves the PAE and gain remarkably, and the die area is also reduced by 20%. The DPA exhibits more than 18 dBm saturated output power, with a peak gain of 7 dB. The peak PAE and 6-dB PBO PAE are 23% and 17%, respectively. The quarter wavelength TL at the input of the auxiliary amplifier is replaced by an amplifier in Agah et al. (2013), in order to increase the gain of DPA and reduce the occupied area. The gain and PAE at 6-dB PBO are 8 dB and 20%, respectively, showing remarkable improvement compared to the work in Agah et al. (2012).

In the Si-based technology process, it is convenient to implement a transformer architecture by utilizing the strong coupling between the different metal layers; thus the size of DPA can be reduced by replacing the distributed TL-based divider and combiner with input and output transformers. An E-band DPA using transformers rather than TLs to achieve load modulation is demonstrated in a 40 nm CMOS process (Kaymaksut et al., 2015). An asymmetrical DPA structure with an asymmetrical transformer combiner is employed to improve the back-off efficiency. A high output impedance of the auxiliary PA is achieved by inserting an additional inductor and capacitor (LC) tuning network. The implementation with a cascode PA units demonstrates 21-dBm output power with a peak PAE of 13.6% and a 6-dB PBO PAE of 7% at 72 GHz.

A 28-GHz DPA based on a 45-nm CMOS SOI process is presented in Rostomyan et al. (2018). State-of-the-art performance is achieved due to the use of a low-loss combiner synthesis technique. Impedance inversion and parasitic compensation are integrated in one compact network, leading to a loss reduction of 1 dB. The fabricated DPA exhibits a peak PAE of 40%, and a 6-dB PBO PAE of 28%, as shown in Fig. 10.14. A high output power of 22.4 dBm is achieved by using two stacked power devices. Moreover, the PAE at 6-dB PBO is higher than 19% in the frequency range of 25–31 GHz.

Multiple mmWave frequency bands have been opened for 5G deployment. Multiband operations will greatly facilitate future cross-network/international roaming and enable ultra-compact

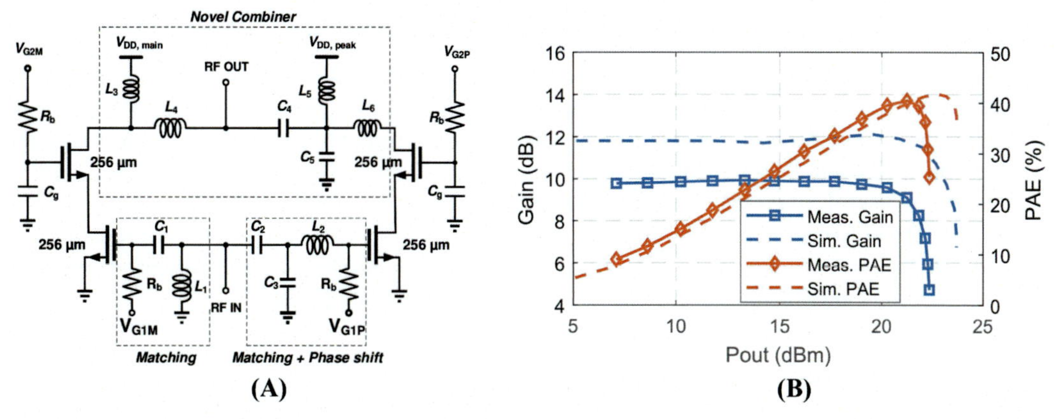


**FIGURE 10.14** 28-GHz Doherty power amplifier (DPA) in CMOS SOI. 28-GHz DPA in CMOS SOI: (A) schematic, (B) experimental results. *Source: From Rostomyan, N., Ozen, M., & Asbeck, P. (2018). 28 GHz Doherty power amplifier in CMOS SOI with 28% back-off PAE. IEEE Microwave and Wireless Components Letters, 28(5), 446–448. https://doi.org/10.1109/LMWC.2018.2813882.*

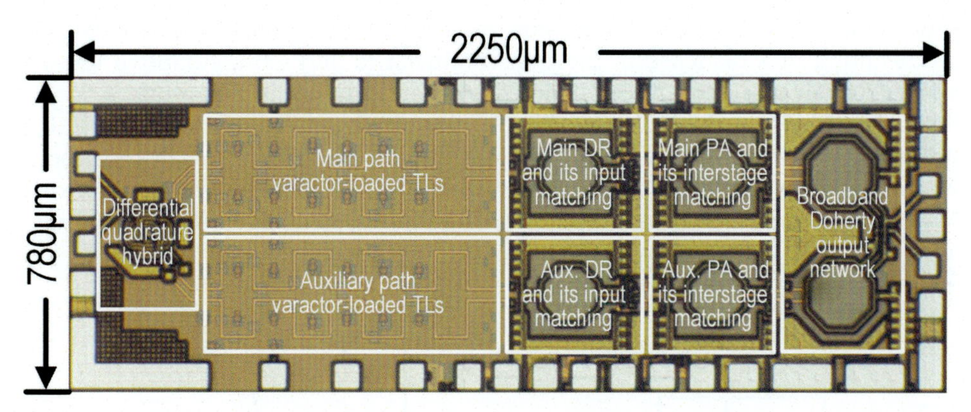


**FIGURE 10.15** 28/37/39 GHz multiband Doherty power amplifier (DPA). Chip photo of the 28/37/39 GHz multiband DPA. *Source: From Hu, S., Wang, F., & Wang, H. (2017). A 28 GHz/37 GHz/39 GHz multiband linear Doherty power amplifier for 5G massive MIMO applications. In* Digest of technical papers - IEEE international solid-state circuits conference *(Vol. 60, pp. 32–33). Institute of Electrical and Electronics Engineers Inc. https://doi.org/10.1109/ISSCC.2017.7870246.*

mMIMO 5G systems. In this situation, a fully integrated 28/37/39 GHz multiband DPA is demonstrated using 0.13-um SiGe BiCMOS in Hu et al. (2017). It achieves + 16.8/ + 17.1/ + 17 dBm peak output power, 18.2/17.1/16.6-dB peak power gain, 29.4/27.6/28.2% peak collector efficiency (CE), and CE of 25.4%@5.9 dB/ 26%@6 dB/ 20.6%@6.7 dB PBO at 28/37/39 GHz, respectively. A transformer-based output network with reduced impedance transformation ratios in PBO is employed to broaden the bandwidth of the DPA. The relative phase of the main path relative to the auxiliary path is adjusted using 9-section varactor-loaded TLs. In consequence, the bandwidth is extended significantly by using different varactor settings for 28 and 37/39 GHz. Fig. 10.15 presents the chip photo for clarification.

## 10.5 Linearity improvement from circuit design

In order to relieve the linearization burden of the digital predistortion, many researchers are seeking novel analog or low-power consumption linearization schemes to enhance PA linearity. In mmWave hybrid mMIMO systems, the linearity requirements are relatively low than the case in sub 6 GHz due to the lower power output of each PA, the circuit-level linearization design will be useful. Even if DPD is mostly employed in sub 6 GHz to compensate the nonlinearity, linearity enhancement design methods are also called to reduce the DPD complexity and its associated power consumption. The analog linearization methods can be divided into two main categories of nonlinearity compensation and enhancement circuits.

### 10.5.1 Nonlinearity compensation

Analog nonlinearity compensation methods mainly include a linear gain/attenuation unit, a predistortion unit for AM-AM and amplitude modulation-phase modulation (AM-PM). The early-stage analog predistortion method used a linear bias offset to stabilize the DC operating point (Yamauchi et al., 1997; Yoshimasu et al., 1998). The conventional bias circuit is constructed by a resistance network, due to the nonlinearity of the base-emitter diode of the input amplification transistor, the DC operating point will decrease at large signals, resulting in a decrease in the transconductance and gain. Adaptive linearization bias can effectively compensate gain compression and phase distortion under large signal conditions by stabilizing the DC operating point as shown in Fig. 10.16 (Wang et al., 2019).

Another alternative solution is to compensate the PA nonlinearities with cascaded analog predistorters as shown in Fig. 10.17A (Tsai et al., 2006). A common method for predistortion of AM-AM characteristics is to use the cold-mode FET (Kao et al., 2013; Tsai et al., 2006, 2011).

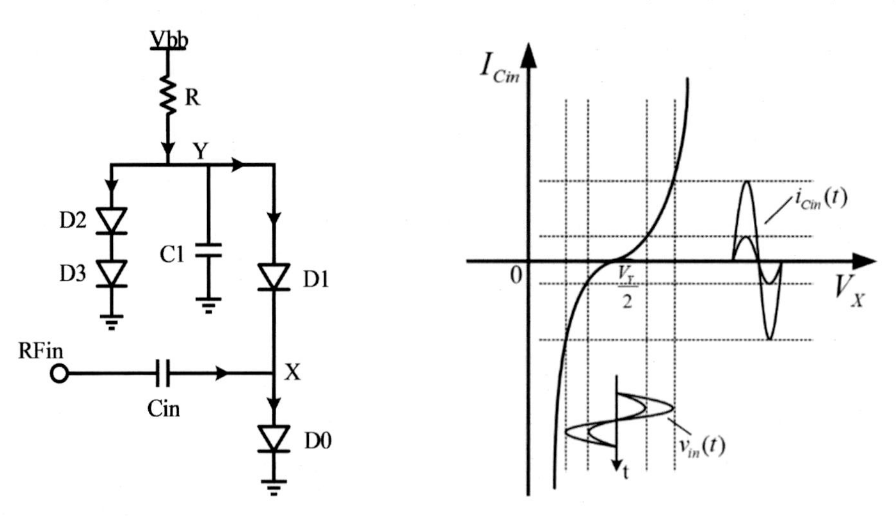


**FIGURE 10.16** Equivalent circuit. The equivalent circuit of the linearization bias circuit and its input IV characteristic curve. *Source: From Wang, D., Chen, W., Chen, L., Liu, X., & Feng, Z. (2019). A Ka-band highly linear power amplifier with a linearization bias circuit. In* IEEE MTT-S international microwave symposium digest *(Vols. 2019-, pp. 320–322). Institute of Electrical and Electronics Engineers Inc. https://doi.org/10.1109/mwsym.2019.8701069.*

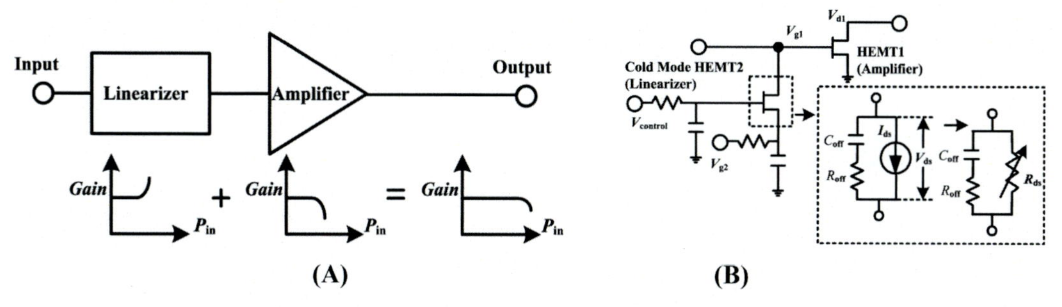


FIGURE 10.17 Block diagram of analog predistortion. Block diagram of analog predistortion. (A) Principle of analog predistorter. (B) Equivalent circuit model of cold-mode FET. *Source: From Tsai, J.-H., Chang, H.-Y., Wu, P.-S., Lee, Y.-L., Huang, T.-W., & Wang, H. (2006). Design and analysis of a 44-GHz MMIC low-loss built-in linearizer for high-linearity medium power amplifiers. IEEE Transactions on Microwave Theory and Techniques, 54(6), 2487–2496. https://doi.org/10.1109/TMTT.2006.875800.*

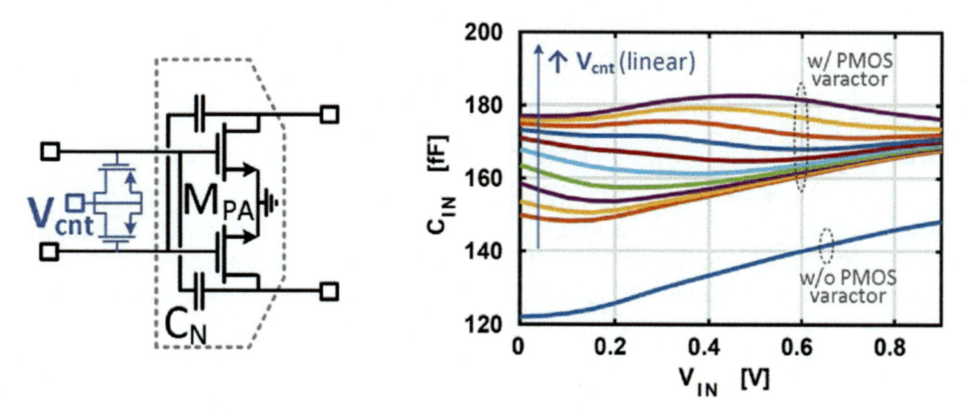


FIGURE 10.18 Architecture of the CS power amplifiers (PA). Architecture of the CS PAs with input PMOS varactors for linearization. *Source: From Vigilante, M., & Reynaert, P. (2018). A wideband class-AB power amplifier with 29–57-GHz AM-PM compensation in 0.9-V 28-nm bulk CMOS. IEEE Journal of Solid-State Circuits, 53(5), 1288–1301. https://doi.org/10.1109/JSSC.2017.2778275.*

The equivalent circuit of the cold-mode FET can be expressed as the parallel combination of a capacitor series with a small resistor and a current source, as shown in Fig. 10.17B. When the input power increases, the slope of the DC IV curve decreases, which results in increased Rds. So that the cold-mode FET linearizer can achieve positive gain and compensate the gain of the PA.

Many works in the literature also focus on using positive channel metal oxide semiconductor (PMOS) transistors to compensate the AM-PM distortion caused by the nonlinear input capacitance of an negative channel metal oxide semiconductor (NMOS) transistor (Kulkarni & Reynaert, 2014, 2016; Vigilante & Reynaert, 2018; Wang et al., 2004; Xi et al., 2017). A PMOS varactor is used to compensate the nonlinear of the input capacitance as shown in Fig. 10.18; thus, the net AM-PM distortion is reduced. According to the experimental results, the analog AM-PM correction method is a promising solution for mmWave PA linearization, which is expected to satisfy the PA specifications in 5G mMIMO.

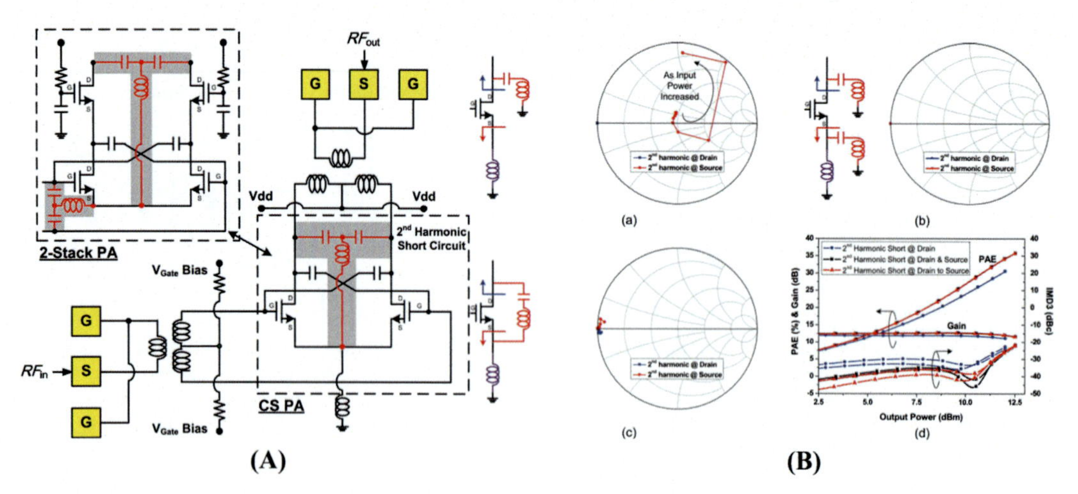


FIGURE 10.19 Linearity enhancement. Linearity enhancement by harmonic short circuit: (A) schematic of the designed power amplifier, (B) simulated results. *Source: From Park, B., Jin, S., Jeong, D., Kim, J., Cho, Y., Moon, K., & Kim, B. (2016). Highly linear mm-Wave CMOS power amplifier. IEEE Transactions on Microwave Theory and Techniques, 64(12), 4535–4544. https://doi.org/10.1109/TMTT.2016.2623706.*

## 10.5.2 Linearity enhancement circuits

The primary method for using linearity enhanced circuits to improve the linearity of amplifiers include antiphase techniques, feedback, multi-gate transistors, third-order intercept cancellations, and harmonic short circuits. The antiphase technique is a method for cancellation of the third order intermodulation distortion intermodulation distortion (IMD)3 between the driver and power stages of the PA (Choi et al., 2010; Joo et al., 2013; Lu et al., 2007; Park, Lee & Yoo, & Park, 2017; Park, Lee, & Park, 2017). This is achieved by the opposite sign $gm^3$ of each stage. The driver stage generates a positive-sign IMD3s, which can pass through the interstage matching network because they appear closely to the signals to be transmitted. At this time, the positive-sign IMD3s are still quite small, but after amplification in the power stage with the signals, the positive-sign IMD3s reach a sufficient size to cancel the negative-sign IMD3s generated in the power stage. Thus, the linearity of the PA can be improved significantly.

The control of the second harmonic impedances can be a practical solution to reduce the nonlinearity because the memory effect is mainly caused by the reflections of the envelope frequency and second harmonics at the device terminals (Francois & Reynaert, 2015; Jin et al., 2013; Kang et al., 2006; Park et al., 2016). A highly linear PA with harmonic short circuit is achieved in Park et al. (2016) as shown in Fig. 10.19. By achieving second harmonics short both at the source and drain, the PA achieves good linearity.

# 10.6 Conclusions

In this chapter, energy-efficient integrated DPA MMICs and linearization techniques are reviewed both for the sub 6 GHz and mmWave 5G mMIMO systems. Different semiconductor technologies and architectures are compared and analyzed, and the enhancement is also investigated. Since the PA specifications for mMIMO are still under consideration and not yet finalized, it is worth proposing novel design methods to further improve their efficiency and their linearity performance.

# References


Agah, A., Hanafi, B., Dabag, H., Asbeck, P., Larson, L., & Buckwalter, J. (2012). A 45GHz Doherty power amplifier with 23% PAE and 18dBm output power, in 45nm SOI CMOS. *IEEE MTT-S International Microwave Symposium Digest*. Available from https://doi.org/10.1109/MWSYM.2012.6259632, United States.

Agah, A., Dabag, H. T., Hanafi, B., Asbeck, P. M., Buckwalter, J. F., & Larson, L. E. (2013). Active millimeter-wave phase-shift doherty power amplifier in 45-nm SOI CMOS. *IEEE Journal of Solid-State Circuits, 48*(10), 2338–2350. Available from https://doi.org/10.1109/JSSC.2013.2269854.

Ali, S. N., Agarwal, P., Mirabbasi, S., & Heo, D. (2017). A 42–46.4% PAE continuous class-F power amplifier with Cgd neutralization at 26–34GHz in 65nm CMOS for 5G applications. *Digest of papers - IEEE radio frequency integrated circuits symposium* (pp. 212–215). United States: Institute of Electrical and Electronics Engineers Inc. Available from https://doi.org/10.1109/RFIC.2017.7969055.

Ali, S. N., Agarwal, P., Renaud, L., Molavi, R., Mirabbasi, S., Pande, P. P., & Heo, D. (2018). A 40% PAE frequency-reconfigurable CMOS power amplifier with tunable gate-drain neutralization for 28-GHz 5G radios. *IEEE Transactions on Microwave Theory and Techniques, 66*(5), 2231–2245. Available from https://doi.org/10.1109/TMTT.2018.2801806, http://ieeexplore.ieee.org/xpl/tocresult.jsp?isYear = 2009&isnumber = 4747395&Submit32 = View + Contents.

Ayad, M., Byk, E., Neveux, G., Camiade, M., & Barataud, D. (2017). Single and dual input packaged 5.5–6.5GHz, 20W, quasi-MMIC GaN-HEMT Doherty power amplifier. *IEEE MTT-S international microwave symposium digest* (pp. 114–117). Canada: Institute of Electrical and Electronics Engineers Inc. Available from https://doi.org/10.1109/MWSYM.2017.8058804.

Bai, T. & Heath, R. W. (2014). Asymptotic coverage and rate in massive MIMO networks. In *IEEE global conference on signal and information processing, GlobalSIP 2014* (pp. 602–606), United States: Institute of Electrical and Electronics Engineers Inc. Available from https://doi.org/10.1109/GlobalSIP.2014.7032188.

Camarchia, V., Fang, J., Moreno Rubio, J., Pirola, M., & Quaglia, R. (2013). 7GHz MMIC GaN Doherty power amplifier with 47% efficiency at 7dB output back-off. *IEEE Microwave and Wireless Components Letters, 23*(1), 34–36. Available from https://doi.org/10.1109/LMWC.2012.2234090.

Camarchia, V., Rubio, J. J. M., Pirola, M., Quaglia, R., Colantonio, P., Giannini, F., Giofre, R., Piazzon, L., Emanuelsson, T., & Wegeland, T. (2013). High-efficiency 7GHz doherty GaN MMIC power amplifiers for microwave backhaul radio links. *IEEE Transactions on Electron Devices, 60*(10), 3592–3595. Available from https://doi.org/10.1109/TED.2013.2274669.

Campbell, C. F., Tran, K., Kao, M. Y., & Nayak, S. (2012). A K-band 5W doherty amplifier MMIC utilizing 0.15μm GaN on SiC HEMT technology. *Technical Digest - IEEE Compound Semiconductor Integrated Circuit Symposium, CSIC*. Available from https://doi.org/10.1109/CSICS.2012.6340057, United States.

Chen, D., Zhao, C., Jiang, Z., Shum, K. M., Xue, Q., & Kang, K. (2018). A V-band Doherty power amplifier based on voltage combination and balance compensation marchand balun. *IEEE Access, 6*, 10131–10138. Available from https://doi.org/10.1109/ACCESS.2018.2795379, http://ieeexplore.ieee.org/xpl/RecentIssue.jsp?punumber = 6287639.

Chen, S., Wang, G., Cheng, Z., Qin, P., & Xue, Q. (2017). Adaptively biased 60-GHz Doherty power amplifier in 65-nm CMOS. *IEEE Microwave and Wireless Components Letters, 27*(3), 296–298. Available from https://doi.org/10.1109/LMWC.2017.2662011.

Chen, X., Chen, W., Ghannouchi, F. M., Feng, Z., & Liu, Y. (2016). A broadband doherty power amplifier based on continuous-mode technology. *IEEE Transactions on Microwave Theory and Techniques, 64*(12), 4505–4517. Available from https://doi.org/10.1109/TMTT.2016.2623705, http://ieeexplore.ieee.org/xpl/tocresult.jsp?isYear = 2009&isnumber = 4747395&Submit32 = View + Contents.

Choi, K., Kim, M., Kim, H., Jung, S., Cho, J., Yoo, S., Kim, Y. H., Yoo, H., & Yang, Y. (2010). A highly linear two-stage amplifier integrated circuit using InGaP/GaAs HBT. *IEEE Journal of Solid-State Circuits, 45*(10), 2038–2043. Available from https://doi.org/10.1109/JSSC.2010.2061612.

Curtis, J., Pham, A. V., Chirala, M., Aryanfar, F., & Pi, Z. (2013). A Ka-Band doherty power amplifier with 25.1dBm output power, 38% peak PAE and 27% back-off PAE. *Digest of Papers - IEEE Radio Frequency Integrated Circuits Symposium*, 349–352. Available from https://doi.org/10.1109/RFIC.2013.6569601, United States.

Francois, B., & Reynaert, P. (2015). Highly linear fully integrated wideband RF PA for LTE-advanced in 180-nm SOI. *IEEE Transactions on Microwave Theory and Techniques*, *63*(2), 649–658. Available from https://doi.org/10.1109/TMTT.2014.2380319, http://ieeexplore.ieee.org/xpl/tocresult.jsp?isYear = 2009&isnumber = 4747395&Submit32 = View + Contents.

Gao, X., Edfors, O., Rusek, F., & Tufvesson, F. (2015). Massive MIMO performance evaluation based on measured propagation data. *IEEE Transactions on Wireless Communications*, *14*(7), 3899–3911. Available from https://doi.org/10.1109/TWC.2015.2414413, http://ieeexplore.ieee.org/xpl/RecentIssue.jsp?puNumber = 7693.

Gao, X., Dai, L., Han, S., Chih-Lin, I., & Heath, R. W. (2016). Energy-efficient hybrid analog and digital precoding for MmWave MIMO systems with large antenna arrays. *IEEE Journal on Selected Areas in Communications*, *34*(4), 998–1009. Available from https://doi.org/10.1109/JSAC.2016.2549418, http://ieeexplore.ieee.org/xpl/tocresult.jsp?isnumber = 5678773.

Giofre, R., Piazzon, L., Colantonio, P., Giannini, F., Camarchia, V., Quaglia, R., Pirola, M., & Ramella, C. (2015). GaN-MMIC Doherty power amplifier with integrated reconfigurable input network for microwave backhaul applications. *IEEE MTT-S international microwave symposium, IMS 2015*. Italy: Institute of Electrical and Electronics Engineers Inc. Available from https://doi.org/10.1109/MWSYM.2015.7166763.

Giofre, R., Colantonio, P., & High, A. (2017). Efficiency and low distortion 6W GaN MMIC Doherty amplifier for 7GHz radio links. *IEEE Microwave and Wireless Components Letters*, *27*(1), 70–72. Available from https://doi.org/10.1109/LMWC.2016.2629972.

Giofre, R., Del Gaudio, A., & Limiti, E. (2019). A 28GHz MMIC Doherty power amplifier in GaN on Si technology for 5G applications. *IEEE MTT-S International Microwave Symposium Digest*, *2019*, 611–613. Available from https://doi.org/10.1109/mwsym.2019.8700757, Institute of Electrical and Electronics Engineers Inc. Italy.

Giofré, R., Colantonio, P., & Giannini, F. (2015). A design approach for two stages GaN MMIC pas with high efficiency and excellent linearity. *IEEE Microwave and Wireless Components Letters*, *26*(1), 46–48. Available from https://doi.org/10.1109/LMWC.2015.2505634.

Guo, R., Tao, H., & Zhang, B. (2018). A 26GHz Doherty power amplifier and a fully integrated $2 \times 2$ PA in 00.15μm GaN HEMT process for heterogeneous integration and 5G. *IEEE MTT-S international wireless symposium, IWS 2018 – Proceedings* (pp. 1–4). China: Institute of Electrical and Electronics Engineers Inc. Available from https://doi.org/10.1109/IEEE-IWS.2018.8401017, http://ieeexplore.ieee.org/xpl/mostRecentIssue.jsp?punumber = 8395091.

Gustafsson, D., Cahuana, J. C., Kuylenstierna, D., Angelov, I., Rorsman, N., & Fager, C. (2013). A Wideband and compact GaN MMIC doherty amplifier for microwave link applications. *IEEE Transactions on Microwave Theory and Techniques*, *61*(2), 922–930. Available from https://doi.org/10.1109/TMTT.2012.2231421.

Gustafsson, D., Cahuana, J. C., Kuylenstierna, D., Angelov, I., Fager, C., & GaN, A. (2014). MMIC modified Doherty PA with large bandwidth and reconfigurable efficiency. *IEEE Transactions on Microwave Theory and Techniques*, *62*(12), 3006–3016. Available from https://doi.org/10.1109/tmtt.2014.2362136.

Gustafsson, D., Andersson, K., Leidenhed, A., Malmstrom, M., Rhodin, A., & Wegeland T. (2016). A packaged hybrid doherty PA for microwave links. In *European microwave week 2016: \microwaves everywhere\, EuMW 2016 - Conference proceedings; 46th European microwave conference, EuMC 2016* (pp. 1437–1440). Sweden: Institute of Electrical and Electronics Engineers Inc. Available from https://doi.org/10.1109/EuMC.2016.7824624.

Han, S., Xu, C. L. I. Z., & Rowell, C. (2015). Large-scale antenna systems with hybrid analog and digital beamforming for millimeter wave 5G. *IEEE Communications Magazine*, *53*(1), 186–194. Available from https://doi.org/10.1109/MCOM.2015.7010533.

Harris, P., Malkowsky, S., Vieira, J., Bengtsson, E., Tufvesson, F., Hasan, W. B., Liu, L., Beach, M., Armour, S., & Edfors, O. (2017). Performance characterization of a real-time massive MIMO system with LOS mobile channels. *IEEE Journal on Selected Areas in Communications*, *35*(6), 1244–1253. Available from https://doi.org/10.1109/JSAC.2017.2686678, http://ieeexplore.ieee.org/xpl/tocresult.jsp?isnumber = 5678773.

Heath, R. W., Gonzalez-Prelcic, N., Rangan, S., Roh, W., & Sayeed, A. M. (2016). An overview of signal processing techniques for millimeter wave MIMO systems. *IEEE Journal on Selected Topics in Signal Processing*, *10*(3), 436–453. Available from https://doi.org/10.1109/JSTSP.2016.2523924, http://www.ieee.org/products/onlinepubs/news/0806_01.html.

Hu, S., Wang, F., & Wang, H. (2017). A 28GHz/37GHz/39GHz multiband linear Doherty power amplifier for 5G massive MIMO applications. In *Digest of technical papers - IEEE international solid-state circuits conference* (Vol. 60, pp. 32–33). United States: Institute of Electrical and Electronics Engineers Inc. Available from https://doi.org/10.1109/ISSCC.2017.7870246.

Huang, C., He, S., & You, F. (2018). Design of broadband modified class-J doherty power amplifier with specific second harmonic terminations. *IEEE Access*, *6*, 2531–2540. Available from https://doi.org/10.1109/access.2017.2784094.

Indirayanti, P., & Reynaert, P. (2017). A 32GHz 20 dBm-PSAT transformer-based Doherty power amplifier for multi-Gb/s 5G applications in 28nm bulk CMOS. *Digest of papers - IEEE radio frequency integrated circuits symposium* (pp. 45–48). Belgium: Institute of Electrical and Electronics Engineers Inc. Available from https://doi.org/10.1109/RFIC.2017.7969013.

Ishikawa, R., Takayama, Y., & Honjo, K. (2018). Fully integrated asymmetric doherty amplifier based on two-power-level impedance optimization. In *EuMIC 2018 - 2018 13th European microwave integrated circuits conference* (pp. 253–256). Japan: Institute of Electrical and Electronics Engineers Inc. Available from https://doi.org/10.23919/EuMIC.2018.8539899, http://ieeexplore.ieee.org/xpl/mostRecentIssue.jsp?punumber = 8521654.

Jee, S., Lee, J., Son, J., Kim, S., Kim, C. H., Moon, J., & Kim, B. (2015). Asymmetric broadband Doherty power amplifier using GaN MMIC for femto-cell base-station. *IEEE Transactions on Microwave Theory and Techniques*, *63*(9), 2802–2810. Available from https://doi.org/10.1109/TMTT.2015.2442973, http://ieeexplore.ieee.org/xpl/tocresult.jsp?isYear = 2009&isnumber = 4747395&Submit32 = View + Contents.

Jin, S., Park, B., Moon, K., Kwon, M., & Kim, B. (2013). Linearization of CMOS cascode power amplifiers through adaptive bias control. *IEEE Transactions on Microwave Theory and Techniques*, *61*(12), 4534–4543. Available from https://doi.org/10.1109/TMTT.2013.2288206.

Joo, T., Koo, B., & Hong, S. (2013). A WLAN RF CMOS PA with large-signal MGTR method. *IEEE Transactions on Microwave Theory and Techniques*, *61*(3), 1272–1279. Available from https://doi.org/10.1109/TMTT.2013.2244228.

Kang, J., Yoon, J., Min, K., Yu, D., Nam, J., Yang, Y., & Kim, B. (2006). A highly linear and efficient differential CMOS power amplifier with harmonic control. *IEEE Journal of Solid-State Circuits*, *41*(6), 1314–1322. Available from https://doi.org/10.1109/JSSC.2006.874276.

Kao, K. Y., Hsu, Y. C., Chen, K. W., & Lin, K. Y. (2013). Phase-delay cold-FET pre-distortion linearizer for millimeter-wave CMOS power amplifiers. *IEEE Transactions on Microwave Theory and Techniques*, *61*(12), 4505–4519. Available from https://doi.org/10.1109/TMTT.2013.2288085.

Kaymaksut, E., Zhao, D., & Reynaert, P. (2015). Transformer based Doherty PA mmWave applications in 40nm CMOS. *IEEE Transactions on Microwave Theory and Techniques*, *63*(4).

Kim, C. H., Jee, S., Jo, G. D., Lee, K., & Kim, B. (2014). A 2.14-GHz GaN MMIC doherty power amplifier for small-cell base stations. *IEEE Microwave and Wireless Components Letters*, *24*(4), 263–265. Available from https://doi.org/10.1109/LMWC.2014.2299536.

Kulkarni, S., & Reynaert, P. (2014). A push-pull mm-Wave power amplifier with $<0.8^{\circ}$ AM-PM distortion in 40nm CMOS. In *Digest of technical papers - IEEE international solid-state circuits conference* (Vol. 57, pp. 252–253). Belgium: Institute of Electrical and Electronics Engineers Inc. Available from https://doi.org/10.1109/ISSCC.2014.6757422.

Kulkarni, S., & Reynaert, P. (2016). A 60-GHz power amplifier with AM-PM distortion cancellation in 40-nm CMOS. *IEEE Transactions on Microwave Theory and Techniques*, *64*(7), 2284–2291. Available from https://doi.org/10.1109/TMTT.2016.2574866, http://ieeexplore.ieee.org/xpl/tocresult.jsp?isYear = 2009&isnumber = 4747395&Submit32 = View + Contents.

Larsson, E. G., Edfors, O., Tufvesson, F., & Marzetta, T. L. (2014). Massive MIMO for next generation wireless systems. *IEEE Communications Magazine*, *52*(2), 186–195. Available from https://doi.org/10.1109/MCOM.2014.6736761.

Lee, H., Lim, W., Bae, J., Lee, W., Kang, H., Hwang, K. C., Lee, K. Y., Park, C. S., & Yang, Y. (2017). Highly efficient fully integrated GaN-HEMT Doherty power amplifier based on compact load network. *IEEE Transactions on Microwave Theory and Techniques*, *65*(12), 5203–5211. Available from https://doi.org/10.1109/TMTT.2017.2765632, Institute of Electrical and Electronics Engineers Inc. South Korea, http://ieeexplore.ieee.org/xpl/tocresult.jsp?isYear = 2009&isnumber = 4747395&Submit32 = View + Contents.

Lee, H., Lim, W., Lee, W., Kang, H., Bae, J., Park, C. S., Hwang, K. C., Lee, K. Y., & Yang, Y. (2017). Compact load network for GaN-HEMT Doherty power amplifier IC using left-handed and right-handed transmission lines.

*IEEE Microwave and Wireless Components Letters*, *27*(3), 293–295. Available from https://doi.org/10.1109/LMWC.2017.2661706.

Lee, J., Lee, D. H., & Hong, S. (2014). A doherty power amplifier with a GaN MMIC for femtocell base stations. *IEEE Microwave and Wireless Components Letters*, *24*(3), 194–196. Available from https://doi.org/10.1109/LMWC.2013.2292926.

Li, S. H., Hsu, S. S. H., Zhang, J., & Huang, K. C. (2018). Design of a compact GaN MMIC Doherty power amplifier and system level analysis with X-parameters for 5G communications. *IEEE Transactions on Microwave Theory and Techniques*, *66*(12), 5676–5684. Available from https://doi.org/10.1109/TMTT.2018.2876255, http://ieeexplore.ieee.org/xpl/tocresult.jsp?isYear = 2009&isnumber = 4747395&Submit32 = View + Contents.

Liu, B., Mao, M., Boon, C. C., Choi, P., Khanna, D., Fitzgerald, E. A., & Fully, A. (2018). Integrated class-J GaN MMIC power amplifier for 5-GHz WLAN 802.11ax application. *IEEE Microwave and Wireless Components Letters*, *28*(5), 434–436. Available from https://doi.org/10.1109/LMWC.2018.2811338.

Lu, C., Pham, A. V. H., Shaw, M., & Saint, C. (2007). Linearization of CMOS broadband power amplifiers through combined multigated transistors and capacitance compensation. *IEEE Transactions on Microwave Theory and Techniques*, *55*(11), 2320–2328. Available from https://doi.org/10.1109/TMTT.2007.907734.

Lv, G., Chen, W., & Feng, Z. (2018). A compact and broadband Ka-band asymmetrical GaAs Doherty power amplifier MMIC for 5G communications. *IEEE MTT-S International Microwave Symposium Digest*, *2018*, 808–811. Available from https://doi.org/10.1109/MWSYM.2018.8439219, Institute of Electrical and Electronics Engineers Inc. China.

Lv, G., Chen, W., Chen, L., & Feng, Z. (2019). A fully integrated C-band GaN MMIC Doherty power amplifier with high gain and high efficiency for 5G application. *IEEE MTT-S International Microwave Symposium Digest*, *2019*, 560–563. Available from https://doi.org/10.1109/mwsym.2019.8701103, Institute of Electrical and Electronics Engineers Inc. China.

Lv, G., Chen, W., Chen, X., Ghannouchi, F. M., & Feng, Z. (2019). A compact Ka/Q dual-band GaAs MMIC Doherty power amplifier with simplified offset lines for 5G applications. *IEEE Transactions on Microwave Theory and Techniques*, *67*(7), 3110–3121. Available from https://doi.org/10.1109/TMTT.2019.2908103, http://ieeexplore.ieee.org/xpl/tocresult.jsp?isYear = 2009&isnumber = 4747395&Submit32 = View + Contents.

Lv, G., Chen, W., Liu, X., & Feng, Z. (2019). A dual-band GaN MMIC power amplifier with hybrid operating modes for 5G application. *IEEE Microwave and Wireless Components Letters*, *29*(3), 228–230. Available from https://doi.org/10.1109/LMWC.2019.2892837.

Lv, G., Chen, W., Liu, X., Ghannouchi, F. M., & Feng, Z. (2019). A fully integrated C-band GaN MMIC Doherty power amplifier with high efficiency and compact size for 5G application. *IEEE Access*, *7*, 71665–71674. Available from https://doi.org/10.1109/ACCESS.2019.2919603, http://ieeexplore.ieee.org/xpl/RecentIssue.jsp?punumber = 6287639.

Maroldt, S., & Ercoli, M. (2017). 3.5-GHz ultra-compact GaN class-E integrated Doherty MMIC PA for 5G massive-MIMO base station applications. In *12th European microwave integrated circuits conference, EuMIC 2017* (Vol. 2017, pp. 196–199). France: Institute of Electrical and Electronics Engineers Inc. Available from https://doi.org/10.23919/EuMIC.2017.8230693.

Marzetta, T. L. (2016). *Fundamentals of massive MIMO*. Cambridge University Press.

Nakatani, K., Yamaguchi, Y., Komatsuzaki, Y., Sakata, S., Shinjo, S., & Yamanaka, K. (2018). A Ka-band high efficiency Doherty power amplifier MMIC using GaN-HEMT for 5G application. *IEEE MTT-S international microwave workshop series on 5G hardware and system technologies, IMWS-5G 2018*. Japan: Institute of Electrical and Electronics Engineers Inc. Available from https://doi.org/10.1109/IMWS-5G.2018.8484612, http://ieeexplore.ieee.org/xpl/mostRecentIssue.jsp?punumber = 8470025.

Nguyen, D. P., & Pham, A. V. (2016). An ultra compact watt-level Ka-band stacked-FET power amplifier. *IEEE Microwave and Wireless Components Letters*, *26*(7), 516–518. Available from https://doi.org/10.1109/LMWC.2016.2574831.

Nguyen, D. P., Pham, B. L., & Pham, A. V. (2017). A compact 29% PAE at 6dB power back-off E-mode GaAs pHEMT MMIC Doherty power amplifier at Ka-band. *IEEE MTT-S international microwave symposium digest* (pp. 1683–1686). United States: Institute of Electrical and Electronics Engineers Inc. Available from https://doi.org/10.1109/MWSYM.2017.8058964.

Nguyen, D. P., Curtis, J., Pham, A. V., & Doherty, A. (2018). Amplifier with modified load modulation scheme based on load-pull data. *IEEE Transactions on Microwave Theory and Techniques*, *66*(1), 227–236. Available from https://doi.org/10.1109/TMTT.2017.2734663, http://ieeexplore.ieee.org/xpl/tocresult.jsp?isYear = 2009&isnumber = 4747395&Submit32 = View + Contents.

Nguyen, D. P., Pham, T., & Pham, A. V. (2018). A 28-GHz symmetrical Doherty power amplifier using stacked-FET cells. *IEEE Transactions on Microwave Theory and Techniques, 66*(6), 2628–2637. Available from https://doi.org/10.1109/TMTT.2018.2816024, Institute of Electrical and Electronics Engineers Inc. United States, http://ieeexplore.ieee.org/xpl/tocresult.jsp?isYear = 2009&isnumber = 4747395&Submit32 = View + Contents.

Nguyen, H. T., Chi, T., Li, S., & Wang, H. (2018). A 62-to-68GHz linear 6Gb/s 64QAM CMOS doherty radiator with 27.5%/20.1% PAE at peak/6dB-back-off output power leveraging high-efficiency multi-feed antenna-based active load modulation. In *Digest of Technical Papers - IEEE international solid-state circuits conference* (Vol. 61, pp. 402–404). United States: Institute of Electrical and Electronics Engineers Inc. Available from https://doi.org/10.1109/ISSCC.2018.8310354.

Niu, Y., Su, L., Gao, C., Li, Y., Jin, D., & Han, Z. (2016). Exploiting device-to-device communications to enhance spatial reuse for popular content downloading in directional mmWave small cells. *IEEE Transactions on Vehicular Technology, 65*(7), 5538–5550. Available from https://doi.org/10.1109/TVT.2015.2466656.

Özen, M., Rostomyan, N., Aufinger, K., & Fager, C. (2017). Efficient millimeter wave Doherty PA design based on a low-loss combiner synthesis technique. *IEEE Microwave and Wireless Components Letters, 27*(12), 1143–1145. Available from https://doi.org/10.1109/LMWC.2017.2763739.

Park, B., Jin, S., Jeong, D., Kim, J., Cho, Y., Moon, K., & Kim, B. (2016). Highly linear mm-Wave CMOS power amplifier. *IEEE Transactions on Microwave Theory and Techniques, 64*(12), 4535–4544. Available from https://doi.org/10.1109/TMTT.2016.2623706, http://ieeexplore.ieee.org/xpl/tocresult.jsp?isYear = 2009&isnumber = 4747395&Submit32 = View + Contents.

Park, J., Lee, C., & Park, C. (2017). A quad-band CMOS linear power amplifier for EDGE applications using an anti-phase method to enhance its linearity. *IEEE Transactions on Circuits and Systems I: Regular Papers, 64*(4), 765–776. Available from https://doi.org/10.1109/TCSI.2016.2620559, http://ieeexplore.ieee.org/xpl/RecentIssue.jsp?punumber = 8919.

Park, J., Lee, C., Yoo, J., & Park, C. (2017). A CMOS antiphase power amplifier with an MGTR technique for mobile applications. *IEEE Transactions on Microwave Theory and Techniques, 65*(11), 4645–4656. Available from https://doi.org/10.1109/TMTT.2017.2709304, http://ieeexplore.ieee.org/xpl/tocresult.jsp?isYear = 2009&isnumber = 4747395&Submit32 = View + Contents.

Park, Y., Lee, J., Jee, S., Kim, S., Kim, C. H., Park, B., & Kim, B. (2015). GaN HEMT MMIC Doherty power amplifier with high gain and high PAE. *IEEE Microwave and Wireless Components Letters, 25*(3), 187–189. Available from https://doi.org/10.1109/LMWC.2015.2390536.

Pi, Z., & Khan, F. (2011). An introduction to millimeter-wave mobile broadband systems. *IEEE Communications Magazine, 49*(6), 101–107. Available from https://doi.org/10.1109/MCOM.2011.5783993.

Piazzon, L., Colantonio, P., Giannini, F., & Giofré, R. (2014). 15% bandwidth 7GHz GaN-MMIC Doherty amplifier with enhanced auxiliary chain. *Microwave and Optical Technology Letters, 56*(2), 502–504. Available from https://doi.org/10.1002/mop.28108.

Probst, S., Martinelli, T., Seewald, S., Geck, B., & Manteuffel, D. (2017). Design of a linearized and efficient doherty amplifier for C-band applications. In *12th European microwave integrated circuits conference, EuMIC 2017* (Vol. 2017, pp. 121–124). Germany: Institute of Electrical and Electronics Engineers Inc. Available from https://doi.org/10.23919/EuMIC.2017.8230675.

Quaglia, R., Camarchia, V., Jiang, T., Pirola, M., Donati Guerrieri, S., & Loran, B. (2014). K-band GaAs MMIC Doherty power amplifier for microwave radio with optimized driver. *IEEE Transactions on Microwave Theory and Techniques, 62*(11), 2518–2525. Available from https://doi.org/10.1109/TMTT.2014.2360395, http://ieeexplore.ieee.org/xpl/tocresult.jsp?isYear = 2009&isnumber = 4747395&Submit32 = View + Contents.

Quaglia, R., Camarchia, V., Pirola, M., Rubio, J. J. M., & Ghione, G. (2014). Linear GaN MMIC combined power amplifiers for 7-GHz microwave backhaul. *IEEE Transactions on Microwave Theory and Techniques, 62*(11), 2700–2710. Available from https://doi.org/10.1109/TMTT.2014.2359856, http://ieeexplore.ieee.org/xpl/tocresult.jsp?isYear = 2009&isnumber = 4747395&Submit32 = View + Contents.

Quaglia, R., Greene, M. D., Poulton, M. J., & Cripps, S. C. (2019). A 10.8-3.2-GHz Doherty power amplifier in quasi-MMIC technology. *IEEE Microwave and Wireless Components Letters, 29*(5), 345–347. Available from https://doi.org/10.1109/LMWC.2019.2904883, https://ieeexplore.ieee.org/servlet/opac?punumber = 7260.

Rappaport, T.S., & Deng, S. (2015). 73GHz wideband millimeter-wave foliage and ground reflection measurements and models. In *IEEE international conference on communication workshop, ICCW* (pp. 1238–1243).

United States: Institute of Electrical and Electronics Engineers Inc. Available from https://doi.org/10.1109/ICCW.2015.7247347.

Rappaport, T. S., Sun, S., Mayzus, R., Zhao, H., Azar, Y., Wang, K., Wong, G. N., Schulz, J. K., Samimi, M., & Gutierrez, F. (2013). Millimeter wave mobile communications for 5G cellular: It will work!. *IEEE Access, 1*, 335–349. Available from https://doi.org/10.1109/ACCESS.2013.2260813, http://ieeexplore.ieee.org/xpl/RecentIssue.jsp?punumber = 6287639.

Roh, W., Seol, J. Y., Park, J. H., Lee, B., Lee, J., Kim, Y., Cho, J., Cheun, K., & Aryanfar, F. (2014). Millimeter-wave beamforming as an enabling technology for 5G cellular communications: Theoretical feasibility and prototype results. *IEEE Communications Magazine, 52*(2), 106–113. Available from https://doi.org/10.1109/MCOM.2014.6736750.

Rostomyan, N., Ozen, M., & Asbeck, P. (2018). 28GHz Doherty power amplifier in CMOS SOI with 28% back-off PAE. *IEEE Microwave and Wireless Components Letters, 28*(5), 446–448. Available from https://doi.org/10.1109/LMWC.2018.2813882.

Sarkar, A., Aryanfar, F., & Floyd, B. A. (2017). A 28-GHz SiGe BiCMOS PA with 32% efficiency and 23-dBm output power. *IEEE Journal of Solid-State Circuits, 52*(6), 1680–1686. Available from https://doi.org/10.1109/JSSC.2017.2686585, http://ieeexplore.ieee.org/Xplore/home.jsp.

Shakib, S., Park, H. C., Dunworth, J., Aparin, V., & Entesari, K. (2016). A highly efficient and linear power amplifier for 28-GHz 5G phased array radios in 28-nm CMOS. *IEEE Journal of Solid-State Circuits, 51*(12), 3020–3036. Available from https://doi.org/10.1109/JSSC.2016.2606584; http://ieeexplore.ieee.org/Xplore/home.jsp.

Tsai, J. H., Chang, H. Y., Wu, P. S., Lee, Y. L., Huang, T. W., & Wang, H. (2006). Design and analysis of a 44-GHz MMIC low-loss built-in linearizer for high-linearity medium power amplifiers. *IEEE Transactions on Microwave Theory and Techniques, 54*(6), 2487–2496. Available from https://doi.org/10.1109/TMTT.2006.875800.

Tsai, J. H., Wu, C. H., Yang, H. Y., & Huang, T. W. (2011). A 60GHz CMOS power amplifier with built-in pre-distortion linearizer. *IEEE Microwave and Wireless Components Letters, 21*(12), 676–678. Available from https://doi.org/10.1109/LMWC.2011.2171929.

Valenta, V., Davies, I., Ayllon, N., Seyfarth, S., & Angeletti, P. (2018). High-gain GaN Doherty power amplifier for Ka-band satellite communications. In *PAWR 2018 - Proceedings 2018 IEEE topical conference on power amplifiers for wireless and radio applications* (Vol. 2018, pp. 29–31). Netherlands: Institute of Electrical and Electronics Engineers Inc. Available from https://doi.org/10.1109/PAWR.2018.8310059.

Vigilante, M., & Reynaert, P. (2018). A wideband class-AB power amplifier with 29–57-GHz AM-PM compensation in 0.9-V 28-nm Bulk CMOS. *IEEE Journal of Solid-State Circuits, 53*(5), 1288–1301. Available from https://doi.org/10.1109/JSSC.2017.2778275, http://ieeexplore.ieee.org/Xplore/home.jsp.

Wang, C., Vaidyanathan, M., & Larson, L. E. (2004). A capacitance-compensation technique for improved linearity in CMOS class-AB power amplifiers. *IEEE Journal of Solid-State Circuits, 39*(11), 1927–1937. Available from https://doi.org/10.1109/JSSC.2004.835834.

Wang, D., Chen, W., Chen, L., Liu, X., & Feng, Z. (2019). A Ka-band highly linear power amplifier with a linearization bias circuit. *IEEE MTT-S International Microwave Symposium Digest, 2019*, 320–322. Available from https://doi.org/10.1109/mwsym.2019.8701069, Institute of Electrical and Electronics Engineers Inc. China.

Xi, T., Huang, S., Guo, S., Gui, P., Huang, D., & Chakraborty, S. (2017). High-efficiency E-band power amplifiers and transmitter using gate capacitance linearization in a 65-nm CMOS process. *IEEE Transactions on Circuits and Systems II: Express Briefs, 64*(3), 234–238. Available from https://doi.org/10.1109/TCSII.2016.2563698. http://www.ieee-cas.org.

Yamauchi, K., Mori, K., Nakayama, M., Mitsui, Y., & Takagi, T. (1997). Microwave miniaturized linearizer using a parallel diode. *IEEE MTT-S International Microwave Symposium Digest, 3*, 1199–1202, IEEE Japan.

Yoshimasu, T., Akagi, M., Tanba, N., & Hara, S. (1998). An HBT MMIC power amplifier with an integrated diode linearizer for low-voltage portable phone applications. *IEEE Journal of Solid-State Circuits, 33*(9), 1290–1296. Available from https://doi.org/10.1109/4.711326.

# Index

Note: Page numbers followed by "*f*" and "*t*" refer to figures and tables, respectively.

## C

## F

## I

## N

## Q



## R

## S

## T

## U



## V